PRINCIPLES OF MACHINING

by cutting, abrasion and erosion

by

J. KACZMAREK

Member of the Polish Academy of Sciences

Professor at the Cracow Technical University

Translated by A. Voellnagel, M.Sc. (Eng.) (Parts 1–3)
and E. Lepa (Part 4)

Peter Peregrinus Limited, Stevenage

Wydawnictwa Naukowo-Techniczne, Warsaw

Peter Peregrinus Limited

Southgate House, Stevenage,
Herts. SGI IHQ, England

Originally published under the title
Podstawy obróbki wiórowej, ściernej i erozyjnej

ISBN: 0 901223 66 2

Printed in Poland

Contents

Introduction

1 Machining by means of cutting and erosion as part of machine technology and machine production engineering

Production is the sequence of functions leading to the creation of parts by processing or joining various materials and then to the assembly of these parts to obtain the final product.

Production engineering is the selection of production means and methods in order to optimize their application as judged by technological criteria.

Technology is the complex of synthetic* sciences whose aim and substance consists in the research concerning production means and processes, the discovering and formulating of laws and principles of those processes and the preparation of the base for optimization of production processes.

Technology creates and develops the theoretical bases of production engineering, whose task is to take advantage of technological i.e. scientific bases and production principles instead of empiricism and improvisation. Technology and production engineering are interconnected, just as theory and practice generally used to be.

In common language the difference between those notions often disappears and the word 'technology' is used not only to designate the science concerning production but also as a synonym of production engineering.

In the strict sense narrowed down to the production of machines, tools and equipment, a distinction should be made between manufacturing technology and production engineering.

Manufacturing technology in its actual stage of development comprises synthetic sciences (or fields of knowledge) constituting its branches and creating a theoretical basis for practical engineering:

— foundry work,
— powder metallurgy,
— plastic working,
— cutting and eroding,

* By 'synthetic sciences' are meant those sciences which are based on a creative synthesis of basic sciences like physics, mathematics, chemistry, logics etc.

— heat treatment,
— coating and plating,
— measurement and inspection,
— permanent joining (flame welding, resistance welding, soldering, gluing),
— detachable joints (assembly),
— technological planning of production.

The manufacturing industry develops dynamically and in accordance with the principles of economy as long as the principles of manufacturing technology are adhered to. These are:

(a) The principle of consistent aims and trends of the development of all individual branches of manufacturing technology with the latter's general aims and trends,
(b) The principle of harmonic development of manufacturing technology branches, i.e. the principle of correct proportions of the speed of development of those branches and of correct percentage shares of particular engineering techniques in production.

Thus the development principles stipulate that the problems of any particular branch of manufacturing technology should be considered in connection with the entire set of problems of the latter.

Among the various branches of manufacturing technology, machining by cutting and machining by erosion show a similarity of characteristics and applications. Although different forms of energy are used: mechanical energy in cutting and electrical, chemical and thermal energies in eroding, still it is the common feature of both processes that machine parts and other components are shaped by removing a certain portion of their material, called the *machining allowance*. In addition, in most methods of machining by erosion the same problems of tool wear arise as in machining by cutting.

Cutting has a very considerable share both in production engineering and in the technology of machines and related products. It plays an important role because of widespread applications, due to its versatility and to the possibility of achieving the highest working accuracy.

Machining by erosion, in its present stage of development, widens and complements the working possibilities, particularly in the case of materials which are hard to machine by cutting or which cannot be machined at all by the latter method.

2 Classification of the methods of machining

It has been agreed that by the term 'machining' all those methods will be meant in which the process of shaping parts and products is connected with the necessity to remove a certain part of material.

Machining (see Table I.1) comprises both machining by cutting and machining by erosion.

It must be stressed here that the removal and, consequently, loss of material are not an exclusive feature of the cutting and erosion processes, but are also present in all methods of shaping machine parts. Therefore, the methods of material removal should not be identified with methods yielding a low output, i.e. resulting in high loss of material.

Table I.1 *Classification of the methods of machining*

Code	Method	Code	Method
M	Machining		
MC	Machining by cutting	ME	Machining by erosion
C	Chip removing processes	ED	Spark machining
A	Abrasion processes	EC	Chemical milling
		EB	Beam-erosion processes Micro-machining

The measure of the actual technological waste of material in production is the so-called *material utilization factor* U_M

$$U_M = \frac{V_N}{V_G}\,100\% \tag{I.1}$$

where V_N—net volume or weight after the technological process in question, V_G—gross volume or weight before this process.

On the other hand, the measure of the technological merits of the given production method, as judged from the point of view of the utilization of material, is the so-called *technological factor of material utilization* U_T

$$U_T = \frac{V_N}{V_{GT}}\,100\% \tag{I.2}$$

where V_{GT} is the technically justified minimum volume or weight of the material necessary for the shaping of the final form of the given product with the desired accuracy and other qualities.

Whenever it is not possible to obtain the final shape of the product by using a certain method, technologically justified material allowances for an additional method, by which the product will have to be finished, must be taken into account in the calculation of material waste.

The index of the technical, economical and organizational efficiency of an industrial enterprise, in the field of material utilization, is the so-called *material savings factor* e_M

$$e_M = \frac{U_M}{U_T}\,100\% = \frac{V_{GT}}{V_G}\,100\% \tag{I.3}$$

The technological factor of material utilization U_T is important for comparing the technological merits of various production methods from the point of view of efficient utilization of material. For most machine cutting methods this factor is relatively high, over 85%, and in the case of abrasion $U_T > 95\%$.

However, the material utilization factor U_M for machine cutting is mostly within the limits of 60–75%. This shows that the main causes of the high waste of material are not inherent in the principle of shaping by means of cutting.

Machining by means of cutting, or *machine cutting*, consists in removing a certain volume of material (machining allowance) by mechanical action, using tools harder than the machined material and provided with a wedge-shaped cutting point.

Within the field of machine cutting, the chip removing processes (C) and abrasion processes (A) are to be distinguished.

Chip removing processes (C) consist in machining by means of tools having a definite number and shape of cutting points. The machining allowance is removed here in the form of chips visible with the naked eye.

Abrasion processes (A) consist in machining by means of tools having many small points of indefinite number and shape, while the machining allowance is removed under the form of minute chips, invisible to the naked eye, and in the form of particles torn away by the force of friction. Both the chips and particles may, under certain conditions, become subject to partial or total oxidation (burning) or fusion.

Eroding, or machining by erosion (E) consists in removing a certain volume of material (machining allowance) by means of the process of erosion.

By the *process of erosion* is meant the removal of successive surface layers of the material as a result of: chipping and spalling of very small parts of material in its solid phase, dissolution or melting and evaporation of the material being machined.

In the field of machining by erosion the following methods have been developed so far.

— *Spark machining*, mainly based on the application of electrical erosion caused by discharges evoked by suitable electric impulses.
— *Chemical milling*, mainly based on the application of chemical erosion resulting from chemical processes accompanying the flow of current between electrodes and electrolytes.
— *Beam-erosion processes*, or micro-machining, mainly based on the erosion by means of beams of particles with a high degree of energy concentration, e.g. electron, plasma or photon beams.

3 Development trends in the field of machining by cutting and by erosion

It is the purpose of machining by cutting and by erosion to give the workpiece the desired shape and dimensions of necessary accuracy, as well as to ensure that the surface layer of the workpiece has the desired quality. This purpose should be achieved in the most economical way.

Therefore, the main development trends in the field of machining by cutting and by erosion are:

(a) Reduction of material losses—and this concerns both the machined material and the tools. In connection with this aim there arises the task of reducing the machining allowances to the limits of a theoretical minimum, determined by the thickness of surface layer damaged by the preceding technological process (e.g.

casting crust) and the probable dimensional errors. The problem of reducing the wear of cutting tools and eroders* also plays its part.

(b) Improvement of the quality of machined products. This means in particular the attainment of the necessary accuracy of shape and dimensions, as well as the optimization of surface layer properties from the point of view of the intended use of the product.

(c) Raising of the output and reduction of the labour consumption by the machining process, improvement of work safety and comfort. In particular, the task of mechanization and automation belongs here, especially the introduction of programme-controlled machining. The rational fulfilment of these tasks is conditioned by the solution of a number of organizational and technical problems, such as the standardization of design and technology, specialization, co-operation, application of the principles of human engineering etc. A higher degree of mechanization and automation means higher consumption of energy. Thus there arises the problem of rational and economical use of energy, in particular electrical energy. This problem becomes acute in the case of processes consuming large amounts of energy, such as the erosion processes.

(d) The widening of machining possibilities. This requires, on the one hand, the improvement of the machinability of materials and, on the other hand, the development of new methods, ways and types of machining and improvement of the existing ones, as well as the optimization of machining conditions.

The science of metal cutting and erosion or the technology of machining by means of cutting and erosion, should create and furnish the theoretical basis for production engineering applying those techniques. In this process the main development trends should be followed and the most advantageous economic solutions aimed at.

LIST OF MANUALS AND BOOKS on the machining of metals, analysed and taken into consideration in the preparation of this book

I/1 ARTINOV, W. A., and ALEKSEEV, Ch. A.: 'Rezanie metallov' (Metal Cutting), Mashgiz, Moskva 1953

I/2 BAKER, H.: 'Modern Workshop Technology', Cleaver-Hume Press Ltd., London 1960

I/3 BIERNAWSKI, W.: 'Teoria obróbki metali skrawaniem' (Theory of the Machining of Metals by Cutting), PWN, Warszawa 1956

I/4 BLACK, P. H.: 'Theory of Metal Cutting', McGraw-Hill, New York 1961

I/5 CLAIR, L. J. St.: 'Étude et emploi de l'outil de coupe', Dunod, Paris 1956

I/6 DANEL'YAN, A. M.: 'Rezanie metallov i instrument' (Metal Cutting and Tools), Mashgiz, Moskva 1950

I/7 DMOCHOWSKI, J.: 'Podstawy obróbki metali skrawaniem' (Elements of Metal Machining by Cutting), I, Lectures for the students of correspondence courses of the Warsaw Institute of Technology 1960

I/8 GRANOVSKII, G. I. *et al.*: 'Rezanie metallov' (Metal Cutting), Mashgiz, Moscow 1954

I/9 HIRSCHFELD, E.: 'Teoria rezania kovov' (Theory of Metal Cutting), Vyd. Slov. Akad. Ved, Bratislava 1963

I/10 HORNING, A.: 'Forgácsolás—Elmélet és Forgácsoló Szerszámok', Tankönyvkiadó, Vállalat, Budapest 1953

I/11 KACZMAREK, J.: 'Podstawy skrawania metali' (Elements of Metal Cutting), PWT, Warsaw 1956

I/12 KLUSHIN, M. I.: 'Rezanie metallov' (Metal Cutting), Mashgiz, Moskva 1953

* *Eroder, working electrode*—terms used to describe generally the tools applied in erosion processes.

I/13 KRONENBERG, M.: 'Grundzüge der Zerspanungslehre', 1. Band—'Einschneidige Zerspanung', 2nd ed., Springer-Verlag, München 1954, 2. Band—'Mehrschneidige Zerspanung (Stirnfräsen, Bohren)', 2nd ed., Springer-Verlag, Heidelberg 1963
I/14 MERCHANT, M. E., and ERNST, H.: 'Principles of Metal Cutting and Machinability', Tool Engineer's Handbook, Section 17, McGraw-Hill, New York 1959
I/15 PRIKRYL, Z.: 'Nauka o obrábeni' (The Science of Machining), SNTL, Praha 1959
I/16 SALMON, P., and CAROUGEAU, M.: 'Le travail de métaux par abrasion', Soc. de Publ. Méc., Paris
I/17 SHAW, M. C.: 'Metal Cutting Principles', 3rd ed., The M.I.T. Press, Cambridge, Mass. 1960
I/18 STANKOWICZ, P. P.: 'Maszine ałatkie i industriska proizwodna maszina', Cz. I, 'Obrada mietała riezanijem' (Machine Tools and Industrial Production Machinery, Part I, Metal Machining by Cutting), 3rd ed., Naucznaja Kniga, Beograd 1958
I/19 VIEREGGE, G.: 'Zerspanung der Eisenwerkstoffe', Verlag Stahleisen mbH, Düsseldorf 1959
I/20 WILSON, F. W. (Ed.): 'Machining with Carbides and Oxides', McGraw-Hill, New York 1962

PART 1

Fundamentals of the theory of cutting

CHAPTER 1

Stereometric and kinematic elements of machining by cutting

1.1 Basic and general notions

The machining of materials by cutting is carried out by means of cutting tools. The characteristic feature of all such tools is the wedge-shaped working part, called the *tool point*.

Tools with wedge-shaped points may be used for *parting off* (Figs. 1.1a and b) or for *generating shapes* (Fig. 1.1c). The difference between these two operations is that in parting off the layer separated from the material is usually small and is totally, or for the most part of it, plastically deformed.

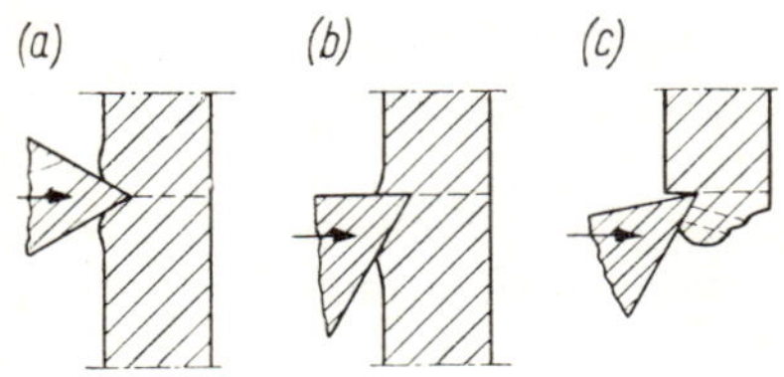

Fig 1.1 Separation of material by means of wedge-shaped tool points

The workpiece being machined and the cutting tool must preserve their correct positions during the machining process and must carry out suitable motions in relation to each other. The workpiece and tool are fixed on a machine, called *machine tool,* by means of jigs and fixtures. The machine tool also provides the necessary motions of the tool and workpiece.

The machining process is carried out in air, but in the cutting zone liquid or gaseous *cutting fluids* may be used.

Thus the course and results of the machining process depend on the reciprocal action of the workpiece (w), tool (t), machine tool (m) with its equipment (e) and the cutting fluid (f). The whole set of these factors is called the machining system, designated by symbols m–e–w–t–f. In practice, this designation is reduced to three letters m–w–t (machine tool–workpiece–tool).

The theory of machining by cutting depends upon several basic sciences which are often interconnected.

At present the following machining theories—independent, though interconnected—have been developed:

(a) cutting of materials, particularly of metals, in various media; this theory has considerable fundamental importance,
(b) machine tool design (kinematics and dynamics),
(c) cutting tool design and the study of tool materials,
(d) design of machining jigs and fixtures,
(e) dimensioning and workshop measurements (workshop metrology).

The theory of metal cutting comprises:

— research and scientific induction concerning the phenomena of the cutting process, based on the principles of physics and other basic sciences,
— research on important practical characteristics of cutting, comprising: cutting forces and energy, cutting heat and temperature, tool wear, geometrical and physical properties of the surface layer being machined, workpiece accuracy and the kind of waste produced (chip),
— preparation of instructions for a scientifically and technically justified choice of machining conditions, in order to optimize the industrial machining process.

1.2 Kinematics of cutting

1.2.1 Definitions of motions

Preparatory and basic motions are necessary for the machining to be carried out.

Preparatory motions, also known as setting motions, comprise all motions connected with the setting of tool or workpiece, approaching the tool to the workpiece or vice-versa and removing the tool from the workpiece or vice-versa. The aim of these movements is the preparation for movements resulting in the cutting process.

Basic motions, also known as working motions, are those motions of the workpiece and tool, which are necessary for the actual machining operation.

A basic motion may be of the primary or auxiliary type.

The *primary basic motion*, shortly called the primary motion, allows only a single contact or a simple removal of a layer of material from a certain spot. The primary motion has a higher speed than the auxiliary one.

The *auxiliary basic motion*, shortly called the auxiliary motion or feed motion, in combination with the primary motion allows repeated or continuous removal of the machined layer. The feed motion is slower than the primary one.

The feed motion may be a resultant motion, but at least one of its components must be rectilinear.

Basic motions are usually divided between the workpiece and the tool, but there are cases, too, where both basic motions are carried out by the tool only. The way of dividing the basic motions between the tool and workpiece is the basis for the classification of various cutting methods.

1.2.2 Definitions of basic motion speeds

The speed of a primary motion is designated by v_m, and that of a feed motion by v_f.

Geometrical composition of the speed vectors of the primary and auxiliary basic motions results in the cutting speed defined by the formula

$$\bar{v} = \bar{v}_m + \bar{v}_f \tag{1.1}$$

If the primary motion speed is much higher than that of the feed motion, which is usually the case, it may be accepted with an accuracy sufficient for technical applications that the primary motion speed equals the cutting speed

$$v_m \approx v \tag{1.2}$$

The feed motion speed is defined as:

(a) the speed whose value equals the ratio of distance travelled to time, or travel distance covered in unit time (e.g. a minute); thus defined, feed motion speed is generally called the *time feed* and in particular cases, e.g. the *minute feed* (designated by v_f);

(b) the speed whose value is expressed by the distance covered during one motion cycle (e.g. one revolution or stroke). The feed motion speed thus defined is called the *feed per revolution* or *feed per stroke* (designated by s);

(c) in the case of multipoint tools as the speed whose value is expressed by the distance covered between the contacts of two consecutive tool points with the material. The feed motion speed thus defined is called the *feed per point* or *feed per tooth* (s_p).

In practice, instead of the term 'feed motion speed' the shorter term 'feed' is used.

The following relationship exists between various kinds of feeds

$$v_f = n\, s = n\, i\, s_p \tag{1.3}$$

where: n—rotational speed (e.g. in rev/min) or stroke frequency (e.g. in strokes/min), i—number of points (teeth).

1.2.3 Definition of cutting time and machine time

The time of simultaneous duration of the primary and auxiliary motions, which is necessary for the machining operation, is called the *machine time.*

The machine time t_m is calculated in the same way as the time of a uniform rectilinear motion, or

$$t_m = \frac{L}{v_f} \tag{1.4}$$

where: L—tool travel distance in relation to the workpiece, or vice versa, in the direction of the rectilinear component of feed motion; it is called shortly the *action distance*; v_f—the speed of rectilinear feed motion.

The time of contact of the point or points with the material machined is called the *cutting time.*

The cutting time t is usually a fraction of the machine time, or

$$t \leqq t_m \tag{1.5}$$

and is calculated according to the formula

$$t = \frac{l}{v_f} \tag{1.6}$$

The action distance is always greater than the cutting distance.

In a general case illustrated by Fig. 1.2, the action distance consists of:

— tool approach l_a—the path of mechanized relative motion of the tool, during which cutting may not occur at first, while later the tool enters into the machined material (path l_e);

— cutting distance l—the path of mechanized relative motion of the tool, during which the tool cuts the material; the length of this path equals that of the machined surface;

— tool exit l_x—the path of mechanized relative motion of the tool, during which the tool comes out of the material after the distance l_c at which point the cutting ceases.

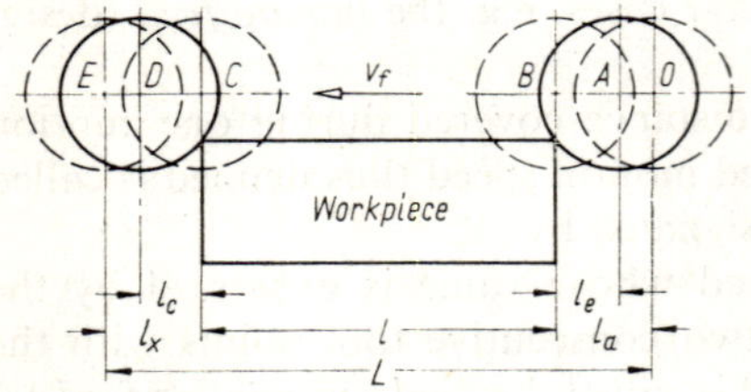

Fig. 1.2 Designations of action distance, cutting distance, approach and exit

Thus the action distance is determined by the formula

$$L = l_a + l + l_x \tag{1.7}$$

The approach distance should in principle equal that of the entering of the tool into the machined material. In practice, because of the difficulties of accurately setting the tool point so that it should just contact the material and because of the risk of damage to the former, the approach distance is usually slightly greater than the entering distance, or $l_a > l_e$.

Depending on the machining method and tool shape, the exit distance may equal zero, e.g. when the tool is being retracted after having passed the length of the machined surface and having finished the cutting process. In other cases the exit distance should equal that of the tool coming out of the material machined. In practice this length may be increased because of the need to provide free motions for setting the tool relative to the workpiece, and then $l_x > l_c$.

1.3 Classification of cutting methods [2/1], [1/2]

Regarding the degree of mechanization of the cutting process there can be distinguished:

(a) manual cutting, carried out with tools operated by man,

(b) manual cutting, carried out with mechanized tools, electrically or pneumatically driven,

(c) machine cutting, carried out on machine tools controlled by man,

(d) machine cutting, carried out on machine tools automatically controlled according to a pre-set programme (this kind of cutting is designated by the symbol NC).

Machine cutting is called *machining* in everyday language.

Regarding the way of removing the machining allowance, single-pass and multi-pass machining may be distinguished.

A single travel of the work tool in relation to the entire machined surface is called a *pass*.

Single-pass machining takes place, when the whole machining allowance is removed in a single pass, and *multi-pass machining*—when more than one pass is used for this purpose.

Regarding the way the tool enters the material and comes out of it, through cutting and blind cutting are distinguished.

Through cutting means a cutting operation in which the tool has an approach and exit after each pass, *blind cutting* when each pass ends without tool exit.

Regarding the number of tool points participating in the cutting of the machining allowance in each pass, single-point and multi-point machining are distinguished.

Similarly, as regards the number of tools working simultaneously, single-tool and multi-tool machining are distinguished.

Regarding the continuity of contact between tool points with the material machined during the cutting process, continuous and intermittent cutting are distinguished.

In *continuous cutting* the tool point maintains contact with the material machined during the entire pass. In *intermittent cutting* the tool point or points periodically enter and come out of the material machined.

The most important criteria for the classification of cutting processes are the kinematics of the primary motions of tool and workpiece and the tool design.

Depending on the kind of combination of these motions—and of the kind of cutting tool—there can be distinguished the cutting methods and, within these, the varieties of cutting methods, shortly called the cutting varieties.

The *cutting methods* are illustrated in Table 1.1. Out of 16 possible combinations four should be rejected (those designated with numbers 1, 3, 9, 11), as none of the component primary motions is rectilinear, and thus it would be impossible to continue

Table 1.1 *Classification of chip removing methods*

Primary motions of workpiece		Primary motions of tool: ●	↗	↻	↻↗
	●	(1) ✕	(2) Shaping Broaching	(3) ✕	(4) Drilling
	↗	(5) Planing	(6)	(7) Milling	(8)
	↻	(9) ✕	(10) Turning	(11) ✕	(12)
	↻↗	(13)	(14)	(15) Hobbing	(16)

Explanation of symbols: ● —stationary; ↗—rectilinear motion; ↻—rotary motion; ↻↗—resultant motion: helical, cycloidal or spiral the cutting process. Out of the remaining 12 only a few are practically useful (2, 4, 5, 7, 10) and one of those, No. 2, comprises two cutting methods: shaping and broaching. The difference between these two methods consists in tool design, machine tool design and in the choice of machining conditions.

Beside the motion combinations marked in Table 1.1 and the corresponding machining methods, several other combinations, not contained in Table 1.1 (e.g. those used in the machining of gear teeth), are used.

Regarding the necessity of applying care and effort to achieve the desired shape accuracy, and the quality of the machined surface, various types of machining are distinguished.

There are four conventional types of machining: rough, fairly accurate, accurate and highly accurate. The ranges of the different types of machining for the individual machining methods regarding dimensional accuracy and surface quality are given in Table 1.2. Taking into consideration the succession of operations or cuts and the principal aim of the machining, preliminary machining, profiling and finishing are distinguished.

Preliminary machining consists in cutting operations, or cuts, aiming at the removal of the so-called damaged layer—the result of such technological processes as casting or forging. Generally, preliminary machining is called *roughing* or *skinning*. Preliminary machining also comprises cutting off to size and other cuts preparing the workpiece for further machining operations.

Profiling comprises the sequence of operations and cuts whose purpose it is to give the basic shape to the workpiece, while leaving a small machining allowance for the finishing operations.

Finishing consists of a set of operations and cuts whose purpose it is to give the final shape to the workpiece, while maintaining the desired accuracy and surface quality.

In this respect terms are sometimes used whose meaning does not quite correspond to the notion of finishing. For instance smoothing operations have the only purpose of imparting a suitable smoothness to the surface, and surface treatment aims at imparting suitable geometrical and physical properties to the surface layer.

This review of classifications of machining by cutting and particularly of chip removing processes contains only the most frequently used classification criteria, leaving out others which may also be used when necessary.

1.4 Tool point stereometry [1/3]

1.4.1 General description of a tool point

In a cutting tool two functional parts may be distinguished: the shank and the working part, called the point.

The *cutting point* is a trihedral or polyhedral corner characterized by certain geometric elements.

The cutting point of a tool is limited by the following surfaces (Fig. 1.3):
— the *rake face* against which the chip is removed and along which it moves,
— the *flank*, situated on the side of the surface machined and of the plane tangent to it during machining.

The line of intersection of the rake face with the flank is called the *cutting edge*.

The cutting edge, depending on the shape of the flank and of the rake face, may consist of rectilinear and curvilinear sections. Between the rectilinear sections there are usually arc-shaped ones.

The *leading cutting edge* is that portion of the edge which is intended to do the major part of the cutting work.

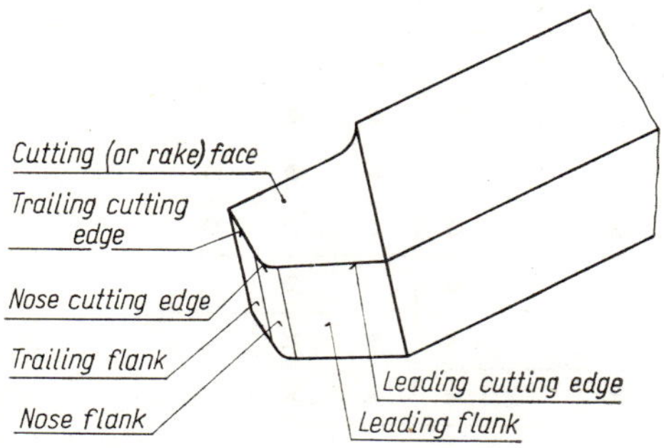

Fig. 1.3 Cutting tool point

The *trailing edge* is that portion of the cutting edge which carries out the minor part of the cutting work.

The *nose cutting edge*, usually arc-shaped, is the portion of the cutting edge, connecting the leading and trail edges.

Similarly, the flank may be divided into:

— *leading flank*, shortly called the flank; it is this portion of the flank which, as a result of intersection with the rake face, forms the leading cutting edge,

— *trailing flank*—it is this portion of the flank whose line of intersection with the rake face constitutes the trailing edge,

— *nose flank*—it is this portion of the flank which, as a result of intersection with the rake face constitutes the nose cutting edge.

The most protruding portion of the cutting point is called the *nose*. In the geometrical sense the nose is the point situated on the cutting edge and reaching deepest into the workpiece material.

1.4.2 Reference systems for tool point characteristics

In order to define unequivocally the geometrical elements of the tool point, the following reference systems are used:

(a) *stationary systems*

(i) basic (main and auxiliary) system, shortly called the tool point system. It is constituted by the reference plane, cross-section plane and projection plane;

(ii) longitudinal-transverse system, consisting of the reference plane, longitudinal plane and transverse plane. It is also called the tool-in-hand system;

(b) *setting system*, connected with the position of the workpiece in relation to the tool;

(c) *motion* or *working system*, connected with the directions of basic motions.

Stationary systems are used to characterize the tool point independently of its position in relation to the workpiece or of the working motions. The common feature

Table 1.2 *Classification of cutting methods according to accuracy and the desired surface quality*

Machining method	Classes (according to ISO)																									
	of accuracy												of surface quality													
	1 to 3	4	5	6	7	8	9	10	11	12	13	14 to 16	1	2	3	4	5	6	7	8	9	10	11	12	13	14
I. CHIP REMOVING PROCESSES																										
Turning				●	●	o	o	o	×	×	—	—	—	—	—	×	×	o	o	●						
Boring				●	●	o	o	o	×	×	—	—	—	—	—	×	×	o	o	●						
Drilling								●	o	×	×	—	—	—	×	×	o	●								
Reaming			●	●	o	o	×	×	—	—							—	×	o	o	●	●				
Peripheral milling					●	●	o	o	×	×	—	—	—	—	×	×	o	o	●	●						
Face milling					●	●	o	o	×	×	—	—	—	—	×	×	o	o	●	●						
Planing and shaping						●	o	o	o	×	×	—	—	—	×	×	o	o	●	●						
Broaching			●	●	o	o	×	×	—	—	—	—				—	×	o	o	●	●					
II. ABRASION PROCESSES																										
A. *Using abrasive tools*																										
Centre-type cylindrical grinding			●	o	o	×	×	—	—								—	×	×	o	●	●				
Centreless cylindrical grinding			●	o	×	×	×	—	—								—	×	×	o	o	●				

Internal grinding			●	o	o	x	x	—	—								—	x	x	o	●	●				
Surface grinding			●	o	o	x	x	—	—	—							—	x	x	o	o	●				
Abrasive belt grinding				●	o	o	x	x	—	—						—	x	x	o	o	●					
Surface honing		●	●	o	o	x	—	—										—	—	x	x	o	●			
Shaft and internal honing	●	●	●	o	o	x	—											—	—	x	x	o	●	●		
Superfinish	●	●	o	o	x	x	—												—	—	x	o	o	●	●	
B. *Using loose abrasive*																										
Lapping		●	●	o	o	x	x	—									—	—	x	o	o	●	●			
Mechanical polishing	●	●	o	o	x	x	—	—										—	—	x	x	o	o	●	●	●
Vibratory and barrel finishing					●	●	●	o	o	x	x	—			—	—	x	x	o	●						
Abrasive-blast treatment				●	●	●	o	o	x	x	—	—			—	—	x	o	o	●	●					
Ultrasonic machining				●	●	o	o	x	x	—						—	x	o	o	●	●					

Accuracy
— rough
x fairly accurate
o accurate
● highly accurate

Surface quality
— rough
x fairly smooth
o smooth
● very smooth

of both stationary systems is that they have one plane in common, viz. the basic plane.

The basic stationary system (Fig. 1.4) is used exclusively to determine unequivocally and to define the shape of tool point elements. This system is connected with the position of projections of cutting edges upon the basic plane. The line *BC* in Fig. 1.4 is the projection of the leading cutting edge and *CD* that of the trailing edge.

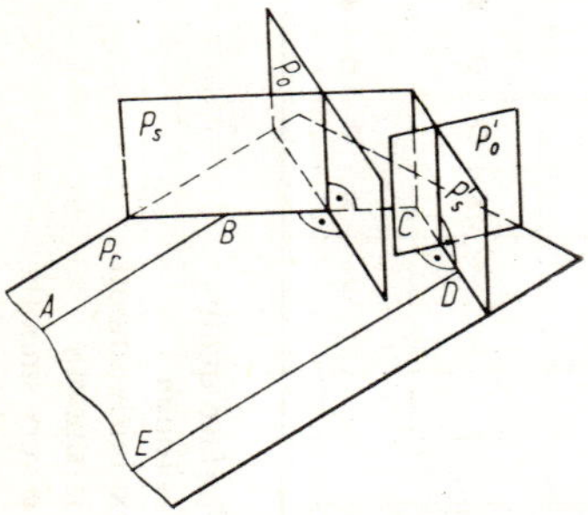

Fig. 1.4 Stationary basic system, main and auxiliary

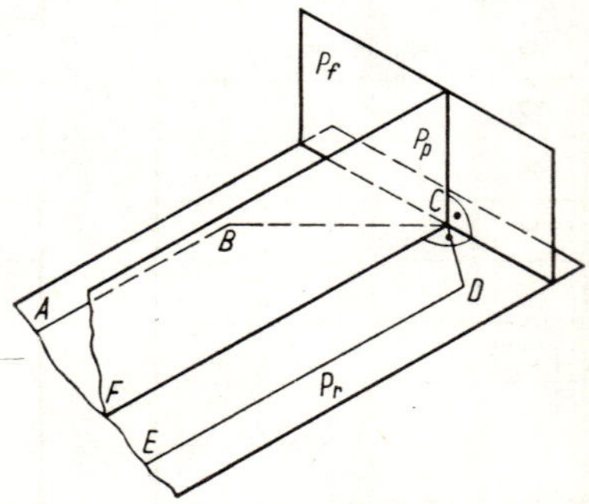

Fig. 1.5 Stationary longitudinal-transverse system

The longitudinal-transverse stationary system (Fig. 1.5) is used to characterize and define not only the point, but also the entire tool. The positioning of this system corresponds to directions characteristic of the given type of tool, e.g. with its axis of symmetry, direction of length etc. Figure 1.5 shows the longitudinal-transverse system of a lathe tool, corresponding to the direction of length *CF*.

The setting system enables one to determine the geometrical elements of a tool point with reference to the workpiece position. This is useful for determining the methods for fastening the tools in their holders.

The motion system allows one to determine the actual magnitudes of tool point angles during the cutting process.

1.4.3 The basic tool point reference system (tool point system)

The *basic stationary reference system* is called in short the basic reference system or tool point system. The following three planes are assumed to form the basic reference system:

P_r (π_1)*—reference plane defined conventionally for each particular kind of tool, e.g. in the case of cutting tools the reference plane coincides with the shank base plane. Examples of reference plane positions for various types of tools are shown in Fig. 1.6,

P_0 (π_3, π_n)—cross-section plane (or normal plane) which is perpendicular (or normal) to the projection of the cutting edge upon the reference plane in a selected point of this projection,

* P_r, P_0 P_s—plane designations accepted by ISO, designations used previously are in parentheses.

Tool sketch	Description of reference plane
Cutting tools	reference plane is the tool shank base plane
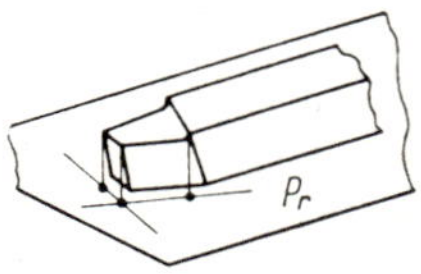	
Drills	reference plane is the plane passing through drill axis, parallel to the cutting edges
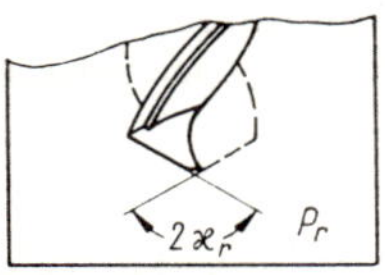	
Milling cutters	reference edge is the plane passing through milling cutter axis and a point on the cutting edge
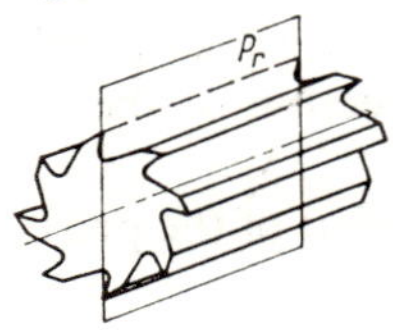	
Pull broaches	reference plane is the plane perpendicular to the direction of broach length and passing through a point on the cutting edge
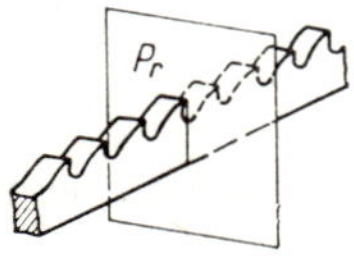	

Fig. 1.6 Examples of reference planes of various tools

P_s (π_2, π_s)—projection plane (also called tangential plane), i.e. a plane perpendicular to both planes mentioned above and passing through a selected point or rectilinear section of the cutting edge.

The *main reference system* constitutes a case of the basic reference system in which the cross-section plane and the projection plane are passing through a characteristic section or point of the leading cutting edge. The cross-section plane belonging to such a system is called the *orthogonal plane.*

The *auxiliary reference system* constitutes a case of the basic reference system in which the cross-section plane and the projection plane are passing through a characteristic section or point of the trailing edge.

On the reference plane of the basic system there are to be defined (Fig. 1.7):

— included angle ε_r, between rectilinear sections of cutting edge projections or between tangents at selected points of cutting edge projections,

— curvatures characterized by rounding radii, particularly by the nose (nose cutting edge) radius designated by r_ε.

The designations on the cross-section plane (in the example shown in Fig. 1.7—on the orthogonal plane) are:

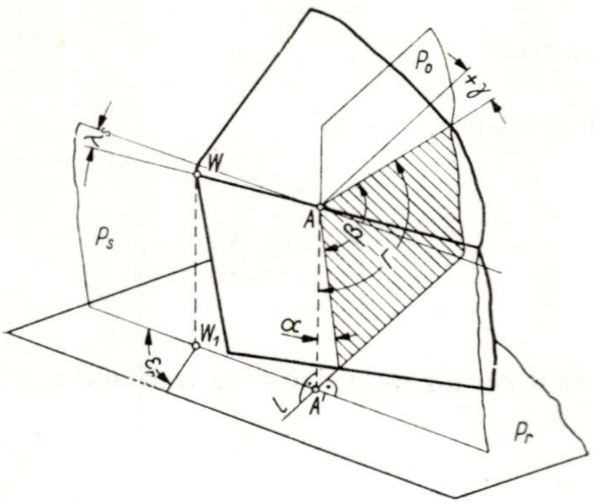

Fig. 1.7 An example of a main reference system and of the designation of geometrical elements of the tool point

— *wedge angle* (formerly called tool point angle) β included between the traces of intersection with cutting face and flank,
— *rake angle* γ included between the traces of rake face and of a plane parallel to the reference plane and passing through the point of intersection of the cutting edge with the cross-section plane,
— *clearance angle* α included between the trace of the flank and that of the projection plane,
— *cutting angle* Γ included between the trace of the rake face and that of the projection plane. Thus the cutting angle

$$\Gamma = \alpha+\beta$$

On the projection plane the *cutting edge inclination angle* λ_s is defined as being included between the cutting edge and the trace of a plane parallel to the reference plane at the point of intersection of the cutting edge with the cross-section plane.

The terms and designations of the angles given above refer to the main reference system. In the auxiliary reference system the corresponding angles are designated by the same symbols, but with a superscript ′ or subscript 1 and to angle names the word 'auxiliary' is added. For example, γ' or γ_1 the auxiliary rake angle.

There exist conventions concerning the algebraic signs of those angles which can have positive or negative values, such as the rake angle and cutting edge inclination angle.

The rake angle is positive when the trace (or a tangent to it at the point of intersection with a cutting edge) of the cutting face on the cross-section face is declining, starting from the point of intersection of the cutting edge with cross-section plane; it is negative when the trace of the rake face is rising and it is equal to zero when this trace is parallel to the reference plane.

This convention may also be noted by means of formulae:
— the rake angle is positive when

$$\alpha+\beta = \Gamma < 90° \quad \text{because} \quad \gamma = 90-\Gamma = 90-\alpha-\beta > 0 \tag{1.8}$$

— the rake angle is equal to zero when

$$\alpha+\beta = \Gamma = 90° \quad \text{because} \quad \gamma = 90-\Gamma = 90-\alpha-\beta = 0 \tag{1.9}$$

— the rake angle is negative when

$$\alpha+\beta = \Gamma > 90° \quad \text{because} \quad \gamma = 90-\Gamma = 90-\alpha-\beta < 0 \tag{1.10}$$

The cutting edge inclination angle is shown in Fig. 1.8.

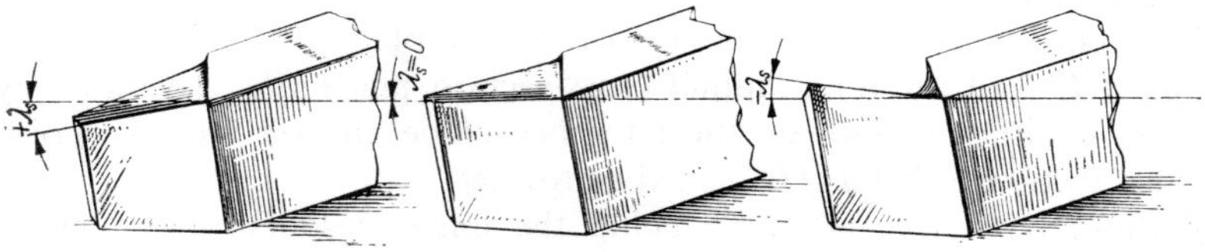

Fig. 1.8 Determination of the positive or negative value of the cutting edge inclination angle

1.4.4 Relations between the values of angles in the basic and longitudinal-transverse system

It is possible to establish trigonometric relations between the rake and clearance angle in the tool point system and in the tool-in-hand system.

Figure 1.9 will be used to derive the formulae. It is assumed that the tool-in-hand (longitudinal-transverse) system has been obtained by rotating the basic system clock-wise through the angle $\varkappa_r$. Rake and clearance angles in the longitudinal-transverse system are designated by subscripts p and f (or y and x), respectively.

The *angle of rotation* $\varkappa_r$ is measured in the reference plane of the stationary longitudinal-transverse system* (Fig. 1.10). It is called the *cutting edge angle* and is

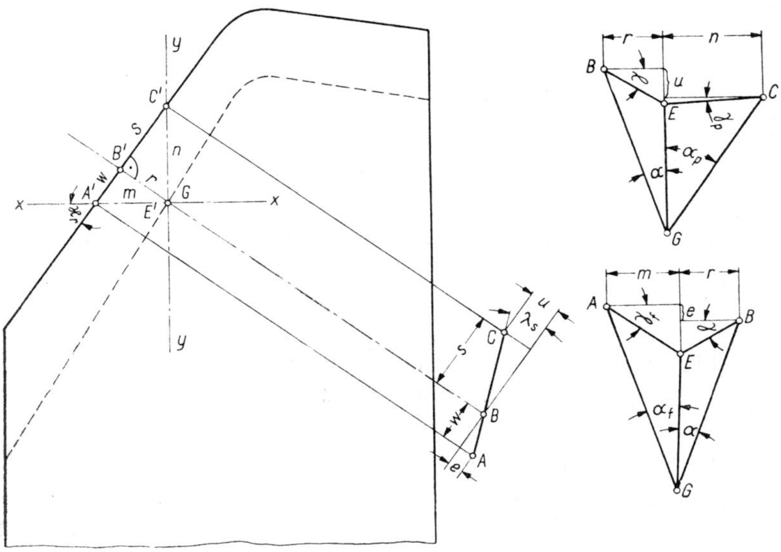

Fig. 1.9 Designation used to determine the relations between angles in the tool point system and in the tool-in-hand system

* Cutting edge angles in a motion system are established on the reference plane of that system, between the projection of the respective edge or tangent to it at a given point and the projection of the direction of feed rate vector.

included between the projection of the rectilinear portion of cutting edge or the tangent to curvilinear edge at a given point and the trace of transverse plane or of a plane parallel to it.

This leads to distinguishing the leading cutting edge angle corresponding to the leading edge, and the trail edge angle corresponding to the trail edge.

The tool point is cut across (Fig. 1.9) by the orthogonal plane $B'G$, the transverse cross-section plane $A'E'$ and the longitudinal cross-section plane $C'E'$ so that they all intersect at point G situated both on the reference plane and on the tool point base edge. Triangles BEG, CEG and AEG are thus formed.

By grouping those triangles and introducing the angle designations given in Fig. 1.7, the following equations can be obtained:

$$u = r\tan\gamma - n\tan\gamma_p \quad \text{and} \quad u = r\cot\alpha - n\cot\alpha_p \tag{1.11}$$

$$e = m\tan\gamma_f - r\tan\gamma \quad \text{and} \quad e = m\cot\alpha_f - r\cot\alpha \tag{1.12}$$

and also:

$$r = n\cos\varkappa_r = m\sin\varkappa_r \tag{1.13}$$

$$n = m\tan\varkappa_r; \quad m = n\cot\varkappa_r$$

$$u = s\tan\lambda_s; \quad s = n\sin\varkappa_r; \quad u = n\sin\varkappa_r\tan\lambda_s \tag{1.14}$$

$$e = w\tan\lambda_s; \quad w = m\cos\varkappa_r; \quad e = m\cos\varkappa_r\tan\lambda_s \tag{1.15}$$

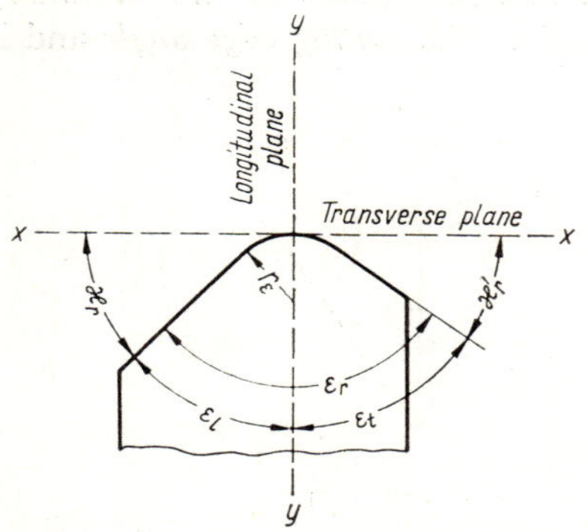

Fig. 1.10 Designations of cutting edge angles

Equations (1.11)–(1.15) may be used for calculating rake and clearance angles in the longitudinal-transverse system, as a function of system rotation, rake, clearance and edge inclination angle in the basic system.

In the longitudinal cross-section

$$\begin{aligned} \tan\gamma_p &= \cos\varkappa_r\tan\gamma - \sin\varkappa_r\tan\lambda_s \\ \cot\alpha_p &= \cos\varkappa_r\cot\alpha - \sin\varkappa_r\tan\lambda_s \end{aligned} \tag{1.16}$$

In the transverse cross-section

$$\begin{aligned} \tan\gamma_f &= \sin\varkappa_r\tan\gamma + \cos\varkappa_r\tan\lambda_s \\ \cot\alpha_f &= \sin\varkappa_r\cot\alpha + \cos\varkappa_r\tan\lambda_s \end{aligned} \tag{1.17}$$

The relation between rake and clearance angles in the longitudinal and transverse cross-section is obtained from eqns. (1.16) and (1.17), thus

$$\begin{aligned} \tan\gamma_p &= \cot\varkappa_r \tan\gamma_f - \frac{\tan\lambda_s}{\sin\varkappa_r} \\ \cot\alpha_p &= \cot\varkappa_r \cot\alpha_f - \frac{\tan\lambda_s}{\sin\varkappa_r} \end{aligned} \qquad (1.18)$$

$$\begin{aligned} \tan\gamma_f &= \tan\varkappa_r \tan\gamma_p + \frac{\tan\lambda_s}{\cos\varkappa_r} \\ \cot\alpha_f &= \tan\varkappa_r \cot\alpha_p + \frac{\tan\lambda_s}{\cos\varkappa_r} \end{aligned} \qquad (1.19)$$

For calculating rake and clearance angles in the basic system as functions of rake angles in the longitudinal-transverse system and of the system rotation angle, the following equations are applied:

$$\tan\gamma = \sin\varkappa_r \tan\gamma_f + \cos\varkappa_r \tan\gamma_p \qquad (1.20)$$

$$\cot\alpha = \sin\varkappa_r \cot\alpha_f + \cos\varkappa_r \cot\alpha_p \qquad (1.21)$$

The equations derived above may be used for determining angles in any cross-section of the tool point, if angle $\varkappa_r$ is measured in the reference plane and is included between this cross-section and the trace of the projection plane.

1.4.5 Relationship between angles in the longitudinal-transverse system and in the setting and working systems

Angle differences in the setting system, as compared with those in the longitudinal-transverse system, result from a change of the reference plane.

The position of the reference plane in the setting system is determined by a characteristic direction of the workpiece, e.g. the axis of symmetry (Fig. 1.11) and by a point selected on the cutting edge (e.g. tool point nose). Consequently, angle ξ is formed between both planes. For the purpose of determining the rake and clearance angles this is equivalent to a rotation of the reference system about the x–x axis.

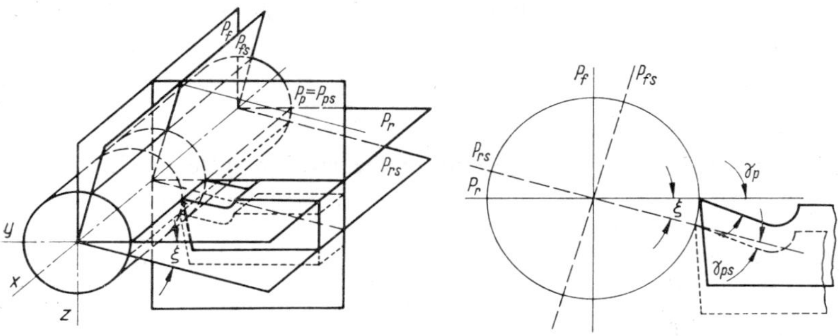

Fig. 1.11 The setting system as a transformation of the longitudinal-transverse system

Therefore the values of rake and clearance angles in the longitudinal cross-section of the setting system are expressed as follows

$$\begin{aligned} \gamma_{ps} &= \gamma_p \pm \xi \\ \alpha_{ps} &= \alpha_p \pm \xi \end{aligned} \tag{1.22}$$

where the subscripts p correspond to the longitudinal-transverse system and the subscripts ps to the setting system.

It is assumed conventionally that the value of angle ξ is positive when the longitudinal-transverse system is rotated anti-clockwise. It follows that the angle of rotation of the system shown in Fig. 1.11 has a negative value.

The working (motion) system can be considered as a transformation of the longitudinal-transverse system.

This transformation results from a different position of the transverse plane (Fig. 1.12) or of the longitudinal plane, which is equivalent to the rotation of the system about the x or y axis. As a consequence, the reference plane is also rotated from the position P_r to the position P_{re}.

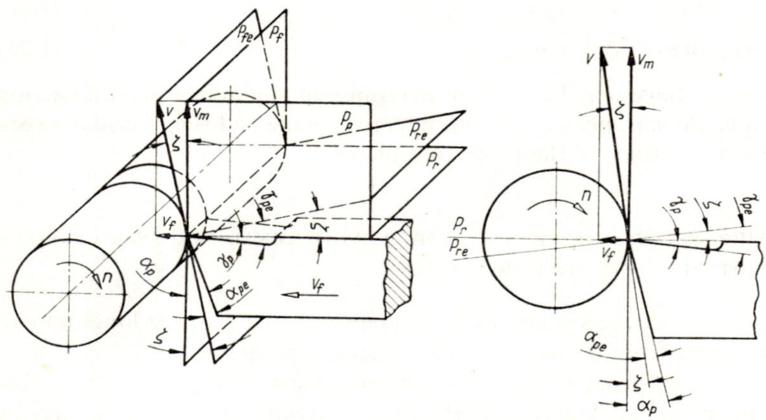

Fig. 1.12 The motion system as a transformation of the longitudinal-transverse system

In the case shown in Fig. 1.12 the position of the transverse cross-section plane is determined by the cutting speed vector v resulting from the addition of the primary motion speed vector v_m to the auxiliary motion speed vector v_f. The angle of rotation ζ thus obtained has the value

$$\zeta = \arctan \frac{v_f}{v_m} \tag{1.23}$$

and the angles in the longitudinal cross-section of the motion system will be determined by

$$\begin{aligned} \gamma_{pe} &= \gamma_p + \arctan \frac{v_f}{v_m} \\ \alpha_{pe} &= \alpha_p - \arctan \frac{v_f}{v_m} \end{aligned} \tag{1.24}$$

It is evident from the second equation above that with feed rate v_f sufficiently high in relation to primary motion speed v_m the motion system clearance angle may become equal to zero. This will happen when

$$\alpha_p = \arctan \frac{v_f}{v_m} \tag{1.25}$$

In practice, for a given type of tool and a given machining method and its variety, a minimum value of this angle, $\alpha_{p\,\min}$, is established, so that excessive friction between the flank and material is prevented.

1.4.6 Type of tool points

The classification of tool points is based on the differences in shape of the projection upon the reference plane and of the orthogonal cross-section.

A very wide differentiation of tool point shapes as projected upon the reference plane makes a systematic classification of tool points difficult. A classification according to this criterion is connected with tool type distinction.

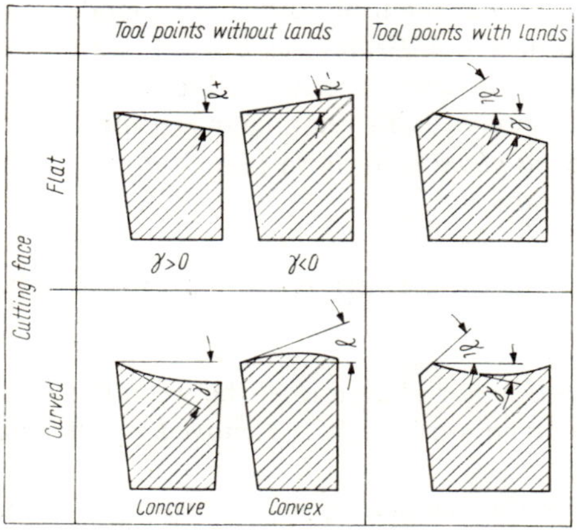

Fig. 1.13 Type of tool points depending on the shape of cutting face

Depending on the tool point shape in the orthogonal cross-section plane (Fig. 1.13):

— tool points with a flat rake face and
— tool points with a curved rake face

are distinguished.

Rake faces can be made with or without lands.

A *land* means a high chamfer of the rake face running along the cutting edge under angle γ_s which is always smaller than the rake angle γ.

The tool point classification shown in Fig. 1.13 is simplified; there are many intermediate and combined varieties, depending on the type of operation.

1.5 Characteristics of undeformed chip [1/4]

1.5.1 Work surface, machined surface and cut surface

The surface of the workpiece before it is machined (Fig. 1.14) is called the *work surface* and after the machining is finished the *machined surface*. The term *cut surface* describes the surface reproduced on the workpiece by that portion of the cutting edge which is active in the cutting process.

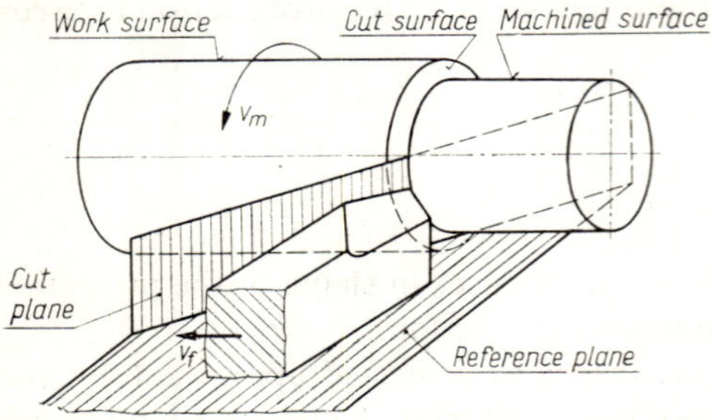

Fig. 1.14 The cut surface, work surface and machined surface

It follows from these definitions that the cut surface joins the work surface with the machined surface. In the case of turning by means of a tool point with angle $\lambda_s = 0$ and angle $\varkappa_r \neq 0°$ or $90°$, and assuming that the tool nose is not rounded and there is no feed motion, the cut surface would constitute the curved surface of a cone frustum. If the feed motion is taken into account, the cut surface becomes an involute helical surface.* In the case of tool points with rectilinear leading cutting edge, a cut plane may be singled out. By the *cut plane* is meant a plane tangent to the cut surface and including the leading cutting edge.

1.5.2 Undeformed chip dimensions and the tool point portions taking part in machining

The *undeformed chip* is the part of workpiece material contained between the work surface, the future cut surface and the machined surface. It is removed by the tool point and transformed into the actual chip.

The undeformed chip is characterized by the cross-section perpendicular to the cutting speed vector passing through the tool point nose and called the characteristic cross-section of the undeformed chip.

In the characteristic cross-section (Fig. 1.15) are distinguished:

— *width of cut b* assumed to be the length of one of the sides of the geometric figure approximating the shape of chip cross-section (in the case shown in Fig. 1.15 it is the side *AC* of the parallelogram *ACDG*),
— *undeformed chip thickness h* measured perpendicularly to width *b*,
— *depth of cut a*, dimension of undeformed chip, measured perpendicularly to the primary motion speed vector and to the auxiliary motion speed vector,
— *feed s*, dimension of undeformed chip in the direction of auxiliary motion speed,
— *leading cutting edge angle* $\varkappa_r$ included between the direction of auxiliary motion

* A surface obtained by unwinding a straight line from a helix on a cylinder.

speed and an extension of the rectilinear section of leading cutting edge or—if the latter is curvilinear—a tangent to it at the given point,

— *trailing edge angle* $\varkappa_r'$ included between the direction of auxiliary motion and an extension of the rectilinear section of the trailing edge or—if the latter is curvilinear—a tangent to it at the given point,

— *tool point nose radius* (or radius of curvature of nose cutting edge) r_ε.

The area of undeformed chip section A is the area of a geometrical figure approximating the area of undeformed chip in its characteristic cross-section.

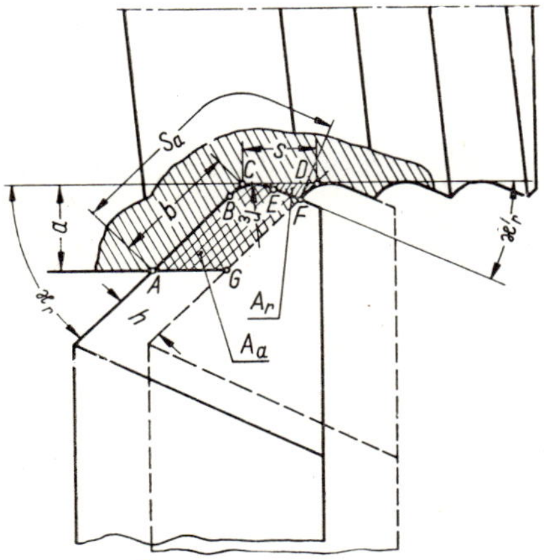

Fig. 1.15 Geometrical dimensions of undeformed chip

In the case when this figure constitutes a parallelogram (Fig. 1.15) the following relationships are valid:

$$a = b \sin \varkappa_r; \qquad s = \frac{h}{\sin \varkappa_r} \tag{1.26}$$

$$A = h\,b = s\,a \tag{1.27}$$

The actual area of undeformed chip section A_a is usually smaller than the nominal one.

The residual area of undeformed chip section A_r is the difference

$$A_r = A - A_a \tag{1.28}$$

The residual area usually constitutes a small percentage of the nominal area and in multi-pass machining the residual areas of the preceding pass are added to the actual areas of the following pass. Therefore, in most cases it can be assumed that

$$A \approx A_a \tag{1.29}$$

The shape of the undeformed chip section area is characterized by the ratio: $a/s = m$ or $b/h = \delta$. An indication of the shape of undeformed chip section area can be the unit length of active cutting edge, determined by the ratio:

$$j = \frac{S_a}{A} = \frac{S_a}{a\,s} = \frac{S_a}{h\,b} \tag{1.30}$$

where S_a is the active length of the cutting edge.

REFERENCES

1/1 GRANOVSKII, G. I.: 'Kinematika rezaniya' (Kinematics of Machining), Mashgiz, Moskva 1948

1/2 ETIN, A.O.: 'Kinematicheskii analiz metodov obrabotki metallov rezaniem' (Kinematic Analysis of the Methods of Machining Metals by Cutting), Izd. Machinostroenie, Moskva 1964

1/3 KUNSTETTER, S.: 'Narzędzia skrawające do metali' (Metal Cutting Tools), WNT, Warszawa 1961

1/4 BIERNAWSKI, W.: 'Teoria obróbki metali skrawaniem' (Theory of Machining Metals by Cutting), PWN, Warszawa 1956

CHAPTER 2

Physical basis of the cutting process

2.1 Mechanics of the cutting process

2.1.1 PRELIMINARY INFORMATION

The removal of the undeformed chip, its transformation into actual chips, and the production of the machined surface, is a complex physical (and, partly, chemical) process. During elastic and plastic deformation several complex processes such as heat generation and propagation, internal and external friction, wear, hardening and tempering etc. are simultaneously taking place.

Research in cutting takes place in order to understand the nature and course of all those phenomena which lead to a rational application and control of the cutting process. The study of the nature and course of the complex cutting process is usually achieved by means of observation, research and analysis of phenomena and processes using simplified models of the actual cutting process.

2.1.2 ORTHOGONAL CUTTING AS A CASE OF PLANE PLASTIC DEFORMATION

A form of cutting which can be considered as a case of plane deformation is called *orthogonal cutting*. It takes place when cutting is carried out by means of a single rectilinear edge which is longer than the width of cut and the width of cut is sufficiently great and considerably greater than the undeformed chip thickness.

The mechanics of the orthogonal cutting process is considered using the following assumptions:

— the properties of the machined material are isotropic,
— the temperature does not change in the whole cutting area,
— the machined material hardens (becomes stronger) as a result of permanent deformation,
— the condition of plasticity is the appearance of maximum shear stresses,
— the portion of the tool point, which adjoins the flank, does not take part in the process of plastic deformation, and the radius of the cutting edge is equal to zero.

The orthogonal cutting process can be considered analogous to the process of pressing a punch into a piece of material (Fig. 2.1). In such a case deformation lines appear in the pressed material. If it is assumed that the compression is not accompanied by friction, then the orthogonal network of deformation lines will form an angle of 45° with the flat surface of the punch. In the area *OBC* the deformation network will consist of a number of radii starting from point *O* and of concentric arcs. Such a system is formed when point *O* is considered as a curve whose radius is equal to zero.

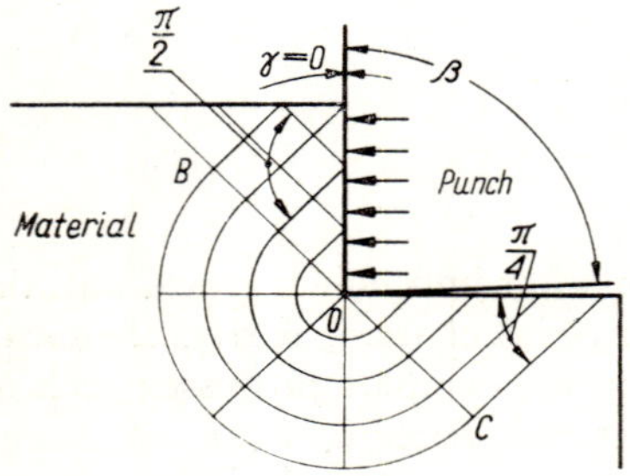

Fig. 2.1 Plastic pressing of a flat punch into material

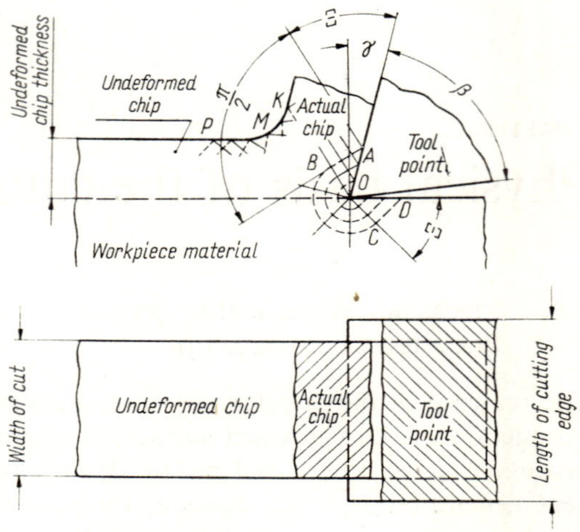

Fig. 2.2 Deformation lines in chip formation area

In the case of orthogonal cutting with no external friction (Fig. 2.2), the tool point exerting pressure upon the material generally has a rake angle $\gamma \neq 0$. Consequently, the deformation lines will constitute a network over a larger area *OBC*.

Plastic deformation stretches over a certain area, limited on the outside by section *PK*. This section is reached by lines of deformation forming angles $\Xi = 45°$ with tangents to curve *PMK*.

If external friction is taken into consideration, then the deformation lines in the area near the tool point nose will change their position. If the angle of external friction is designated by Θ, the angle of inclination of the deformation lines, marked by continuous lines, will be

$$\Xi = \frac{\pi}{4} + \Theta \tag{2.1}$$

Thus the area *OBC* will become narrower by the angle 2Θ.

Of the deformation lines running in two principal directions, those marked by continuous lines are more important. In most cases of cutting the intensity of deformation in the direction marked by intermittent lines is small and can be neglected.

A simplified diagram of the area and lines of plastic deformation in the orthogonal cutting process is shown in Fig. 2.3. This area is limited by:

— line *OP* called the initial boundary of plastic deformation,

— line PK called the external boundary of plastic deformation,
— line KO called the final boundary of plastic deformation.

The area within these boundaries is the area where the undeformed chip becomes the actual chip, or the area of chip formation.

The position of the boundary curves, and particularly that of the initial boundary of plastic deformation, was established by theoretical analysis and confirmed experimentally by a number of research workers including Krivoukhov [2/1]. It is important to note that the initial curve OP in its lower part is situated below the line of cutting. This indicates that the area of plastic deformation reaches the material not transformed into the chip (i.e. to the upper layer of the machined material).

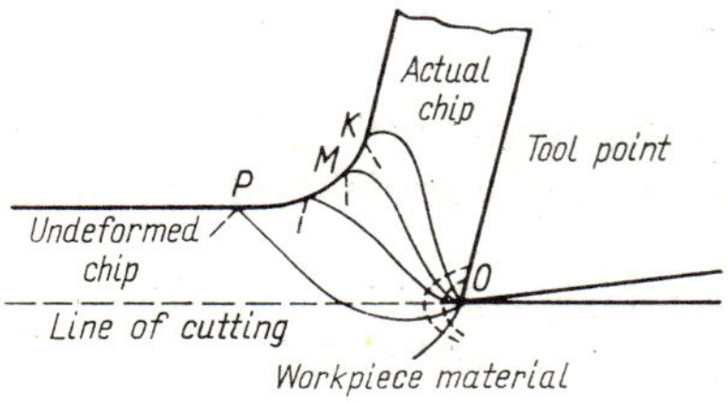

Fig. 2.3 Marking of the plastic deformation zone *POK* or chip formation zone

It follows from the system of deformation lines shown that shear does not take place on a single plane. Compressive stresses change within the chip formation area from the value corresponding to the initial yield stress of the material to the value corresponding to the yield stress of the hardened (strengthened) chip material. Plastic deformation also increases correspondingly. The boundaries of the chip formation area and the chip deforming process itself are not stable and uniform. A study of the chip formation process indicates a periodic increase in stress and deformation, accompanied by slip. As a result, on the external surface of the chip there appear creases and shoulder-like irregularities. This phenomenon is the more distinct when the deformed material is less plastic. The periodic increase in stress, deformation and slip produces the elementary chip structure observed by the early researchers on the cutting process, like Time, Taylor and many others.

In order to establish the relationship between the parameters characterizing the cutting process, a simplified model is used, as shown in Fig. 2.4.

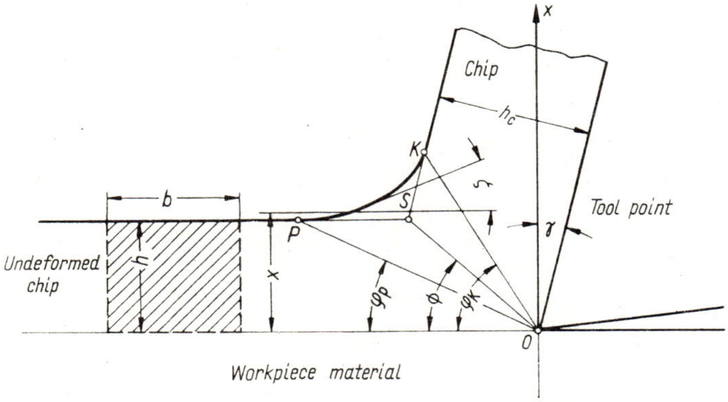

Fig. 2.4 Simplified diagram of orthogonal cutting

The main feature of this simplified diagram of orthogonal cutting is the replacement of curved slip lines by straight ones. Consequently, the chip formation area is contained between the initial boundary *OP* and the final boundary *OK*. Co-ordinates of point *P* are x_P and φ_P and those of point *K* x_K and φ_K. To establish the position of the successive lines, the same reference system is used and the angle of inclination of the slip line in question is given.

To complete the diagrammatic representation of the chip formation process, the curvilinear external boundary *PK* is replaced by the broken line *PSK*. The co-ordinates of point *S* are designated by Φ, x_S.

2.1.3 Shear strain during the cutting process

In order to determine the shear strain ε during orthogonal cutting, it is assumed that a measure of ε is the tangent of the angle of rotation of a straight line, perpendicular to the gliding plane.

The value of ε at point *K*, i.e. at the final boundary of chip formation area (Fig. 2.4), when the angle of tangent $\zeta = 90° - \gamma$, is determined as follows:

$$\varepsilon = \cot\varphi_K + \tan(\varphi_K - \gamma) \tag{2.2}$$

On the other hand, at the initial boundary of the chip deformation area the tangent angle $\zeta = 0$, and therefore the shear strain $\varepsilon = 0$.

2.1.4 State of stress in the chip deformation area

On the basis of equilibrium (disregarding the moment of the resultant force in relation to the centre of the slip plane) the state of stress on the final boundary plane is given by

$$\tau_K \frac{x_K b}{\sin\varphi_K} = W_\gamma \cos(\varphi_K + \Theta - \gamma) \tag{2.3}$$

where: τ_K—shear stress in the slip (shear) plane, acting at an angle φ_K relative to the *z* axis, x_K—ordinate of the point of intersection of the slip plane with the external

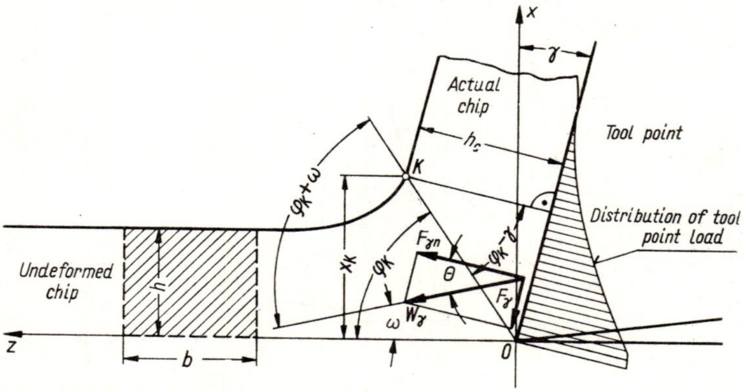

Fig. 2.5 Distribution of forces on the cutting face

boundary curve in the final plane, b—width of cut, W_γ—the resultant force acting on the active cutting face, $W_\gamma = \sqrt{F_{\gamma n}^2+F_\gamma^2}$; Θ—the average friction angle on the cutting face $\left(\Theta = \arctan \dfrac{F_\gamma}{F_{\gamma n}}\right)$, γ—rake angle.

Thus the shear stress on the final boundary plane is

$$\tau_K = \frac{W_\gamma \sin\varphi_K \cos(\varphi_K+\Theta-\gamma)}{b\, x_K} \tag{2.4}$$

The angle $\Theta-\gamma$ is generally designated as

$$\omega = \Theta-\gamma \tag{2.5}$$

Angle ω is called the *angle of action* of the chip forming force. As seen from Fig. 2.5, the angle of action is contained between the direction of the resultant chip-forming force W_γ and the cutting direction.

Taking into account eqn. (2.5), the stress on the final boundary plane is

$$\tau_K = \frac{W_\gamma}{b\, x_K} \sin\varphi_K \cos(\varphi_K+\omega) \tag{2.6}$$

Stress values calculated according to eqn. (2.6) are about 20% higher than those obtained by measurement. The reason for this is that the state of stress in the cutting process is a complex one and consequently the tangential stresses in such a case are lower than in uniaxial stress.

To compensate for these differences, in a suitable slip plane a coefficient which reduces the shear angle can be introduced. The corrected shear angle will then be

$$\varphi_z = \varphi - e\,\varphi_K \tag{2.7}$$

where e—coefficient depending on the state of deformation and stress.

Using eqn. (2.7) it is possible to determine the tangential stress in any slip plane

$$\tau = \frac{W_\gamma}{b\, x} \sin\varphi_z \cos(\varphi_z+\omega) \tag{2.8}$$

and in the final boundary plane

$$\tau_K = \frac{W_\gamma}{b\, x_K} \sin(k\varphi_K) \cos(k\varphi_K+\omega) \tag{2.9}$$

where $k = 1-e$, and therefore $\varphi = \varphi_K$.

Using eqn. (2.8) it is possible to establish the limits of the chip formation area:

— the *initial boundary* of the chip formation area can be determined taking into account that:

$$\tau = \tau_0 \qquad x = h \qquad \text{and} \qquad \varphi = \varphi_p$$

thus

$$\varphi_p = f(h, \tau_0, k, \varphi_K, \omega) \tag{2.10}$$

— the *external boundary* of the chip formation area can be determined, assuming that

$$\tau = A\,\varepsilon^m \tag{2.11}$$

where: A—a constant characterizing the yield stress of the material, m—a constant characterizing the ability of the material to strengthen (harden); from eqns. (2.8) and (2.11)

$$x = f(h, m, x_K, \varphi_K, k, \gamma, \omega) \tag{2.12}$$

— the *final boundary* of the chip formation area may then be established, assuming that $\tau = \tau_K$, where τ_K is the yield stress of the strengthened (hardened) chip material. Then, after the substitution of $\tau = \tau_K$ and $x = x_K$ into eqn. (2.8), φ_K is determined as follows:

$$\varphi_K = f(k, \omega, \gamma) \tag{2.13}$$

This method of establishing the boundaries of the chip formation area is given by N. N. Zorev [2/2].

Metallographic studies of the chip and the zone of its formation, or the so-called *chip root*, reveals the existence of elongation lines whose direction does not coincide with that of slip lines. Figure 2.6 shows diagrammatically the direction of chip texture. The angle between the final deformation boundary and the direction of elongation lines is designated by ψ_c and called the *angle of crystal elongation.*

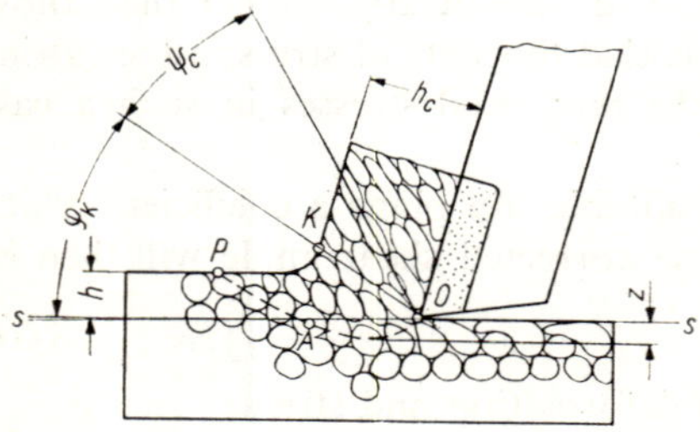

Fig. 2.6 Diagrammatic presentation of the chip elongation lines and angle; z—thickness of permanently deformed layer, other designations are explained in the text

The position of the elongation line may be determined by the *angle of texture*

$$\eta_c = \varphi_K + \operatorname{arccot} \frac{\varepsilon}{2} \pm \sqrt{1 + \frac{\varepsilon^2}{4}} \tag{2.14}$$

The angle of chip texture η_c is measured between the direction of the elongation line and that of the cutting line.

2.1.5 Chip compression

Within the range of stress and deformation appearing in the cutting process it can be assumed that the actual chip volume is equal to that of the undeformed chip.

Designating in Fig. 2.7 by:

b —average width of undeformed chip, in mm,
h —average thickness of undeformed chip, in mm,
l —average length of undeformed chip, in mm,
b_c—average width of actual chip, in mm,
h_c—average thickness of actual chip, in mm,
l_c—average length of actual chip, in mm,

it can be written that

$$b\,h\,l = b_c h_c l_c \quad \text{mm}^3 \tag{2.15}$$

This equation may be transformed as follows:

$$\frac{b}{b_c}\frac{h}{h_c}\frac{l}{l_c} = 1 \tag{2.16}$$

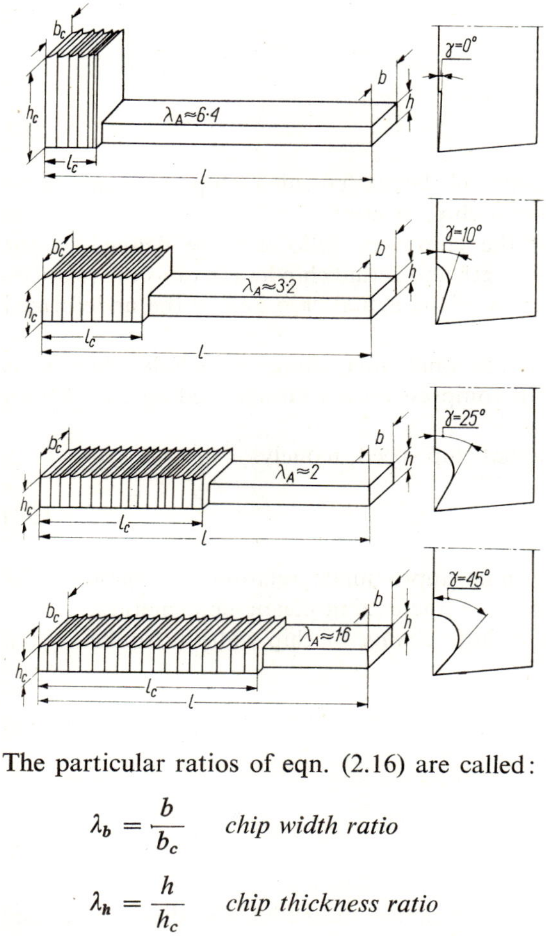

Fig. 2.7 Illustration of the influence of rake angle on chip compression

The particular ratios of eqn. (2.16) are called:

$$\lambda_b = \frac{b}{b_c} \quad \textit{chip width ratio} \tag{2.17}$$

$$\lambda_h = \frac{h}{h_c} \quad \textit{chip thickness ratio} \tag{2.18}$$

$$\lambda_l = \frac{l}{l_c} \quad \textit{chip length ratio} \tag{2.19}$$

Thus

$$\frac{\lambda_l}{\lambda_b \lambda_h} = 1 \tag{2.20}$$

or

$$\lambda_l = \lambda_b \lambda_h \tag{2.21}$$

This means that the chip length ratio—or longitudinal compression of the chip—is equal to the product of the chip thickness and width ratios.

In the literature it is sometimes assumed that $\lambda_l = \lambda_h$. Actual experiments show that in all cases of cutting producing a continuous chip, some widening occurs, and therefore the definition of the chip area ratio by the relation of actual chip thickness to undeformed chip thickness is incomplete and inaccurate.

If chip length ratio λ_l is generally called the *chip area ratio* λ_A, then

$$\lambda_l = \lambda_A \tag{2.22}$$

and

$$\lambda_A = \lambda_b \lambda_h = \frac{b_c h_c}{b\,h} = \frac{A_c}{A}$$

where: A—nominal cross-sectional area of the undeformed chip, in mm², A_c—average cross-sectional area of the actual chip, in mm².

It follows from eqn. (2.22) that the chip area ratio may be defined as the relationship between undeformed chip length and actual chip length or as the relationship between average cross-sectional area of the actual chip and of the undeformed chip.

Because of the non-uniform cross-sectional area along the whole chip, it is generally accepted in practice that chip compression is characterized by chip length ratio.

This allows one to calculate the actual chip speed, namely

$$v_c = \frac{v_m}{\lambda_l} \quad \text{m/min} \tag{2.23}$$

Using Fig. 2.4 it is possible to determine approximate relationships between the chip area factor and the angles and data characterizing plastic deformation.

In the case of orthogonal cutting without change of chip width it can be written that

$$\lambda_A = \frac{h_c}{h} = \frac{OS \sin(90 - \Phi + \gamma)}{OS \sin \Phi} \tag{2.24}$$

or

$$\lambda_A = \frac{\cos(\Phi - \gamma)}{\sin \Phi} \tag{2.25}$$

Thus, after the chip area factor and rake angle are measured, the shear angle may be calculated

$$\tan \Phi = \frac{\cos \gamma}{\lambda_A - \sin \gamma} \tag{2.26a}$$

or

$$\cot \Phi = \frac{\lambda_A}{\cos \gamma} - \tan \gamma \tag{2.26b}$$

Equation (2.25) was established by Time. It concerns a simplified diagram of chip formation, while the curvature of the external boundary of chip formation area was not taken into account.

The relationship between λ_l and other data can also be established while this curvature is taken into consideration.

It is evident from Fig. 2.5 that

$$h_c = \frac{x_K \cos(\varphi_K - \gamma)}{\sin \varphi_K} \tag{2.27}$$

From Zorev's calculations it is known that

$$x_K \approx \frac{h}{0{\cdot}09 - 0{\cdot}05 \cot \varphi_K} \tag{2.28}$$

Therefore

$$A = \frac{h_c}{h} \approx \frac{\cos(\varphi_K - \gamma)}{0{\cdot}9 \sin \varphi_K - 0{\cdot}05 \cos \varphi_K} \tag{2.29}$$

The angle of the final plane of deformation can be determined from eqn. (2.29), if γ and λ_A are known; then

$$\tan \varphi_K \approx \frac{\cos \gamma + 0{\cdot}05 \lambda_A}{0{\cdot}9 \lambda_A - \sin \gamma} \tag{2.30}$$

It follows from eqn. (2.30) that φ_K increases with increasing rake angle, and when the chip area factor decreases.

2.1.6 Relationship between the chip area factor and the degree of chip compression

The chip area factor is related to the degree of plastic deformation or compression.

An evaluation of the degree of plastic deformation in the cutting process may be obtained by comparing the shear strain during cutting ε to that resulting from compression ε_c.

According to Davidenkov [2/4]

$$\varepsilon_c = 1{\cdot}5 \ln \frac{h_0}{h} \tag{2.31}$$

where: h_0—original height of the compressed specimen, h—height of the specimen after compression.

Using eqns. (2.26) and (2.2) and substituting angle Φ for φ_K, the following relation is found between the shear strain during cutting and the chip area factor and rake angle

$$\varepsilon = \frac{\lambda_A^2 - 2\lambda_A \sin \gamma + 1}{\lambda_A \cos \gamma} \tag{2.32}$$

Now the shear strain from eqn. (2.32) compared with that resulting from compression, for the same degree of deformation, is

$$\frac{\lambda_A^2 - 2\lambda_A \sin \gamma + 1}{\lambda_A \cos \gamma} = 1{\cdot}5 \ln \frac{h_0}{h} \tag{2.33}$$

Hence the degree of plastic deformation is

$$\frac{h_0}{h} = e^{\frac{\lambda_A^2 - 2\lambda_A \sin\gamma + 1}{1 \cdot 5\lambda_A \cos\gamma}} \tag{2.34}$$

On the assumption that $\gamma = 0$ and $\lambda_A = 1$

$$\frac{h_0}{h} = e^{2/1 \cdot 5} = e^{4/3} \approx 3 \cdot 7 \tag{2.35}$$

It is evident from eqn. (2.34) and from the example cited above, that even in the case when $\lambda_A = 1$, the degree of chip compression is relatively high.

2.1.7 Chip texture and built-up edge

When cutting dry, the chip moving along the rake face must overcome the frictional force. The latter takes place under specific conditions:

— the normal pressure may reach several thousand kilograms force per square cm,
— mating surfaces, and particularly the cutting surfaces, are clean in the physical and chemical sense,
— contact surfaces are in a non-uniform temperature field and maximum temperature at the contact area may reach the melting temperature of the material being machined.

The frictional force of the chip against the rake face depends on the area of the contact surface and on the distribution of normal pressure and coefficient of friction over that surface.

Experiments conducted by many research workers indicate that the coefficient of friction may considerably exceed 0·5. Principal factors influencing the value of the coefficient of friction are (besides the properties of materials rubbing against each other): the temperature of the contact surface, the medium in which cutting takes place, the normal pressure and tool point rake angle.

As a result of friction on the contact surface between chip and tool point, the movement of chip layers nearest to the contact surface is delayed.

Under certain conditions prevailing on the surface of contact between chip and the rake face, friction may increase so much (on a smaller or larger part of the face) that the movement of further layers of the chip being formed is delayed. This results from the external frictional stress exceeding the yield stress of the nearest layers of the moving chip. A characteristic texture of the bottom layer of the chip then appears, which can be detected by metallographic methods.

Besides chip texture which is always more or less evident, under certain conditions of cutting ductile metals the appearance of the so-called *built-up edge* (BUE) can be observed. This is a wedge-like extension of the edge, consisting of the machined material.

The built-up edge partially takes over the work of the tool point, at the same time not only changing the process of chip formation, but also (which is more important) changing the dimensions and character of the machined surface. Therefore a knowledge of the law governing this building-up process if of great importance.

Beginning with Taylor [2/5] and Usachev [2/6] many research workers have contributed to the formation of modern views upon the building-up process.

The cutting edge is preceded by an area of stress and accompanying deformations. In a certain zone of this stress area there appears plastic deformation and compression of the material, changing the physical properties of the latter. In particular the yield stress of the hardened material increases. This phenomenon is most pronounced in the direction of the resultant stress of chip formation, approximately determined by the working angle ω (Fig. 2.8).

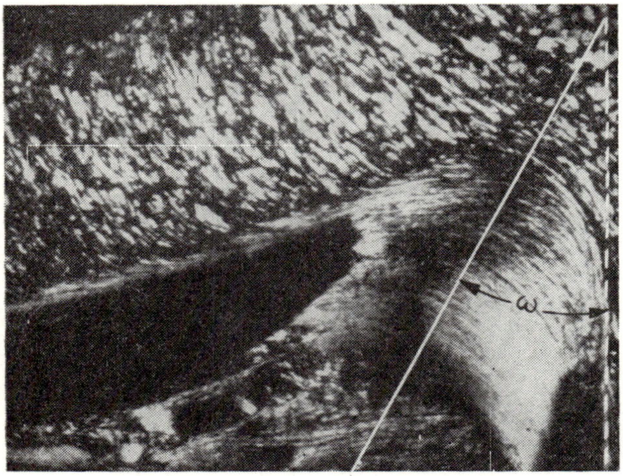

Fig. 2.8 Photomicrograph of built-up edge at the cutting of medium-carbon steel with a speed of 20 m/min. Magnified 60× (acc. to A. HULTGREN and G. KULLBERG: 'Metallographic observations concerning orthogonal metal cutting', *Jernkontorets Annaler*, **148**, 1964)

Plastic deformation and external friction of the chip against the tool point generate a certain amount of heat. This results in the formation of a high temperature field which also comprises the chip formation zone.

The high temperature field is characterized by a high gradient and therefore considerably modifies the physical and mechanical properties of the chip formation area.

According to observations and experiments carried out, conditions may be created in which, notwithstanding the influence of the high temperature field, the state of plasticity cannot be maintained in the area of maximum compression. As a result, the harder, wedge-shaped part of the machined material, adhering to the tool point because of high friction, cuts like an extended edge into the plastic workpiece material. This phenomenon occurs in a continuous way.

Thus the built-up edge gradually becomes thicker (Fig. 2.9). The condition necessary for the BUE to adhere to the tool point is a sufficiently high coefficient of external friction (in principle, higher than 0·5) and such a system of forces acting upon the tool point that does not exceed the strength or the frictional grip of the BUE on the tool point.

As the BUE increases on the tool point, so does the friction of the chip material flowing around it, and so do the forces and moments acting upon the BUE.

If at first the frictional resistance around the built-up edge (in the direction of chip flow) becomes equal to the friction of the BUE against the rake face, then the entire BUE flows along with the chip (Fig. 2.9a). If at first the strength of the BUE is exceeded, which may cause its splitting, breaking or shearing off, then isolated parts of the BUE may be retained by the chip and pressed into the machined material (Fig. 2.9b). The latter case occurs more frequently.

In both cases the formation or growth of the BUE is recommenced after it has broken away.

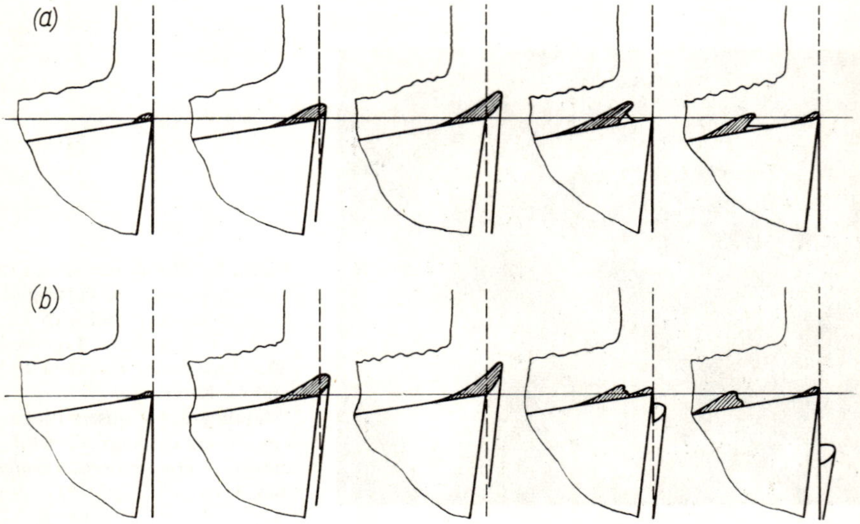

Fig. 2.9 Diagrammatic presentation of the cycle of built-up edge formation and flow: (a) BUE flow with chip only, (b) BUE flow with chip and partial pressing into the machined material

The result of the BUE is as if a new tool point, with a different rake angle, were created. This angle is called the *BUE rake angle* γ_u (Fig. 2.10) or, shortly, the BUE angle. It is greater than the rake angle.

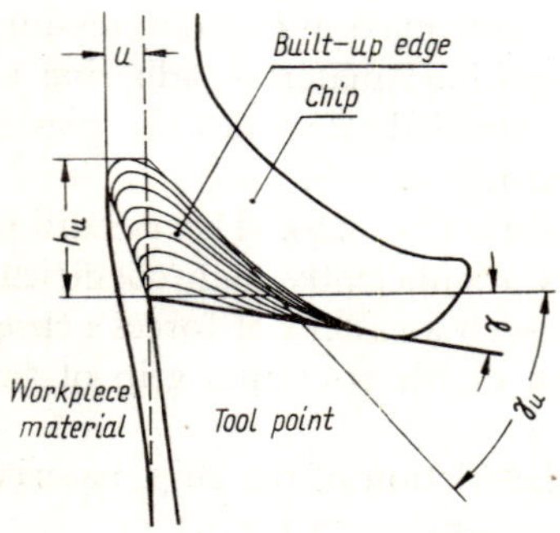

Fig. 2.10 Geometrical characteristics of built-up edge

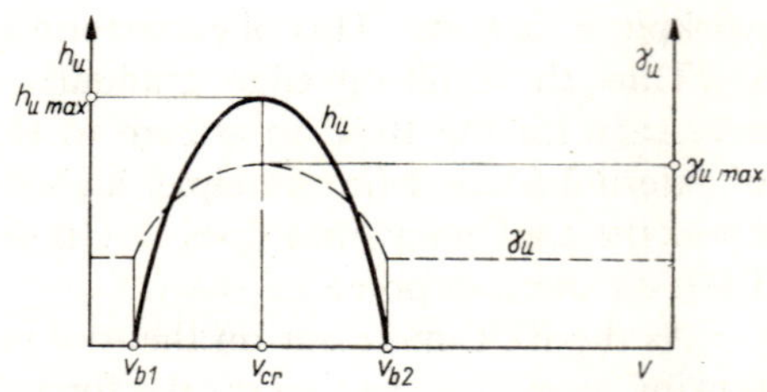

Fig. 2.11 Height and rake angle of built-up edge plotted against the cutting speed

The dimensions of BUE are characterized by:

— BUE height h_u,

— BUE overhang u which to a large extent determines the irregularities of the machined surface.

The boundaries of the BUE may be determined from photomicrographs or by hardness testing. At the boundary of the BUE there is a sudden increase in hardness, i.e. the BUE hardness is higher than that of the chip which, in turn, is harder than the machined material.

Figure 2.11 visually demonstrates the relationship between the dimensions and shape of the BUE and the cutting speed, according to the results of Rozenberg and Eremin [2/11]. It is evident from Fig. 2.11 that for a given machined material and cutting conditions there exists a *critical cutting speed* v_{cr} at which the height of the BUE and its rake angle reach their maximum values. When the speed is increased or reduced with respect to v_{cr}, both the height and rake angle of the BUE decrease. There are certain boundary cutting speeds v_{b1} and v_{b2} at which BUE completely disappears.

The *lower boundary speed* v_{b1} is for most so-called ductile materials so low that it has no importance from a practical point of view.

The *upper boundary speed* v_{b2} depends on the rake angle of the tool point, the material being machined and the dimensions of the undeformed chip.

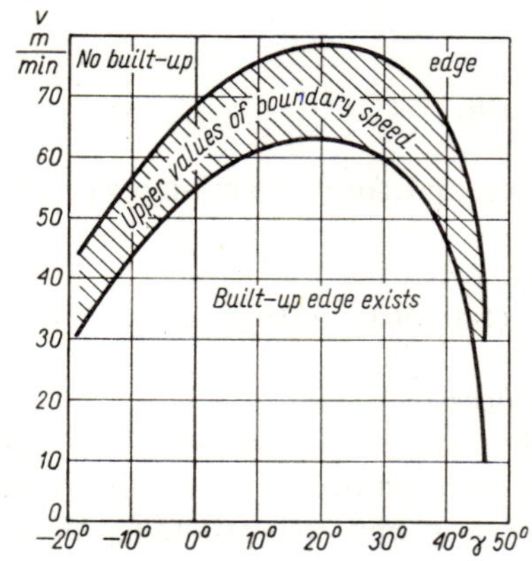

Fig. 2.12 Approximate values of the boundary speed v depending on rake angle γ, for the cutting of structural carbon steels within the range of $h = 0{\cdot}1$–$0{\cdot}3$ mm

In Fig. 2.12 the range of variation of the upper boundary cutting speed is shown, when cutting a structural steel, with various rake angles. Below the curve given in Fig. 2.12 conditions prevail under which the friction of the chip against the rake face is sufficiently high for the built-up edge to form. Its size depends on the time it takes the adherence force of the BUE to tool point to be counterbalanced or the BUE breaking stress to be exceeded.

The dimensions of the undeformed chip influence the formation of BUE mainly as a result of changes in the amount of heat which is generated.

The smaller the undeformed chip thickness, the higher is the upper boundary speed and the critical speed. According to research conducted by Eremin [2/11], when turning a chrome steel, the upper boundary cutting speed is $v_{b2} = 41$ m/min for $h = 0{\cdot}51$ mm. On the other hand, a reduction of undeformed chip thickness to $h = 0{\cdot}08$ mm resulted in a shift of the boundary cutting speed value to $v_{b2} = 118$ m/min.

Undeformed chip width influences the formation of BUE in a similar way as undeformed chip thickness, but to a considerably lesser extent.

Built-up edge has a great and, taking all circumstances into consideration, negative influence upon the course and results of the cutting process. On chip formation and flow, BUE has an influence similar to that of an increase in rake angle. The larger the built-up edge, the smaller is the chip area factor and the smaller becomes the cutting force. BUE also reduces frictional wear of the tool point, but on the other hand, by periodic appearance and disappearance, accompanied by changes of cutting force value, may become a cause of vibrations. The frequency of appearance and disappearance of BUE on the tool point is within the range from a few to hundreds of Hertz units, and therefore mechanical parts of machine tools may vibrate in resonance. This in turn may cause an increase of fatigue wear of the tool point. First of all, however, BUE negatively influences the quality of the machined surface and this is the main reason why cutting under conditions permitting the formation of BUE is considered to be abnormal and disadvantageous.

A knowledge of the laws governing the appearance and disappearance of BUE should be used to the purpose of avoiding the machining conditions favouring its appearance.

2.1.8 Characteristics of non-orthogonal cutting

Orthogonal cutting seldom occurs in industrial practice, and, for production purposes, non-orthogonal cutting is commonly applied.

A non-orthogonal machining process occurs when cutting:

— is carried out simultaneously by two or more rectilinear cutting edges or else by a curvilinear edge of the tool point;

— cannot be considered as a plane deformation process.

Whilst there are differences caused in the non-orthogonal process by simultaneous work of two or several cutting edges, all physical phenomena occur in a qualitatively similar way. Therefore, for the purposes of explaining and understanding those phenomena and the laws governing them, the simpler model of orthogonal cutting is used.

When evaluating the non-orthogonal cutting process in a quantitative manner the same basic definitions should be used, but at the same time the fact that all active sections of the cutting edge may work simultaneously should be taken into consideration.

Quantitative differences between non-orthogonal cutting and those concerning orthogonal cutting increase with the ratio of the length of trail edge to that of leading cutting edge.

Therefore, if this ratio is smaller than 0·1 then, from the point of view of the mechanics of the cutting process, the relations valid for orthogonal cutting approximately apply to the non-orthogonal case.

2.2 Classification and flow of chips

2.2.1 Chip classification according to structure and shape

Chips are a waste product of the cutting process and consequently should not deserve such attention as the problems connected with the workpiece, cutting tools and machine tool. However, chip features characterize the machinability of materials and determine the degree of difficulty involved in their removal. Therefore, a uniform classification and a corresponding system of evaluation of chips is necessary.

The classiffication of chips is based on the characteristics of their structure and shape.

Classification of chips according to their structure which is connected with the method of chip formation, distinguishes:

— *spalled* or *torn-off chips* (Fig. 2.13a) resulting from exceeding the cohesion strength. In practice, this kind of chip is very seldom encountered;
— *shorn chips* (Fig. 2.13b) resulting from exceeding the shear strength. Such chips appear in several varieties under practical machining conditions.

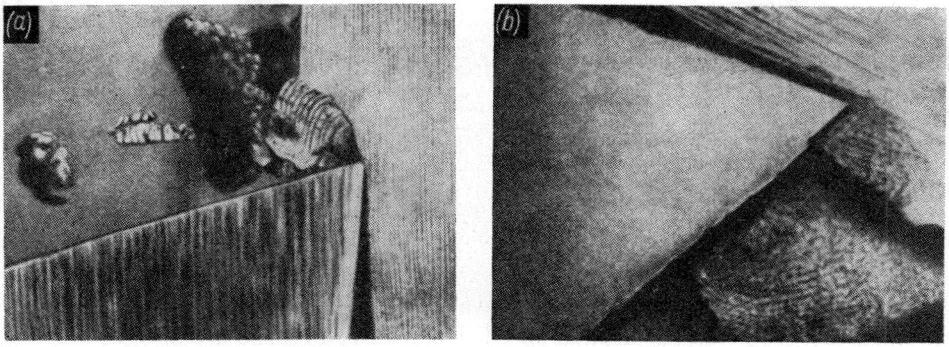

Fig. 2.13 Chip types: (a) spalled (torn-off), (b) shorn

Among shorn chips, the following varieties are distinguished:

(a) discontinuous chips, also called splintering chips (Fig. 2.14). These are single chip particles or chip particle groups loosely connected with each other and scattering as a result of collision;
(b) wavy chips (Fig. 2.15), whose characteristic feature is good cohesion of the chip with clearly defined stair-like lateral parts;
(c) uniform chips (also called ribbon chips or continuous chips), as in Fig. 2.16, characterized by complete cohesion and only light traces of structural elements.

Besides the typical varieties listed above, there are also some intermediate ones.

Each variety of chip indicates the specific conditions under which it was formed. It is known from theoretical assumptions and confirmed by experience that all kinds of chips could be obtained when cutting any material, if suitable cutting conditions could be provided.

For example, when cutting cast iron which under normal conditions yields discontinuous splintering chips, under certain conditions wavy ribbon chips are produced.

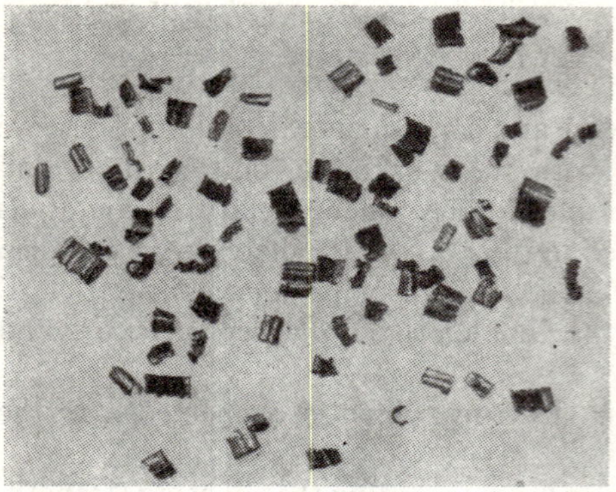

Fig. 2.14 Discontinuous chips

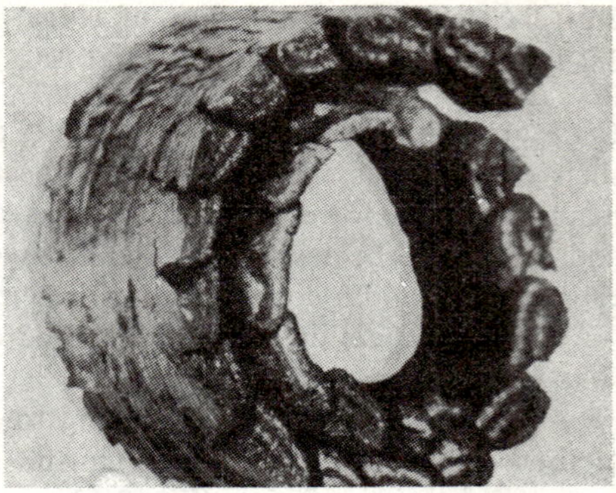

Fig. 2.15 Wavy chip (according to [2/3])

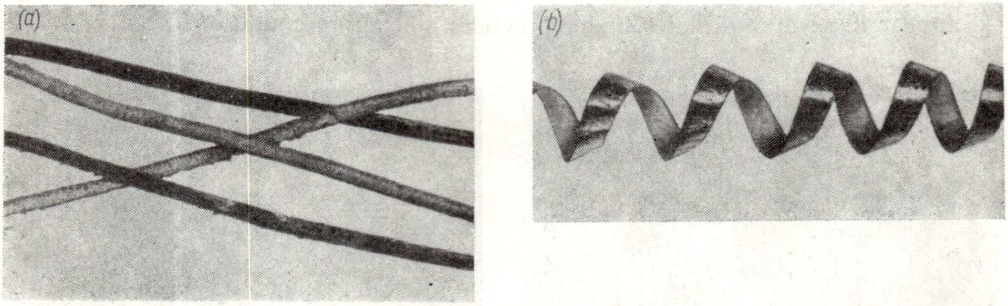

Fig. 2.16 Uniform chips: (a) straight ribbon chip, (b) helical chip (acc. to [2/3])

Figure 2.15 shows an example of a wavy cast-iron chip. It was obtained when modified cast iron was cut at high speed. Irregularity of structure and particle size is caused by free graphite inclusions.

Classification of chips according to their shape is particularly important in practice, since on this feature of the chip mainly depends work comfort and safety of the worker, as well as of chip collection, storing and transport.

In Table 2.1 the classification and characteristics of chip types and varieties are listed, according to their typical, most commonly encountered shapes.

Chips of uniform structure are divided into 4 kinds, i.e. ribbon (R), helical (H), spiral (S) and arc (A) chips.

Ribbon and spiral chips have several varieties.

Each kind may be further subdivided in respect of the degree of chip continuity. Continuous and broken chips are distinguished here, and it depends on the method and sort of machining, on cutting conditions, on tool point design etc. whether the former or the latter chips are obtained.

By continuous chips are meant those which form without interruption as long as the cutting process lasts during one pass of the tool point in relation to the workpiece.

Broken chips are obtained not only by applying methods of intermittent machining, but also by continuous machining, if chip breaking is due to suitable

Table 2.1 *Classification and characteristics of shorn chips*

Chip structure	Chip shape definition: Type	Variety	Symbol	Chip characteristics and evaluation: Chip characteristics and characteristic dimensions		Shape symbol	Evaluation
Uniform and wavy	Ribbon	Straight	Rs	Continuous	Chip thickness and width		Undesirable
				Broken	Chip length, thickness and width		Permissible
		Wavy	Rw	Continuous	Wave length and height		Undesirable
				Broken	Section length, wave length and height		Permissible
		Tangled	Rt	Overall dimensions of tangles			Undesirable
	Helical		H	Continuous	Coil diameter		Suitable
				Broken	Coil length and diameter	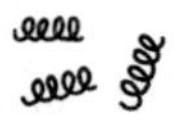	Advantageous
	Spiral	Flat	Sf	Outer diameter of coil		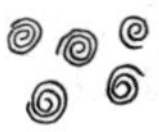	Advantageous
		Conical	Sc	Continuous	Max. diameter of coil		Permissible
				Broken	Max. diameter and height of coil		Advantageous
	Arc		A	Continuous	Arc height and length		Suitable
				Broken	Arc height and length, section length		Advantageous
Discontinuous	Splintering		D	Overall dimensions of elements			Suitable

tool point design, to mass forces or to the proper choice of cutting conditions (resulting in the appearance of a suitable field of chip stress).

In addition to chip form the *flow surface* or so-called *bottom surface* of the chip (i.e. the surface coming in contact with the tool point cutting face) also characterizes

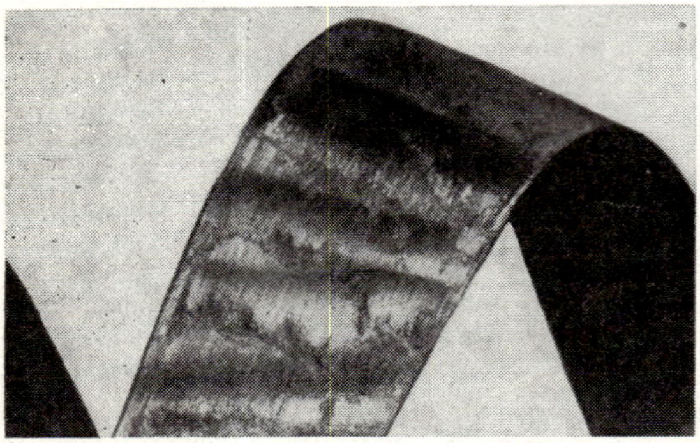

Fig. 2.17 Scales on the surface of chip flow (acc. to [2/3])

the machining process and machinability. Among others, the traces of the existence of built-up edge and of the occurrence of scuffing (Fig. 2.17) may be observed on this surface in the form of characteristic scales. A smooth and shiny chip flow surface is an indication of the correct choice of machining conditions.

2.2.2 Classification of chip running directions

In order to classify chip running directions, four quarters of space, as projected upon a plane parallel to the reference plane, have been designated as in Fig. 2.18. Thus the following chip running directions are distinguished:

- in quarter II, chip running towards the workpiece and forwards, i.e. in accordance with the direction of auxiliary motion (Fig. 2.18a),
- in quarter III, chip running towards the workpiece and backwards, i.e. contrary to the direction of auxiliary motion (Fig. 2.18b),

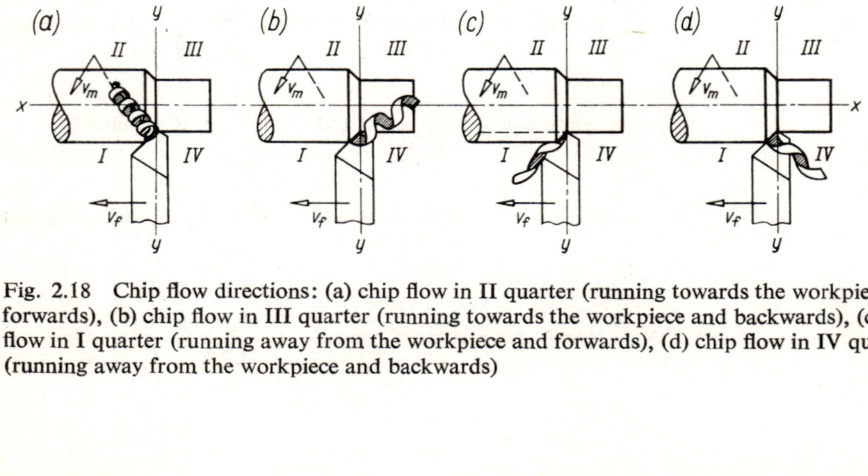

Fig. 2.18 Chip flow directions: (a) chip flow in II quarter (running towards the workpiece and forwards), (b) chip flow in III quarter (running towards the workpiece and backwards), (c) chip flow in I quarter (running away from the workpiece and forwards), (d) chip flow in IV quarter (running away from the workpiece and backwards)

— in quarter I, chip running away from the workpiece and forwards, i.e. in accordance with the direction of auxiliary motion (Fig. 2.18c),
— in quarter IV, chip running away from the workpiece and backwards, i.e. contrary to the direction of auxiliary motion (Fig. 2.18d).

From the point of view of work comfort and safety, chips are evaluated as: undesirable (1), inconvenient (2), permissible (3), suitable (4) and advantageous (5). Such evaluation is influenced not only by the shape of chip, but also by the direction of its running.

Consequently, chips running towards the workpiece and backwards are evaluated as undesirable, as they may cause scratching of the machined surface and in the case of continuous chips they may also wind around the workpiece, and be thrown off by centrifugal force.

Chips running towards the workpiece and forwards are also inconvenient. Although they do not scratch the machined surface, they may still wind around the workpiece.

Chips running away from the workpiece (forwards or backwards) may be permissible, suitable or advantageous depending on their shape. For example, broken helical chips running forwards away from the workpiece are considered advantageous.

2.2.3 Chip flow

Chip running direction should be distinguished from the direction of chip flow along the rake face. As a result of a change of the state of stress and deformation in the chip, the direction of chip flow along the rake face may be changed. These changes may be brought about by correspondingly shaping the rake face or selecting the cutting conditions and the method of chip cooling. Many factors, such as the shape of the undeformed chip, properties of the machined material etc., influence the cooling conditions of a plastic chip.

For the purpose of unequivocal definition, the direction of chip flow along the rake face is characterized by the angle of chip flow ν_c, measured from the trace of intersection of the rake face by a plane perpendicular to the cutting edge.

In the simplest case of non-orthogonal cutting, when $\gamma = \lambda_s = \varkappa_{r1} = 0$ and $\varepsilon_r = 90°$, from the condition of equilibrium of forces [2/2] and suitable transformation we have

$$\sin 2\nu_c = \frac{S_{at}}{S_{al}} \tag{2.36}$$

i.e. in this case the angle of chip flow depends on the ratio of the active length of the trail edge S_{at} to that of the leading cutting edge S_{al}.

Equation (2.36) may also be applied in the general case of non-orthogonal cutting, when $\lambda_s \neq 0°$, $\gamma \neq 0°$, $\varkappa_{r1} > 0°$ and $\varepsilon_r \neq 0°$. For this purpose the inclination of the cutting face λ_f in the plane perpendicular to the direction of chip flow should be considered.

It may be assumed with sufficient accuracy that

$$\nu_n = \nu_c - \lambda_f \tag{2.37}$$

where ν_n—angle of chip flow in the general case of non-orthogonal cutting. The procedure applied when deriving eqns. (1.16)–(1.21) in Chapter 1 permits one to derive a formula for $\tan \lambda_f$ and for the angle of chip flow

$$\nu_n = \arcsin\left(\frac{1}{2}\frac{s}{a}\frac{\sin^2 \varkappa_r}{\sin(\varkappa_r+\varkappa_{r1})}\right) - \arctan(\tan\lambda_s \cos\nu_c + \tan\gamma \sin\nu_c) \qquad (2.38)$$

Equation (2.38) indicates that the chip flow angle in the general case of non-orthogonal cutting depends on the feed and depth of cut, on leading and trail cutting edge angle, on leading cutting edge inclination angle and on rake angle. Namely:

— when the values of feed and leading cutting edge angle increase, so does the chip flow angle,

— when the depth of cut, trail edge angle, leading cutting edge inclination angle and rake angle increase, the chip flow angle decreases.

Also the factors determining the value of shear stresses τ_{shi} may have a certain physical influence upon the chip flow angle, e.g. the cutting speeds or the method and intensities of cooling which modify the temperature field and, consequently, the mechanical properties of undeformed chip.

By improving the knowledge of the dependence of chip flow on all factors which determine it in a geometrical, kinematic and physical sense it is possible to control the flow and running of chips in a purposeful way.

This principle is applied practically in controlling the process of chip curling.

2.2.4 Length of contact between chip and tool point

As a result of the acting forces and of the partially plastic condition of the chip, the latter adheres to the rake face over a certain length called the *length of contact* between chip and tool KB_0.

Figure 2.19 shows a simplified diagram of the zone of chip flow and of the contact between chip and tool. The resultant cutting force F acts in the direction n–n, and so does the resultant cutting resistance R, equal to F, but acting in the opposite sense. This direction intersects the cutting face at point C and the gliding plane at point B.

The projection of the resultant cutting force upon the direction perpendicular to the rake face yields component $F\gamma_n$. Similarly, the projection of the cutting resistance upon the direction perpendicular to the gliding plane yields component R_s. Both these components form with the direction n–n angles Θ and Θ_s, respectively. They are called the *angles of friction*, as their tangents are mean values of the coefficient of friction.

From the condition of equilibrium of moments with reference to point W [2/2] it can be written that

$$F\gamma_n e_n = R_s e_s \qquad (2.39)$$

Considering that

$$F\gamma_n = F\cos\Theta \quad \text{and} \quad R_s = R\cos(90 - \Phi_i - \Theta + \gamma)$$

and the fact that, in absolute values, $F = R$,

$$e_n \cos\Theta = e_s \sin(\Phi_i + \Theta - \gamma) \qquad (2.40)$$

It can be assumed that the length of contact between chip and tool point

$$KB_0 = m\, e_n \tag{2.41a}$$

and the length of the shear plane

$$KC = \omega\, e_s \tag{2.41b}$$

Besides, it is obvious from Fig. 2.19 that

$$h_c = KC \cos(\Phi_i - \gamma) \tag{2.42}$$

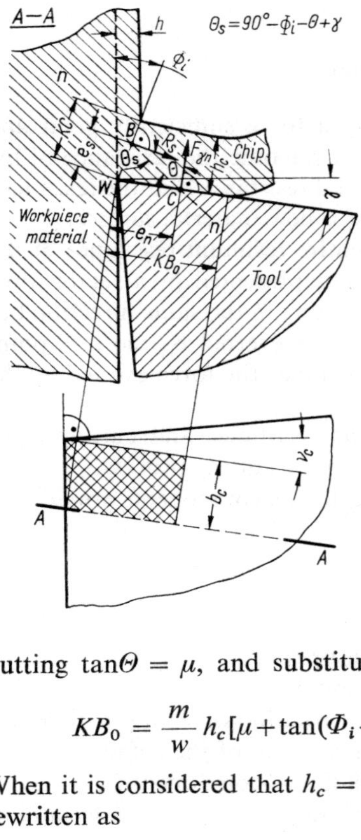

Fig. 2.19 Length of contact between chip and cutting face

Putting $\tan\Theta = \mu$, and substituting eqns. (2.41) and (2.42) into eqn. (2.40) gives

$$KB_0 = \frac{m}{w}\, h_c[\mu + \tan(\Phi_i - \gamma)] \tag{2.43}$$

When it is considered that $h_c = \lambda_h h = (\lambda_A/\lambda_b)h$ and that $\lambda_b \approx 1$, eqn. (2.43) can be rewritten as

$$KB_0 = \frac{m}{w}\, \lambda_A h[\mu + \tan(\Phi_i - \gamma)] \tag{2.44}$$

or, using the relationship between λ_A, Φ_i and γ [eqn. (2.25)]

$$KB_0 \approx \frac{m}{w}\, h\, \frac{\cos(\Phi_i - \gamma_0)}{\sin\Phi_i}\, [\mu + \tan(\Phi_i - \gamma)] \tag{2.44a}$$

It follows from eqn. (2.44a) that the length of contact between chip and tool is influenced by:

— angle of friction of the tool point,

— undeformed chip thickness,
— coefficient of friction on the rake face, and thus all factors influencing this coefficient,
— coefficients m and w, depending on material properties and on the conditions under which the cutting process takes place. According to Zorev [2/2], the m/w ratio when turning carbon steel, over wide range of undeformed chip thicknesses, is about 4.

The length of chip contact has an important influence upon the distribution of loads and upon the character of tool point wear.

2.3 Cutting forces

2.3.1 Definition of cutting forces and resistances

To start the cutting process it is necessary to apply a force sufficiently great and suitably situated in space. This force must be able to overcome the resistance of the material to elastic and plastic deformation, the frictional resistance and the resistance to the destruction of cohesion and to the creation of a new material surface.

Such a force is called the *cutting force* and the entire resultant resistance of the machined material is called the *cutting resistance*. It is represented by a vector force of a magnitude equal to that of the cutting force and acting in the same direction but in the opposite sense. Thus the cutting force is the force exerted by the tool point upon the material being machined and the cutting resistance the force exerted by the material upon the tool point.

The magnitude and direction of both the cutting resistance and cutting force change in space depending on many factors involved in the cutting process.

In order to determine the cutting force unequivocally, its components are defined in an orthogonal reference system.

The most frequently used reference systems are:
— the system involving the machine tool,
— the system involving the tool.

The system involving the machine tool (Fig. 2.20) consists of:
(a) the vertical direction, that of the vertical component S,
(b) the horizontal direction coinciding with the axis of rotation or with the direction of one of the basic motions (e.g. with the axis of rotation of the spindle), determining the direction of the horizontal—or longitudinal—component Q,
(c) the horizontal direction, perpendicular to both those listed above, determining the direction of the transverse component R.

The system involving the tool consists of:
(a) the straight line z, parallel to the direction of the primary motion. The projection of the cutting force on this line is called the *tangential* or *primary cutting force* F_v;
(b) the straight line y, parallel to the reference plane and perpendicular to the vector of primary motion velocity. The projection of the cutting force on this line is called the *thrust force* F_p;
(c) the straight line x, perpendicular to both the former ones. The projection of the cutting force on this line is called the *axial cutting force* F_f. This component is also called the *feed force*, if its direction coincides with that of the feed motion.

Fig. 2.20 The reference system involving the machine tool (on the example of turning work)

Components in the x–y–z system directly characterize the cutting process itself.

The cutting force is a resultant of forces exerted upon the rake flank. This can be explained in a simplified way be means of Fig. 2.21.

The actual cutting edge always has a nose radius $r_\beta > 0$. Considering the simplified model of chip formation by orthogonal cutting (Fig. 2.4), it may be assumed that the imaginary shear surface will start from a certain point K, situated on the curved part of the tool point profile (Fig. 2.21). Thus it may be assumed that point K divides the zones of action of the rake and flank.

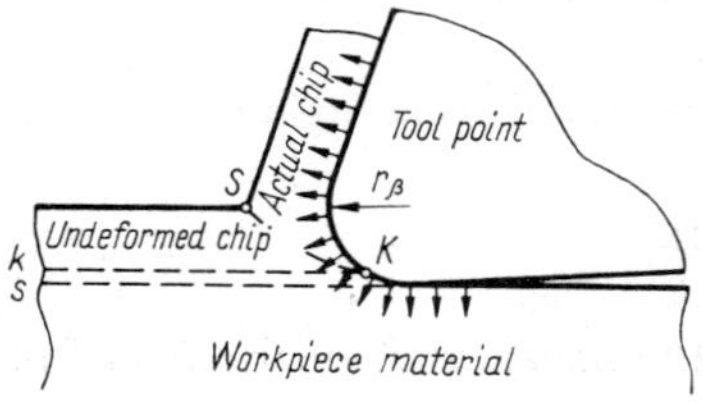

Fig. 2.21 Diagram of the division into zones of action of face and flank

Under this assumption the forces exerted by the rake will mainly do the chip formation work and the frictional work of the chip against the rake face.

Forces exerted upon the flank are mainly carrying out the work of elastic and plastic deformation of the layer situated below the K–k line, as well as the frictional work of the machined surface against the flank.

The model presented above is not quite accurate, since in reality the zone of action of the forces on the rake face reaches below the K–k line. Thus there exists an interconnection between both systems of forces. However, when certain problems are dealt with, the approximate assumption is made that there are two separate systems of forces on the rake face and on the flank. Such an assumption enables one to explain more clearly some phenomena connected with the cutting process.

2.3.2 Forces in orthogonal cutting

In the simplest case of cutting, i.e. in orthogonal cutting, when $\lambda_s = 0$, the distribution of forces exerted by the tool point is as illustrated by Fig. 2.22.

On the rake face acts the resultant force Q_c conventionally called the *chip forming force.*

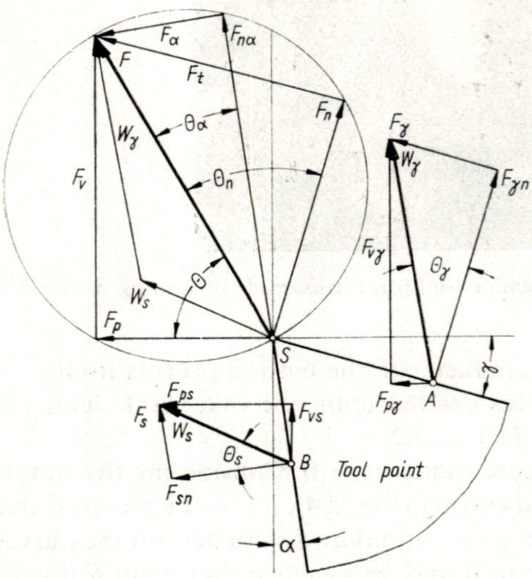

Fig. 2.22 Cutting forces exerted by the tool point in the case of orthogonal cutting ($\lambda_s = 0$)

This force may be decomposed into:

— the force $F_{\gamma n}$ perpendicular to the cutting plane (generally normal to the cutting face),

— the force F_γ tangential to the rake face, or the frictional force against the rake face, or into:

— partial primary component $F_{v\gamma}$ (on the rake face),

— partial reaction component $F_{p\gamma}$ (on the rake face).

The angle formed by the directions of $F_{\gamma n}$ and Q_c is the *mean angle of friction of the chip* against the rake face and is defined by

$$\mu_n = \tan\Theta_c = \frac{F_\gamma}{F_{\gamma n}} \tag{2.45}$$

where μ_n—the mean value of the coefficient of friction on the rake face.

On the flank acts the resultant force Q_s, conventionally called the *force of compression of the surface layer.*

This force may be decomposed into:

— the force F_{sn} normal to the flank,

— the force F_s tangential to the flank,

or into:

— partial primary component F_{vs} (on the flank),

— partial reaction component F_{ps} (on the flank).

The angle formed by F_{sn} and Q_s is the *mean angle of friction of the tool point* against the machined surface and is defined by

$$\mu_\alpha = \tan\Theta_s = \frac{F_s}{F_{sn}} \tag{2.46}$$

Considering the entire tool point, the resultant force F exerted by it may be determined as the vectorial sum

$$\bar{F} = \bar{W}_\gamma + \bar{W}_s \tag{2.47}$$

The resultant cutting force may be decomposed in the directions mentioned before, and thus e.g. the cutting force components F_v and F_p, F_n and F_t or $F_{n\alpha}$ and F_α can be distinguished.

The conditions of the equilibrium of forces are as follows:

$$\begin{aligned} F_v &= F_{v\gamma} = F_{vs} \\ F_p &= F_{p\gamma} = F_{ps} \\ F_{\gamma n} l_\gamma - F_{sn} l_s &= 0 \end{aligned} \tag{2.48}$$

where

$$l_\gamma = SA \quad \text{and} \quad l_s = SB$$

When relationships evident from Fig. 2.22 and from eqns. (2.45) and (2.46) are used, eqn. (2.48) may be rewritten in the form

$$F_v = F_{\gamma n}(\mu_\gamma \sin\gamma + \cos\gamma) - F_{sn}(\sin\alpha - \mu_\alpha \cos\alpha) \tag{2.49}$$

$$F_p = F_{\gamma n}(\mu_\gamma \sin\gamma - \cos\gamma) + F_{sn}(\cos\alpha + \mu_\alpha \sin\alpha) \tag{2.50}$$

The components F_v and F_r may also be presented in a different form, using the geometry of Fig. 2.22 and eqn. (2.4) and assuming that $x_K = h$, $\varphi_K = \Phi_i$ and $\Theta = \Theta_\gamma$. This gives

$$F_v = \frac{\tau_{shi}\, h\, b}{\sin\Phi_i}\left[\frac{\cos(\Theta_c - \gamma)}{\cos(\Phi_i + \Theta_c - \gamma)} + K\frac{\sin(\Theta_s - \alpha)}{\cos(\Phi_i + \Theta_c - \gamma)}\right] \tag{2.51}$$

$$F_p = \frac{\tau_{shi}\, h\, b}{\sin\Phi_i}\left[\frac{\sin(\Theta_c - \gamma)}{\cos(\Phi_i + \Theta_c - \gamma)} + K\frac{\cos(\Theta_s - \alpha)}{\cos(\Phi_i + \Theta_c - \gamma)}\right] \tag{2.52}$$

where $K = W_s/W_\gamma$.

It is evident from eqns. (2.51) and (2.52) that the component cutting forces depend on:

(a) the dimensions h and b of the undeformed chip,
(b) stereometric parameters α and γ of the tool point,
(c) properties of the material being machined and a number of other factors exerting a physical influence upon the magnitudes of τ_{shi}, Θ_c and K.

When interpreting and applying eqns. (2.51) and (2.52), it should be taken into account that

— μ_γ and μ_α may not be considered as equal, since they vary in different ways under the influence of the same factors,
— friction coefficients may not be considered as constant, since they depend more

or less on a number of factors. E.g. as the flank angle increases, μ_α decreases, but μ_γ hardly changes at all. On the other hand, when the rake angle increases, μ_γ decreases, but μ_α remains almost constant,

— the coefficient K may not be considered as constant. Its magnitude depends, *inter alia*, on the hardness and strength of the material being machined and on the dimensions of the undeformed chip. As the hardness of the material increases, so does the coefficient K, and as the dimensions of the undeformed chip are reduced, the coefficient K decreases.

It follows that the dependence of τ_{shi}, Θ_c and K on many factors is of a complicated character and is therefore inconvenient for practical application.

In order to simplify eqns. (2.51) and (2.52), statistically and empirically determined coefficients are substituted for the terms in brackets, Z in eqn. (2.51) and Y in eqn. (2.52). It has also been assumed that

$$\frac{Y}{Z} = w_y \tag{2.53}$$

Therefore

$$F_v = h\, b\, \tau_{shi} Z \tag{2.54}$$

and

$$F_p = h\, b\, \tau_{shi} Z\, w_y \tag{2.55}$$

The product $\tau_{shi} Z$ is called the *specific cutting resistance* or *cutting stress*. In other words the cutting stress or specific cutting resistance k_s is the quotient of the main cutting force and the cross-sectional area of the undeformed chip

$$k_s = \tau_{shi}\, Z = \frac{F_v}{h\, b} \tag{2.56}$$

Thus the equations for *component cutting forces* may be written as follows:

$$F_v = h\, b\, k_s \tag{2.57}$$

$$F_p = w_y\, h\, b\, k_s \tag{2.58}$$

2.3.3 Forces in non-orthogonal cutting

In the case of non-orthogonal cutting, when $\lambda_s \neq 0$, the distribution of forces on the tool point may be illustrated as in Fig. 2.23.

Plane $\pi_{xy}(ABCD)$ is parallel to the reference plane and plane $\pi_y(ABEF)$ is the rake face plane. The cutting edge makes the cutting edge angle $\varkappa_r$ with the direction of the f axis.

At the centre of the *fpv* system, situated on the cutting edge, another reference system, *lmn*, is established. It was created in the following way: first of all the *fpv* system was rotated about the v axis through the cutting edge angle $90° - \varkappa_r$, so that the p axis changed its position to l to coincide with the cutting edge, whilst the f axis changed its position to m. Then the *lvm* system was rotated about the l axis through the rake angle γ so that the m' axis took the position m and the v axis changed its position to n.

Forces F_n and P (see Fig. 2.22), components of the resultant cutting resistance act on the rake surface. Component P coincides with the direction of chip flow and forms the chip flow angle ν_c with the axis m.

The following equations for the components F_f, F_p and F_v may be derived from the relationships evident in Fig. 2.23:

$$F_v = F_{nv}+M_v = F_n \cos\gamma + P\cos\nu_c \sin\gamma \quad (2.59)$$

$$F_p = -F_{np}+L_p+M_p = -F_n \sin\gamma \cos\varkappa_r + P\sin\nu_c \sin\varkappa_r + P\cos\nu_c \cos\gamma \cos\varkappa_r \quad (2.60)$$

$$F_f = -F_{nf}-L_f+M_f = -F_n \sin\gamma \sin\varkappa_r - P\sin\nu_c \cos\varkappa_r + P\cos\nu_c \cos\gamma \sin\varkappa_r \quad (2.61)$$

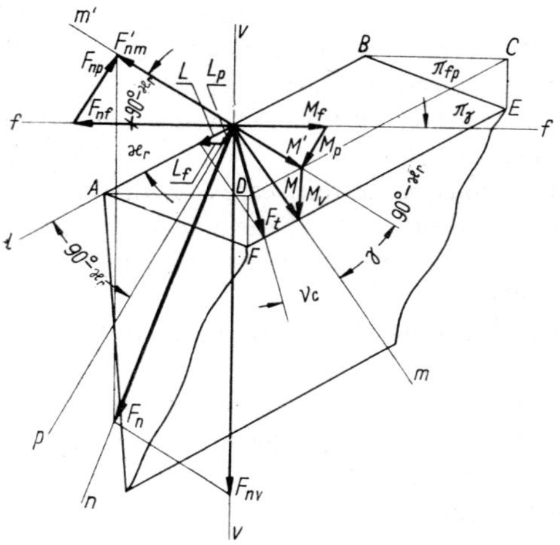

Fig. 2.23 Distribution of forces on the tool point in the case of non-orthogonal cutting

Proceeding in a similar way as in the case of orthogonal cutting, it is possible to designate $F_t/F_n = \mu$ and

$$\begin{aligned} Z &= \cos\gamma + \mu\cos\nu_c \sin\gamma \\ Y &= \mu(\sin\nu_c \sin\varkappa_r + \cos\nu_c \cos\varkappa_r \cos\gamma) - \sin\gamma \cos\varkappa_r \\ X &= \mu(\cos\nu_c \sin\varkappa_r \cos\gamma - \sin\nu_c \cos\varkappa_r) - \sin\gamma \sin\varkappa_r \end{aligned} \quad (2.62)$$

Furthermore, on the assumption that $w_y = Y/Z$ and $w_x = X/Z$

$$\begin{aligned} F_v &= F_n Z = h\,b\,k_s \\ F_p &= F_n Z_{wy} = w_y h\,b\,k_s \\ F_f &= F_n Z_{wx} = w_x h\,b\,k_s \end{aligned} \quad (2.63)$$

It follows from eqns. (2.59)–(2.61) that the cutting force components in the case of non-orthogonal cutting depend on the angle of chip flow, and, consequently, on all factors influencing the magnitude of that angle.

2.3.4 Empirical and statistical determination of cutting forces

In order to calculate the magnitude of the cutting force components according to eqn. (2.63), the values of k_s, w_y and w_x are determined in an empirical and statistical way. The dependence of these magnitudes upon a number of factors, involving the machine tool, tools, material and medium, is represented graphically and then suitable equations are chosen (approximated), which define those relationships with sufficient accuracy.

The empirical and statistical equations do not always reflect the physical side of the cutting process in a correct way. However, they furnish useful practical relationships between the cutting resistance and various cutting conditions.

For the user's convenience nearly all the relationships are approximated by products of power functions whose bases are constituted by cutting parameters, and the exponents, consistent within certain ranges, characterize the intensity of influence of the given parameter upon the cutting resistance. An exception from this rule is the rake angle, approximated by a function of the type represented by $F = C/A^{\gamma}$. This is due to the fact that the angle γ may have positive, zero and negative values. The normally accepted type of power function would, in the case when $\gamma = 0$, yield an infinitely large value of the cutting resistance.

The approximation by a power function may in certain cases be the cause of difficulties as far as true presentation of the actual influence of the given parameter is concerned. It is due to the fact that in some equations the so-called constant coefficients and exponents may in actual fact vary.

Therefore, when empirical and statistical equations based on such functions are used, it should be remembered that:

— they may be applied only within a range which has been studied previously; extrapolation of empirical and statistical equations is not permissible because of the risk of introducing errors of unknown magnitude;

— they are built on the assumption that the particular factors are independent of one another; actually, there exists an indirect or direct interdependence of those factors, and therefore the equations based on statistical means yield only approximate values;

— out of the many cutting conditions, only those whose influence is most important or which vary most frequently are shown separately in the equations; the influence of the remaining conditions is taken account of in the form of so-called constants or correction factors, the values of which are strictly established only for a certain range of machining processes.

For example, the empirical and statistical method was used to determine the equation for specific cutting force per unit cross-sectional area of the undeformed chip as

$$k_s = \frac{C_z}{h^{m_z} b^{n_z}} \tag{2.64}$$

where: C_z—coefficient taking care of the influence of all factors not accounted for separately in the equation, m_z and n_z—exponents established in an empirical way for the given range.

In view of this relationship (2.64), the equations for the cutting force components (2.63) may be rewritten in the form

$$F_v = C_z b^{e_z} h^{u_z}; \quad F_o = C_y b^{e_y} h^{u_y}; \quad F_f = C_x b^{e_x} h^{u_x} \tag{2.65}$$

In eqn. (2.65)

$$C_y = w_y C_z; \quad C_x = w_x C_z$$

and

$$u_z = 1 - m_z; \quad u_y = 1 - m_y; \quad u_x = 1 - m_x$$

$$e_z = 1 - n_z; \quad e_y = 1 - n_y; \quad e_x = 1 - n_x$$

Particular values of the coefficients C_z, C_y and C_x as well as of exponents u_z, u_y, u_x, e_z, e_y and e_x depend, among other factors, on the kinematics of cutting. Therefore, these values will be given when the cutting methods and their varieties are discussed.

2.3.5 Influence of shock and vibration on cutting forces

2.3.5.1 *Dynamic coefficient*

The equations for calculating cutting forces suggest that under constant cutting conditions those forces also remain constant. However, recordings of cutting force values by sufficiently sensitive, inertialess methods (e.g. resistance strip method, with observation on oscillograph screen) indicate cutting force magnitude variations during the cutting process.

In order to characterize those variations generally, the so-called *dynamic coefficient* K_d has been adopted, defined by the ratio

$$K_d = \frac{F_{max}}{F_{min}} \tag{2.66}$$

where: F_{max}—maximum value of the force during the period under study, F_{min}—minimum value of the force during the same period.

In addition, to characterize the dynamics of the process more unequivocally, the *force variation coefficient* K_v may be adopted, defined by the ratio

$$K_v = \frac{F_{max} - F_{min}}{F_m} \tag{2.67}$$

where F_m— mean value of the cutting force during the period under study.

Combining eqns. (2.66) and (2.67) gives the relationship between K_v and K_d, i.e.

$$K_v = (K_d - 1)\frac{F_{min}}{F_m} \tag{2.68}$$

Cutting force variations may result from shock or vibration.

Shock may be caused by both the starting or stopping of the tool motion under full load, by sporadic heterogeneities of the material etc. In this case the dynamic coefficient is called the *coefficient of impact* (Fig. 2.24) and is defined by the ratio

$$K_i = \frac{F_i}{F_{min}} \tag{2.69}$$

The external effect of cutting force variations during the cutting process are vibrations. Their importance in the cutting process was realized only a relatively short time ago.

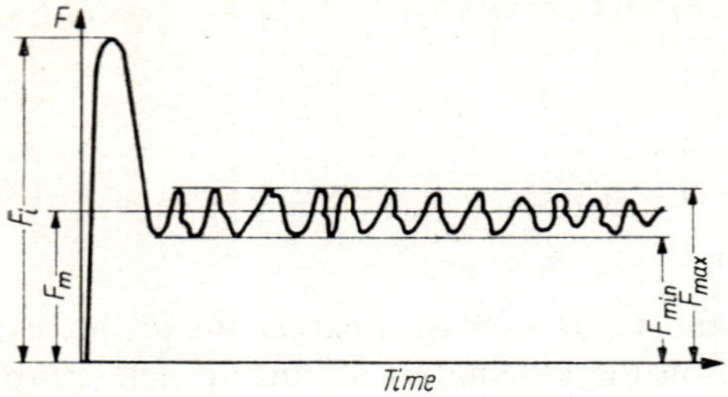

Fig. 2.24 Variation of cutting force in the course of time

2.3.5.2 *Classification and cause of vibrations during cutting*

Vibrations which occur during cutting may be generally divided into two groups:
— vibrations independent of the cutting process,
— vibrations dependent on the cutting process.

The first group includes:

(a) vibrations forced by: other machines (vibrations transmitted by foundations), unbalance of rotating parts, inertia forces of reciprocating parts, kinematic inaccuracies of the drive (caused by faulty production or assembly) etc.,
(b) vibrations caused by periodically varying rigidity of machine tool parts (e.g. of shafts with splineways),
(c) relaxation vibrations, occurring in feed mechanisms of low rigidity in the case of high friction resistance of those mechanisms.

To the second group of vibrations, connected with the cutting process, belong:

(a) vibrations caused by the varying cutting force,
(b) self-excited vibrations.

Only those vibrations which are connected with the cutting process will be discussed below.

The cause of forced vibrations are variations of the cutting force, these in turn being the result of changes in undeformed chip dimensions. Thus, for example, when excentrically chucked shafts or ones with a non-uniform machining allowance (e.g. in roughing operations) are turned, a periodical force appears which produces vibration. The principal feature of these vibrations is their frequency equal to that of the varying exciting forces. The magnitude of the influence of forced vibrations depends mainly on:

(a) the amplitude of variations of the vibromotive force,
(b) the ability of the machine tool–workpiece–tool system to disperse the vibration energy,
(c) the degree of synchronization of the angular frequency of vibromotive force variations ω_f with the angular frequency of free vibrations of the parts and assemblies of the machine tool–workpiece–tool system ω_0.

In the case when $\omega_f = \omega_0$ the well-known phenomenon of resonance occurs and its intensity then depends only on the ability of the given object to damp vibrations. The diagram in Fig. 2.25 illustrates this relationship. The magnification factor of the amplitude of forced vibrations, i.e. the ratio of this amplitude A_1 to the displacement A_0 of the object statically loaded by the exciting force, is plotted against the ratio of the variation frequency of the vibromotive force ω_f to natural frequency ω_0.

Self-excited vibrations are those vibrations which occur without any action on the part of external exciting forces. Such vibrations occur in the machine tool–workpiece–tool system able to produce non-damped vibrations, at the expense of an external source of energy, in most cases—of an electric motor. Self-excited vibrations are characterized by the independence of their angular frequency of the cutting

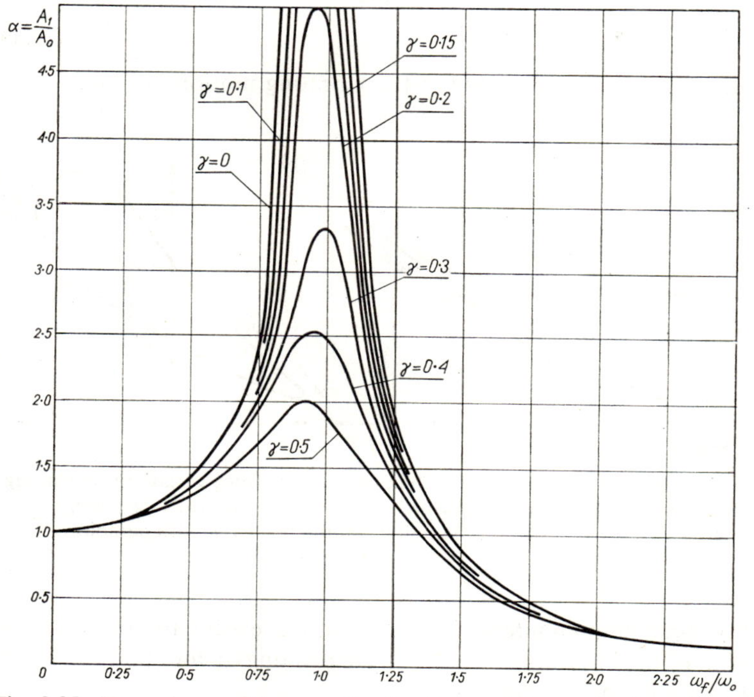

Fig. 2.25 Dependence of the magnification factor of the amplitude of forced vibrations α on the ratio of variation frequency of the exciting force ω_f to natural frequency ω_0 (γ damping coefficient)

conditions (a, s, v_m) and of the tool point shape. On the other hand, the frequency of these vibrations is influenced by the mass and rigidity of the machine tool–workpiece–tool system. The frequency of self-excited vibrations is similar to one of the harmonic frequencies of free vibration.

For self-excited vibrations to occur, a situation must be created in which, under the action of a single external impulse, the system will begin a vibrating motion resulting in varying forces that will support the continuation of these vibrations. As during this motion various resistances have to be overcome, a source of energy able to compensate for the resulting losses is also necessary.

The initial impulse in the cutting process is easily obtained even as a result of vibrations not connected with the process itself (such as instantaneous changes of the cutting resistance as a result of non-homogeneity of material).

Circumstances under which forces supporting the vibrations may appear as a result of the vibrating motion itself, may be explained by the example of vibrations with a single degree of freedom, e.g. vibration occurring in the direction of action of the radial component F_p (Fig. 2.26).

In the case of orthogonal vibrationless cutting the magnitude of force F_{po}* is determined by the equation

$$F_{po} = C_y b^{e_y} h_0^{u_y} \quad \text{kg} \tag{2.70}$$

where: b—width of undeformed chip in mm, h_0—thickness of undeformed chip in mm.

As a result of vibrations, the instantaneous thickness h will be

$$h = h_0 - y \quad \text{mm} \tag{2.71}$$

where y—displacement of the tool point due to the vibrating motion, in mm.

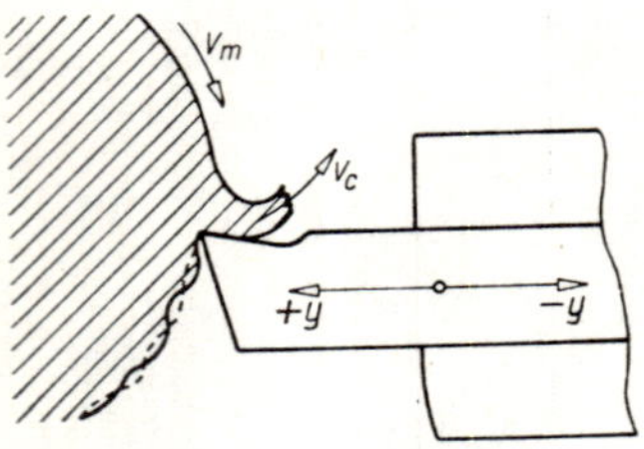

Fig. 2.26 Diagram of vibrating motion, with a single degree of freedom, caused by force F_p (orthogonal cutting)

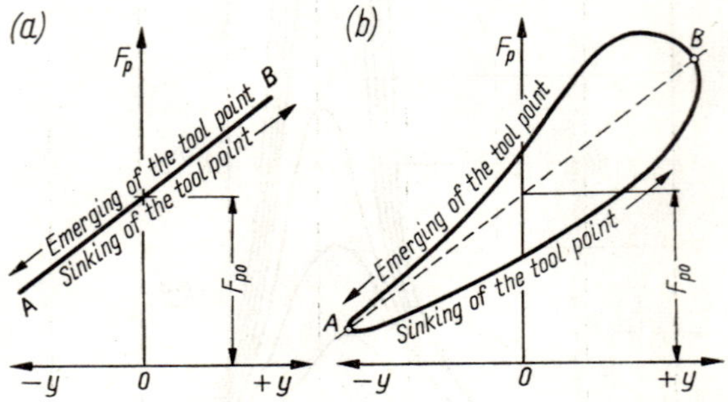

Fig. 2.27 Characteristics of force variations depending on undeformed chip thickness: (a) the ideal case, without damping, (b) the real case, with damping

If there were no resistances counteracting this motion, except for the force of elasticity, the variation of the instantaneous force F_p depending on tool point position could be presented by means of the equation

$$F_p = F_{po}\left(1 - \frac{y}{h_0}\right)^{u_y} \quad \text{kg} \tag{2.72}$$

A graphical interpretation of this equation is given in Fig. 2.27a.

In reality resistance to motion does exist. Thus a continuous process occurs involving the loss of energy. The equation of the equilibrium of forces in the process of self-excited vibrations is, therefore,

$$m\,y'' + \beta\,y' + j\,y = F_0$$

where: m—the vibrating mass, β—coefficient characterizing the resistance counteracting the motion and proportional to the speed of the vibrating motion, j—rigidity characterizing the elastic resistance, $F_p - f(y, y', y'')$—the variable cutting force supporting the self-excited vibrations.

The variations of force F_p depending on tool point position, in the case of damped self-excited vibrations, are illustrated in Fig. 2.27b.

* Subscript $_0$ designates this magnitude under conditions of vibrationless cutting or, as it were, 'static' cutting.

The area limited by both curves AB shows the work of a single vibration cycle. For this work to be positive (according to the designations adopted in the vibration theory), the cutting force when the tool point sinks must be smaller than when it emerges. Besides, and this follows from the vibration theory, for the work of a single vibration cycle W to be positive ($W > 0$) the condition $0 < \varphi \leqq \pi$ must be fulfilled.

At the same time

$$W = F_{p0} A \pi \sin\varphi \tag{2.73}$$

where: A—amplitude, φ—angle of phase shift.

This means in practice that cutting forces should be phase-shifted through angle φ in respect to the vibrating motion of the tool point. Measurements of vibrations during turning, carried out by Sokolovski and Amosov [2/8] established the magnitude of this angle as $\varphi = 60$–$85°$.

It follows from the condition that force F_p must be smaller at the sinking than at the emerging of the tool point, that in the process of self-excited vibrations must exist forces decreasing with the increase of cutting speed which accompanies the sinking of the tool point. The cutting speed increases since the relative speed of chip displacement along the tool point v_c is augmented by the sinking speed (Fig. 2.26).

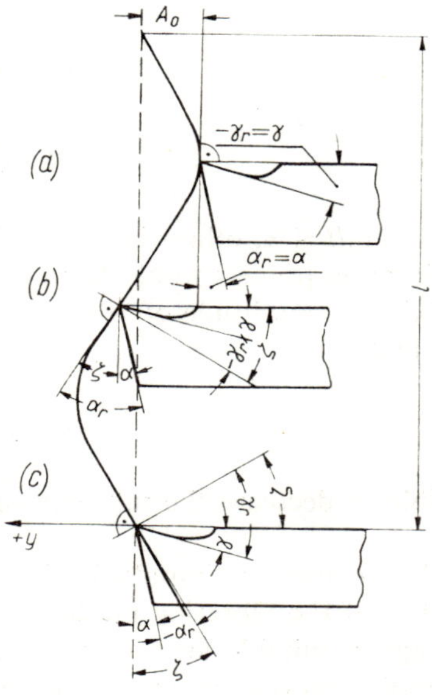

Fig. 2.28 Variation of tool point angles as a result of vibrating motion

Changes in tool point geometry may constitute a factor supporting the vibrations. This follows from the variations of rake and clearance angles, and is evident from Fig. 2.28. When, due to vibrations, the tool sinks into the material, the rake angle increases by ζ and the clearance angle decreases by the same amount.

When the tool point is sinking (provided that $\zeta \leqq \alpha_r$), the radial force component decreases and thus the basic condition is fulfilled. When the tool point emerges, the resistance increases because of the reduction of angle γ_r.

It is assumed that the tool point carries out a sinusoidal motion, so that $y = A_0 \sin\omega t$, and that

$$y' = \tan\zeta = A_0\,\omega \cos\omega\, t \tag{2.74}$$

whence

$$(\tan\zeta)_{max} = A_0\omega = 2\pi\,\frac{A_0}{l} \tag{2.75}$$

where $2\pi/l$ angular frequency of vibration, A_0 amplitude of vibration, l wavelength on the cut surface (as indicated in Fig. 2.28).

For small ζ values $\tan\zeta \approx \zeta$, then

$$\zeta_{max} = 2\pi\,\frac{A_0}{l} \tag{2.76}$$

Using eqns. (2.74), (2.75) and (2.76) the following results are obtained:

$$\gamma_{r\,max} = \gamma + 2\pi\,\frac{A_0}{l} \tag{2.77}$$

$$\gamma_{r\,min} = \gamma - 2\pi\,\frac{A_0}{l} \tag{2.78}$$

$$a_{r\,max} = \alpha + 2\pi\,\frac{A_0}{l} \tag{2.79}$$

$$a_{r\,min} = \alpha - 2\pi\,\frac{A_0}{l} \tag{2.80}$$

The conclusion may be drawn from eqn. (2.80) that $2\pi\, A_0/l \leqq \alpha$, otherwise the tool point would be pushed out of the material and the vibration process would be interrupted. Another conclusion follows, i.e. that the maximum amplitude of vibrations depends on the angle α and is determined by the formula

$$A_{0\,max} = \frac{l}{2\pi}\,\alpha \tag{2.81}$$

This means that by reducing the angle α it is possible to decrease the intensity of vibrations.

The occurrence of self-excited vibrations as a result of cutting an 'impulse-excited' surface is conditioned by the existence of a phase shift of the vibrating motion, in respect of the wave on the machined surface, through a suitable angle. Its value must be such that F_p be smaller at the sinking of the tool point than at its emergence. Experience shows that such cases occur for phase-shift angles between 7° and 120°.

2.3.5.3. *Influence of cutting conditions upon vibrations and the methods of preventing self-excited vibrations*

Vibrations occur in the cutting process very frequently. However, their intensity is not always sufficiently great to justify their definition as a harmful phenomenon or paying attention to them at all.

The magnitude of vibrations depends on factors involving:
(a) the machine tool and the clamping of the workpiece,
(b) the cutting tool, the material being machined and the cutting parameters.
Vibrations accompanying the cutting process are three-dimensional. To define them in space, projections of the vibratory motion on the direction of cutting force components F_f, F_p and F_v are used.

So far vibrations in the direction of force F_p have been studied most effectively. Therefore, vibrations in this direction will be dealt with below in the discussion of various factors influencing vibrations.

However, it must be stressed here that the influence of certain parameters of vibrations in various directions may be different, as indicated by some experiments involving vibrations in the direction of the circumferential cutting force.

Of the factors involving machine tool and workpiece clamping, the following have a particular importance: rigidity and vibration damping ability of the machine tool–workpiece–tool system.

Increasing rigidity results in the reduction of vibration range so that in the case of sufficient rigidity vibrations do not occur at all. Greater rigidity causes vibration frequency to increase and the amplitude to decrease.

Of the tool point parameters studied, the following exert an important influence upon the amplitude of vibrations: rake angle γ, clearance angle α and nose radius r_β. Vibration frequency is practically independent of tool point parameters.

A reduction of rake angle is conducive to a widening of vibration range and an increase of amplitude. The influence of rake angle on amplitude is very strong as it may cause an 8- to 10-fold magnification or reduction of the latter. A decrease of clearance angle is conducive to a reduction of vibrations in the direction of force F_p. An increase of tool point nose radius causes a widening of the range of occurrence of vibrations and an increase of their amplitude.

Of the properties of the material being machined the following have the greatest influence on vibrations: ductility and the vibration damping ability. Experiments indicate that vibrations occur to a higher degree when ductile metals are being machined than in the case of brittle materials.

The influence of the particular cutting conditions is not identical. Feed increase causes a reduction of the range of occurrence of radial vibrations, while an increase of the depth of cut—vice-versa—causes a widening of that range and an increase of vibration amplitude. This does not involve any change of vibration frequency.

As the cutting edge angle is increased, both the range of occurrence and the amplitude of vibrations decrease. This is particularly evident within the range of $\varkappa_r$ from 45 to 90°.

Vibrations in the directions of components F_v and F_f have not been studied sufficiently so far. It follows from observations of machining by cutting under industrial conditions that increasing wear (blunting) of the tool point results in greater intensity of vibrations in the direction of force F_v.

Prevention of self-excited vibrations which are most closely connected with the cutting process may consist, first of all, in a change of cutting conditions. This, however, is not always feasible, both for technological and economic reasons. Only the cutting edge angle may be varied within a considerable range.

Therefore, changes of tool point parameters are more useful in preventing vibrations. Such changes follow from the previous discussion of their influence upon vibrations. As far as the reduction of clearance angle is concerned, it should be kept in mind that an excessive decrease of this angle may cause vibrations in the direction of force component F_v (thus acting in a similar way as blunting of the tool point). In many cases self-excited vibrations are prevented from occurring by the application of cutting fluids and improvement of tool point surface finish. Satisfactory results are also obtained by increasing machine toll rigidity and, in certain cases, by damping vibrations by means of vibration dampers of various kinds.

2.4 Cutting energy and heat

2.4.1 Cutting energy components

The entire* cutting energy consists of:
(a) plastic deformation energy E_{pl},
(b) external friction energy e_x,
(c) internal friction energy due to elastic deformations and vibrations E_i.
The latter component of the cutting energy may be completely omitted under normal cutting conditions without vibrations, as its magnitude is negligible in comparison with the first two components. Thus practically the entire *cutting energy* is defined by

$$E \approx E_{pl}+E_x \quad \text{kg mm} \tag{2.82}$$

Since the energy of external friction consists of the energy of friction on the cutting face $E_{x\gamma}$ and the energy of friction on the flank $E_{x\alpha}$,

$$E = E_{pl}+E_{x\gamma}+E_{x\alpha} \quad \text{kg mm} \tag{2.83}$$

The cutting process may be characterized by the *specific cutting energy*, also called the *unit cutting energy*. Both terms mean the amount of energy needed in order to transform unit volume or mass of undeformed chip into actual chips.

The *volumetric specific cutting energy* $E_{v\,sp}$ is the amount of energy necessary to transform 1 mm³ of undeformed chip into actual chips†.

The *gravimetric specific cutting energy* $E_{g\,sp}$ is the amount of energy necessary to transform 1 g of undeformed chip into actual chips.

In practice the notion of volumetric specific cutting energy is used much more frequently. It may be expressed by

$$E_{v\,sp} = E_{sp\,pl}+E_{sp\,x\gamma}+E_{sp\,x\alpha} \quad \text{kg mm/mm}^3 \tag{2.84}$$

The specific cutting energy is one of the most characteristic and sensitive indices of the cutting process. According to Gordon's research [2/9] the total volumetric specific cutting energy and its components depend on a number of factors.

* Strictly speaking, in addition to the energy components listed above, the energy of formation of the new surface and the energy of chip bending should be taken into consideration. They are being neglected in this case, since they are very small (relatively). Together with internal friction energy due to elastic deformations, those energies to not exceed 3% of the total cutting energy.

† It is easy to perceive that the value of the specific cutting energy is equal to that of specific cutting resistance. Since $E = F_v l$ and $V = A\, l$, it follows that $E_{v\,sp} = \dfrac{E}{V} = \dfrac{F_v l}{A\, l} = k_s$, where: l—cutting path, V—volume and other designations as used before.

The influence of undeformed chip thickness h is illustrated in Fig. 2.29, where the volumetric specific cutting energy $E_{v\,sp}$ is plotted against undeformed chip thickness h. It can be seen that an increase of h causes a decrease of $E_{v\,sp}$. As a result of this, the percentage share of specific deformation energy and specific friction energy on the flank is reduced, while such share of the specific friction energy on the rake face increases.

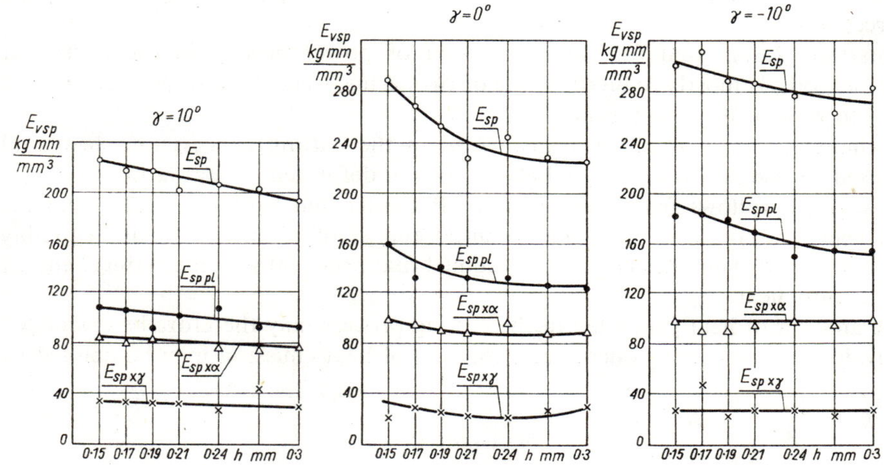

Fig. 2.29 Influence of undeformed chip thickness h on the total volumetric specific cutting energy $E_{v\,sp}$ and its components in the case of machining structural steel of 0·35% carbon content, with various rake angles

An increase in rake angle causes a reduction of specific cutting energy. This particularly concerns the specific energy of plastic deformation, and to a slightly lesser degree the specific energy of friction on the cutting face. On the other hand, the energy of friction on the flank slightly increases.

2.4.2 Cutting power

Generally, *cutting power* is defined as

$$N_c = \frac{dE}{dt} \tag{2.85}$$

where t—cutting time.

In numerous cases of cutting N_c is constant, e.g. when the speeds of basic motions are constant and the cutting force varies so slightly that its mean value may be considered as constant. Then

$$N_c = \frac{E}{t} \tag{2.86}$$

The total cutting energy consists of the energies of all components of the cutting force. Thus, if the cutting power in kilowatts is to be determined, then

$$N_c = \frac{F_v v_m}{6120} + \frac{F_p v_p}{6120} + \frac{F_f v_f}{6120} \tag{2.87}$$

where: F_v, F_p and F_f—mean values of cutting force components, v_m, v_p and v_f—mean values of speeds in the direction of action of the corresponding components.

Denominator value is the product $60 \times 75 \times 1{\cdot}36$ when speeds are given in m/min and forces in kg.

Equation (2.87), used for the calculation of power needed in the particular cutting methods, is mostly reduced to one or two terms, since the rest may be omitted as equal to zero or negligibly small.

Mechanical energy is partly transformed in the cutting process into heat and partly used to increase the potential energy of the deformed crystal lattice.

However, it follows from the results of Rebinder and Epifanov [2/10] that the part of energy used to increase the potential energy of crystal lattice is negligibly small (less than 1%) and thus it can be assumed that practically all mechanical energy is turned into heat.

The amount of heat generated in the cutting process may therefore be calculated as equivalent to the cutting energy and the rate of heat emission may be defined by

$$Q = \frac{F_v v_m}{427} \quad \text{cal/min} \tag{2.88}$$

where 427 kg m/cal—the mechanical equivalent of heat.

The heat emitted is absorbed by the chip (Q_c), and by the tool (Q_s), transmitted to workpiece material (Q_w) and absorbed by the cutting fluid and ambient air (Q_a). Radiation is omitted as negligible. This can be written as (Fig. 2.30)

$$Q = Q_c + Q_s + Q_w + Q_a \tag{2.89}$$

A knowledge of heat distribution among chip, tool point, material being machined

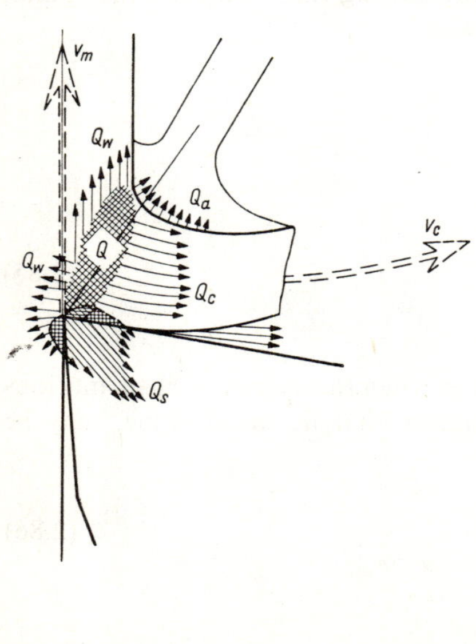

Fig. 2.30 Diagram of heat generation and distribution during the cutting process

and the cutting fluid and ambient air is of great practical importance. Heat distribution is dependent on cutting conditions, kind of chip, cooling method, properties of workpiece material and on tool point material and shape.

On the basis of studies conducted so far, the following conclusions may be drawn:

(a) one of the factors having most important influence on heat distribution is cutting speed,

(b) increase of cutting speed causes a reduction of the heat transmitted into the material,

(c) increase of cutting speed causes an increase of the percentage share of heat absorbed by the chip, and reduces heat transmitted to the tool point.

From a practical point of view, however, the important factor is not the amount of heat absorbed by the tool point and material being machined, but, in the first place, the temperature level and distribution.

2.4.3 Distribution of heat and temperature during the cutting process

Heat generated during cutting is not uniformly distributed in the entire mass of deformed chip. Also, the distribution of heat is not uniform in the material being machined. This also concerns the tool point and the actual chip. Consequently, temperatures at various points of the chip, workpiece material and tool point differ from one another very considerably. The whole cutting process depends to a decisive degree on temperature distribution, and therefore the knowledge of temperature level and distribution is of fundamental importance.

Temperature will undoubtedly be highest in the spots of greatest heat concentration. These are: the area of maximum chip deformations and the friction surface of chip and tool point.

In view of the results of studies presented, it is obvious that no single temperature can be said to determine unequivocally the whole cutting process, because it is rather the temperature field that is decisive in this case. However, since the character of this field is generally constant, and only the temperature level is changing, determination of the temperature in a given spot on the tool point gives some notion about temperature levels in other spots.

Therefore, in this sense the cutting temperature may be discussed, as long as simultaneously the point is given at which the temperature was measured or (in some cases) the measuring method is given (as certain measuring methods directly determine the kind of temperature measured, e.g. maximum or mean temperature).

In practice, the greatest importance is attached to the knowledge of maximum temperature. In the spots where it occurs, the tool point wear is most rapid. Therefore, if no further explanation is given as to which temperature is being meant, by cutting temperature the maximum temperature is understood.

REFERENCES

2/1 Krivoukhov, V. A.: 'Deformirovanie poverkhnostnykh sloev metalla v protsesse rezaniya' (Deformation of Surface Layers of Metal in the Cutting Process), Mashgiz, Moskva 1956

2/2 Zorev, N. N.: 'Voprosy mekhaniki protsessa rezaniya metallov' (Problems of the Mechanics of Metal Cutting Process), Mashgiz, Moskva 1956

2/3 BIERNAWSKI, W.: 'Teoria obróbki metali skrawaniem' (Theory of Metal Machining by Cutting), PWN, Warszawa 1958

2/4 DAVIDENKOV, N. N.: 'O krivoi techenii Ludwiga' (On Ludwig's Flow Curve), *Izvestiya AN SSSR*, 4–5, 1944

2/5 TAYLOR, W.: 'On the art of cutting metals', *American Machinist*, 1902

2/6 USACHEV, Ya. Ch.: 'Yavleniya proiskhodyashchie pri rezanii metallov' (Phenomena occurring during the cutting of metals), *Izvestiya Petersburskogo Politekhnicheskogo Instituta*, 1915

2/7 EREMIN, A. N.: 'Fizicheskaya sushchnost yavlenii pri rezanii stali' (Physical Nature of Phenomena Accompanying the Cutting of Steel), Mashgiz, Moskva 1951

2/8 AMOSOV, I. S.: 'Eksperimentalnoe i tekhnicheskoe issledovanie vibratsii pri tochenii' (Experimental and Technological Research on Vibration during Turning), Mashgiz, Moskva 1951

2/9 KLUSHIN, M. I.: 'Rezanie metallov' (Cutting of Metals), Mashgiz, Moskva 1953

2/10 REBINDER, P. A., and EPIFANOV, G. I.: 'Ob energeticheskom balanse protsessa rezaniya metallov' (On energy balance of the metal cutting process), *Doklady AN SSSR*, **16**, 4, 1949

2/11 ROZENBERG, A. M., and EREMIN, A. N.: 'Elementy teorii protsessa rezaniya metallov' (Elements of Theory of the Metal Cutting Process), Mashgiz, Moskva 1956

CHAPTER 3

Cutting fluids and their importance

3.1 Basic functions of cutting fluids

The usual medium in which the cutting process takes place is air. It exerts an influence upon the course and results of the cutting process. This can be established experimentally by comparison with a cutting process taking place in a high vacuum.

Exchanging air for another fluid influences the cutting process to a higher or lower degree. As cutting in air constitutes the most frequent case, the influence of other media is usually defined by comparison with results obtained in air, or with so-called dry cutting.

The medium in which the cutting process takes place is called the *cutting fluid*, it may consist of a liquid or gas. However, solids may also be introduced into this fluid and they influence the cutting process in their own way.

Cutting fluids (given the popular name of coolants) should fulfil the following basic functions:

(i) cooling,
(ii) lubricating,
(iii) washing,
(iv) so-called surface loosening (in certain cases).

The cutting fluid is supposed to be non-corrosive (anticorrosive), harmless to man and economical.

The cooling action of the cutting fluid consists in abstracting the heat generated during the cutting process. It is important that heat is abstracted from the correct zones in the cutting process. For example, heat abstraction from the undeformed chip zone is generally not advantageous, whilst the cooling of the tool point, work surface and actual chip is desirable.

Thus the cooling properties of cutting fluid depend to a large extent on the cooling method. The cooling efficiency increases with heat conductivity and specific heat of the cutting fluid, with its heat of fusion and of vaporization, and also with its viscosity. Cooling effect increases, too, with the flow rate of the fluid, the area of application and with decreasing temperature of the cutting fluid.

The lubricating action of the cutting fluid reduces the external friction between rake face and chip and between flank and machined surface. Effectiveness of the lubricating action depends, therefore, in the first place on the degree of penetration of the coolant between those surfaces. It follows from the magnitude of specific cutting forces that very often, i.e. when machining steel, pressures between chip and tool point and between workpiece and tool point may reach 20 000–30 000 kg/cm². Notwithstanding such high pressures, the use of suitable cutting fluids results in the reduction of friction and wear.

There are several hypotheses on the subject of penetration of cutting fluids into the zones of external friction. It is assumed in all these hypotheses, however, that in the zone of contact between chip and tool point there exist very low pressure points (momentary vacuum spots). The ambient fluid penetrates into these gaps, overcoming even very high resistance to flow. The appearance of vacuum spots is explained in the particular hypotheses by:

— a smaller or larger preceding gap appearing before the tool point (Fig. 3.1a),
— contact between chip and tool point—as a result of built-up edge, discontinuous structure or non-uniform cohesion of chip—is not uniform but wavy, and therefore gaps form during chip flow (Fig. 3.1b),

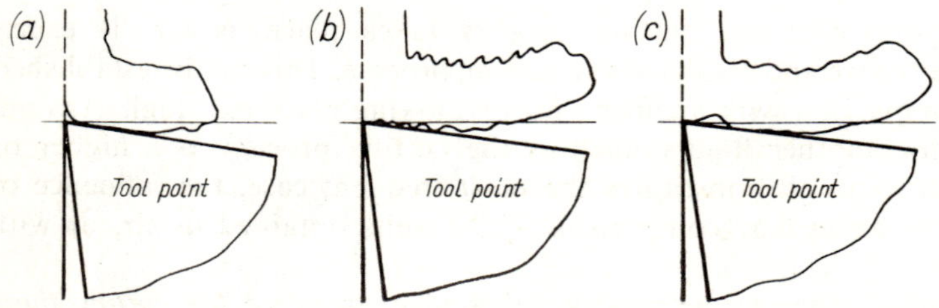

Fig. 3.1 Causes of penetration of cutting fluid between chip and tool point: (a) preceding gap, (b) wavy surface of contact between chip and tool point, (c) gaps caused by chip adhesion to tool point

— contact between chip and tool point is momentarily interrupted as a result of the formation of discontinuous chip and of the clinging of chip to tool point (e.g. because of the so-called *friction-type built-up edge*, Fig. 3.1c).

The particular causes listed above are valid under different machining conditions. Thus e.g. the appearance of the preceding gap was established only within the range of very low cutting speeds and when cutting brittle materials.

The washing action of a cutting fluid consists in removing the products of abrasion, chip fragments and tool wear particles. In practice, cutting fluids exhibit good washing properties having high wettability and because particles immersed in the liquid move downstream.

This action has an advantageous influence on the smoothness of the machined surface and on the reduction of tool point wear.

Surface loosening action of the cutting fluid, also called the *cutting action* of this fluid, results from the so-called *Rebinder effect* [3/1]. Adsorbed molecules of cutting fluid, particularly of liquids containing surface-active additives (e.g. organic acids reducing surface tension), disintegrate, diffuse and penetrate into microgaps resulting from defects in the crystal lattice. Because of the wedge-like action of the adsorbed liquid, a very thin surface layer of the machined material becomes 'loosened'.

According to Rebinder's research, the thickness of this layer may reach 0·1 mm. Since the machined surface 2 in the next pass becomes the work surface 1 (Fig. 3.2), plastic deformation appearing in the chip formation area and slip of the chip elements occur more easily and more frequently. This phenomenon is called *internal lubrication.* As a result of the appearance of smaller and more densely spaced chip elements, the formation of a ribbon chip is facilitated.

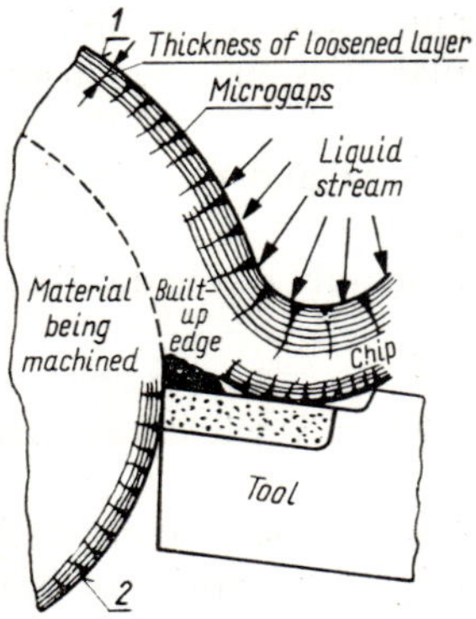

Fig. 3.2 Surface loosening action of cutting liquids

The efficiency of activated liquids increases with a reduction in undeformed chip thickness, since in this case the percentage share of the thickness of loosened layer is greater in relation to the total thickness.

Increasing the cutting speed decreases the efficiency of activated liquids. This is due to a more ductile chip, having a much smaller number of larger gaps of all sorts (lack of preceding gap and of built-up edge). Consequently, much fewer local vacuum spots are formed and the liquid does not penetrate everywhere, and dry friction may result.

It follows from the experiments of Rebinder and his collaborators that the efficiency of cutting liquids with surface-active additives depends, too, to a large extent on the selection of the liquid and of the machined material.

An influence similar to that exerted on the material being machined is also exerted by the cutting liquid with surface-active additives on the tool point. However, due to the latter's greater hardness and its being subject to abrasion only and not to plastic deformation, this influence is by far less important. Thus the advantages of the use of coolants (abstraction of heat and reduction of friction) outweigh the disadvantages and the useful life of a suitably cooled tool point is always longer than that of a dry cutting one.

3.2 Classification of cutting fluids

Cutting fluids may be divided into the following groups:
— gases,
— mixtures of liquids and gases (liquids sprayed in a stream of gas),
— liquids,
— liquids with suspended solids,

— pastes (mixtures of solids).
Of these groups of cutting fluids, liquids are most commonly used.

If ambient air is omitted in this classification, it can be said that gases are applied as cutting fluids under pressure. Under normal conditions gases have a comparatively low specific heat and thermal conductivity. On the other hand, when rapidly expanding, they intensively abstract heat from the surroundings.

The use of gases as cutting fluids is limited by the high price of compressed gases, the cost of special installation, unpleasant hissing of gas leaving the nozzle and also by the difficulties of extracting those gases which, in large quantities, may be harmful to men or machines.

For these reasons, of all gases only compressed air is used and only in those cases where the use of a liquid is impossible or troublesome.

More and more frequently gas–liquid mixtures are used, mainly of the air–liquid type. The effectiveness of their action is brought about by:
— expanding air abstracting heat from the surroundings,
— the stream of compressed air having a high speed (up to 300 m/s) which improves the cooling effect,
— highly atomized liquid falling in the form of minute droplets upon high-temperature surfaces and evaporating, thus abstracting large quantities of heat,
— the high speed of the stream, causing the liquid droplets to have a high kinetic energy which permits their penetration into even very small gaps.

As a result of these properties of air stream with liquid spray, a very high cooling efficiency is obtained, since—depending on the degree of compression and atomization—at the spot of stream incidence upon the cooled surface, sub-zero temperatures of the order of -2 to -20°C can be obtained.

The lubricating action of the liquid spray may be arbitrarily adjusted by the addition of lubricants.

In Fig. 3.3 a diagram of a liquid spray–compressed air cooling system is shown. Compressed air flows through filter 1, then through pressure regulator 2 and valve 3 to Venturi tube 4. Passing the pressure regulator 9 the air also exerts pressure on the liquid in tank 10. This liquid is forwarded through filter 11 and nozzle 6 to a low pressure zone in tube 4. Liquid–air proportions are adjusting the air pressure (pressure

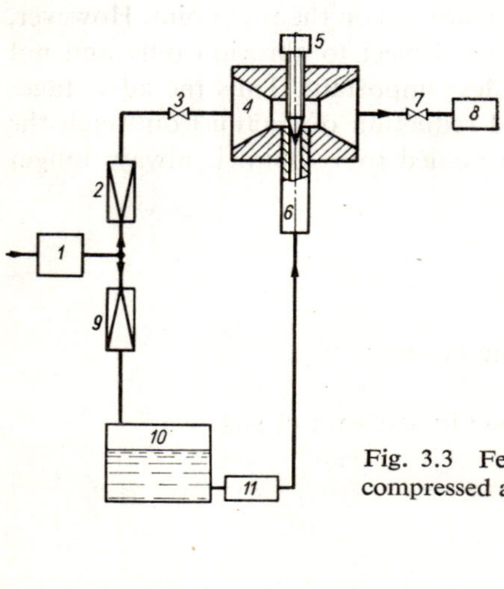

Fig. 3.3 Feed diagram of a liquid spray–compressed air cooling system

regulator 9), metering needle 5 and the speed of air flow in the tube (i.e. by means of air pressure regulator 2). The gas–liquid mixture flows through valve 7 to spraying pipe 8.

Liquids used as cutting fluids are divided into water solutions and mixtures, emulsions and oils.

Water solutions of various salts, e.g. sodium nitrite ($NaNO_3$) and potassium or sodium chromate, are used mainly to obtain a good cooling effect and in order to passivate the surface and to protect it from corrosion (e.g. sodium nitrate is an inhibitor of corrosion). On the other hand, water solutions of salts have a poor lubricating effect.

In the case of finishing operations, surface-active additives (e.g. soaps) are added to the water solutions.

A large group of cutting fluids is constituted by water-and-oil emulsions. i.e. mixtures of water and emulsified oil with the addition of a corrosion inhibitor. In addition, emulsions may also contain surface-active additives (activated emulsions). These liquids have good cooling and lubricating properties, the latter being regulated by emulsion content.

Another large group is constituted by oils. Mineral oils and—less frequently—vegetable oils are used. They have very good lubricating properties and poor cooling ability.

To increase their activity or the capacity for surface action, oils are subjected to sulphonation or greasing. Such oils are called *activated oils*.

In order to improve the lubricating properties, liquids with suspended solids are also applied. These are oils with a very fine suspension of graphite or of molybdenum disulphide.

If the main problem consists in reducing friction and wear, while the cooling and washing action is of no real importance, then lubricants in the form of pastes or of pastes mixed with graphite or molybdenum disulphide powder may be used.

3.3 Ways of applying cutting fluids and the influence of those fluids on the machining process

In order to define the application of a cutting fluid unequivocally, the temperature of the fluid, its flow rate (usually in litres per minute) and the method of application must be determined.

The various methods of applying cutting fluids may be divided into two groups:

(a) methods influencing the machining process by cooling the tool only,

(b) methods influencing the machining process by complex action, as the process occurs in a medium consisting of the given cutting liquid.

Influencing the machining process by cooling the tool only is achieved by means of the so-called *internal cooling* with closed circuit. An example of such a solution is given in Fig. 3.4.

The efficiency of internal cooling action depends on the temperature of the liquid and the speed of its flow. A decrease of tool point temperature changes the conditions of chip flow along the rake face, the coefficient of friction also changes. In general, the dependence of the coefficient of friction on the rake face upon temperature is as shown in Fig. 3.5. It follows that within the temperature range from 0 to

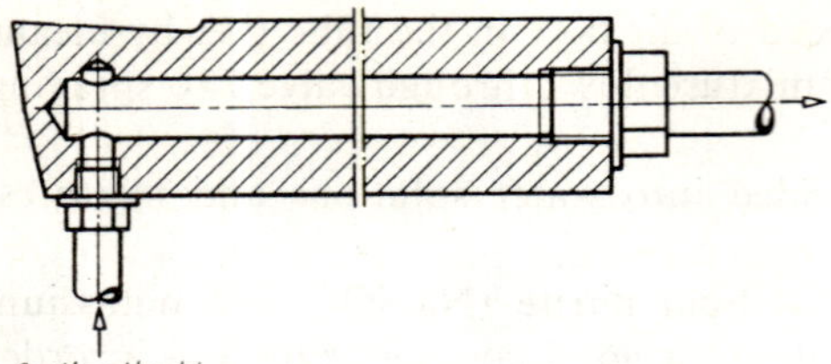

Fig. 3.4 Example of internal cooling with closed circuit

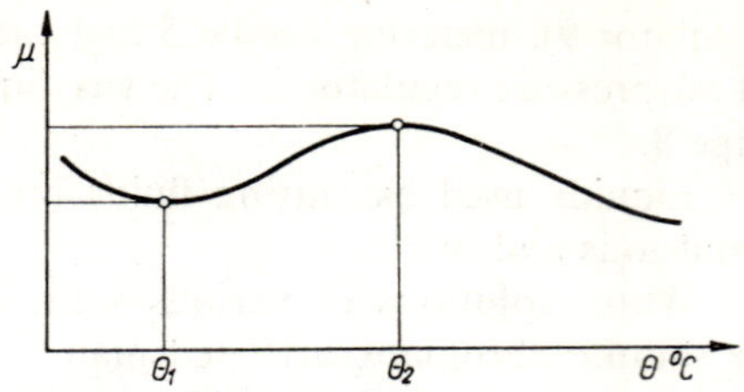

Fig. 3.5 Dependence of the coefficient of friction on the rake face upon the temperature of contact surface between chip and rake face

Θ_1 and for temperatures exceeding Θ_2 the decrease of contact surface temperature causes an increase of friction. This reduces cooling effectiveness as far as tool point wear is concerned. Instead, within the range between Θ_1 and Θ_2 decreasing the contact surface temperature yields an improved effect, as not only does tool point temperature decrease—and, consequently, the strength of tool point material increases—but also friction is reduced.

If the cutting liquid is to be used as the medium in which the cutting process occurs, a suitably ample stream of the fluid is directed on the chip formation zone, on the tool point and the material being machined (Fig. 3.6).

The influence of the medium on the cutting process may be illustrated by comparing the results of research on the application of water as a cutting fluid with the results of cutting in air.

Fig. 3.6 Example of cutting in a medium consisting of cutting fluid [3/3]

The conclusions following from such a comparison are:

(a) in the case of dry cutting (or cutting in air) under certain conditions a built-up edge forms on the cutting face and chip adhesion occurs, while during cutting in a stream of water no built-up edge or chip adhesion is observed;

(b) when cutting in air it is noted that the coefficient of friction varies according to the diagram shown in Fig. 3.5, whilst when cutting in water this coefficient remains constant, or

$$\mu = \frac{q_F}{q_N} = \text{const} \tag{3.1}$$

This is a consequence of the absence of built-up edge and chip adhesion to the rake face;

(c) the water in which the cutting process occurs (involving undeformed chip thicknesses usually applied in practice) does not perceptibly change the magnitude of shear stress and of the specific energy of plastic deformation, but it reduces the coefficient of friction;

(d) the quality of a surface machined in water is better than that obtained in air, which is also a consequence of the absence of built-up edge on the rake face;

(e) in the case of cutting in water the tool point wear is less than in air.

It is evident from these conclusions that the use of water as a cutting fluid is advantageous. Also, many other liquids act in a similar way to water (e.g. carbon tetrachloride). Therefore, water solutions and water-and-oil emulsions, used as a continuous stream or spray are widely applied in practice.

In certain cases the effectiveness of cutting fluids may be improved by directing the stream of fluid in a special way.

One such method consists in directing the fluid stream to the zone of contact between flank and the machined surface under a pressure of several scores of atmospheres.

The influence of cutting fluids upon the course and results of the cutting process depends on the chemical composition of the fluids and on the proportion of their components.

With time, cutting fluids change both in respect to their composition and the proportion of components. The cause of it may be pollution, evaporization of water and other liquids, non-uniform precipitation of components (e.g. through settling of oil drops or chips), growth of bacteria etc.

Consequently, cutting fluids have to be handled properly, and their properties checked and restored. This permits one to take full advantage of a correctly selected cutting fluid.

REFERENCES

3/1 REBINDER, P. N., EPIFANOV, G. I., and PLETNEVA, N. A.: *Doklady AN SSSR* **97**, 1954

3/2 PETRYS, T., and ŚNIECHOWSKI, R.: 'Chłodziwa do obróbki metali skrawaniem' (Coolants for the Machining of Metals by Cutting), PWT, Warszawa 1960

3/3 WAKEFIELD, C. C. and Co. Ltd.: 'Cutting and Metalworking Oils', London

3/4 KLUSHIN, M. I.: 'Smazochno-okhlazhdayushchie zhidkosti pri rezanii metallov i tekhnika ikh primeneniya' (Lubricating and Cooling Fluids in Metal Cutting and the Technology of their Application), Mashgiz, Moskva 1961

CHAPTER 4

Role and importance of the cutting tool

The cutting tool influences in a decisive way the quality of the workpiece, i.e. its accuracy and surface quality. Also, the work output and its cost depend, to a large extent, on the tool. The tool cost relative to the other machining costs increases the less wear resistant is the tool.

Usability of a cutting tool for machining is defined by the term cutting capacity or simply tool performance.

Tool performance is the capacity to machine materials with a high output while maintaining the desirable qualities of the workpiece, minimum wear and lowest possible cost.

The cutting capacity of the tool depends mainly on:
— the type and properties of tool point material,
— tool point shape,
— working conditions.

4.1 Tool wear

4.1.1 TYPES OF TOOL WEAR

General principles and laws concerning friction and abrasion of solids also apply to tool point under conditions of its contact with the chip and with the material being machined. In addition, specific conditions of the cutting process should be taken into consideration.

Specific features of friction and abrasion in the cutting process are:

(a) tool point surfaces take part in the process of friction against a continuously newly formed work surface and newly formed chip. Consequently, wear symptoms of practical importance may appear only at the cutting point,
(b) normal pressures on the tool point surfaces are several times higher than those normally encountered in mechanical engineering,
(c) metal surfaces taking part in the friction process may be almost chemically clean,

(d) the friction process occurs in a non-uniform temperature field, while contact temperature may reach the melting point of the workpiece material,
(e) plastic deformation occurs at the friction surface of chip and machined surface,
(f) as a result of wear at the tool point, its geometrical form changes, continuously varying the conditions under which friction and abrasion take place,
(g) built-up edge may appear on the tool point and influence the distribution of forces and thus, indirectly, the friction and abrasion processes.

The cause of tool wear may be: mechanical forces, adhesion, diffusion and chemical processes.

Mechanical wear of the tool point may take the following forms:

— mechanical abrasion, when the cause of wear is the reciprocal engagement of irregularities of both surfaces,
— ultimate strength wear, when, as a result of high cutting forces, the tool point strength is exceeded locally,
— fatigue strength wear, when the tool point loses cohesion from material fatigue.

Mechanical abrasion is a commom form of tool point wear. A characteristic feature of mechanical abrasion is the scratching of the rubbing part of tool point surface and the uniformity of wear on the friction surface.

Ultimate strength wear should occur only incidentally during normal use of a tool, even if the machining is of the intermittent type. This means that in normal use cutting conditions should be selected in such a way as to avoid exceeding the tool strength.

The following symptoms of strength wear are distinguished:

— dents, small defects of cutting edge within the zone of contact between edge and the material being machined and chip. Overall dimensions of dents are of the order of tenths of a millimetre and therefore for their measurement optical magnifying instruments (e.g. so-called Brinell's magnifier) are needed;
— chippings, larger defects of cutting edge within the zone of contact between the tool edge and the material being machined and chip. Overall dimensions of chippings exceed 0·5 mm and may be determined with the naked eye, and measured with a rule with millimeter scale;
— breaches, larger defects of cutting edge, stretching beyond the zone of contact between the tool edge and the material being machined and chip;
— cracks which usually constitute the beginning of defects, i.e.—depending on their dimensions—dents, chippings or breaches.

The failure of the tool point owing to strength wear depends on the size of defects in relation to tool point dimensions, and also on the defect shape and position.

There is no distinct boundary between strength wear and mechanical abrasion of the tool point. As the dimensions of particular dents decrease, strength wear becomes mechanical abrasion. In this connection the so-called *brittle abrasion* is distinguished, an intermediate form of tool point wear. It occurs when microdents appear frequently and on the entire contact surface.

Symptoms of fatigue strength wear are similar to those of ultimate wear, but do not appear soon after the beginning of the cutting process. They seldom occur at all, since, in general, before material fatigue sets in, the tool point is worn out by abrasion.

Adhesive* wear of the tool point occurs under certain cutting conditions. Its characteristic symptom is the appearance of built-up edge, chip adhesion and welding etc. The condition for adhesive wear to occur is a sufficiently high normal pressure and a not too high contact temperature.

Diffusion wear of the tool point occurs at a sufficiently high temperature (depending on the type of tool point material).

Usually, adhesion and diffusion wear do not occur simultaneously to any great extent, while mechanical abrasion accompanies all other forms of wear.

Chemical wear of the tool point consists in the rapid formation of chemical compounds on the tool point surface. These are loosely connected with the tool point material and are easily abraded. An example of chemical wear of the tool point may be the formation of oxide coatings.

4.1.2 Wearing action and wearability

Independently of the causes and forms of tool point wear, its effect always consists in a loss of volume. Since abrasion is the principal form of wear, it is common practice to describe the phenomenon of mass and volume loss during the friction process by the term abrasion.

The partaking of two bodies in the friction process is characterized by the notions of wearing action and wearability.

By *wearing action* is understood the capacity of the material to change the dimension and shape of the mating body by a gradual loss of its volume and mass during the friction process.

The measure of wearing action is its mean intensity determined by

$$Q_s = \frac{S}{t} \quad \text{mm}^3/\text{min} \tag{4.1}$$

where: S—volume of the abraded material of the mating body in mm³, t—time of friction of both bodies in min.

Wearability of a material means that it changes dimensions and shape as it gradually looses volume and mass during the friction process as a result of the wearing action of the mating body.

The measure of wearability is its mean intensity determined by

$$Q_z = \frac{Z}{t} \quad \text{mm}^3/\text{min} \tag{4.2}$$

where: Z—volume of the abraded material of the given body in mm³, t—time of friction of both bodies in min.

It follows from these definitions that a cutting point is characterized by both wearing action and wearability. The lower is the latter, the better is the tool point.

Wearing action and wearability may be considered as relative notions.

* Adhesion—mutual attraction of particles of adhering materials, cohesion—mutual attraction of particles within the material.

Then, to characterize the abrasion process, the *coefficient of relative wearing action* is used

$$K_s = \frac{Q_s}{Q_z} = \frac{S}{Z} \tag{4.3}$$

as well as the *coefficient of relative wearability*

$$K_z = \frac{1}{K_s} = \frac{Q_z}{Q_s} = \frac{Z}{S} \tag{4.4}$$

Thus the relative intensity of tool point wearability is at the same time the relative intensity of wearing action of the material being machined.

4.1.3 Geometrical features of tool point wear

The most visible results of tool point wear are the changes of point shape and dimension, called the geometrical magnitudes of tool point wear.

The typical *geometrical magnitudes of wear* are shown in Fig. 4.1:

— on the cutting face; *N*—top wear, *KA*—crater length, *KB*—crater width, *KT*—(maximum) crater depth, *KL*—width of crater lip, *KM*—crater centre distance to cutting edge,

— on the flank: *VB*—width of the wear land, *VL*—length of wear land.

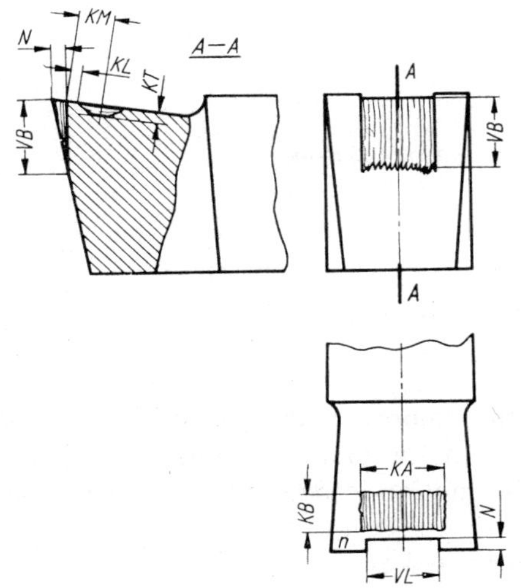

Fig. 4.1 Dimensions of traces of tool point abrasion wear

The *wear land* on the flank is the result of its friction against the machined surface of the workpiece. This symptom of tool point wear always appears during the cutting process, and is dependent on the type of machining, workpiece material and tool point shape.

Wear on the cutting face, although it also always accompanies cutting, does not always attain clearly visible dimensions. The appearance of a crater on the rake face is the result of a certain temperature field and normal pressure. The spot in which crater depth is greatest is that of maximum wearability.

In the case of cutting brittle materials at low cutting speeds, when a discontinuous splintering chip is formed, no tool point wear in the form of a crater on the rake face is observed.

4.1.4 Theoretical calculation of tool point wear land

In order to determine the increase of wear with time of the tool point, the case of orthogonal cutting will be considered, in which wear occurs on the flank.

It is easy to conclude from Fig. 4.2 that:

$$VB = VB_0(1-\tan\alpha\tan\gamma) \tag{4.5}$$

$$N = VB_0\tan\alpha = VB\,\frac{\tan\alpha}{1-\tan\alpha\tan\gamma} \tag{4.6}$$

Besides, mean width of wear land VB can be determined from

$$VB \approx \frac{2V}{VL.N} \tag{4.7}$$

where: V—volume of abraded material, and other designations are as before.

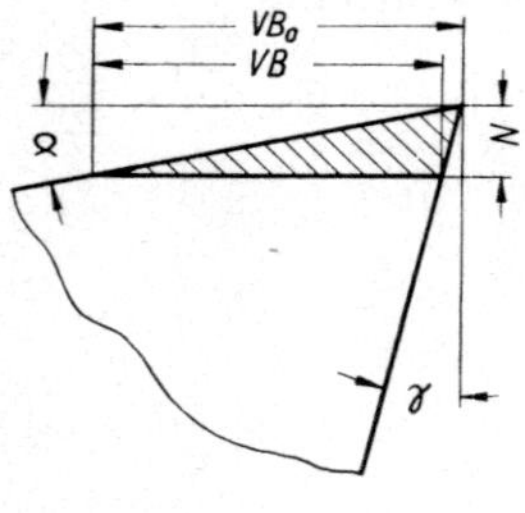

Fig. 4.2 Magnitude of wear on the flank

The volume of abraded material may be determined from

$$V = A_s F_\mu v\, t \quad \text{mm}^3 \tag{4.8}$$

where: A_s—specific output of abrasion, i.e. the volume of material which can be removed by abrasion if the energy has unit value in $\text{mm}^3/(\text{kg mm})$, F_μ—friction force on the flank in kg, v—speed of tool point motion in relation to the machined surface in mm/min, t—abrasion time in min.

The force of friction on the flank may be expressed as

$$F_\mu = \mu_\alpha q_{N\alpha} VB.VL \tag{4.9}$$

where: μ_α—coefficient of friction on the flank, $q_{N\alpha}$—unit pressure on the flank.

Substitution of eqns. (4.6), (4.8) and (4.9) into eqn. (4.7) yields

$$VB = \frac{1-\tan\alpha\tan\gamma}{\tan\alpha}\,2A_s\mu_\alpha q_{N\alpha}\, v\, t \tag{4.10}$$

If the magnitudes contained in eqn. (4.10) were not variable with cutting time and friction, this equation could be used for the calculation of VB. In reality, however, the wear land surface increases with cutting time and so does the amount of heat which in most cases is not abstracted as rapidly as it is generated. An increased contact temperature results. This relation may be expressed empirically as [2/4]

$$\Theta = E\, t^e \tag{4.11}$$

Consequently, changes will also concern

— specific output of abrasion A_s, because with increasing temperature both the hardness and strength of tool point material decrease and thus A_s increases. This relationship may be represented empirically by

$$A_s = A_{so} + B\Theta^b \tag{4.12}$$

— coefficient of friction on the flank μ_α, since temperature influences it in a non-monotonic way. This relation may be approximated by

$$\mu_\alpha = \mu_{\alpha o} - C\Theta^c + D\Theta^d \tag{4.13}$$

— normal pressure on the flank $q_{N\alpha}$, since the resistance to plastic deformation varies under the influence of temperature. This relation is approximated by

$$q_{N\alpha} = \frac{q_{N\alpha o}}{K+\Theta^n} \tag{4.14}$$

Substitution of eqn. (4.11) into eqns. (4.12)–(4.14) allows one to write eqn. (4.10) in a new form, with new constants resulting from the multiplication of those from eqns. (4.11)–(4.14):

$$VB = 2\,\frac{1-\tan\alpha\tan\gamma}{\tan\alpha}\,v(A_{so}+P\,t^z)(\mu_{\alpha o}-M\,t^u+N\,t^w)\,\frac{q_{N\alpha o}\,t}{1+W\,t^m} \tag{4.15}$$

Equation (4.15) shows the dependence of the mean width of flank wear land when all cutting conditions remain constant and the temperature changes only as a result of heat abstraction becoming with time slower than heat generation.

In the case when other machining conditions will vary, too (influencing the cutting temperature in a direct or indirect way) the magnitude of coefficients P, M, N and W and of exponents z, u, w and m will change correspondingly. Generalizing, it may be said that these coefficients make allowance for factors influencing the temperature.

With the use of eqns. (4.15) and (4.6) a similar relationship between the shortening of the tool point with time may be presented.

Equation (4.15) for the calculation of the width of flank wear land indicates that this relationship is complicated and unsuitable for practical application. Thus the value of eqn. (4.15) consists in demonstrating the factors determining the wear land width VB. This equation also indicates that the wear curve may change its character depending on the constants which in turn are functions of almost all cutting conditions.

4.1.5 Tool point wear curves

Experimental research on how wear varies with cutting time confirms the theoretical predictions, since the form of the tool point wear curves varies within wide limits. Figure 4.3 shows examples of wear curve shapes. Curve *a* corresponds to the cutting of a highly abrasive material at low cutting speeds, with very good cooling, i.e. at a rather low, stabilized temperature. Curve b is typical of the machining of steel at medium cutting speeds without cooling or with low intensity cooling. Curve c corresponds to cutting at high cutting speeds, when the temperature grows rapidly with cutting time.

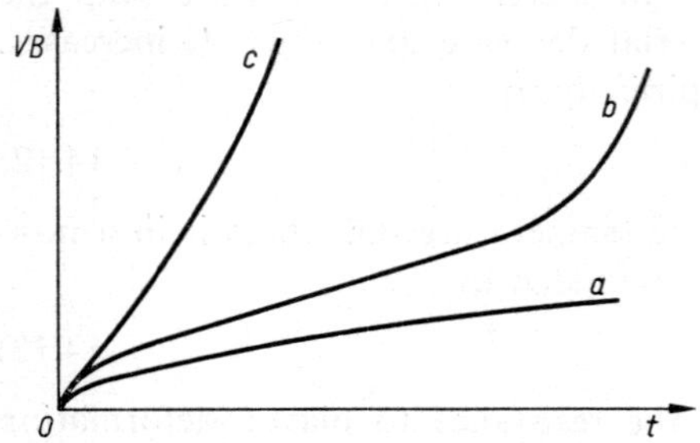

Fig. 4.3 Varieties of tool point wear curves

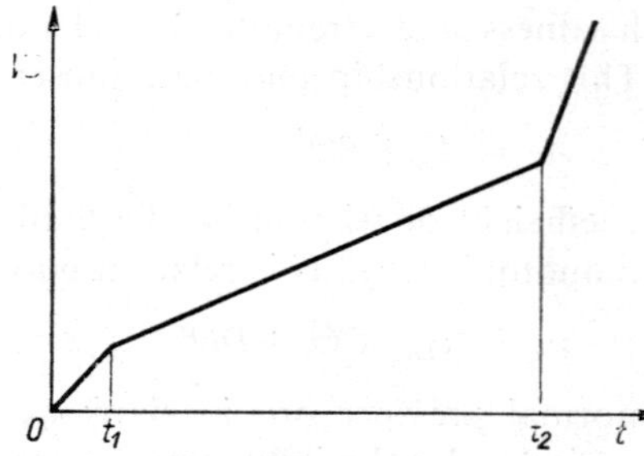

Fig. 4.4 Typical tool point wear curve

The most general form of the wear process is considered to be represented by a curve (Fig. 4.4) which includes:

— a period of initial wear from $t = 0$ to $t = t_1$,
— a period of moderate wear, lasting from $t = t_1$ to $t = t_2$,
— a period of accelerated wear, lasting from $t = t_2$ to the physical loss of cutting capacity.

The period of initial wear may be reduced by suitable shaping of the tool point and its careful lapping. The period of accelerated wear may not occur at all during the practically adopted service life of the tool point. These, however, are considered to be exceptional cases and the wear curve shown in Fig. 4.4 is a typical one.

4.1.6 Influence of some factors on tool point wear

4.1.6.1 *Influence of the material being machined*

On the basis of numerous experiments it has been established that tool point wear increases with specific cutting energy and with the wearing action of the material being machined. In the case of a particular type of material this characteristic depends on the chemical composition (alloy additions) and structure.

Thus, for example, the construction material used most frequently, steel, influences tool point wear in the following way (according to research conducted by E. Z. Feldsztejn):

(a) wearing action of steel increases with carbon content,
(b) wearing action of steel increases with the amount of alloy additions causing the formation of carbides,
(c) steels of ferritic structure show the lowest wearing action,
(d) from among pearlitic steels, lamellar pearlite is characterized by the highest wearing action and globular cementite by the lowest,

(e) wearing action of pearlitic steels decreases with the size of cementite grains,
(f) wearing action of high-speed steels of globular cementite structure depends on the distribution of carbides: irregular distribution tends to reduce wearing action, while network-type distribution increases it considerably,
(g) austenitic steels containing a small amount of carbides are characterized by low wearing action,
(h) as steel hardness increases a tendency for the wearing action to increase may be observed,
(i) the principal structural component causing wearing action increase are carbides, and particularly cementite.

4.1.6.2 *Influence of tool point material*

Resistance of tool point material to wear is characterized by its wearing action, wearability and thermal resistance, i.e. resistance to the loss of mechanical properties at high temperature. This depends on the chemical composition and structure of the tool point material.

4.1.6.3 *Influence of tool point elements and stereometric features*

Amongst the geometrical parameters of tool point, the rake angle γ, cutting edge angle $\varkappa_r$ and nose radius r_β have the greatest influences upon abrasive wear.

Figure 4.5 shows an example of how the width of the flank wear land varies with rake angle. Figure 4.6 shows similar examples involving cutting edge angle and nose radius. These diagrams are based on the results of Danel'yan's research [4/7].

Increase of tool point wear with a reduction of rake angle is explained by the increasing thrust force when rake angle decreases and, consequently, by the simultaneous rise of normal pressure.

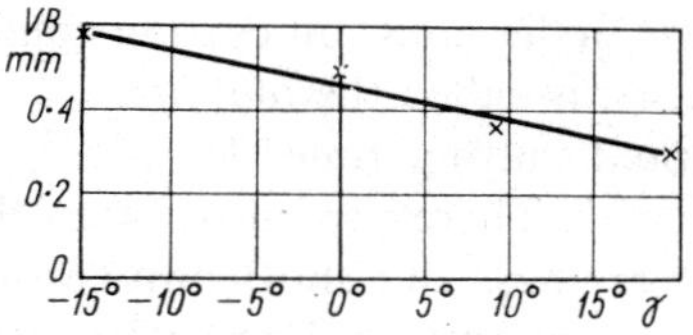

Fig. 4.5 Influence of rake angle on wear land width appearing on the flank after a constant cutting time (according to Danel'yan)

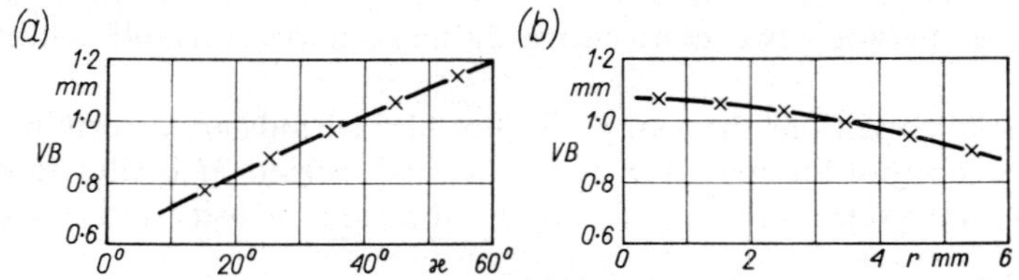

Fig. 4.6 Dependence of wear land width *VB* on the flank: (a) upon cutting angle $\varkappa_r$, (b) upon nose radius *r* (according to Danel'yan)

Increase of tool point wear when cutting edge angle increases and nose radius decreases results from the reduction of corner volume. As a consequence, tool point temperature and wear increase.

4.1.6.4 *Influence of cutting conditions*

Of all cutting conditions the cutting speed (v) and feed (s) have the greatest influence on tool point wear. The influence of the depth of cut a is considerably less important.

On the basis of numerous research works the influence of cutting conditions in the case of sintered carbide tool bits may be generalized and approximated by the statistical and empirical equation

$$VB = C_h v^{(3-5)} s^{(1 \cdot 5-2)} a^{(0 \cdot 15-0 \cdot 25)} \tag{4.16}$$

Exponents of v, s and a indicate the quantitative differences of the influence of the three cutting parameters.

The influence of the medium in which the cutting takes place is the more significant, the worse is the cutting ability of the tool point. Thus relatively greater effects are obtained in the case of tool steel than in the case of sintered carbides.

Interrupted type cutting accelerates abrasive wear of the tool point, although it does not change the relative influences of the particular cutting conditions. The cause of accelerated tool point wear in the case of interrupted cutting is mainly microdents appearing on the cutting edge.

4.2 Total tool life and service life

4.2.1 Determination of tool life and tool work period

Progressing tool point wear leads to the tool losing its ability to achieve the necessary machining quality and output.

The loss of the tool point's capacity to machine workpieces under desired conditions is called *tool point dulling*.

The working time of the tool point, or cutting time under constant cutting conditions until conventional dulling of the tool point occurs, is called the *tool life*.

In principle, this definition concerns uninterrupted cutting time. However, in practice, this condition can be fulfilled only in rare cases. Therefore, most often the term tool life is applied to the sum of several cutting times, and short non-cutting times, until the tool point dulls. The condition of constant cutting parameters must still be fulfilled.

Beside the notion of *tool life period* that of *tool work period* is applied, this is the sum of the cutting times between two consecutive dullings, under variable cutting conditions.

Under most industrial conditions the tool work (not life) periods are determined.

The difference between tool life period T and tool work period M is illustrated in Fig. 4.7. Depending on the selection of v, s and a, the abrasion intensity and cutting times differ.

Thus if the same magnitude of wear land width on the flank is assumed as permissible in both cases, then, in general, $T \neq M$.

Each tool point may be used several times for cutting, either by resharpening, or by using consecutive corners (in the case of tools having multipoint tips).

The sum of all the tool life periods which can be obtained from a given tool point, is called the *total tool life U*.

The sum of all tool work periods, which can be obtained from a given tool point, is called the *total service life W*.

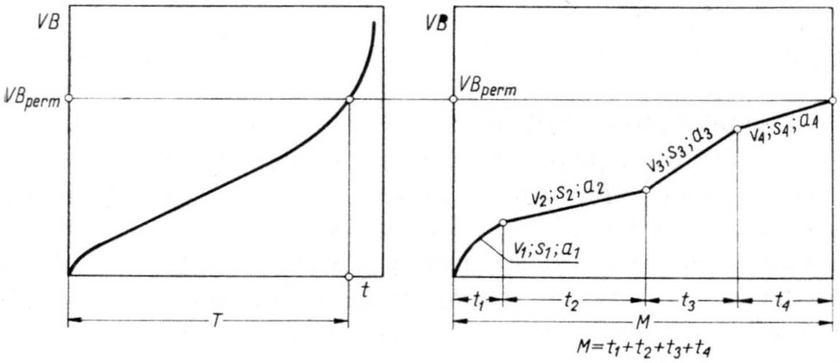

Fig. 4.7 Graphical interpretation of tool life period and tool work period

Those definitions may be written as follows

$$U = \sum_{i=1}^{i=m} T_i \tag{4.17}$$

$$W = \sum_{i=1}^{i=m} M_i = \sum_{i=1}^{i=m} \Big(\sum_{n=1}^{u=z} t_u\Big)i \tag{4.18}$$

where: m—number of tool life periods, z—number of work times in one tool work period.

4.2.2 Tool point wear criteria

The notion of tool point wear is a relative one. Depending on the demands made, the moment of dulling may be determined in various ways.

The basis for the determination of the moment when cutting is to be stopped (i.e. when the tool point is considered to be dulled) is called the *wear criterion.*

There are three groups of wear criteria: physical, technological and economic.

The physical criterion of tool point wear assumes that the tool point is to be considered dulled when the third wear period begins (accelerated wear). Direct determination of this point by measuring the wear land on the flank or any other geometrical parameter of tool point wear is practically impossible. However, this point may be determined on the basis of some phenomena accompanying it, such as the change of tool point shape as a result of abrasion, increase of cutting resistance and power, temperature increase and vibrations.

The increase of cutting power and resistance may be ascertained by means of suitable instruments installed on the machine tool. Thus, for example, the role of an indicator of power increase may be played by an ammeter or wattmeter included into the circuit energizing the motor.

Temperature increase may be determined approximately by temper colours of the chip or by glossy traces on the machined surface of steel workpieces turned with high-speed steel tools.

Physical criteria are seldom applied in practice.

Technological criteria of tool point wear are applied most often. These comprise:
(a) loss of accuracy (exceeding of tolerance limits),
(b) deterioration of surface quality (increase of surface roughness beyond the permissible limit).

It follows from the definition of this type of criterion that the tool point may not have reached the beginning of the third phase of wear, but if point wear in the radial direction in respect of the shaft being turned exceeds the dimensional tolerance of workpiece diameter, then the tool point is considered dulled.*

By analogy, the same concerns the deterioration in surface quality. It is understandable that the frequency of tool point dulling in the technological sense will be the greater, the closer will be the tolerance of dimensions and of surface quality.

An economical criterion of tool point wear assumes that the magnitude of permissible wear should be such that a maximum value of U or W is obtained [eqn. (4.17) or (4.18)].

If it is assumed, for the sake of simplicity, that all tool life periods are equal, then

$$U = m\,T \tag{4.19}$$

The number of possible sharpenings, and thus of tool life periods, is determined by the quotient

$$m = \frac{L}{e+g} \tag{4.20}$$

where: L—tool point dimension limiting the number of sharpenings (e.g. width or thickness of tool bit minus a certain magnitude, making the cutting still possible), e—tool point wear, measured in the direction of the limiting dimension, g—grinding allowance, measured in the direction of the limiting dimension (usually $g = 0{\cdot}1$–$0{\cdot}3$ mm).

In the case when VB is adopted as the tool wear criterion, the value of e may be calculated by means of Fig. 4.8, i.e.

$$e = VB\frac{\sin\alpha}{\cos(\alpha+\gamma)} \tag{4.21}$$

Substitution of (4.21) to (4.20) yields

$$m = \frac{L}{VB\dfrac{\sin\alpha}{\cos(\alpha+\gamma)}+g} = \frac{w\,L}{VB+w\,g} \tag{4.22}$$

* Under the assumption that re-setting of the partly worn tool to correct dimension is either uneconomical or impossible.

when

$$w = \frac{\cos(\alpha+\gamma)}{\sin\alpha} \tag{4.23}$$

If it is assumed that $\frac{\mathrm{d}U}{\mathrm{d}VB} = \frac{\mathrm{d}(m\,T)}{\mathrm{d}VB} = 0$, the value of $\frac{\mathrm{d}T}{\mathrm{d}VB}$ may be found for which U will have its maximum value

$$U = m\,T = \frac{w\,L\,T}{VB+w\,g} \tag{4.24}$$

$$\frac{\mathrm{d}U}{\mathrm{d}VB} = \frac{\mathrm{d}\left(\frac{w\,L\,T}{VB+w\,g}\right)}{\mathrm{d}VB} = 0 \tag{4.25}$$

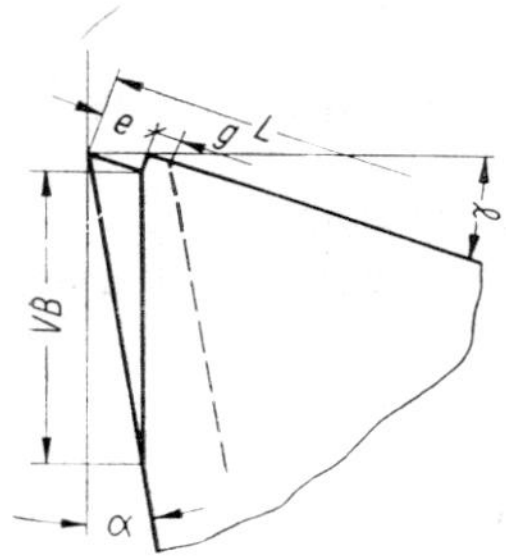

Fig. 4.8 Auxiliary drawing for the calculation of tool wear *e* in the direction of the limiting dimension *L*

After differentiation in which $T = f(VB)$

$$\frac{w\,L\,T}{(VB+w\,g)^2} = \frac{w\,L\,\mathrm{d}t/\mathrm{d}VB}{VB+w\,g} \tag{4.26}$$

and thence

$$\frac{\mathrm{d}VB}{\mathrm{d}T} = \frac{VB+w\,g}{T} \tag{4.27}$$

Equation (4.27) allows one to determine the optimum value of VB by a graphical method (Fig. 4.9) when the relationship $VB = f(t)$ is known.

The value of $w\,g$ is marked on the VB axis from t axis downwards and thus section OA is obtained. The tangent to the curve $VB = f(t)$ is drawn from point A. The point of tangency B determines the optimum wear value VB_{opt} and the corresponding tool life period T.*

This method of optimum tool point life determination indicates that for each set of machining conditions on which the form of the wear curve depends, there is a different VB_{opt} value.

From an industrial point of view this is inconvenient, as it would result in excessive differentiation of recommended wear values.

* The method of determination of optimum wear with a view to obtaining maximum total tool life was developed by M. N. Larin in 1947. The graphical method of determining VB_{opt} was developed by N. N. Zorev in 1949.

Therefore, mean values of recommended tool point wear *VB* have been established for the particular methods, varieties and types of cutting.

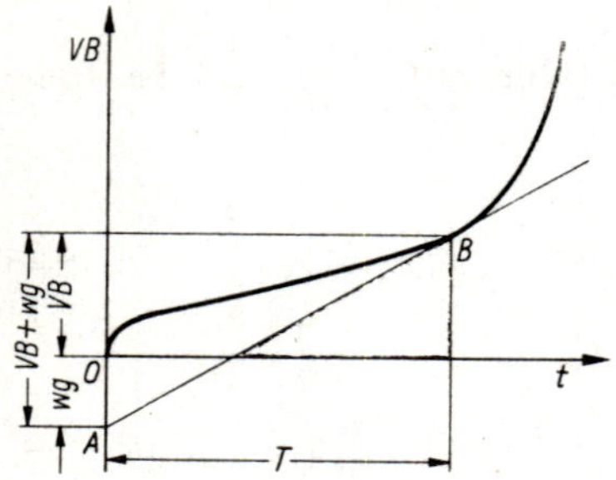

Fig. 4.9 Method of determining VB_{opt} for maximum total tool life $U = U_{max}$

4.2.3 INFLUENCE OF SHARPENING METHODS ON TOTAL TOOL LIFE AND SERVICE LIFE

Restoring the proper tool point shape after dulling is called *sharpening*.

Sharpening may be carried out by means of grinding and other abrasive processes or by erosion machining.

Sharpening may be carried out by:

(a) removing a suitable layer from the flank only,
(b) removing a suitable layer from the rake face only,
(c) removing suitable layers from both the flank and rake face.

The character of tool point wear, positioning and shape of the tool bit constituting the point decide which of the methods listed above would be most useful and economical.

Sharpening from the flank side is recommended in the case of a relatively large wear land appearing on the flank only, of horizontally situated and relatively wide tool bits.

Sharpening from the side of the rake face is justified only if wear appears mainly as a land on the rake face and the tool bits are relatively smooth. The disadvantage of this method is the reduction of tool bit thickness resulting in the reduction of bit strength and thus of total tool life and service life.

Simultaneous sharpening from both the flank side and rake face may be most economical depending upon the positioning and dimensions of the tool bit. This can be demonstrated by means of Fig. 4.10.

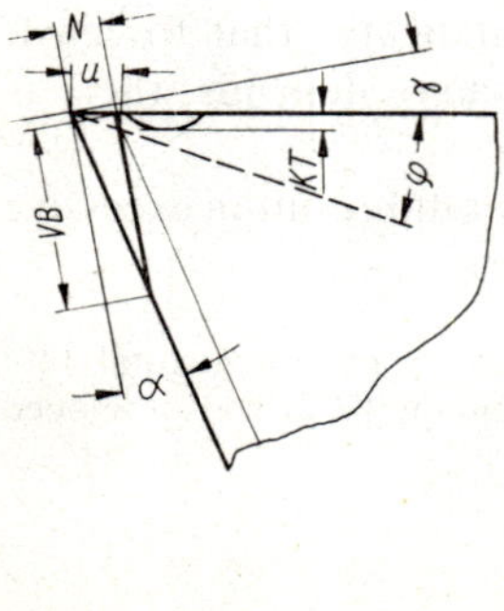

Fig. 4.10 Determination of the direction of tool point nose shifting as a consequence of wear and sharpening

It follows from Fig. 4.10 that

$$\tan\varphi = \frac{KT}{u} = \frac{KT}{N/\cos\gamma + KT\tan\gamma} \tag{4.28}$$

and from eqn. (4.06)

$$\tan\varphi = \frac{KT}{VB\dfrac{\tan\alpha}{\cos\gamma(1-\tan\alpha\,\tan\gamma)} + KT\tan\gamma} \tag{4.29}$$

Angle φ determines the direction of tool point nose shift when in consecutive sharpenings layers are repeatedly removed from the side of the flank and rake face.

It is evident from eqn. (4.29) that:

— if crater depth $KT = 0$, i.e. when there is no wear on the rake face, then $\varphi = 0$, and sharpening should be carried out on the flank only,

— if flank wear $VB \leqq KT/\cos\gamma$, then a layer of thickness KT should be removed. As a result of this, angle $\varphi = 90 - \alpha - \gamma = \beta$. Thus in this case sharpening from the side of the rake face should take place.

This being so, eqn. (4.29) allows one to determine the most advantageous positioning of the tool bit when *VB* and *KT*, and thus wear character and magnitude, are known.

Fig. 4.11 shows the most advantageous positioning of the tool bit for known and constant magnitudes of *KT* and *VB*. Such positioning ensures a constant strength of

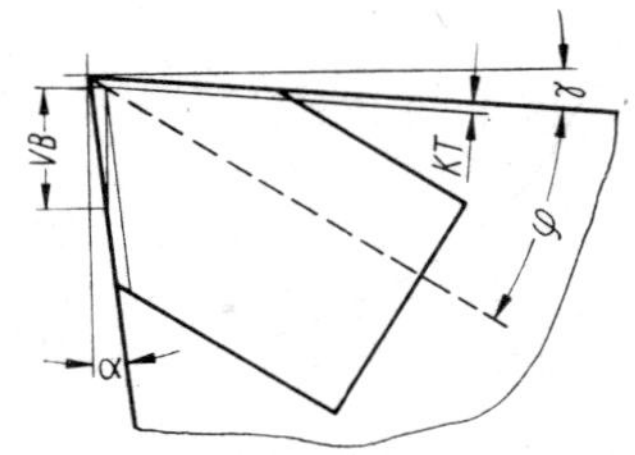

Fig. 4.11 Optimum positioning of tool bit constituting the tool point, with regard to flank and rake face wear rate

the tool bit, notwithstanding the fact that the sharpening proceeds from the rake face, too. At the same time, tool bit position, as determined by angle φ, yields the longest total tool life and service life.

Therefore, angle φ is called the *optimum position angle* of tool bit constituting the tool point [4/15].

4.3 Tool life speed

4.3.1 Relationship between tool life and cutting speed; tool life speed

Of all factors influencing the tool life period, cutting speed has the greatest influence. As cutting speed is varied within a wide range, this may considerably increase or reduce tool life.

For this reason the dependence of tool life period upon cutting speed has long been the subject of research.

The mathematical form of this relationship was given by F. W. Taylor [2/5] as a statistical and empirical equation

$$T = \frac{C_T}{v^z} \tag{4.30}$$

where: C_T—a constant taking into account the influence of all factors not appearing separately in the formula; in the physical sense the constant C_T is equal to the tool life period for cutting speed $v = 1$ m/min; z—exponent determined empirically.

Values of the exponent z vary for sintered carbides within the limits 3–6, and for high-speed steels between 8 and 10. This indicates that doubling the cutting speed causes a reduction in tool life for sintered carbide tool points to about 1/32th and for high-speed steel tool points to about 1/260th of the former value. Such a great influence of cutting speed may be explained by the fact that with its increase the cutting temperature increases very quickly. Consequently, mechanical strength and resistance to abrasion of the tool point material decrease simultaneously to a degree depending on its properties.

Equation (4.30) suggests that the dependence of tool life period on cutting speed is a monotonic hyperbolic function (for $v = 0$ $T = \infty$ and for $v = \infty$ $T = 0$). A graphical interpretation of eqn. (4.30) is given in Fig. 4.12.

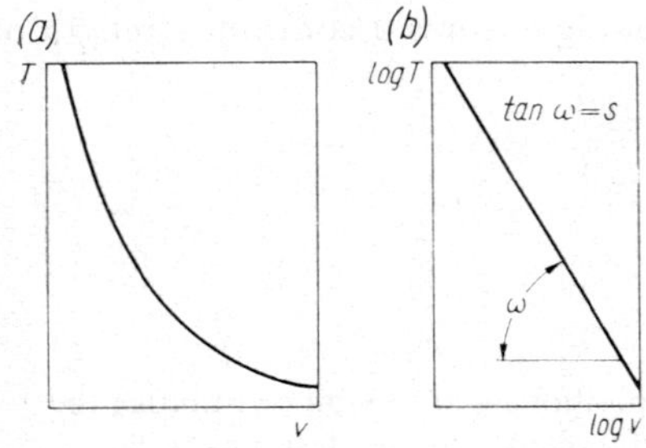

Fig. 4.12 Dependence of tool life on cutting speed according to equation $Tv = C_T$: (a) linear scale, (b) logarithmic scale

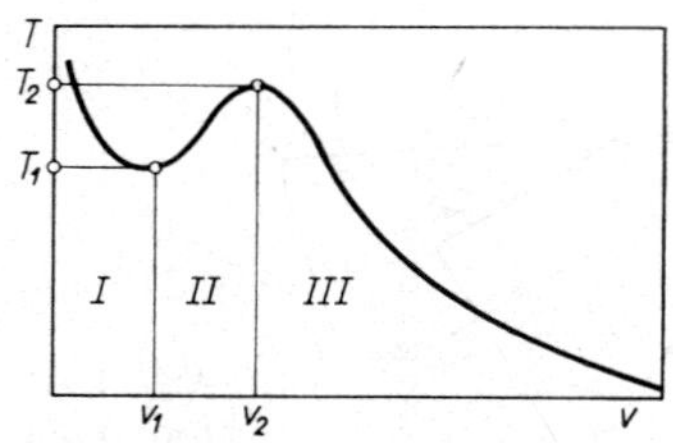

Fig. 4.13 General dependence of tool life on cutting speed

More detailed studies (among others those of Avakov [4/13]) indicate that the dependence of tool life on cutting speed within a wide range of the latter has a non-monotonic character, as shown in Fig. 4.13.

From the form of the relationship $T = f(v)$ it is possible to distinguish:

— range I: reduction of tool life with increase of speed, when $v < v_1$,

— range II: increase of tool life with increase of speed, when $v_1 \leqq v \leqq v_2$,

— range III: monotonic reduction of tool life, when $v > v_2$.

It is evident that the relationship $T = f(v)$ has a complex character and the empirical formula $Tv^z = C_T$ does not express its physical nature.

Many attempts* have been made to evaluate for the generalized relationship $T = f(v)$ a mathematical form which would fully correspond to the physical basis.

* One of the first to make attempts at a physical understanding of the relationship $T = f(v)$ was N. A. Safronov [4/8 .

So far there is no accepted and acknowledged function fulfilling those conditions for a given tool point material and for the complete range of cutting speeds.

Besides, in most cases occurring in practice, cutting speeds $v > v_2$ are applied, i.e. speeds belonging to the range where tool life decreases monotonically with increasing cutting speed.

For these reasons equations similar to (4.30) are used, as they sufficiently approximate the curve $T = f(v)$ in the III range of cutting speeds.

When machining certain materials, as e.g. austenitic steels, speeds belonging to the II or even to I range are applied in practice. In these cases the inapplicability of equations similar to (4.30) should be born in mind. As seen from Fig. 4.13, three values of cutting speed may correspond to a single value of T.

In the III range of cutting speed there is an unequivocal relationship between T and v. Equation (4.30) may be written in the form

$$v = \frac{C_v}{T^m} \tag{4.31}$$

where $C_v = \sqrt[z]{C_T}$—a constant taking into account the influence of factors not appearing separately in the equation; the value of C_v is equal to that of cutting speed for $T = 1$ min; $m = 1/z$—exponent determined empirically.

In the case when the tool life period is being established, the cutting speed corresponding to this period is called the tool life speed and designated as v_T, and thus

$$v = \frac{C_v}{T^m} = v_T \tag{4.32}$$

For example, the speed corresponding to the tool life period $T = 20$ min is called the twenty minutes cutting speed v_{20}. Similarly, there may be an hourly cutting speed v_{60}, four hours cutting speed v_{240}, etc.

4.3.2 INFLUENCE OF CERTAIN FACTORS ON TOOL LIFE SPEED

4.3.2.1 *Influence of the material being machined*

The influence of the material being machined on tool life speed is expressed mainly by the amount of heat equivalent to the amount of specific cutting energy and by the wearing action of the material under given cutting conditions. Both these properties of the material are, of course, the result of its chemical composition and structure. Research conducted on the direct connection between these factors and tool life speed have not yet yielded results permitting one to establish general quantitative relationships, which are valid for various materials.

Therefore, in practice correction factors are used, determined in an empirical and statistical way for various workpiece materials. Tool life speed for any such material may be calculated from the equation

$$v_T = \frac{C_v}{T^m} B_M \tag{4.32a}$$

where: B_M—correction factor, C_v—constant determined for carbon steel of tensile strength $\sigma_B = 75$ kg/mm^2.

Correction factors for various types of steel are given in Table 4.1.

Table 4.1 *Values of correction factors B_M for various types of steel*

Tool point material	Type of material being machined	B_M*
High-speed steel and sintered carbides	Free cutting steel	1·2
	Structural carbon steels (C < 0·6%) and nickel steels	1·0
	Chromium–nickel steels	0·95
	Tool carbon steels (C > 0·6%), manganese steels and chromium–molybdenum, chromium–nickel–molybdenum, chromium–aluminium, chromium–aluminium–molybdenum and chromium–nickel–tungsten steels	0·8
	Chromium–manganese, chromium–silicon, silico–manganese and chromium–silico–manganese steels	0·7
	High-speed tool steels	0·6

* Factors B_M for steel are determined in each group for steel of $\sigma_B = 75$ kg/mm²

It is evident from this table that the highest tool life speeds are obtained when machining free cutting steels and the lowest when machining high-speed tool steels. For other materials—in the case of cast iron, although both its hardness and specific cutting energy are inferior to those of steel, lower tool life speeds are applied than for carbon steels. The reason is mainly the higher wearing action and inferior heat conductivity of cast iron, causing increased cutting temperatures.

Aluminium and its alloys may be turned at speeds 4 to 6 times higher than steel because of the considerably lower specific cutting energy.

With respect to other properties of the various materials being machined, the dependence of tool life speed on hardness and tensile strength (for steel and other ductile materials) has been established, also in an empirical and statistical way. For the same type of material, increased hardness or tensile strength causes a decrease of tool life speed. For ductile materials this relationship may be described by the following equation

$$v_{T2} = v_{T1}\left(\frac{\sigma_{B1}}{\sigma_{B2}}\right)^a \tag{4.33}$$

where: v_{T2}—tool life speed of a material of tensile strength σ_{B2}, v_{T1}—tool life speed of a material of tensile strength σ_{B1}, a—exponent, whose value is different for various materials.

The same relationship for cast iron and other brittle metals is expressed by the equation

$$v_{T2} = v_{T1}\left(\frac{HB_1}{HB_2}\right)^b \tag{4.34}$$

where: v_{T2}—tool life speed of a material of hardness HB_2, v_{T1}—tool life speed of a material of hardness HB_1, b—exponent, different for the various materials being machined.

Values of the exponents *a* and *b* are listed in Table 4.2.

Table 4.2 *Values of exponents in equations for the dependence of tool life speed on tensile strength* σ_B *and hardness HB*

Tool point material	Type of material being machined	*B* kg/mm²	*a*	*b*
High-speed steel	Free cutting steel	—	1·75	—
	Carbon steel	45	1·00	—
	Carbon and chromium steel	45	1·75	—
	Cr–Ni steel, nickel steel, manganese steel, silicon steel etc.	—	1·50	—
	Chromium–nickel–molybdenum steel, chromium–molybdenum steel, high-speed steel	—	1·25	—
Sintered carbides	Non-hardened steels	—	1·50	—
	Hardened steels	—	2·00	—
	Grey cast iron, malleable cast iron	—	—	1·75

Influence upon tool life speed is also exerted by the production engineering methods used in the preparation of semi-finished products (casting, forging, rolling etc.), the condition of the material and of its surface layer. The influence of these factors is expressed in the form of correction factors B_p whose values are listed in Table 4.3. The influence of production engineering methods used in the preparation of semi-finished products and in their heat treatment results from the role played by material structure and crystal size. These factors control to a large extent both the value of specific cutting energy and wearability. The influence of the surface layer in the form of so-called *casting* or *forging scale* is caused by hardening of this layer in comparison with the core of the material. The wearing action of the hardened layer is always several times higher than that of the non-hardened material and this

Table 4.3 *Values of correction factors* B_p *and* C_r *with regard to the type and condition of semi-finished products*

Tool-point material	Material being machined	Semi-finished products: without scale			with scale (of hardness HB): hot rolled			forged			cast		
		cold rolled and forged	hot rolled and forged	cast	160	160–300	200	160	160–200	200	160	160–200	200
High-speed steel and sintered carbides	Steel as rolled, as forged or as cast	1·1	1·0	0·9	0·9	0·9	0·9	0·75	0·80	0·85	0·75	0·80	0·85
	normalized	1·04	0·95	0·86	0·86	0·86	0·86	0·71	0·81	0·71	0·76	0·76	0·81
	annealed	0·99	0·9	0·81	0·81	0·81	0·81	0·67	0·72	0·76	0·67	0·72	0·76
	toughened	0·88	0·8	0·72	0·72	0·72	0·72	0·6	0·64	0·68	0·6	0·64	0·68
	Cast iron	—	—	1·0	—	—	—	—	—	—	0·7	0·75	0·8

results in non-uniform wear of the tool point. The influence of the hardened layer is the greater, the smaller is the depth of cut.

4.3.2.2 *Influence of the tool*

The influence of tool point material upon tool life speed depends on the resistance of the material to high cutting temperatures and the resulting changes of strength, friction and abrasion.

When tool life speeds of high-speed steel SW18 with tungsten content of 18% are taken as a basis for comparison, correction factors B_N for other types of steel indicate that highest life speeds correspond to high-speed steel SK5 with 5% cobalt content. The remaining types of steel differ only slightly from each other in this respect. Correction factors for a few standard high-speed steels are listed in Table 4.4, and their chemical composition (and that of a few other types) according to Polish Standards in Table 4.5.

Table 4.4 *Values of correction factors B_N with regard to the type of tool point material*

Material being machined	Type of machining	Type of high-speed steel SW9	SW18	SK5
		Factor B_N		
Non-hardened steel	rough	1·12	1·0	1·0
	fairly accurate and accurate	1·15	1·0	0·95
Grey cast iron	rough	1·2	1·0	1·0
	fairly accurate and accurate	1·1	1·0	1·0

Table 4.5 *Chemical composition of some high-speed steels, according to Polish Standards*

Type of steel	C	Mn max.	Si max.	P max.	S max.	Cr	W	Ni max.	V	Co	Mo
SW9	0·80–0·95	0·40	0·40	0·03	0·03	3·5–4·5	8·0–10·0	0·40	1·8–2·2	—	—
SW14	0·80–0·95	0·40	0·40	0·03	0·03	3·8–4·8	13·0–15·0	0·40	2·2–2·7	—	—
SW18	0·70–0·85	0·40	0·40	0·03	0·03	3·8–4·8	16·5–19·5	0·40	1·0–1·5	—	—
SK5	0·70–0·85	0·40	0·40	0·03	0·03	3·8–4·8	16·5–19·5	0·40	1·2–1·7	4·5–5·5	—
SW7Mo	0·80–0·88	0·40	0·40	0·03	0·03	3·7–4·5	6·0–7·0	0·35	1·7–2·1	—	4·5–5·5
SW12	1·20–1·30	0·40	0·40	0·03	0·03	3·6–4·8	11·0–13·0	—	3·4–4·2	—	0·70–1·0
SW10K	0·9–1·0	0·40	0·40	0·03	0·03	3·6–4·8	9·0–11·0	—	2·0–2·7	4·4–6·0	max. 0·45

Considerably greater differences of tool life speed occur with various types of sintered carbides.

The influence of tool point material is usually taken account of in the following way: for each type of tool point material separate equations for tool life speed are determined and their constant C_v depends on the influence in question.

The material and dimensions of the tool shank exert an influence upon tool life speed, since the larger are these dimensions (cross-sectional area) and the better is the thermal conductivity of the shank material, the greater amount of heat is abstracted from spots where it is most intensively generated. The influence of the tool shank on tool life speed is only slight.

Among other tool point parameters the following exert an influence on tool life: rake angle, clearance angle, leading cutting edge inclination angle, nose radius and the shape of the rake face.

The influence of rake angle on tool life, and thus also on tool life speed, results, on the one hand, from the improvement of tool point strength and, on the other hand, from a simultaneous increase of specific cutting energy as rake angle decreases. These influences are opposed to each other and experiments show that for each tool–workpiece pair there is an optimum value of rake angle for which tool life speed reaches its maximum.

The influence of clearance angle may be described in a similar way. The greater the clearance angle, the weaker is friction, but also the smaller is heat resistance and strength and the more rapid tool point wear. Reduction of clearance angle leads to opposite results. Consequently, there also exists an optimum value—with respect to tool life—of clearance angle for various workpiece materials and tool points. Experiments indicate that the optimum range of clearance angles is very narrow. For steel and cast iron the clearance angles between 6° and 15° are applicable.

The influence of angles γ and α on tool life speed is illustrated by Fig. 4.14.

The influence of the leading cutting edge inclination angle has been determined differently by various research workers. Those studies were conducted under different conditions. Consequently, their results are not comparable.

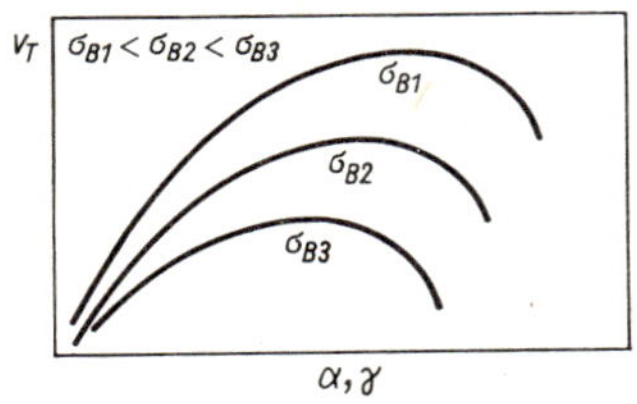

Fig. 4.14 Dependence of tool life speed on rake angle and clearance angle

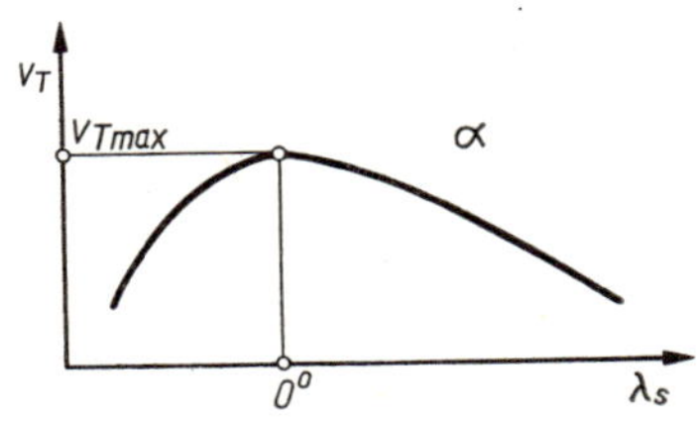

Fig. 4.15 Dependence of tool life speed on leading cutting edge inclination angle

Special studies on the influence of inclination angle conducted by Bobrov [4/9] have indicated that an increase of this angle is accompanied by an increase of specific cutting energy and specific friction energy. On the other hand, reduction of inclination angle causes an increase of the mass of the tool point corner and this, in the case of non-orthogonal cutting, improves the tool point strength and resistance to high temperature.

The result of this complicated influence of inclination angle on tool life speed is shown in Fig. 4.15. The optimum value of inclination angle is usually close to zero. Negative angles influence tool life speed more than positive angles.

The influence of nose radius r_β on tool life and tool life speed is a consequence of a change in the conditions of heat abstraction by the tool point. The greater is radius r_β, the greater are, admittedly, component forces F_v and F_p, but specific cutting energy, and thus heat concentration in tool point is reduced. Therefore, an increase in tool point nose radius results in longer tool life. This influence of nose radius r_β cannot be fully taken advantage of, as excessive increase of this radius tends to cause vibrations.

The influence of the shape of the rake face on tool life follows from changes of friction in chip flow and from the influence or decrease of tool point volume close to the cutting edge. For example, it has been established experimentally that tool points with lands increase tool life speed by an average of about 15% for tool points of high-speed steel and 12% for those of sintered carbides.

Tool point wear exerts an important influence on the value of tool life speed. Increasing the permissible wear causes an increase of tool life or—with a constant tool life—increase of tool life speed. Increasing the recommended permissible extent of tool wear is economical only up to a value corresponding to the maximum total tool life (U_{max}).

In this connection, two curves illustrating the dependence of tool life speed on tool life should be distinguished (Fig. 4.16)

— for a constant mean recommended value of permissible wear VB,

— for optimum wear, corresponding to maximum value of total tool life U_{max}.

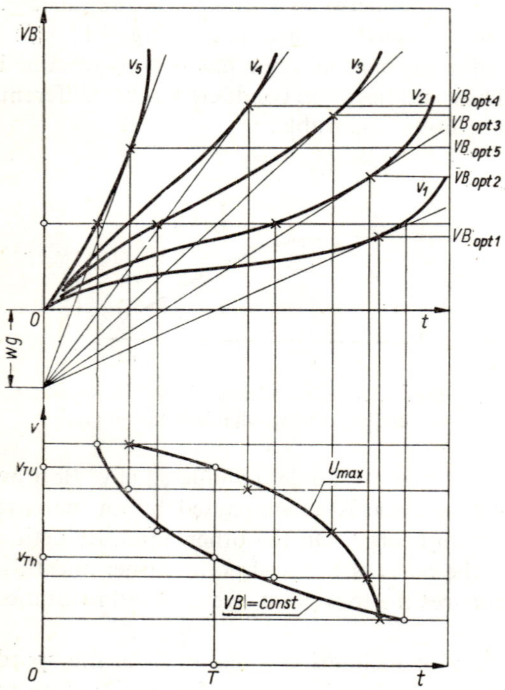

Fig. 4.16 Influence of tool wear criterion upon tool life speed

4.3.2.3 *Influence of cutting conditions*

The influence of the dimensions and shape of the undeformed chip is a result of the effect of depth of cut, feed rate and cutting edge angle which characterizes the shape of the undeformed chip.

The influence of the depth of cut and feed rate on tool life is similar to that of cutting speed (Fig. 4.13), although considerably less. Feed rate has a relatively greater influence on tool life than the depth of cut.

In the III range of Fig. 4.13 the dependence of tool life on feed rate and depth of cut may be approximated with sufficient accuracy by the hyperbolic functions

$$T = \frac{C_{Ts}}{v^z s^{u_T}}; \qquad T = \frac{C_{Ta}}{v^z a^{e_T}} \tag{4.35}$$

where C_{Ts}, u_t and C_{Ta}, e_T are determined in an empirical way.

Taking into consideration the relationships (4.31) and (4.35) it may be shown that

$$v_T = \frac{C_v}{T^m a^{e_v} s^{u_v}} \tag{4.36}$$

where $e_v = e^{mT}$ and $u_v = u^{mT}$.
It has been established by research that approximate mean values are: $e_v = 0{\cdot}2$ and $u_v = 0{\cdot}4$.

It follows that in the case when both the feed rate and depth of cut are increased by the same percentage value, the influence of feed is approximately equal to the influence of the square of the depth of cut.

The influence of leading cutting edge angle is very important. Within the range $30° \leqq \varkappa_r \leqq 90°$ the relationship between tool life speed and cutting edge angle $\varkappa_r$ is, according to experiments, determined by the equation

$$v_T = \frac{C_v}{\varkappa_r^z} \tag{4.37}$$

where z—exponent found in an empirical way ($z \approx 0{\cdot}4$–$0{\cdot}6$).

The wider is the cutting edge angle, the shorter is the length of active cutting edge and the greater the thermal load per unit active cutting edge length. Therefore, in order to maintain the same tool life as before, cutting speed must be suitably reduced. If the influence of angle $\varkappa_r$ on tool life speed is characterized by correction factors and it is assumed that for $\varkappa_r = 45°$ the factor $B_\varkappa = 1$, then for angles within the range of 30–90° it will take on values shown in Fig. 4.17.

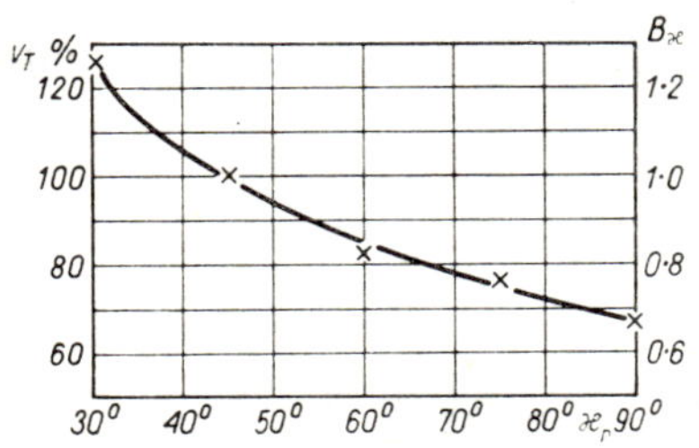

Fig. 4.17 Dependence of tool life speed v_T on leading cutting edge angle $\varkappa_r$

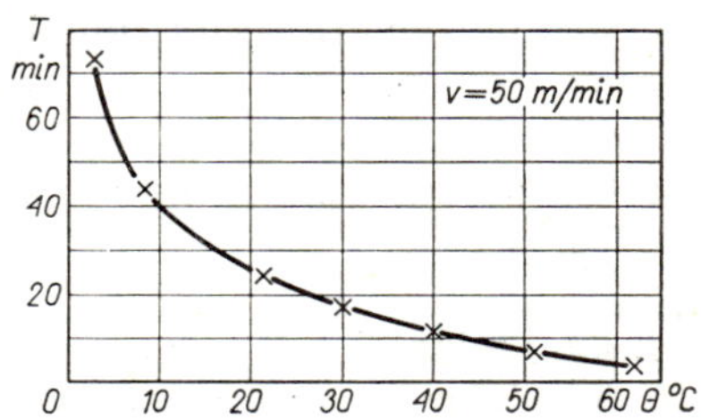

Fig. 4.18 Dependence of tool life T on cutting fluid temperature Θ (according to J. Dmochowski)

The influence of trailing edge angle $\varkappa_r'$ on tool life is twofold. Its reduction causes, on the one hand, an increase of tool point volume and thus also of its life; on the other hand, however, the smaller the angle $\varkappa_r'$, the greater the influence of the trailing edge on the cutting energy, and particularly on friction energy. An increase of trailing edge angle has the opposite effect. Experiments show that, as a result of this double influence of angle $\varkappa_r'$, tool life and thus the tool life speed assume their highest values for a certain magnitude of this angle, namely within 5–10°.

The cooling method and the type of cutting fluid also influence tool life speed, due to the lubricating and cooling action of the fluid. Factors contributing to an improvement of lubrication and cooling include a relatively low temperature of the coolant–lubricant concentration and—in the case of emulsions—a higher concentration.

According to research conducted by J. Dmochowski [4/10] on the influence of coolants on the cutting of steel ($\sigma_B \approx 76$ kg/mm^2) with high-speed steel tools the lower the coolant temperature, the higher is the cutting speed for constant tool life. Figure 4.18 shows the dependence of tool life period on cutting speed at various coolant temperatures.

Research by A. Bułat and R. Sniechowski [4/16] has indicated that, if emulsion is used, its concentration is of considerable importance. The relationship between tool life and emulsion concentration (oil content in percent) depends on the type of machining (rough, accurate) and on the type of material being machined. The use of emulsion with an oil content below the optimum value shortens tool life, since the lubricating qualities of the emulsion deteriorate, causing an increase of friction and, consequently, an acceleration of wear. Application of emulsions of higher oil content also reduces tool life as the cooling capacity of the fluid decreases since oil has a lower conductivity and specific heat than water.

The flow rate and the method of cutting fluid supply play an important role in affecting tool life. No comprehensive research has been conducted so far in this field, but from work already carried out it is seen that tool life increases with cooling flow rate. However, beginning with a certain flow rate value, this increase becomes quite negligible. Therefore, e.g. in the case of a steel turning operation, an emulsion flow rate exceeding 10–12 l/min is uneconomical.

4.3.3 Calculation of tool life speed

On the basis of numerous research works carried out and of a mathematical analysis of their results, the empirical equation for tool life speed may be written in the following form

$$v_T = \frac{C_v}{T^m a^{e_v} s^{u_v}} B_M\, B_p\, B_H\, B_N\, B_r\, B_h\, B_f\, B_k\, B_\varkappa\, B_{\varkappa'}\, B_c \quad \text{m/min} \tag{4.38}$$

where:

C_v —a constant taking into account the influence of factors not appearing separately (e.g. machining variety, type of tool point material and cooling method),

T —tool life period in min,

m —exponent, constant for the given machining variety, workpiece material and tool,

a —depth of cut in mm,
e_v —exponent, constant for the given machining variety, workpiece material and tool,
s —feed in mm/rev,
u_v —exponent, constant for the given machining variety, workpiece material and tool,
B_M—correction factor considering the kind of material being machined; for structural carbon steel $B_M = 1$,
B_p —correction factor considering the kind of preparation of the semi-finished product, the condition of its surface and the structure of material being machined; for hot-rolled steel without scale $B_p = 1$,
B_H—correction factor considering hardness or tensile strength of the material; for carbon steel of $\sigma_B = 75$ kg/mm² and for grey cast iron of $HB = 190$ kg/mm², $B_H = 1$,
B_N—correction factor considering the type of tool point material; for SW18 high-speed steel (see Table 4.5) and for sintered carbides $B_N = 1{\cdot}0$,
B_r —correction factor considering tool point nose radius; for $r_\beta = 2$ mm and a high-speed steel tool point or $r_\beta = 1$ mm and a sintered carbide tool point, $B_r = 1$,
B_h —correction factor considering the extent of permissible wear,
B_f —correction factor considering tool shank dimensions,
B_k —correction factor considering the shape of the rake face; for flat tool points without lands, with a positive rake angle, $B_k = 1$,
$B_\varkappa$ —correction factor considering leading cutting edge angle; for $\varkappa = 45°$, $B_\varkappa = 1$,
$B_{\varkappa'}$—correction factor considering trailing edge angle; for $\varkappa' = 10°$ $B_{\varkappa'} = 1$,
B_c —correction factor considering cooling and lubrication; for dry cutting $B_c = 1$.

In the case when the cutting parameters are selected so that all factors* equal 1, eqn. (4.38) is simplified to the form given in eqn. (4.36).

The tool life period is most frequently assumed on the strength of technological and economic data. Thus, when recommended tool life periods are applied, the equation for tool life speed can be further simplified to the following form

$$v_T = \frac{C_{vT}}{a^{e_v} s^{u_v}} \quad \text{m/min} \tag{4.39}$$

where $C_{vT} = C_v/T^m$.

In a particular case, e.g. when $T = 60$ min or $T = 90$ min, the respective equations may be written in the form

$$v_{60} = \frac{C_{v60}}{a^{e_v} s^{u_v}} \quad \text{and} \quad v_{90} = \frac{C_{v90}}{a^{e_v} s^{u_v}}$$

When a different value of tool life period has been assumed, the conversion of tool life speed for T_2 instead of T_1 is carried out according to equation

$$v_2 = v_1 \sqrt[z]{T_1/T_2} \tag{4.40}$$

Assumption that $B_T = \sqrt[z]{T_1/T_2}$ yields

$$v_2 = v_1 B_T \tag{4.41}$$

* Factor values not given in this chapter may be found in the corresponding chapters of Part II.

REFERENCES

4/1 Orzechowski, S.: Stale narzędziowe. Wiadomości wstępne i katalog (Tool steels. Introductory Information and Catalogue), PWT, Warszawa 1953

4/2 Żmihorski, E.: Stal szybkotnąca (High-speed steel), PWT, Warszawa 1958

4/3 Bryjak, E.: Węgliki spiekane i ich zastosowanie (Sintered Carbides and Their Application), WGH, Katowice 1959

4/4 Imshennik, K. P., and Bukhman, N. A.: Tekhnologiya paiki tverdosplavnogo instrumenta (Technology of the Soldering of Hard-facing Alloy Tools), Mashgiz, Moskva 1954

4/5 Tretyakov, V. I.: Metallokeramicheskie tverdye splavy (Metal-ceramic Hard-facing Alloys), Metallurgizdat, Moskva 1962

4/6 Gibas, T.: Spieki ceramiczne i cermetale (Cermals and Cermets), WNT, Warszawa 1967

4/7 Danel'yan, A. M.: Teplota i iznos instrumentov v protsesse rezaniya metallov (Heat and Wear of Tools in the Metal Cutting Process), Mashgiz, Moskva 1954

4/8 Safronov, N. A.: Termodinamicheskaya teoriya rezaniya metallov (Thermodynamic theory of metal cutting), Vestnik Metallopromyshlennosti, 2, 9 and 10, 1930

4/9 Bobrov, V. F.: Vliyanie ugla naklona glavnoi rezhushchei kromki instrumenta na protsess rezaniya (Influence of the Leading Cutting Tool Edge Inclination Angle on the Cutting Process), Mashgiz, Moskva 1962

4/10 Dmokhovski, G. S.: Vliyanie okhlazhdayushchego deistviya smazochnookhlazhdayushchei zhidkosti v usloviyakh obrabotki metallov rezaniem. Autoreferat dissertatsii (Influence of the cooling action of lubricant-coolant under conditions of metal machining by cutting. Author's review of his own dissertation), Moskva 1951

4/11 Kostetski, B. I.: Stoikost' rezhushchikh instrumentov (Cutting Tool Life), Mashgiz, Moskva 1949

4/12 Danel'yan, A. M.: Teplota instrumentov v protsesse rezaniya metallov (Heat of Tools in Metal Cutting Process), Mashgiz, Moskva 1954

4/13 Avakov, A. A.: Fizicheskie osnovy stoikosti rezhushchikh instrumentov (Physical Basis of Cutting Tool Life), Mashgiz, Moskva 1960

4/14 Żmihorski, E.: Stale narzędziowe i obróbka cieplna narzędzi (Tool Steels and Heat Treatment of Tools), WNT, Warszawa 1967

4/15 Józefik, A.: Analiza racjonalnego doboru płytek i wkładek ze spiekanych węglików na noże tokarskie (Analysis of rational choice of sintered carbide tips and inserts for turning tools), Przegląd Mechaniczny, 1, 1952

4/16 Bułat, A., Sniechowski, R: Wpływ emulsji olejowych na własności skrawne narzędzi skrawających (Influence of oil emulsions on cutting properties of cutting tools), Prace Gł. Inst. Mech., 15, 1965

CHAPTER 5

The role of the machine tool in the cutting process

5.1 Properties and features of a machine tool, which influence the machining process

A machine tool has the following functions in the machining process:

(a) clamping of the tool and workpiece and establishing their correct relative position,

(b) supplying the working motions (primary and auxiliary motion) and maintaining the correct relative position of the workpiece and tool during the machining operation,

(c) supplying the energy for cutting or erosion work.

The method and accuracy of those functions of the machine tool are called its *machining capacity* and exert an influence on the machining process.

In order to use the machine tool to the best advantage it is necessary to know its machining capacity defined by relationships existing between the properties and features of the machine tool and:

— the quality and accuracy of machining,

— machining output,

— unit cost of machining.

When these factors are known and the existing possibilities are made use of, the probability of achieving high output, good quality and low production costs increases.

5.2. Influence of machine tool on machining accuracy

5.2.1 Classification of machine tool influence on machining accuracy

Machining accuracy is evaluated by considering the difference between the desired (nominal) dimensions and properties and those of the finished product. The smaller the machining errors, the greater is accuracy.

The machine tool influences machining errors by:

(a) errors in workpiece positioning in relation to the tool. This may be caused by errors in the production and assembly of machine tool parts and assemblies and by errors in the positioning and clamping of the workpiece and tool relative

to the machine tool. It is called in short the geometric influence on machining accuracy;
(b) errors in the working motions of the tool and workpiece during the machining process, or kinematic influence on machining accuracy;
(c) errors in the displacement of machine tool parts and assemblies, workpiece, and tool as a result of forces appearing in the machining process, called dynamic influence on machining accuracy;
(d) errors in the displacement of machine tool parts and assemblies, workpiece and tool as a result of thermal expansion caused by heat generated in the machining process and as a result of friction in the driving mechanisms. This is called thermal influence on machining accuracy.

The influence of machine tool parts and assemblies on machining accuracy is ascribed to those assemblies and parts on which the relative position of workpiece and tool directly depends. For example, in a lathe such assemblies are considered to be: the headstock assembly, saddle assembly and tailstock assembly. In practice, other assemblies indirectly affect the position of workpiece and tool (e.g. the lathe bed or body), their deformations and displacements may also cause machining errors. However, as this influence results, in turn, in changes of positions and displacements of assemblies having direct contact with workpiece and tool, it has been conventionally agreed that only the displacement errors of the latter assemblies determine the machine tool influence on machining accuracy.

5.2.2 Geometric influence on machining accuracy

The following factors may cause geometric errors of a machine tool:
— errors in the production of particular parts, adding up or balancing each other in the assembly,
— assembly errors, i.e. errors in joining well-made parts,
— deformations and displacements of parts under their own weight,
— deformations and displacements resulting from foundation work and support errors,
— wear and consequent volume reductions and changes of shape and dimensions of wearing parts.

The causes listed above can evoke a number of geometric errors, such as:
(a) out-of-flatness or deviation from the horizontal or vertical (Fig. 5.1 and Fig. 5.2). Verticality errors result in surface waviness or taper of the workpiece;
(b) out-of-parallelism (Fig. 5.3) and out-of-perpendicularity of two or more parts of a machine tool. These errors deform the shape of workpieces and cause their dimensional errors;
(c) out-of-roundness, taper, radial and axial runout of rotating parts. These errors result in irregular or oval holes and in wavy surfaces instead of planes and outer surfaces of revolution;
(d) misalignment of geometric axes and axes of rotation. These errors cause changes of shape and dimensions of workpieces, e.g. taper or deviations from nominal diameter of workpieces representing solids of revolution and deviations from the required surface form of other workpieces;

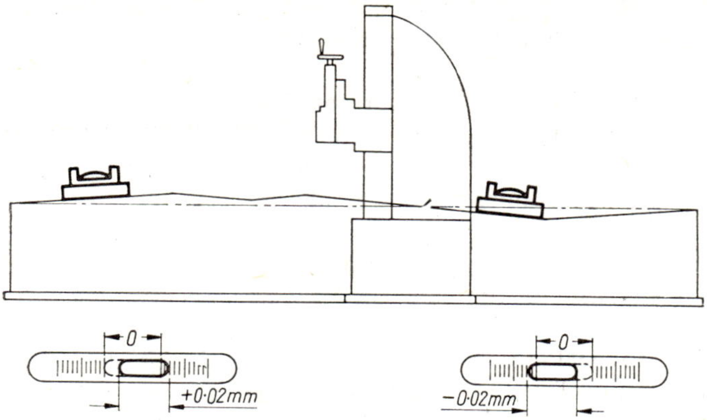

Fig. 5.1 Waviness of planer bed ways

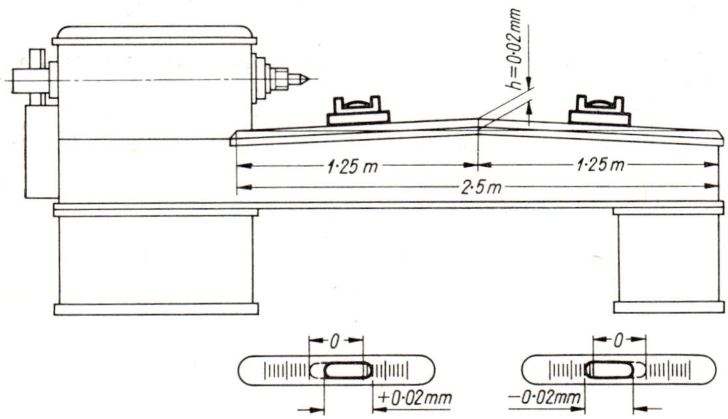

Fig. 5.2 Deviation of lathe bed from the horizontal

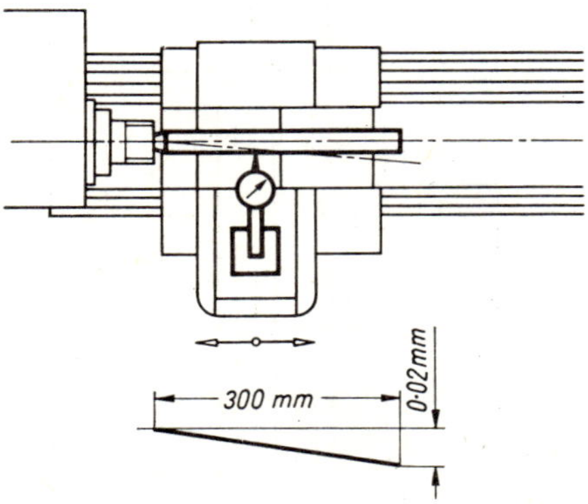

Fig. 5.3 Out-of-parallelism of lathe cross-slide motion with spindle axis in the horizontal plane

(e) variability of pitch on threaded parts, e.g. on lead screws. Errors of this kind are transmitted to the workpiece or periodically change the machining conditions which indirectly affects machining accuracy.

Geometric errors of a machine tool should wherever possible be measured directly. Their evaluation depends on the absolute magnitude of the error and on the sense of deviation from nominal dimensions. Errors whose sense indicates that, as the wear sets in, they will decrease, are considered to be less harmful.

The importance of geometric errors for total machining accuracy is relatively great. This can be demonstrated on the basis of a statistical comparison of errors measured on finished products with errors calculated from geometric deviations only (neglecting kinematic, dynamic and thermal errors). A list of such comparisons is given in Table 5.1 [5/4]. It follows therefrom, that the share of geometric errors in total machining errors varies within a range of 50–75%. Thus a conclusion can be drawn that geometric errors in machine tools should be reduced as far as possible, but still within limits of an economically justified accuracy.

Values of permissible errors should be determined according to the accuracy of part production and assembly obtainable under industrial conditions. The present

Table 5.1 *Percentage share of errors calculated from geometric inaccuracies in total machining error* (acc. to [5/4])

Machining method	Kind of error	Machining error: Calculated (geometric)	Machining error: Actual	Percentage share of geometric errors
Turning	Taper	0·007/200	0·01/150	
	Lead screw pitch	0·015/300	0·02/300	
Grinding	Taper	0·01/1000	0·01/500	
Milling	Work surface deviation from horizontal	0·03/300	0·04/300	
Planing	Work surface deviation from horizontal	0·02/300	0·04/300	
Gear cutting	Out-of-parallelism of teeth	0·02/300	0·01/100	

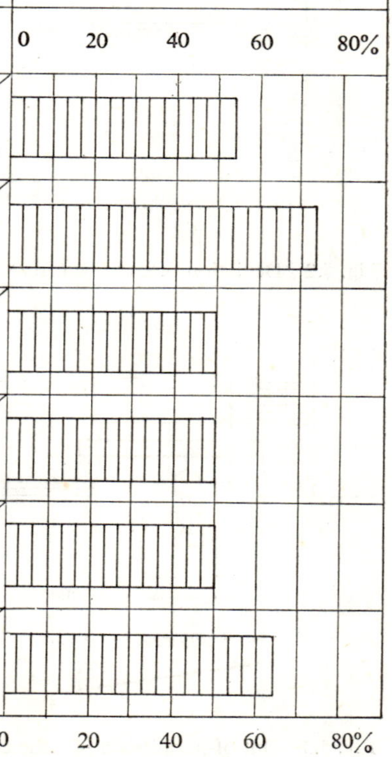

standards of many countries give permissible error values corresponding to machining possibilities existing about 40 years ago. It may be considered that if machine tool parts are properly manufactured and assembled with the use of today's production engineering technology, then the actual geometric errors in machine tools should be reduced by 25–50%. Errors in wearing parts should have a sense opposite to that of dimensional changes resulting from wear.

Wear of mating parts of machine tools is mainly the result of abrasion caused by abrasive dust suspended in the air or in lubricants and cutting fluids. In order to prevent rapid wear of working parts, the latter, e.g. guide ways, should be provided with covers. Filtration systems for lubricants and cutting fluids are very desirable particularly if the latter are supplied to open circuits.

5.2.3 Kinematic influence on machining accuracy

In the case of machine tools with simple working motions, machining errors are mainly caused by geometric errors and by errors resulting from displacements evoked by mechanical forces or thermal expansion.

In the case of machine tools with complex working motions, the kinematic accuracy of the machine tool is of great importance. This can be observed when machining gears, threads and other curved surfaces.

The kinematic accuracy of a machine tool results from the accuracy of coupling of its internal kinematic chains [5/3].

A kinematic error is defined by deviations of the final and initial link of a kinematic chain from nominal positions or displacements.

Kinematic errors may be caused by:

(a) manufacturing errors in kinematic chain elements,
(b) errors in the assembly of kinematic chain elements,
(c) errors in the selection of gear ratios,
(d) errors resulting from the total (torsional, bending, tensile and compressive) deformation of a kinematic chain.

Kinematic errors may be calculated or measured by means of special instruments called *kinematometers*.

Checking of kinematic errors has not yet become a widespread practice and there are no standards establishing their permissible values, just as there are no standards concerning geometric errors. One of the main reasons why the measurement of kinematic accuracy has not spread is that relatively complicated and expensive equipment is necessary, together with time-consuming operations.

In the case of machine tools shaping complicated and accurate curved surfaces, a study of their kinematic errors is justified and should be recommended.

5.2.4 Influence of displacements of machine tool assemblies on machining accuracy

Machine assemblies consisting of many parts, when subjected to load, are displaced in several directions, depending on the direction and method of loading as well as on the design, dimensions and the material of the parts used.

Displacement of particular points of an assembly in relation to each other is the result of:

— play between particular parts, particularly between moving parts,

— contact deformations on the crests of surface irregularities (roughness and waviness—see Chapter 6).

— elastic shear strain of loaded parts.

Figure 5.4 shows examples of displacements caused by play (Fig. 5.4a), contact deformation (Fig. 5.4b) and shear strain (Fig. 5.4c).

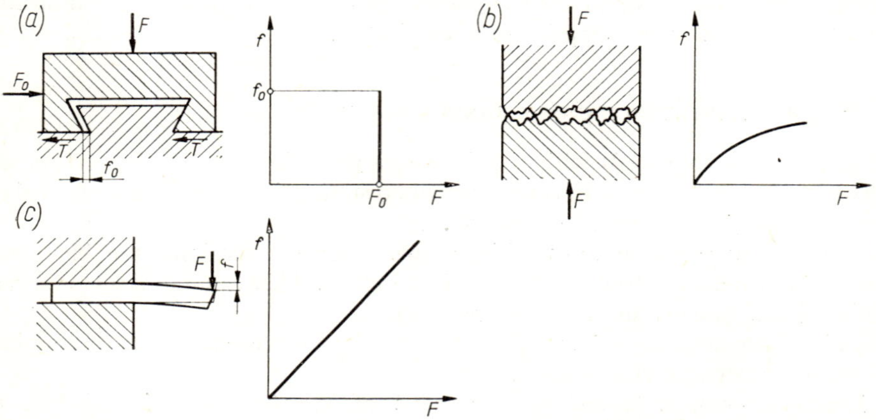

Fig. 5.4 Displacements as a result of: (a) play, (b) contact deformation, (c) shear strain

Under conditions existing in practice, the action of force F causes simultaneous contact deformation and shear strain. When the load attains a value equal to that of friction at the contact of parts with some play between them (when $F = F_0 = T$), gliding occurs.

The resulting curve representing the dependence of displacement f on load F may have various courses. In Fig. 5.5 a few examples of actual displacement curves for the loading and relieving processes are given.

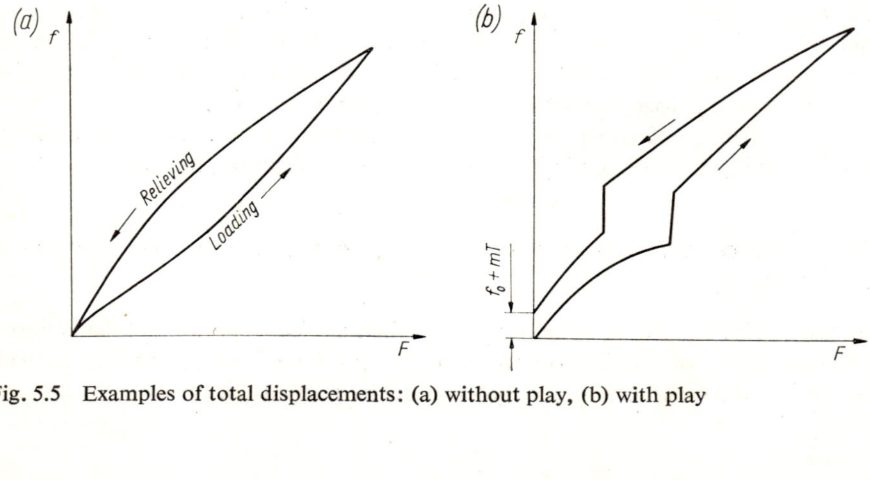

Fig. 5.5 Examples of total displacements: (a) without play, (b) with play

In the case of displacements in the absence of play, when the load is taken off, all points of the assembly return to their original positions. The loading and relieving curves do not coincide because of external friction (on contact surfaces) and internal friction. This phenomenon is known as *friction hysteresis*.

In the case when there is some play in the system being loaded, not all its points return to their original positions after the load is taken off. The cause of residual displacement after complete relief ($F = 0$) is the play (f_0) and the force of friction at contact surfaces which is greater than the force of resilience of the system still partially stressed.

For the evaluation of resistance or susceptibility of a given system to displacements under load, the terms stiffness and compliance are used.

Stiffness is the force increment necessary to cause an increment of linear displacement equal to 1 micrometre or of angular displacement equal to 1 centiradian. According to this definition, linear stiffness will be determined by the equation

$$j = \frac{dF}{df} \quad \text{kg}/\mu\text{m} \tag{5.1}$$

$$j = \frac{dF}{d\alpha} \quad \text{kg/crad} \tag{5.2}$$

where: F—load in kg, f—linear displacement caused by force F, in μm, α—angular displacement in crad.

Compliance m means the increment of linear or angular displacement caused by 1 kg increment of load

$$m = \frac{df}{dF} \quad \mu\text{m/kg} = \frac{1}{j} \tag{5.4}$$

or $$m = \frac{d\alpha}{dF} \text{crad/kg} = \frac{1}{j} \tag{5.3}$$

A graphical interpretation of linear stiffness and compliance (Fig. 5.6) will be, respectively, the tangent and cotangent of the angle formed by the line tangent to

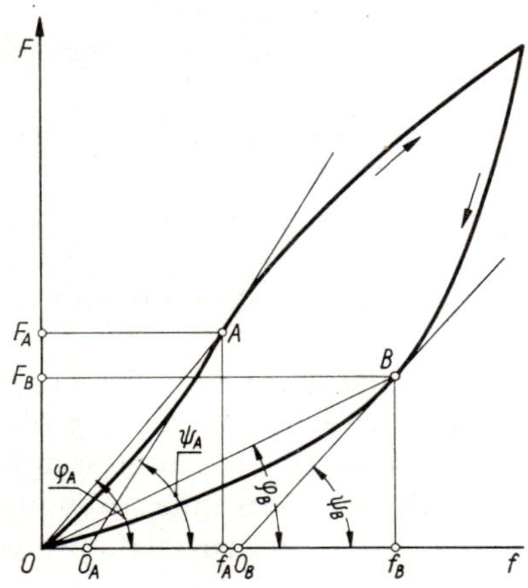

Fig. 5.6 Designations for the determination of mean and local values of static stiffness

loading curve (at point A) or to the relieving curve (at point B). According to the diagram in Fig. 5.6,
— at point A:

$$j_A = \frac{F_A}{f_A - \overline{OO_A}} \quad \text{and} \quad m_A = \frac{f_A - \overline{OO_A}}{F_A}$$

— at point B: (5.5)

$$j_B = \frac{F_B}{f_B - \overline{OO_B}} \quad \text{and} \quad m_B = \frac{f_B - \overline{OO_B}}{F_B}$$

The magnitudes defined in this way are called *local stiffness* and *local compliance.*

In practice, another equation is used for the determination of stiffness and compliance values, viz.:

$$j_A = \frac{F_A}{f_A} \quad \text{and} \quad m_A = \frac{f_A}{F_A}$$
$$j_B = \frac{F_B}{f_B} \quad \text{and} \quad m_B = \frac{f_B}{F_B} \qquad (5.6)$$

The magnitudes defined according to eqn. (5.6) are called *mean stiffness* and *mean compliance.* If stiffness without further definition is mentioned, it is understood to be mean stiffness.

Stiffness and compliance values determined on the basis of measurement of forces and displacements during the machining process are called dynamic, and those determined under conditions of a static load similar to that appearing during the machining process are called static.

When *static stiffness* and *compliance* are being determined, the following conventions are accepted:
— the loading conditions are simplified by taking into consideration only the most important forces and moments, i.e. those which influence the displacements most decisively. Thus, for example, in the measurement of stiffness and compliance of machine tools only the components F_v and F_p are taken into account;
— displacement measurements are carried out in directions and at points which are most characteristic of the machining errors;
— the relieving rather than the loading process is decisive (except for special purposes);
— the range of loads should correspond to the range of permissible machine tool loads (as foreseen by the designer).

Conventionally, the characteristic of static stiffness is given in a diagram with the load on the y axis and the displacement on the x axis.

An example of static stiffness diagrams for some lathe assemblies (height of centres 250 mm), for the case of a two-component load, is given in Fig. 5.7, together with a diagram of loading forces. Angle φ was assumed to be 60°. It can be seen in Fig. 5.7 that the stiffnesses of spindle, saddle and tailstock assemblies differ from one other and that they are not identical in the directions v and p. Similar relationships exist for other types of machine tools.

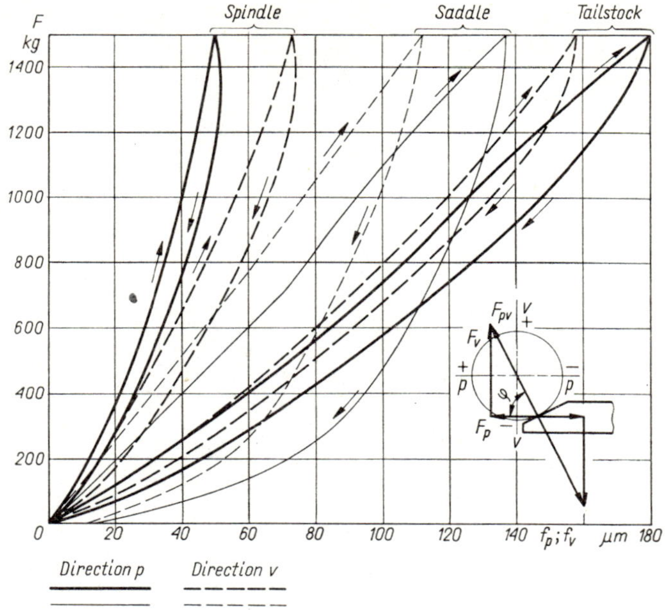

Fig. 5.7 Examples of machine tool stiffness in the case of a two-component load (no moment)

5.2.5 Example of the influence of displacements of machine tool parts and assemblies on accuracy errors

There exists a definite relationship between the compliance of machine tool assemblies which directly influence the relative positions of workpiece and tool and the machining errors in shape and dimensions.

The influence of compliance on machining errors can be demonstrated by considering a lathe.

Figure 5.8 shows diagrams of displacements in the direction of resultant constant load acting on spindle assembly (spindle axis displacement f_{sp}), tailstock assembly (displacement of tailstock centre axis, f_t) and saddle assembly (F_s) and also of elastic deformation of a constant diameter shaft (f_w), depending on the tool position along the machined length between centres.

Adding up ordinates corresponding to all displacements at each point along the machined length yields the total theoretical shaft displacement error. If diameter error is designated by Δd, it may be expressed by the equation

$$\Delta d = 2(f_{sp}+f_t+f_s+f_w) \tag{5.7}$$

where: f_{sp}, f_t, f_s and f_w are functions of the position of force F, while

$$\begin{aligned} f_{sp} &= f_{spo}\frac{x}{l} = m_{sp}F\left(\frac{x}{l}\right)^2 \\ f_t &= f_{to}\frac{l-x}{l} = m_t F\left(\frac{l-x}{l}\right)^2 \end{aligned} \tag{5.8}$$

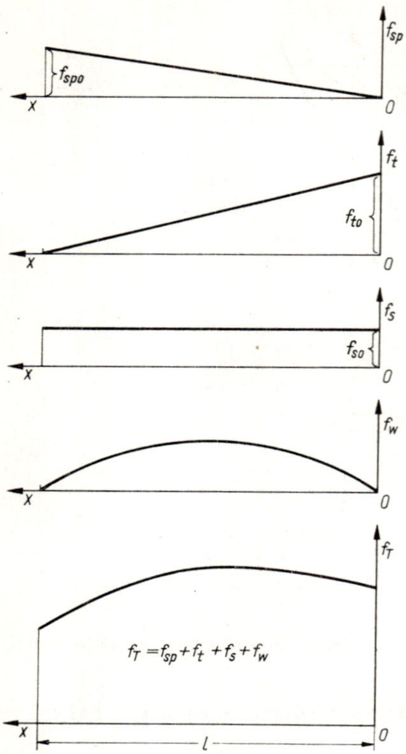

Fig. 5.8 Curves representing the dependence of spindle (f_{sp}), tailstock (f_t), saddle (f_s), workpiece (f_w) and total displacement (f_T) on tool position

where: f_{spo}—spindle displacement when force F is situated at $x = l$, l—workpiece length, m_{sp}—spindle compliance, f_{to}—tailstock displacement when force F is situated at $x = 0$, m_t—tailstock compliance.

Saddle compliance is constant and independent of load position (independent of x), and thus

$$f_s = m_s F \tag{5.9}$$

Deflection of a beam supported at two points, calculated for the point at which the load is applied, is determined by the well-known formula

$$f_w = \frac{F}{3EI} \frac{(l-x)^2}{l} x^2 \tag{5.10}$$

where: E—Young's modulus, I—moment of inertia.

Taking into account eqns. (5.8), (5.9) and (5.10) yields eqn. (5.7) in the form

$$\Delta d = 2F\left[m_{sp}\left(\frac{x}{l}\right)^2 + m_t\left(\frac{l-x}{l}\right)^2 + m_s + \frac{1}{3EI} \frac{(l-x)^2}{l} x^2\right] \tag{5.11}$$

When the functional relationship is known between stiffness and compliance and also the position of the load in relation to the origin of coordinates which is assumed to be situated at the starting point of the turning operation (at tailstock centre), then the diameter error can be determined depending on the actual load position.

It may be assumed that the mean compliance of tailstock and headstock is constant and changes of displacements occur only as a result of force variations acting on both assemblies as load position varies. Headstock and tailstock load changes may be approximated by reactions of a beam supported at two points and acted on by a moving force.

Equation (5.11) indicates that in the case of a constant cutting force—when turning a uniform (constant diameter) shaft—the increment of diameter as compared with the value established at the starting point (i.e. at tailstock centre) will vary.

Two types of errors in shaft form and dimensions, resulting from machine tool and workpiece compliance, may be distinguished:

(a) when workpiece compliance is greater than that of the spindle and tailstock, *barrelling* occurs (Fig. 5.9). If $m_t > m_{sp}$ then the diameter is smaller at the spindle (Fig. 5.8a), when $m_t = m_{sp}$, the convexity is symmetrical (Fig. 5.9b) and if $m_t < m_{sp}$, then the diameter is larger at the spindle (Fig. 5.9c);

(b) when workpiece compliance is smaller than that of spindle and tailstock, a shape error called *concavity* (contraction) occurs. If $m_t > m_{sp}$ then the diameter is smaller at the spindle (Fig. 5.9d), when $m_t = m_{sp}$, the concavity is symmetrical (Fig. 5.9e) and if $m_t < m_{sp}$, then the diameter is larger at the spindle (Fig. 5.9f).

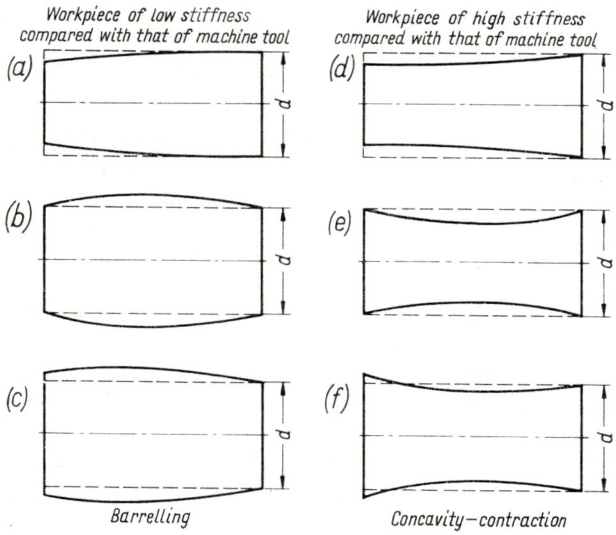

Fig. 5.9 Errors in shaft shape and dimensions, resulting from machine tool and workpiece compliance

The specification of shape errors in Fig. 5.9 demonstrates that depending on the stiffness of machine tool assemblies, for a given tool and workpiece position, and on the relation between their stiffness and that of the workpiece, many varieties of shape errors may occur.

From the point of view of the risk of producing rejects, concavity errors are most disadvantageous, i.e. when the stiffness of machine tool assemblies is not sufficient in relation to that of the workpiece, and also when the stiffness of the tailstock assembly is inferior to that of the headstock assembly (the cases in Fig. 5.9a and d).

Machine tools built so far have, unfortunately, in most cases a tailstock of lower stiffness. Thus the possibility arises of errors consisting in the diameter being reduced at the spindle as compared with the diameter set at the tailstock.

In a similar way as exemplified by the case of a lathe, analysis of the influence of stiffness or compliance on accuracy can be made for other types of machine tools.

Studies on the accuracy of products machined by cutting indicate that there is a considerable agreement between error distribution on products and the distribution of total displacements, determined on the strength of calculation of machine tool and workpiece stiffness or compliance. Therefore, stiffness of machine tool assemblies constitutes one of the features characterizing the accuracy obtainable on the given machine tool.

For an approximate, comparative evaluation of stiffness and accuracy obtainable on machine tools, the notions of *conventional stiffness* and *conventional compliance* of machine tool are introduced. They should be determined in a relative position of workpiece and tool, in which maximum displacements might be expected.

Thus, for example, in the case of a lathe, its conventional compliance m_T is determined by omitting f_w in eqn. (5.11) and substituting $0{\cdot}5l$ for x, then

$$m_T = \frac{1}{2}(m_{sp}+m_t)\,t\,m_s \tag{5.12}$$

The corresponding conventional stiffness j_T of a lathe is expressed by

$$j_T = \frac{\Delta j_s}{j_s(m_{sp}+m_t)+1} \tag{5.13}$$

In eqns. (5.12) and (5.13) l is the centre distance of the given machine tool. Therefore, in order to compare lathes with different turning lengths, an identical conventional length l should be adopted.

5.2.6 Influence of vibrations

Both separately excited (forced) and self-excited vibrations influence the course and results of the machining process.*

As a result of vibrations, disadvantageous symptoms appear in the form of:

- tool life reduction, mainly by the occurrence of microcracks and accelerated abrasion on the flank face,
- increase in roughness surface, often accompanied by waviness,
- reduction of the service life of some machine tool parts, particularly rotating parts (bearings etc.).

Besides, in certain vibration ranges, viz. in the range of audio frequencies, whistling occurs which increases the machine tool noise. This affects the worker's concentration and augments his fatigue, thus indirectly contributing to the deterioration of machining quality.

Of the effects listed above, tool wear and increase in roughness surface are directly connected with the machining process.

Accelerated tool wear as a result of vibrations can be explained by two causes.

The first cause results from periodic reduction of clearance angle (see Fig. 2.28),

* Classification of vibrations was discussed in Chapter 2.

which increases the friction surface. As a result of this phenomenon the tools subjected to vibrations perform more friction work on the flank.

The second cause is an increase in the dynamic coefficient [see eqn. (2.66)]. As a result of this phenomenon, local exceeding of the limit of both the ultimate strength and fatigue strength occurs on the edge.

Exceeding the local strength in the vicinity of tool point edge results not only from periodic changes of cutting forces, but also from changes in the system of forces acting on the tool point. This is a consequence of periodic sinking of the tool point in the material, causing periodic increases of unit pressure acting on the tool point from the side of the cutting face.

The magnitude of tool life reduction, caused by vibrations, depends on the vibration amplitude and frequency. Increase of amplitude causes an increase of dynamic coefficient and thus greater stresses in the tool point. Increase of frequency speeds up the appearance of fatigue symptoms. Particularly disadvantageous are frequencies approximating those of free vibrations of the machine tool parts and assemblies, as the vibration amplitude rises in such cases.

Deterioration of surface finish also depends on vibration amplitude and frequency. Tool point vibrations in relation to the material are three-dimensional, so that their traces are visible in all directions on the machined surface. Generally, however, as a result of vibrations, surface roughness increases most distinctly in the direction of tool marks, since vibration amplitude is usually highest in this direction.

Criteria according to which vibrations should not be permitted to occur in the cutting process are relative and depend on surface quality requirements. It is generally accepted that vibrations which leave traces on the machined surface visible to the naked eye are not permissible. In cases of high surface quality requirements this criterion is not adequate. Vibration meters are then necessary to establish the presence of vibrations and their permissible intensity.

In practice, a method [5/7] of establishing a vibrationless machining range for a machine tool is being introduced, which satisfies high surface quality requirements. This method has the following features:

— a study of vibrations is conducted by means of a vibration meter recording amplitude and frequency, or by measuring the resulting roughness of a machined surface,
— measurement and observation of vibrations are carried out at each step of rotational speed series within the feed range used to obtain the desired range of surface quality and within the depth of cut range permissible considering the available motor power,
— vibration measurements are conducted during workpiece machining, in positions and on spots characteristic for the machining method in question,
— tools selected for these studies have a tool geometry such that the probability of vibration appearance is increased,
— as a reliable criterion for non-permissibility of vibrations, the traces visible with the naked eye are considered, or the surface quality is compared with the quality obtained under similar conditions, but in the absence of vibrations,
— results of the study are presented graphically in the form of diagrams whose coordinates are depth of cut and feed rate, and on which the areas of work in the absence and in the presence of vibrations are marked separately.

In Fig. 5.10 examples of such diagrams are shown, illustrating the results of a study concerning the range of vibrationless work on a toolmaker's centre lathe with 250 mm height of centres. Studies were conducted on the parting of a centre-supported shaft by means of a cut-off tool with a cutting edge angle 0° and various widths *a*. Measurements were carried out at the headstock, in the middle of the shaft length and at the tailstock.

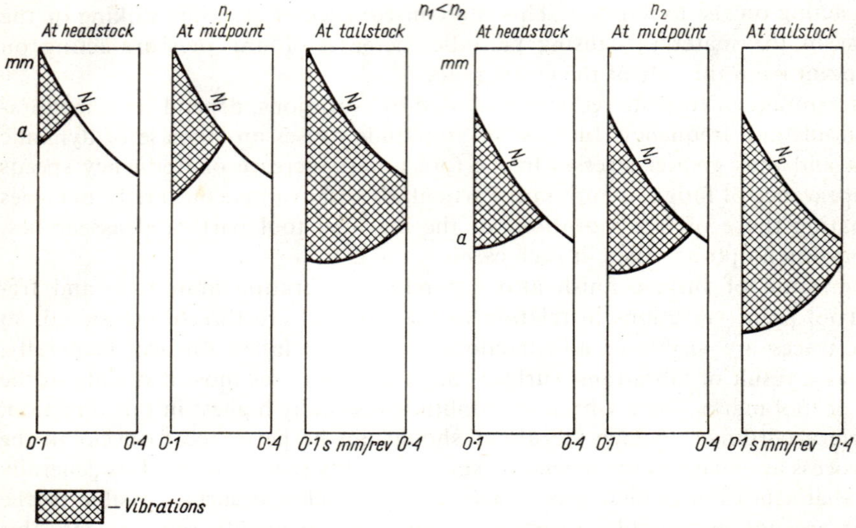

Fig. 5.10 Areas of vibrationless machining on a lathe (parting by means of a tool with $\varkappa_r = 0$)

In Fig. 5.10 curves N_p mark the limitations resulting from the available motor power. Below the shaded areas are those of vibrationless machining.

From the example cited above the following conclusions may be drawn:

— susceptibility to vibrations increases with decreasing stiffness, since at the tailstock, whose stiffness was the lowest in this case, the vibration area is largest,
— an increase of rotational speed and cutting speed reduces the area of vibrationless work,
— an increase of feed rate permits machining without vibrations with greater depth of cut.

Conclusions drawn from the example shown in Fig. 5.10 may not be generalized for all cases of machining and all types of machine tools. The areas of vibrationless work largely depend on the dynamic characteristic of the m–w–t (machine tool–workpiece–tool) system and should be checked for each concrete set of machining conditions.

5.2.7 Thermal displacements caused by heating

Heat sources in the machine tool–workpiece–tool system are: cutting heat, frictional heat in the motion system of the machine tool and heat from outer sources received in the form of radiation or convection heat.

The latter source does not generally play an important role, as machine tools should be installed in such a way that they will not be exposed to solar radiation nor to that of radiators, nor to the flow of heated air stream from blowers. On the other hand, the first two heat sources listed above are connected with the m–w–t system and always accompany machine tool work.

Within the machining range comprising low cutting speeds and large areas of undeformed chip cross-section or under conditions of rough and fairly accurate machining, heat generated by cutting work is more important. In this machining range thermal deformations of machine tools are of second-rate importance, since they are generally much inferior to permissible deviations of shape and dimensions.

Within the range of accurate and highly accurate machining, usually performed at high rotational speeds and with small areas of undeformed chip cross-section, the frictional heat in the machine tool mechanisms plays an equally or more important role than the cutting heat.

Figure 5.11 shows an example of a study on temperature distribution at various points of a grinding machine running idle, so that cutting heat has no part nor influence here [5/8]. It is evident from this example that even without cutting heat, maximum temperature differences reach 30°C.

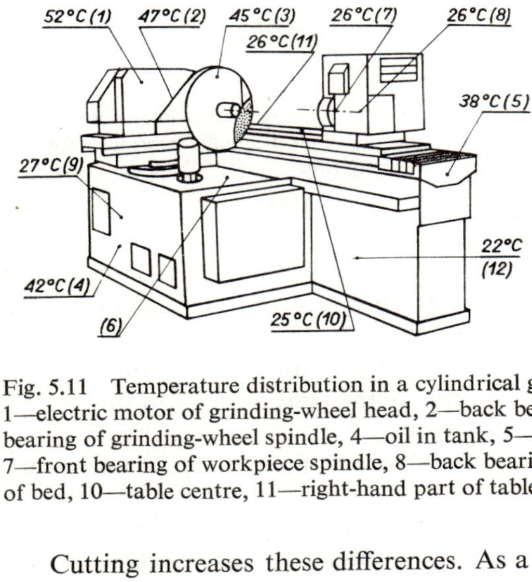

Fig. 5.11 Temperature distribution in a cylindrical grinding machine running idle. Designations: 1—electric motor of grinding-wheel head, 2—back bearing of grinding-wheel spindle, 3—front bearing of grinding-wheel spindle, 4—oil in tank, 5—oil in table drive cylinder, 6—coolant tank, 7—front bearing of workpiece spindle, 8—back bearing of workpiece spindle, 9—back wall of bed, 10—table centre, 11—right-hand part of table, 12—wall of front part of bed (acc. to [5/8])

Cutting increases these differences. As a result displacements may occur caused by:

— thermal expansion of warmed-up machine tool parts,
— deformations of non-uniformly heated parts and stresses evoked in some parts by thermal expansion of other parts,

In some cases the latter cause exerts a greater influence on resultant displacements than just the thermal expansion of heated parts.

Coefficient of linear expansion of steel is approximately 11×10^{-6} 1/°C. Thus a part about 320 mm long, heated by 32°C, will expand some 11 micrometres. If, in addition, possible deformations of unheated parts, caused by thermal expansion of other parts, as well as deformation of non-uniformly heated parts are considered, then the displacements may increase many times.

The heating of particular machine tool parts is an effect of heat conduction, depending on the thermal conductivity under existing conditions, on contact surface area, temperature gradient along heat flow and time.

In the case when the amount of heat generated in time unit is constant, the temperature gradient will also be a function of time. Under these conditions, if thermal conductivity and contact surface area will not be subject to change, it should be expected that a time will come, when heat inflow will counter-balance heat lost by the m–w–t system to the ambient medium, and the temperature will practically remain constant. It is then called the *stabilization temperature*.

This term is applied to a temperature level above which the temperature increase does not exceed the rate of 1°C per hour. Stabilization level is established experimentally under existing conditions.

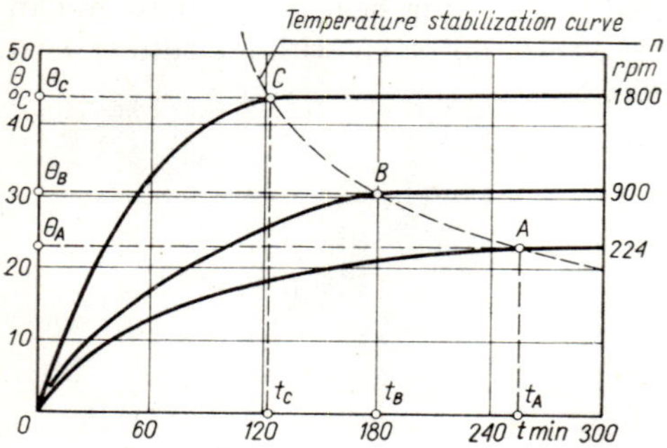

Fig. 5.12 Increase of temperature of the housing of lathe spindle bearings with working time

In Fig. 5.12 an example is given of stabilization temperature measurements on a tool-maker's lathe at various rotational speeds. The following conclusions may be drawn from this diagram:

— the higher the rotational speed, the higher the stabilization temperature,
— the higher the rotational speed, the shorter is the time needed to attain the stabilization temperature.

In the case given, at a maximum speed of $n = 1800$ rpm about 2 hours are necessary for the temperature field to become stable.

It follows that in the case of accurate and highly accurate machining, warming-up of the machine tool to stabilization temperature should be recommended, since during the warming-up period dimensional errors may occur. This has a particular importance in automated quantity production and in the checking of geometrical accuracy of machine tools during their technical inspection.

Thermal displacements of machine tool parts and assemblies, having a direct effect on the position of workpiece and tool, are of decisive importance for machining accuracy. The method of measuring displacements in one of such assemblies can be explained by the example of changes in the position of spindle axis of a toolmaker's lathe with 250 mm height of centres, depending on the temperature measured on the bearing housing.

In Fig. 5.13a the positioning of the adopted reference system is presented. Displacements in the direction of the y axis from the work-stand and also from the tool point have been assumed to be positive, as they cause diameter increase. For the same reason axis displacement upwards, along z axis, has been accepted as positive.

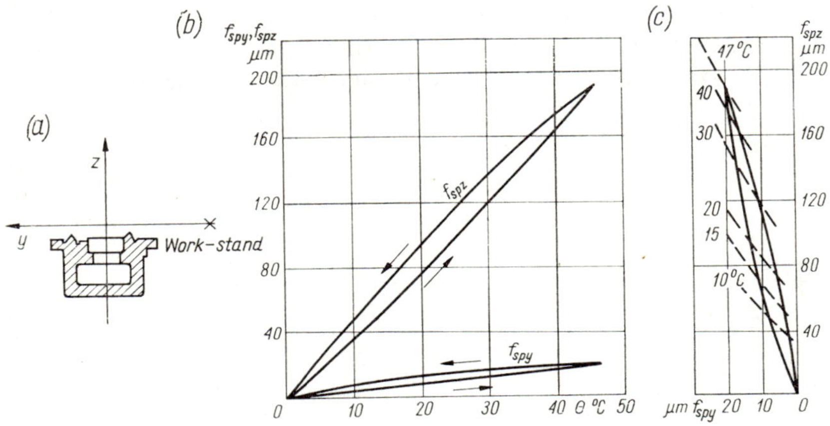

Fig. 5.13 Displacements of spindle axis of a lathe with 250 mm height of centres, as a result of temperature increase (measured on bearing housing) during work at maximum rotational spindle speed

Figure 5.13b shows spindle axis displacement for various temperatures when machining at maximum rotational speed (ambient temperature was assumed to be 0°C).

In Fig. 5.13c the measurement results have been illustrated in a different way, namely showing the curve of spindle axis displacement in the y–z system when heated to 47°C in relation to ambient temperature and then cooled-off to the original temperature.

It is evident from Fig. 5.13 that maximum displacement in the z axis is about 10 times greater than that in the y axis.

In order to determine which of these displacements is more important for the dimensional error, i.e. diameter increase, an equation for radius R is derived with the aid of Fig. 5.14.

As a result of warming-up, the spindle axis was displaced from position O_1 into position O_2. If the tool point nose is at point A, then axis displacement will result in an increase of turning radius from r to R. Then

$$R = \sqrt{(r+f_{spy})^2+f_{spz}^2} \tag{5.14}$$

or, designating $R_y = r+f_{spy}$ and $f_{spz} = a f_{spy}$

$$R = R_y \sqrt{1+a^2\left(\frac{f_{spy}}{R_y}\right)^2} \tag{5.15}$$

The ratio f_{spy}/R_y is very small and the following approximation can be used

$$\sqrt{1+a^2\left(\frac{f_{spy}}{R_y}\right)^2} \approx 1+\frac{a^2}{2}\left(\frac{f_{spy}}{R_y}\right)^2 \tag{5.16}$$

If the right-hand side of (5.16) is designated as E then

$$R = E\,R_y = E(r+f_{spy}) \tag{5.17}$$

Equations (5.16) and (5.17) allow one to calculate the turning radius resulting from displacement of the axis of rotation and to determine whether the influence of

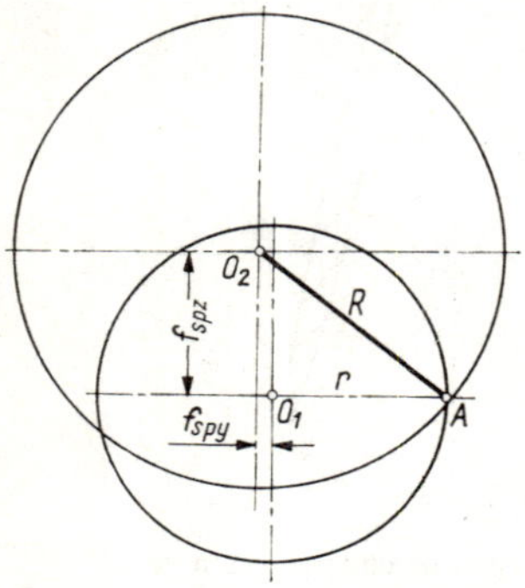

Fig. 5.14 Change of diameter of a shaft being turned, resulting from displacement of the axis of rotation

displacement in the direction of the z axis, represented by E, has a significant effect on the magnitude of the turned radius. In the case of the maximum deformations shown in Fig. 5.13, $E \approx 1{\cdot}0002$. This means that a vertical displacement 10 times greater than the transverse one (in the direction of y axis) creates a negligibly small change of radius R_y (by 0·2°/₀₀).

The conclusion follows (valid also for displacements caused by the action of forces, or displacements resulting from statical deformability) that the increase of turned radius is influenced considerably by transverse displacement, i.e. in the direction of the y axis. Displacement in the direction of the z axis may thus be neglected.

This conclusion concerns the magnitudes normally encountered in practice. In extreme cases calculations according to eqn. (5.17) should be applied.

5.3 Influence of machine tool on working range and production rate

5.3.1 Factors determining the working range of a machine tool and its production rate

Machining productivity* depends on the machine tool, for the following reasons:

(a) on every machine tool only products within a certain range of dimensions and weight can be machined, limited by the so-called nominal dimensions of the machine tool,

* Definition and detailed discussion of machining productivity and derived notions will be found in Chapter 7.

(b) rotational speeds and feed rates are contained within a certain range and in most cases are stepped, which does not permit a free choice of any arbitrary feed rate or rotational speed value,
(c) motor power and gear efficiency limit the possibilities of selection of machining conditions,
(d) the given strength of machine tool parts, particularly of its driving assemblies, also limit the freedom of choice of machining conditions,
(e) the obtainable accuracy of the machine tool causes limitations in the choice of cutting conditions (discussed in the preceding section).

Limitations caused by the range of nominal dimensions are evident and do not require further discussion. These dimensions are normally contained in the technical data of the machine tool, together with rotational speed and feed rate series, which permits one to choose a suitable machine tool for the given machining work.

5.3.2 Limitations resulting from a stepped driving gear

Limitations resulting from the stepping of feed and rotational speed driving gear depend on the kind and density of stepping.

In most cases so-called *geometrical stepping* is applied, i.e. the values of rotational speeds and feed rates constitute a geometrical progression with an established common ratio. However, there are also driving gears stepped according to an arithmetical progression (feed drives).

With regard to the influence of the kind and density of stepping on the selection of cutting conditions, both the geometrical and arithmetical stepping may be characterized by the so-called decrease of cutting speed or of feed rate.

Decrease of cutting speed is the percentage value of the ratio of the difference between two cutting speed values which are obtained for the same workpiece diameter in two consecutive steps of rotational speed, to the minuend of this difference, i.e. to the higher cutting speed.

Decrease of feed rate is defined in a similar way.

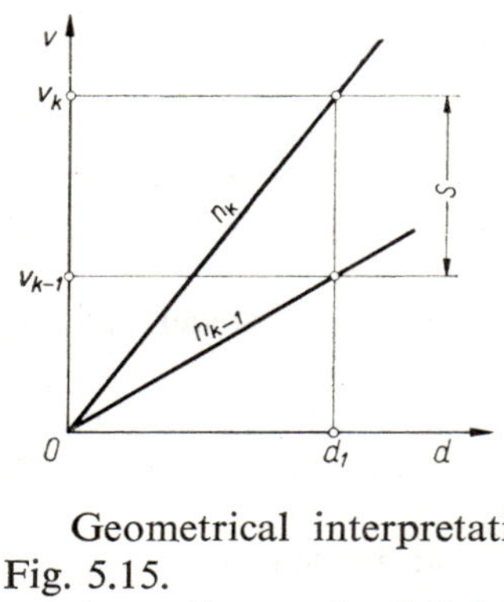

Fig. 5.15 Geometrical interpretation of the decrease of cutting speed

Geometrical interpretation of the decrease of the cutting speed is given in Fig. 5.15.

According to the definition given above, the decrease of cutting speed

$$S = \frac{v_k - v_{k-1}}{v_k} \; 100\% \tag{5.18}$$

In the case of arithmetical progression, eqn. (5.18) may be rewritten in the form

$$S_a = \frac{n_k - n_{k-1}}{n_k} 100\% = \frac{q}{n_k} 100\%$$

$$= \frac{q}{n_1 + (k-1)q} 100\% = \frac{q}{n_1 B_k} 100\% \tag{5.19}$$

where: q—common difference of the arithmetical progression, k—number of the successive terms of the progression, n_1—the first term of the progression, $B_k = n_k/n_1$—span of the progression up to the kth term.

It is evident from eqn. (5.19) that speed decrease in the case of arithmetical progression will be variable, viz. it will be:

— inversely proportional to the value of the kth term of progression, or to the product of the first term and the span of progression up to the kth term,
— growing with an increase of common difference q,
— smaller, the greater is the first term of progression.

In the case of geometrical progression the equation for speed decrease will be

$$S_g = \frac{n_k - n_{k-1}}{n_k} 100\% = \left(1 - \frac{1}{\varphi}\right) 100\% \tag{5.20}$$

where φ—common ratio of geometrical progression.

It is evident from eqn. (5.20) that the decrease of speed in the case of geometrical stepping is constant and depends only on the value of the common ratio. For widely applied geometrical progression with common ratio $\varphi = 1{\cdot}25$, the theoretical maximum deviation of cutting speed from the desired value can be 20%.

In practice such deviations do not exceed half the theoretical value, as principally that speed is selected which is nearer to the desired one. Therefore possible losses in machining productivity or deviations from most economical machining conditions are relatively small and this justifies the application of stepped driving gears in machine tools.

5.3.3 Power and efficiency of machine tool

Power used for the performance of cutting work is always lower than the electric power consumed by the motor. This is caused by losses occurring in the equipment and in the mechanisms transmitting the power.

Figure 5.16 diagramatically presents the flow and losses of power driving a machine tool.

Electric power N_{mc} consumed from the mains is transformed in the motor into motion power. In this process losses L_m (both mechanical and electric) occur, called *motor losses*.

Motor power N_m is the power delivered by the motor and defined by

$$N_m = N_{mc} - L_m \tag{5.21}$$

In the case when motor power equals the power foreseen by the designer as maximum power for continuous work, it is called *nominal motor power* N_{mn}.

Motor power is usually transmitted by belt drive. As a consequence of friction, slips and belt bending work, further losses L occur, called *transmission losses.*

Machine tool power N_{bt} is the power delivered by belt transmission to drive the machine tool mechanisms

$$N_{mt} = N_m - L_b = N - L_m - L_b \tag{5.22}$$

Machine tool losses L_t are power losses occurring in driving mechanisms.

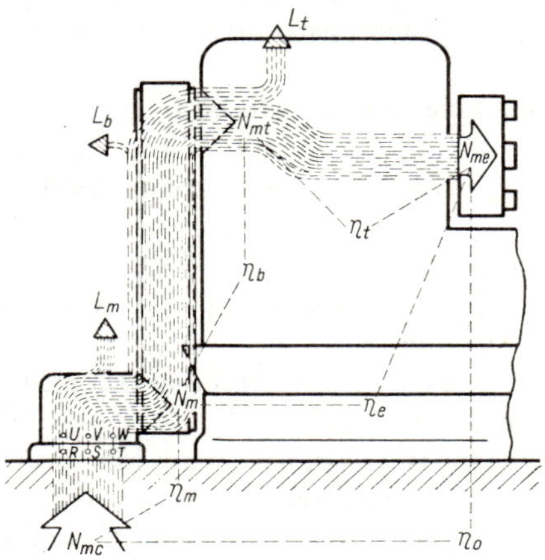

Fig. 5.16 Flow of power in a lathe

Useful (effective) power N_{me} or *cutting power* is the power delivered by the machine tool and used to perform cutting work

$$N_{me} = N_{mt} - L_t = N_m - L_b - L_t = N - L_m - L_b - L_t \tag{5.23}$$

Idle power N_{mo} is the power used to overcome resistance to motion of the motor and machine tool mechanisms without useful load

$$N_{mo} = L_m + L_b + L_t \tag{5.24}$$

With the use of the general definition of efficiency as the ratio of power delivered to power absorbed, the following efficiencies may be determined

— motor efficiency

$$\eta_m = \frac{N_m}{N_{mc}} \tag{5.25}$$

— belt transmission efficiency

$$\eta_b = \frac{N_{mt}}{N_m} = \frac{N_{mt}}{\eta_m N_{mc}} \tag{5.27}$$

— machine tool efficiency

$$\eta_t = \frac{N_{me}}{N_{mt}} = \frac{N_{me}}{N_{mc}\eta_m\eta_b} \tag{5.28}$$

— useful efficiency

$$\eta_e = \frac{N_{me}}{N_m} = \eta_b \eta_t \tag{5.29}$$

— overall efficiency

$$\eta_o = \frac{N_{me}}{N_{mc}} = \eta_m \eta_b \eta_t \tag{5.30}$$

In the case when the motor load is nominal, the adjective 'nominal' should be added to the definitions of efficiencies and the subscript n to their designations, for example: nominal useful efficiency η_{en}.

Power losses in the mechanical drive of a machine tool are caused by:
- friction of working surfaces,
- fluid resistance (hydrodynamic losses); mixing, overcoming of viscosity resistance and splashing,
- air resistance (aerodynamic losses).

The most important role is played by friction losses, while aerodynamic resistance is negligible within the range of speeds applied in practice.

Friction resistance depends on a number of factors, e.g. on the kind and condition of lubricants, on the load and on the speed of motion.

The higher the lubricant viscosity, the greater is resistance and idle power. However, viscosity of any kind of lubricant changes with increasing temperature. An example confirming this is the dependence of idle power on the time of idle running of a machine tool (Fig. 5.17).

Idle power at the moment of starting (point A) is about 1·5–2 times greater (in this example 1·75 times) than idle power after the stabilization of temperature. This occurs after time t_s. Under conditions when the temperature falls considerably below the normal level (20°C), friction resistance resulting from increased viscosity may cause a higher increase of idle power in the starting period.

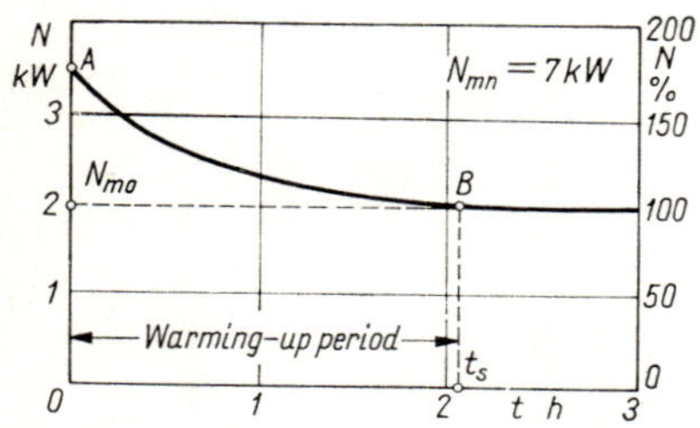

Fig. 5.17 Influence of the idle running time of a machine tool on its idle power

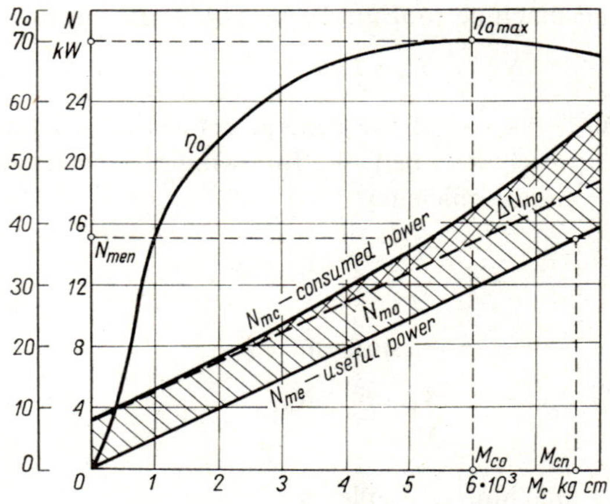

Fig. 5.18 Dependence of consumed power, useful power and overall efficiency on useful load, for a lathe of nominal power N_{mn} = 20 kW

Increasing the useful load of a machine tool causes increased losses for the motion of its mechanisms. This results in an increase of power loss by ΔN_{mo}.

$$\Delta N_{mo} = N_{mc} - N_{me} - N_{mo} \tag{5.31}$$

Figure 5.18 presents an example of the relationship between useful power, consumed power and power loss, the latter being split into idle power and the power of motion of the mechanisms loaded by cutting moment $(N_{mo} + \Delta N_{mo})$.

With increasing load the overall efficiency increases up to a maximum value $\eta_{o\,max}$. The load under which the maximum overall efficiency is obtained is called *optimum load* of the machine tool with regard to the use of motor power and designated by M_{co}.

It is characteristic that optimum load for most machine tools driven by asynchronous motors is smaller than nominal load M_{cn}.

Such a relationship repeats itself at every step of speed series. An example illustrating this in a different coordinate system, for various rotational speeds, is given in Fig. 5.19. By means of this diagram another typical relationship may also be presented, viz. that the higher the rotational speed, the smaller is the optimum useful load and maximum useful power.

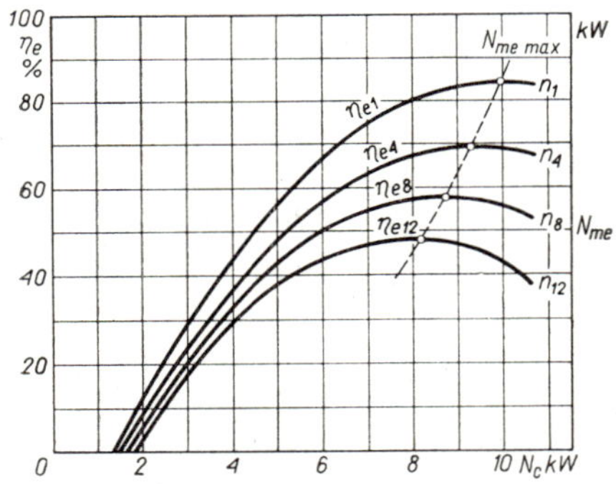

Fig. 5.19 Dependence of useful efficiency and maximum useful power on load at various rotational speeds

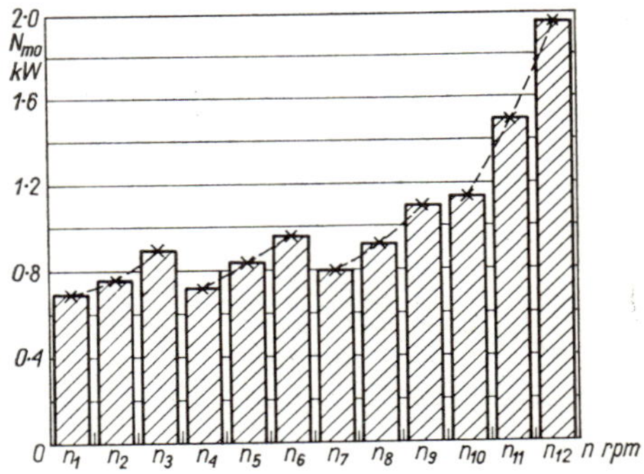

Fig. 5.20 Dependence of idle power on the engagement of particular spindle speeds of a lathe

The influence of rotational speed on idle power is not monotonic. This can be observed in the example shown in Fig. 5.20. The increase of idle power resulting from engagement of successive spindle speeds is not uniform but shows certain irregularities.

Non-uniformity of idle power increase is explained by the fact that in the engagement of successive speeds unequal numbers of various intermediary elements take part. These have different relative speeds and working conditions (influence may also be exerted by errors in the machining and assembling of these elements). Thus the equation for the sum of friction power loss in the machine tool driving gear can be written as follows

$$L_t = C \sum_{i=1}^{i=z} R_i d_i n_i \approx N_{tmo} \quad \text{kW} \tag{5.32}$$

where: C—calculation coefficient needed to obtain the result in kW, z—number of working parts, R—resistance of friction of the given part, d—reduced diameter of contact surface, n—rotational speed of the given part, N_{tmo}—theoretical idle power.

It follows from eqn. (5.32) that the value of L_t may increase by jumps and in a non-monotonic way, depending on the components of the sum.

Figure 5.21 shows the relationship between theoretical and actual idle power (N_{tmo} and N_{mo}) and sum of products $n_i d_i$ at various rotational speeds, for two lathes of different nominal power and different design. If it is assumed that friction resistance is approximately equal for all working parts, then, according to eqn. (5.32), the values of N_{mo} and those of $\sum n_i d_i$ should be proportional.

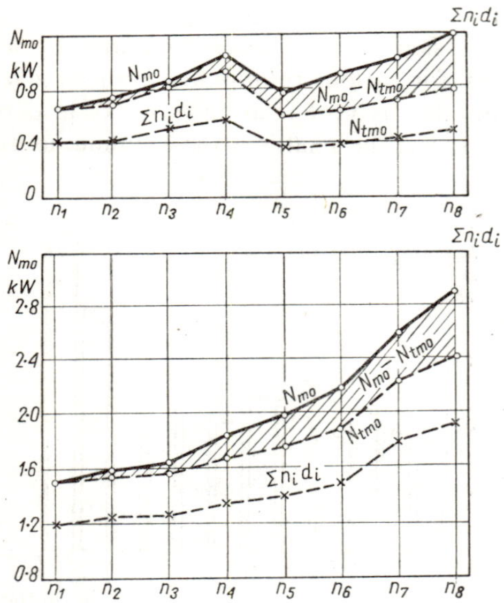

Fig. 5.21 Relationship between idle power and the sum of products $\sum n_i d_i$

The form of the curves in Fig. 5.21 does not confirm such an assumption, as the difference between N_{mo} and N_{tmo} grows with rotational speed.

An analysis of Figs. 5.20 and 5.21 leads to the following generalized conclusions:

— there is a conformity of character in the form of the idle power curves, actual and calculated (according to friction resistance),
— in some driving gear designs there is a cyclic repeatability of idle power increments, while they grow with rotational speed,
— the increase of rotational speed increases the difference between the actual and theoretical idle power, which is the result of growing friction resistance.

Hydraulic resistance depends on several factors. The effect of working time and lubricant temperature influencing its viscosity has already been discussed. Besides, hydraulic resistance is influenced by the lubrication system and the amount of oil being circulated. The least hydraulic losses occur in oil-mist lubrication and the greatest in oil-bath-and-splash lubrication. The latter losses largely depend on the level of oil in the headstock. Raising this level above a certain optimum height causes increased losses of driving power. This is illustrated by Fig. 5.22, containing the results of study on a turret lathe [5/9].

It follows from these considerations that the power and efficiency of a machine tool play an important role in the determination of its machining capacity.

In most cases both the depth of cut and feed are determined for other reasons, so that the only parameter that may be limited by the available power is cutting speed. Maximum cutting speed permissible with regard to useful power is calculated according to formula

$$v_M = \frac{6120}{C_z\, a^{e_z}\, s^{u_z}} N_c\, \eta_o = \frac{6120}{C_z\, a^{e_z}\, s^{u_z}} N_c\, \eta_m\, \eta_b\, \eta_t \tag{5.33}$$

where: C_z—coefficient taking care of the influences of all factors not accounted for separately in the equation, e_z, u_z,—exponents, determined in an empirical way.

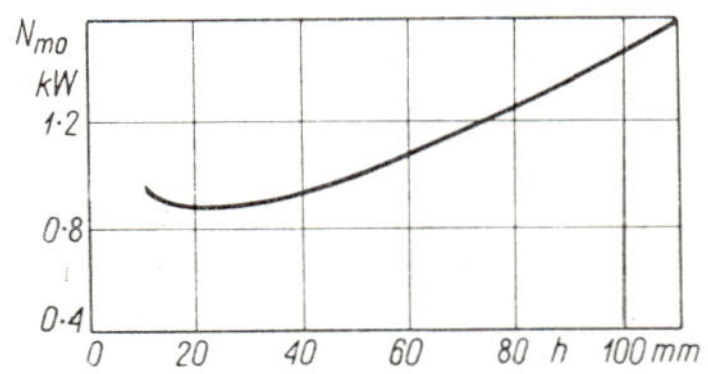

Fig. 5.22 Dependence of idle power of a turret lathe on the level of oil in its headstock (acc. to [5/9])

In practice, another version of the equation for v_M is used

$$v_M = \frac{6120}{C_z\, a^{e_z}\, s^{u_z}} N_{mn}\, \eta_{en} \quad \text{m/min} \tag{5.34}$$

This equation is more practical as for its application it is sufficient to know the nominal motor power and the mean nominal value of useful efficiency.

Mean nominal values of useful efficiency of cutting machine tools, correctly designed and built, vary within the limits

— for the lower half of speed range $\eta_{en} = 0{\cdot}7\text{–}0{\cdot}85$,

— for the upper half of speed range $\eta_{en} = 0{\cdot}55\text{–}0{\cdot}7$.

5.3.4 Strength limitations

Every machine tool is calculated to withstand a determined permissible force and a permissible torque. In view of this, the choice of machining parameters is limited, on the part of machine tool, by conditions expressed by

$$F_n = C_z\, a^{e_z}\, s^{u_z} \leqq F_{n\,\text{perm}} \tag{5.35}$$

$$M_c = 97\,400\, \frac{N_{me}}{n} = 97{\cdot}4\pi\, C_z\, d\, a^{e_z}\, s^{u_z} \leqq M_{c\,\text{perm}}$$

$$M_c = 306 C_z\, d\, a^{e_z}\, s^{u_z} \leqq M_{c\,\text{perm}} \tag{5.36}$$

Figure 5.23 presents diagrammatically a case of depth of cut and feed rate limitation caused by permissible moment $M_{c\,\text{perm}}$ and permissible force $F_{n\,\text{perm}}$.

Adopting minimum depth of cut value a_{min}, it is possible to use with it all feed rates from s_{min} to s_{max} (section *AB*) without strength limitations. On section *BC* limitation exists only as a result of the available feed range (from s_{min} to s_{max}). On section *CD* limitation is constituted by permissible force $F_{n\,perm}$ and s_{min}. On section *DE* the choice of a and s is limited by permissible moment $M_{c\,perm}$ and s_{min}.

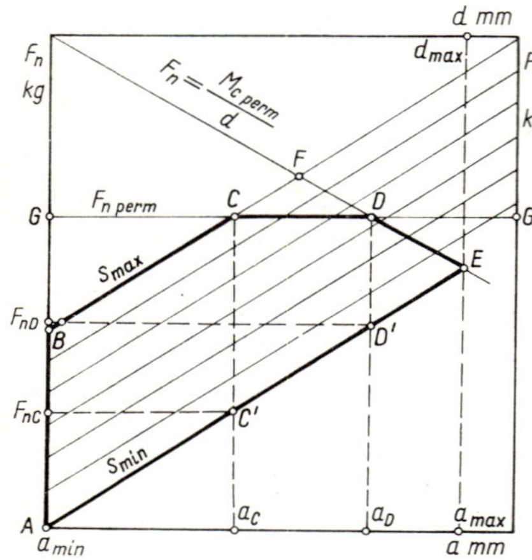

Fig. 5.23 Influence of the permissible moment $M_{c\,perm}$ and permissible force $F_{n\,perm}$ on feed rate and depth of cut

It is evident from the diagram in Fig. 5.23 that the selection of feed rate and depth of cut values may be performed within the area marked out by points *ABCDED'C'A*.

Lines $F_{n\,perm}$ and $F_n = M_{c\,perm}/d$ may acquire various positions in relation to each other and to the feed range and consequently the area of selection of feed rate and depth of cut values may have various shapes.

In most cases machine tools are so designed that strength limitations do not occur within the range of practically applied depths of cut and available feed rates, or may occur only in the case of roughing, and even then, only for the largest cutting diameters attainable on the given machine tool.

REFERENCES

5/1 Polska Norma PN/M-55650. Obrabiarki do metali. Sprawdzanie dokładności. Uwagi ogólne (Polish Standard PN/M-55650. Metal cutting machine tools. Checking of accuracy. General remarks), 1949

5/2 Perzyna, F.: 'Dokładność obrabiarek do metali i sposoby jej uzyskania' (Accuracy of Metal Cutting Machine Tools and the Ways to Obtain It), PWT, Warszawa 1953

5/3 Gwiazdowski, W.: 'Kinematyka obrabiarek' (Kinematics of Machine Tools), WNT, Warszawa 1965

5/4 Troshenskii, S. P.: 'Raschety tochnosti obrabotki na metallorezhushchikh stankakh' (Calculations of the Accuracy of Machining on Metal Cutting Machine Tools), Izd. Mashinostroenie, Moskva 1964

5/5 Projekt normy pomiaru sztywności statycznej obrabiarek (Draft standard on measurement of static rigidity of machine tools)

5/6 'Drgania samowzbudne obrabiarek' (Self-excited Vibrations of Machine Tools), Collective work, WNT, Warszawa 1963

5/7 KACZMAREK, J.: 'Etudes et bases d'appréciation des machines outils', La Machine Outil Française, 11, 1963

5/8 PIETKIEWICZ, M.: 'Zagadnienia odkształceń cieplnych w obrabiarkach' (Problems of thermal deformations in machine tools). Mechanik, 4, 1965

5/9 ENIMS—Raschet poter na trenie w priwodach glawnogo wrashchatel'nogo dwizheniya metallorezhushchikh stankow (Calculation of friction losses in driving gears of primary rotational motion of metal cutting machine tools), CBTI, 1956

CHAPTER 6

Machined surface and surface layer

6.1 Evaluation of product, machined component and its surface layer quality

6.1.1 EVALUATION OF PRODUCT AND MACHINED COMPONENT QUALITY

It is the aim of every branch of production engineering and, among others, of the machining by cutting or by erosion, to form products of the necessary quality.

The quality of a product is evaluated according to:

— the degree of reliability with which it functions and fulfils its tasks,
— its total life or service life, i.e. the time during which the useful features of the machine do not change,
— outer appearance, and particularly the so-called surface finish,
— the suitability and the quality of materials of which the product was made.

Product quality depends on the designer, as he chooses the material, determines the outer shape and useful properties of the product.

A very important influence is also exerted by the producer, through the production engineering technique he has chosen, which determines to what degree the designer's aims will be attained.

The useful features of the product, constituting an important criterion of quality evaluation, depend on the conditions and the way in which this product is being used, too. This also means that the user plays his part in influencing the quality evaluation.

At the moment when the given product has been manufactured, it is impossible to foresee exactly the conditions, the way and range in which it will be used. However, the need for *quality evaluation* already exists. Therefore the useful quality and technological quality have to be distinguished.

The *useful quality* of the product (also called its *usability*) is evaluated on the basis of its behaviour in use.

The *technological quality* of a product is evaluated on the basis of its features which can be defined immediately after it is finished.

The quality of a complex product, e.g. a machine tool or other machine, depends on correct assembling and joining of its parts and on the quality of these parts, most of which were machined by cutting or, possibly, by erosion.

The quality of the particular parts of a product or, speaking generally, the quality of a machined component, is influenced by:

— the state of the semi-finished product or the kind and properties of its material and several other features,

— conditions of the machining process.

The state of the machined component, a resultant effect of the influence of factors belonging to both these groups, i.e. the state after machining, is called the technological quality of the machined component.

Thus the useful quality of a machined component is influenced by:

— its state after machining, or technological quality,

— conditions of use.

Many useful features of components may be listed, such as:

(a) low wearability or high resistance to abrasion,

(b) low coefficient of friction or low resistance offered to sliding on another part,

(c) high fatigue strength (resistance to deformations and surface strength),

(d) good corrosion resistance,

(e) high load capacity or the ability to transmit or carry loads without showing plastic deformations or excessive elastic deformations,

(f) high reflectivity, and several others.

Between the useful features of a component and its state after machining exist certain relationships, provided that the conditions of use are appropriate.

Therefore, it is the aim of particular branches of production engineering to create the desired state of machined components.

The state of a component after machining, or its technological quality, is defined by:

(a) properties of the material,

(b) accuracy of shape and dimensions,

(c) properties of the surface layer.

The accuracy of shape and dimensions depends completely on production engineering and particularly on machine tool properties. This was discussed in Chapter 5.

The properties of the material do not depend on machining by cutting. On the other hand, in this process the properties of the material in the surface layer of the machined component may change.

6.1.2 Structure and evaluation of surface layer and machined surface

The surface layer is the layer of material limited by the actual geometrical surface of the machined component and comprising that part of material—from the geometrical surface inward—which shows changed physical and sometimes also chemical properties as compared with those of the material before machining.

A simplified diagram of surface layer structure after machining is given in Fig. 6.1.

The superficial film, i.e. that situated directly under the geometric surface, constitutes a zone of adsorbed gas, liquid or solid particles (e.g. air particles, particles of tool material, contaminations etc.). In addition, this zone also contains a film of oxides and other compounds of the metal being machined and chipped particles of crystal lattice. The total thickness of this zone is very small, as it usually does not exceed 80 Å (this justifies joint discussion of both films). However, the superficial zone plays a very important role, e.g. in wear processes.

The basic part of surface layer is constituted by the plastically deformed zone of machined material (the compressed zone) resulting from the action of forces occurring

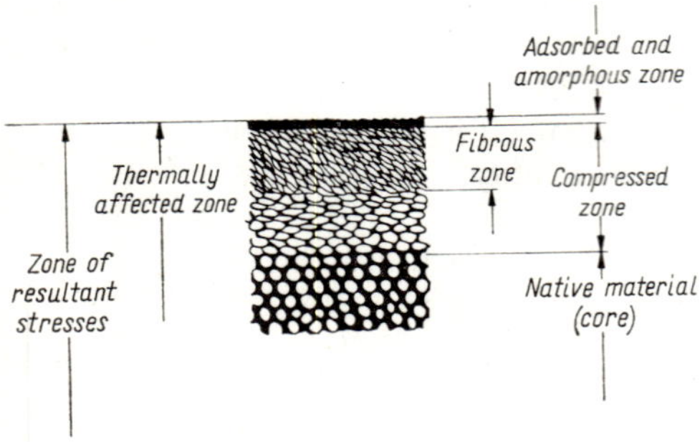

Fig. 6.1 Simplified diagram of surface layer structure after mechanical working

in the technological process. The thickness of this zone may vary from a few hundredths to several tenths of a millimetre, depending on the machining method and conditions.

In many cases, particularly after chip-removing processes, a distinctly fibrous structure of the compressed zone is visible. It results from the action of forces of friction between the material being machined and the cutting tool.

In addition to the plastic deformations caused by mechanical forces, the distribution of the temperature field may also influence the microstructure. As a result of heat generated in the technological process, various transformations of the surface layer structure may occur, like hardening and tempering processes, phase changes, grain size changes etc.

The depth of thermal effects in a mechanically worked surface layer usually reaches only a part of the compressed layer. On the other hand, in processes such as electric spark machining, the thermally affected layer extends almost through the entire surface layer and its thickness may even exceed 1 mm.

The surface layer of a machined component and the quality of this layer is characterized by its physical properties and the geometric properties of the machined surface.

The physical properties of the surface layer are affected by:

— microstructure and fibrous structure, its distribution and size within the surface layer,
— plastic deformation or compression and the accompanying changes of hardness and strength (surface hardening),
— resultant stresses (also called final or after-machining stresses) and their distribution within the surface layer and in the native material.

The reasons for the changes of native material properties and of surface layer formation are:

— field of stresses created by cutting forces,
— temperature field created as a result of heat generation by cutting energy,
— external and internal friction.

The geometric properties of the machined surface are affected by
(a) surface irregularities, from which the following are distinguished:
— surface roughness, evaluated in a conventional way according to mean total height,
— waviness, differing from surface roughness by considerably wider spacing of irregularities;
(b) geometric structure of the surface, including tool mark array and possible surface defects like scratches, cracks, chippings and gaps.

Geometric properties of the machined surface result from:
— geometric and kinematic reproduction of tool point shape in the material being machined (taking into account vibrations),
— external friction between tool point and workpiece material,
— built-up edge formation in the machining of certain materials and within certain ranges of machining conditions.

6.2 Geometric properties of machined surface

6.2.1 Classification, definition and designation of irregularities

6.2.1.1 *Surface area and profile*

The surface to be shaped according to the drawing is called *nominal surface* or, according to standard terminology, *geometric surface*.

The *real surface*, or the surface constituting the boundary between the component and the ambient medium, differs from the nominal surface by deviations directed inward or outward in relation to the latter. The real surface is perceived only approximately, since every observation, being a case of measurement, is weighted with a certain error.

Thus the *observed* or *effective surface* means an approximated image of the real surface.

In order to obtain an image of the surface, so-called *surface profile* is used, i.e. outline of surface resulting from its intersection with a plane.

The position of the intersecting plane, called *profile plane*, may be adopted conventionally. If its direction in relation to the surface is not specified, it is understood to be a plane normal to the nominal surface. Profiles are distinguished in the same way as surfaces (Fig. 6.2):
— nominal,
— effective,
— real.

The more accurate the observation (measuring) methods used, the smaller the differences are between the effective profile and the real one.

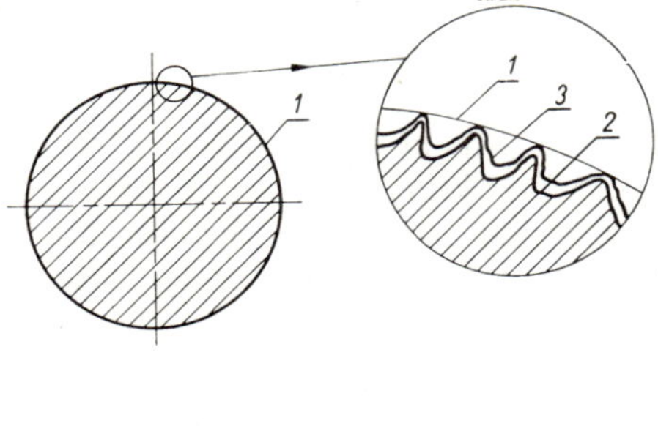

Fig. 6.2 Kinds of surface profiles:
1—nominal (geometric) profile,
2—real profile, 3—effective profile

6.2.1.2 *Types of surface irregularities and their definitions*

Deviations of the real surface outline from the nominal profile are called *surface irregularities.* Among them valleys and peaks are distinguished. Valley depth is measured from the highest point of the peak adjoining the valley, to valley bottom. Correspondingly, peak height is measured from the lowest point of the valley adjoining the peak, to the peak top.

Depending on the ratio of irregularity spacing S to mean total height R_{tm}, there are distinguished:

— surface roughness when $S/R_{tm} < 50$;
— surface waviness, when $50 \leqq S/R_{tm} \leqq 1000$;
— shape error, when $S/R_{tm} > 1000$.

By S is meant the conventional spacing of irregularities. Its conventional character results from the fact that surface profile does not always show a uniform repeatability of irregularities. An exact analysis of such a profile would require the application of harmonic analysis. In practice, spacing of irregularities is determined in an approximate way on the basis of observations and measurements from which mean values are drawn.

In the case of waviness an additional condition must be fulfilled, viz. that at least three waves must appear on the length of the given surface. If the number of waves is less than that, such irregularities are regarded not as waviness but as shape errors.

In order to determine mean total height it is necessary to take measurements of irregularities (valleys or peaks) on a sufficiently long section of surface profile. Therefore, a conventional length of profile is adopted on which mean total height of the given kind of irregularity can be measured with the desired dependability. This is:

— sampling length l_e—or a standardized length of profile which permits one to measure surface roughness without taking into account other kinds of irregularities, and
— measuring length l_g—minimum length measured by the given instrument, permitting one to determine surface roughness without taking into account other irregularities. Measuring length is a standardized multiple of sampling length.

Sampling lengths constitute a standardized series: 25; 8; 2·5; 0·8; 0·25; 0·08 mm. The choice of one of these sampling lengths depends on the size of irregularities and on the machining method.

Recommended sampling lengths are listed in Table 6.1.

An unequivocal determination of mean total height also requires an established reference line.

A *reference line* is a straight line parallel to the general direction of effective profile within the sampling or measuring length.

A particular case of the reference line is the mean line.

The *mean line* is a reference line dividing the profile in such a way that the sum of squares of deviations constituted by valleys and peaks in relation to the mean line acquires a minimum value. In practice, the mean line is determined so as to make the sum of cross-sectional areas of peaks equal to the sum of cross-sectional areas of valleys (Fig. 6.3).

Table 6.1 *Recommended sampling lengths depending on the machining method and roughness number*

<table>
<tr><th rowspan="3">Group of machining methods</th><th rowspan="3" colspan="2">Sort of machining</th><th colspan="14">Roughness number</th></tr>
<tr><th>1</th><th>2</th><th>3</th><th>4</th><th>5</th><th>6</th><th>7</th><th>8</th><th>9</th><th>10</th><th>11</th><th>12</th><th>13</th><th>14</th></tr>
<tr><th colspan="14">Sampling length l_e, mm</th></tr>
<tr><td rowspan="8">Chip removing processes</td><td colspan="2">planing</td><td colspan="3">8</td><td colspan="2">2·5</td><td colspan="9"></td></tr>
<tr><td rowspan="2">milling</td><td>plain</td><td colspan="2"></td><td colspan="2">8</td><td colspan="3">2·5</td><td colspan="2">0·8</td><td colspan="5"></td></tr>
<tr><td>face</td><td colspan="2"></td><td colspan="2">8</td><td colspan="3">2·5</td><td>0·8</td><td>0·25</td><td colspan="5"></td></tr>
<tr><td colspan="2">reaming</td><td colspan="3"></td><td>2·5</td><td colspan="4">0·8</td><td>0·25</td><td colspan="5"></td></tr>
<tr><td colspan="2">spotting</td><td colspan="5"></td><td colspan="3">8</td><td colspan="2">0·25</td><td colspan="4"></td></tr>
<tr><td rowspan="2">turning and boring</td><td>ordinary</td><td colspan="2"></td><td colspan="2">8</td><td colspan="2">2·5</td><td colspan="2">0·8</td><td>0·25</td><td colspan="5"></td></tr>
<tr><td>diamond</td><td colspan="6"></td><td colspan="2">0·25</td><td colspan="3">0·08</td><td colspan="3"></td></tr>
<tr><td colspan="16"></td></tr>
<tr><td rowspan="5">Abrasion processes</td><td colspan="2">grinding</td><td colspan="3"></td><td colspan="2">2·5</td><td colspan="5">0·8</td><td colspan="4"></td></tr>
<tr><td rowspan="2">lapping</td><td>manual</td><td colspan="6"></td><td colspan="2">0·8</td><td colspan="4">0·25</td><td colspan="2">0·08</td></tr>
<tr><td>machine</td><td colspan="4"></td><td colspan="8">0·25</td><td colspan="2">0·08</td></tr>
<tr><td colspan="2">honing</td><td colspan="7"></td><td colspan="3">0·8</td><td colspan="2">0·25</td><td colspan="2">0·08</td></tr>
<tr><td colspan="2">superfinish</td><td colspan="7"></td><td colspan="4">0·25</td><td colspan="3">0·08</td></tr>
</table>

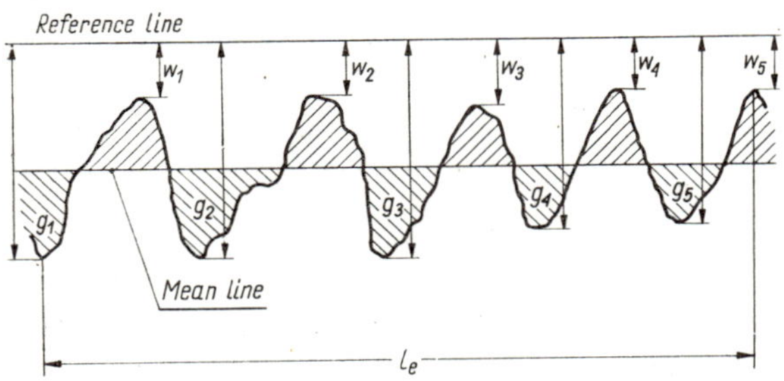

Fig. 6.3 Simplified determination of mean line position: l_e—sampling length

6.2.1.3 *Measurement of surface irregularities*

The following definitions are used in order to define surface irregularities:

(a) mean total height R_{tm} (Fig. 6.3) which is the mean distance between five highest and five lowest points of the profile on the sample length, in relation to the reference line

$$R_{tm} = \frac{(w_1+w_2+w_3+w_4+w_5)-(g_1+g_2+g_3+g_4+g_5)}{5} \tag{6.1}$$

(b) arithmetical mean deviation from the mean line R_a, or the mean of the absolute values of the profile point distances from the mean line on the measuring length

$$R_a = \frac{1}{l_g} \int_{i=l_{g1}}^{i=l_{g2}} y_i \, dx \tag{6.2}$$

In practice, eqn. (6.2) is replaced by the approximately equivalent expression

$$R_a \approx \frac{1}{n} \sum_{i=1}^{i=n} y_i \tag{6.3}$$

(c) maximum height R_{max}, or the distance between reference lines passing through the highest and the lowest point of the profile on the sample length.

With regard to surface quality surfaces have been divided into 14 roughness numbers depending on the dimensions R_{tm} and R_a. Height ranges for R_{tm} and R_a corresponding to the particular roughness numbers are given in Table 6.2.

Table 6.2 *Surface roughness numbers*

Roughness number	Mean total height R_{tm} µm (not more than)	Arithmetical mean deviation from the mean line R_a µm (not more than)	Roughness number	Mean total height R_{tm} µm (not more than)	Arithmetical mean deviation from the mean line R_a µm (not more than)
1	320	80	8	3·2	0·63
2	160	40	9	1·6	0·32
3	80	20	10	0·8	0·16
4	40	10	11	0·4	0·08
5	20	5	12	0·2	0·04
6	10	2·5	13	0·1	0·02
7	6·3	1·25	14	0·05	0·01

6.2.1.4 *Geometric structure of a surface*

Irregularities appearing on a machined surface have in most cases a definite, characteristic lay. The character of marks and their lay are called the *geometric texture* of the given surface.

The lays of tool marks forming surface irregularities have been classified and this classification is listed in Table 6.3.

The character of irregularities on a machined surface is defined by giving the machining method. Thus, for example, among surfaces without a definite lay (point-type texture) a different character is shown by surfaces after electric spark machining or after abrasive-blast treatment or by as-cast surfaces.

Table 6.3 *Classification, terminology and designations of lay*

Kind of lay	Variety of lay			Lay designation	Examples of machining methods
Single lay	Parallel	recti-linear	parallel to surface edge	=	planing, broaching
			perpendicular to surface edge	‖	planing, straight turning
		curvilinear		C	face milling, with part of face working
	Concentric			◎	facing
Multiple lay	Orderly		radial	⊘	face grinding of bush face
			criss-cross	X	honing
	Disorderly			⊕	spotting
Without lay (point-type texture)				⊙	electric spark hollowing

In the case of single lay or multiple lay texture of an orderly character, irregularity measurements depend on the position of the profile plane in relation to the lay of tool marks. The following are distinguished (Fig. 6.4):

— irregularities, and particularly roughness across the lay when the profile plane is perpendicular to the lay of tool marks,

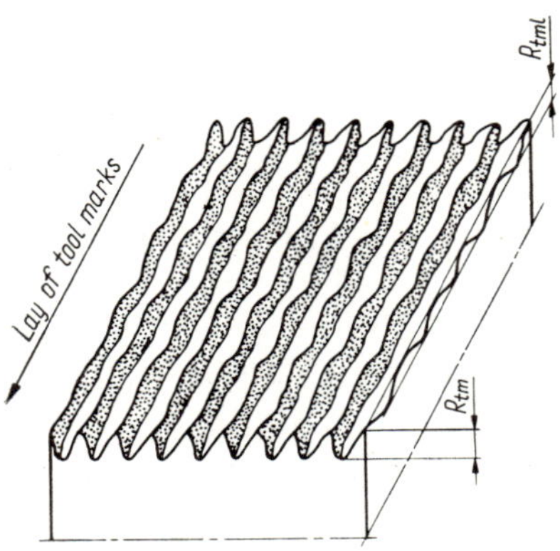

Fig. 6.4 Roughness across lay (R_{tm}) and roughness along lay (R_{tml}) of a geometrically anisotropic surface ($R_{tm} \neq R_{tml}$)

— irregularities, and particularly roughness along the lay when the profile plane is parallel or at a tangent to the lay of tool marks.

Surfaces showing irregularities of the same order and character in every direction are called geometrically isotropic. Other surfaces are geometrically anisotropic.

6.2.1.5 *Load carrying capacity of a surface*

If it were possible to make cross-sections of the real surface by several planes parallel to the nominal surface, then a series of counterfaces would be obtained, i.e. planes determining surfaces of contact between truncated irregularities and the nominal surface.

Such contact surfaces are also called *load carrying surfaces*, as they may result, approximately, from the wearing off of irregularity peaks and may transmit loads.*

The load carrying surface percentage (C_s), commonly called the load carrying capacity of the surface, is the ratio of load carrying surface area to the entire nominal surface area.

In the case of geometrically isotropic surfaces the load carrying surface may be assumed to be proportional to the so-called load carrying length of the surface profile. This is the sum of profile sections being in contact with the nominal counterface (Fig. 6.5).

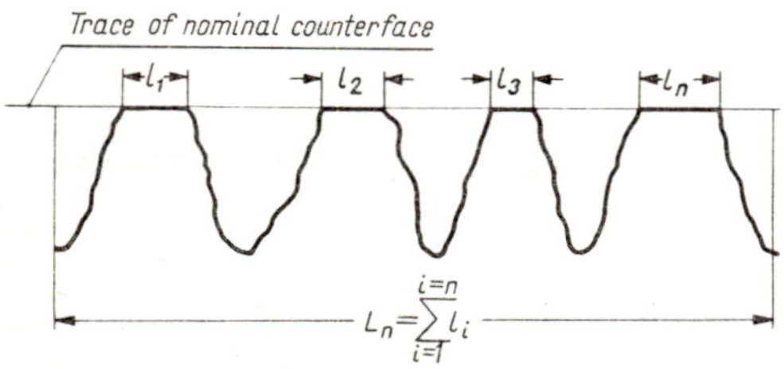

Fig. 6.5 Determination of load carrying length of surface profile

By analogy with the load carrying surface percentage there also exists a definition of load carrying line percentage C_l being the ratio of the load carrying length of the surface profile to the entire length of this profile

$$C_l = \frac{L}{L_n} = \frac{\sum_{i=1}^{i=n} l_i}{L} \tag{6.4}$$

By marking the load carrying lengths of the profile at various heights, from the highest peak to the deepest valley (Fig. 6.6), a so-called curve of load carrying capacity of the profile is obtained.

In the case of isotropic surfaces, the curve of the load carrying capacity of the profile may also be considered as the curve of the load carrying capacity of the surface.

* Strictly speaking, the term *load carrying surface* is not accurate. Load carrying capacity is associated with loading of the surface and then surface deformations should be taken into account. Otherwise, it would be more accurate to speak of *geometric contact surface*.

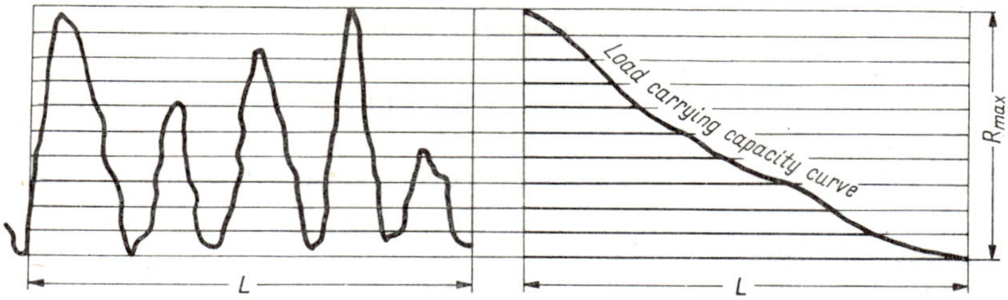

Fig. 6.6 Determination of the curve of the load carrying capacity of the surface profile

Among load carrying capacity curves are distinguished those with capacity growing proportionately, degressively and progressively with the distance from the highest peak. Correspondingly, capacities are shortly called

— proportional (Fig. 6.7a),
— degressive (Fig. 6.7b),
— progressive (Fig. 6.7c),
— mixed (e.g. degressively-progressive or progressively-degressive, Fig. 6.7d, e).

In general, similar interrelationships are again established between the machining methods (whose consequence is a definite character of tool marks), mean total height

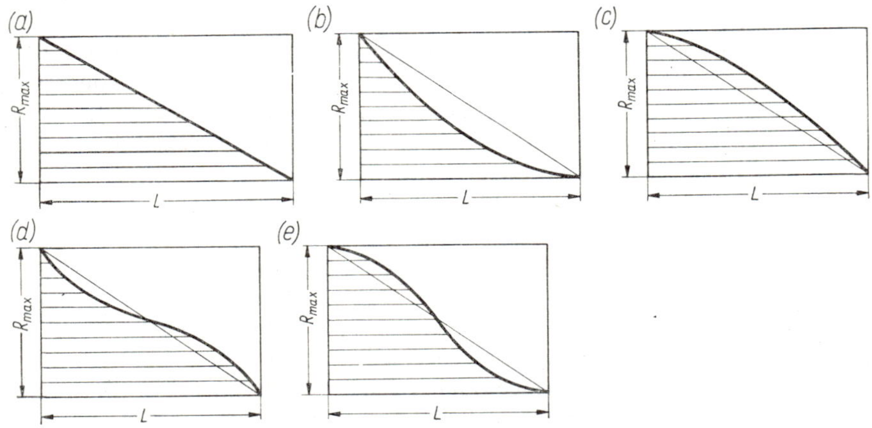

Fig. 6.7 Kinds of curves of load carrying capacity: (a) proportional capacity, (b) degressive capacity, (c) progressive capacity, (d) degressive and progressive capacity, (e) progressive and degressive capacity

and load carrying capacity of the surface. Thus, when the machining method and roughness number are given, the kind of curve of load carrying capacity is approximately determined.

The ratio of the area situated below the load carrying capacity curve to the area of the rectangle with sides L and R may be adopted as a measure for load carrying line percentage (and in the case of geometrically isotropic surfaces also for load carrying surface percentage).

With regard to useful qualities of surfaces (wearability), surface profiles with progressive load carrying capacities are desirable.

6.2.1.6 *Designation of surface quality and geometric texture*

In order to fully designate the requirements concerning surface quality of a machined object, it is necessary to determine:
(a) the geometric texture of the surface (and thus also, indirectly, its load carrying capacity),
(b) limitations concerning surface roughness.
Requirements concerning the geometric texture of the surface are designated by the lay symbol and by specifying the machining method to be used.

Requirements concerning surface roughness should be given while taking the following principles into account:
(a) mark the surfaces to be machined, specifying surface quality requirements. Three possibilities are distinguished here:
  (i) a machined surface, without roughness number specification, is designated by the symbol $\sim$,
  (ii) a machined surface with roughness number specification is designated by an equilateral triangle standing with its apex directed toward the surface to be machined,
  (iii) all machined surfaces, not marked separately with roughness number specification, are designated by an equilateral triangle over the blueprint title block;
(b) determine which roughness measure is toleranced, while it is established that:
  (i) if no symbol for roughness measure is given, requirements concerning one or both measures specified in the standard should be fulfilled,
  (ii) if roughness measure symbol is given, surface quality requirements concerning this measure must be fulfilled;
(c) determine the value of the given roughness measure; it is assumed that
  (i) roughness measure may be toleranced unilaterally or bilaterally,
  (ii) roughness measure may be toleranced by values corresponding to roughness number limits or by any arbitrary values.
Examples of full designations of requirements— concerning surface quality and geometric texture, with explanations, are contained in Table 6.4.

In principle, designations of surface quality requirements placed in the drawing over the given surface should not be smaller than 2·5 mm, and those placed above the blueprint title block not smaller than 5 mm.

6.2.2 SURFACE ROUGHNESS AS A RESULT OF GEOMETRIC AND KINEMATIC REPRODUCTION OF TOOL POINT (THEORETICAL MEAN TOTAL HEIGHT)

A machined surface is created as a result of geometric and kinematic reproduction of the tool point, accompanied by some physical phenomena.

Either the former or the latter group of factors prevails, depending on machining conditions, the kind of material and tool point shape.

Ridges appearing on the machined surface as a result of geometric reproduction depend on tool point shape and on the speed of tool motion in relation to the workpiece. In addition, when the cutting edge is dented, its dents are reproduced, too.

Table 6.4 *Example of designations of surface quality requirements*

Designation	Explanation
∼	Machined surface without roughness number specification
Shot blast ∼	Machined surface without roughness number specification but with indication of machining method (and thus, approximately, of the character of tool marks)
▽9	Machined surface with arbitrary texture, arbitrary machining method, surface roughness unilaterally toleranced to upper limit of roughness number 9, i.e. $R_{tm} \leqq 1{\cdot}6$ μm or $R_a \leqq 0{\cdot}32$ μm
▽8–9	Machined surface with arbitrary texture, arbitrary machining method, surface roughness bilaterally toleranced from lower limit of roughness number 8 ($R_{tm} \leqq 3{\cdot}2$ μm or $R_a \leqq 0{\cdot}63$ μm) to upper limit of roughness number 9 ($R_{tm} \leqq 1{\cdot}6$ μm or $R_a \leqq 0{\cdot}32$ μm)
Grind ▽8–9	As above, with indication of machining method (grinding)
▽R_z18	Machined surface with arbitrary texture, arbitrary machining method, surface roughness unilaterally toleranced to R_{tm} value not constituting a limit of any roughness number range
Electric spark hollow ⊙ ▽R_z15–30	Machined surface without definite lay, electric spark hollowing, surface roughness bilaterally toleranced, R_{tm} to fulfil condition $15 \leqq R_{tm} \leqq 30$

It is to be assumed that under proper conditions of tool application, the cutting edge is not dented or its dents are so small in comparison with the resulting surface roughness that their influence may be neglected. Therefore, in further discussion only the case of smooth edges will be considered.

Tool point shape is characterized, in the plane parallel to the base plane in which roughness across the lay is determined, by:

— nose radius r_ε,

— leading cutting edge angle $\varkappa_r$ and auxiliary cutting edge angle $\varkappa_{r1}$.

The factor characterizing the tool point and workpiece kinematics and having an influence on surface roughness is feed per tool point which, in the case of single-tool turning, is equivalent to feed per revolution.

It is easy to perceive that in the case when $r_\varepsilon > 0$ there are three possibilities (cases) of geometric and kinematic reproduction of the tool point. These cases are illustrated in Fig. 6.8.

When it is assumed that r_ε, $\varkappa_r$ and $\varkappa_{r1}$ are constant, which is true in machining with a definite tool, the particular cases of reproduction depend on the feed rate.

Feed rate values constituting limits between the particular cases of reproduction are called *limit feeds*.

The first type of reproduction (I) takes place when transverse reproduction marks result from the intersection of the arc-shaped portions of the tool point edge. It is easy to perceive that this case occurs when

$$\begin{aligned} & s \leqq s_{12} = 2r_\varepsilon \sin\varkappa_{r1} && \text{when} \quad \varkappa_r > \varkappa_{r1} \\ \text{or} \quad & s \leqq s_{12} = 2r_\varepsilon \sin\varkappa_r && \text{when} \quad \varkappa_r < \varkappa_{r1} \end{aligned} \tag{6.5}$$

The first type of reproduction is called *arc-shaped reproduction* and is designated by the symbol shown in Fig. 6.8.

The second type of reproduction (II) occurs when transverse reproduction marks result from the intersection of the arc-shaped portion of the tool point edge with its rectilinear portion. It can be demonstrated that this type takes place when

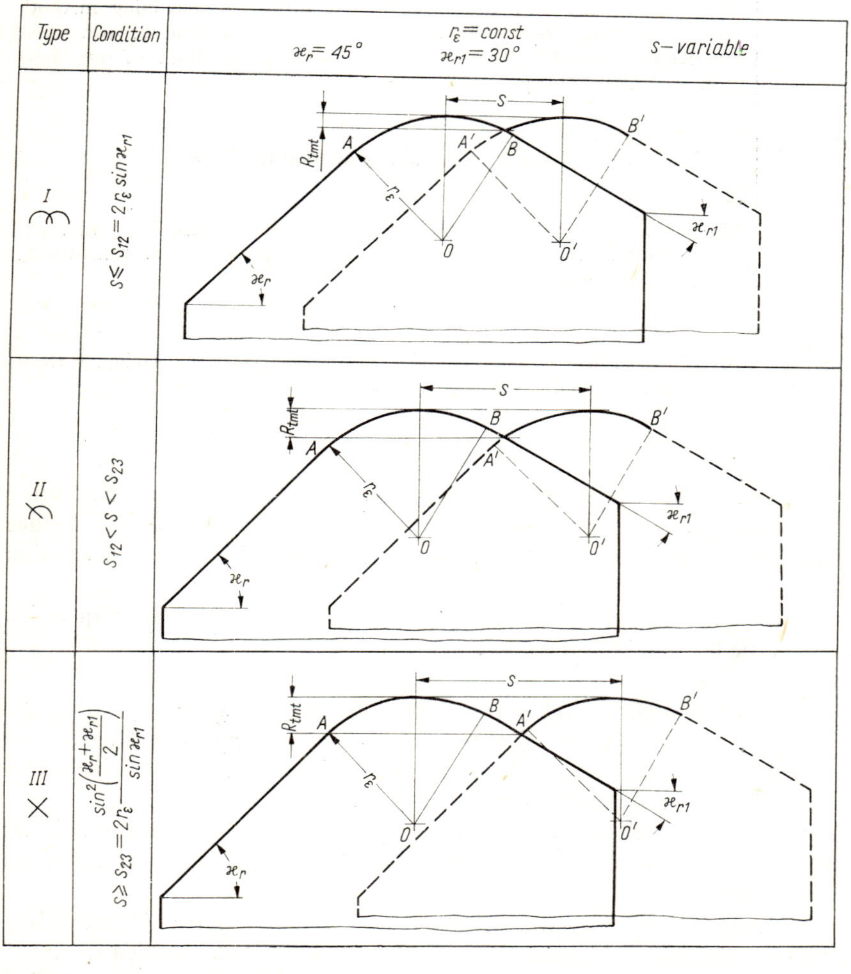

Fig. 6.8 Types of geometric and kinematic reproduction of tool point on material being machined

$$2r_\varepsilon \sin\varkappa_{r1} < s < 2r_\varepsilon \frac{\sin^2\left(\frac{\varkappa_r+\varkappa_{r1}}{2}\right)}{\sin\varkappa_{r1}} \quad \text{for} \quad \varkappa_r > \varkappa_{r1}$$

$$\text{or} \quad 2r_\varepsilon \sin\varkappa_r < s < 2r_\varepsilon \frac{\sin^2\left(\frac{\varkappa_r+\varkappa_{r1}}{2}\right)}{\sin\varkappa_{r1}} \quad \text{for} \quad \varkappa_r < \varkappa_{r1} \tag{6.6}$$

The second type of reproduction is called *arc-shaped-and-rectilinear reproduction* and is designated by the symbol given in Fig. 6.8.

The third type of reproduction (III) takes place when transverse reproduction marks result from the intersection of the rectilinear portions of the tool point edge. This type occurs when

$$s \geqq 2r_\varepsilon \frac{\sin^2\left(\frac{\varkappa_r+\varkappa_{r1}}{2}\right)}{\sin\varkappa_{r1}} \quad \text{for} \quad \varkappa_r > \varkappa_{r1}$$

$$\text{or} \quad s \geqq 2r_\varepsilon \frac{\sin^2\left(\frac{\varkappa_r+\varkappa_{r1}}{2}\right)}{\sin\varkappa_r} \quad \text{for} \quad \varkappa_r < \varkappa_{r1} \tag{6.7}$$

The third type of reproduction is called *rectilinear reproduction* and designated by the symbol given in Fig. 6.8.

Table 6.5 contains a list of conditions (limit feeds and feed ranges), for the theoretical mean total height R_{tmt}, as well as symbols and designations for the particular type of reproduction of tool points having $\varkappa_r > \varkappa_{r1}$ or $\varkappa_r < \varkappa_{r1}$ (the latter possibility is realised very rarely in practice).

In Fig. 6.9 an example is presented of theoretical dependence of mean total height on feed for all types of reproduction. This example concerns a tool point having $r_\varepsilon = 1$ mm, $\varkappa_r = 45°$ and $\varkappa_{r1} = 10°$.

Table 6.5 *Formulae for the theoretical mean total height of irregularities in particular type of tool point reproduction*

For $\varkappa_r > \varkappa_{r1}$	R_{tm}	$r_\beta - 0{\cdot}5\sqrt{4r_\beta^2 - s^2}$	$s\left(\frac{\sin^2\varkappa_{r1}}{2} - \sin^2\varkappa_1 \times \sqrt{\frac{2r_\beta}{s\sin\varkappa_{r1}} - 1}\right) + 2r_\beta\sin^2\left(\frac{\varkappa_{r1}}{2}\right)$	$s\frac{\sin\varkappa_r\sin\varkappa_{r1}}{\sin(\varkappa_r+\varkappa_{r1})} - r_\beta\left[\frac{\sin\varkappa_r+\sin\varkappa_{r1}}{\sin(\varkappa_r+\varkappa_{r1})} - 1\right]$
	Feed range	$s \leqq s_{12} = 2r_\beta\sin\varkappa_{r1}$	$s_{12} < s < s_{23}$	$s \geqq s_{23} = 2r_\beta \frac{\frac{\sin^2(\varkappa_r+\varkappa_{r1})}{2}}{\sin\varkappa_{r1}}$
	Symbol			

Type of reproduction		I	II	III
For $\varkappa_r < \varkappa_{r1}$	Symbol			X
	Feed range	$s < s_{12} = 2r_\beta \sin\varkappa_r$	$s_{12} < s < s_{23}$	$s \geqq s_{23} = 2r_\beta \dfrac{\dfrac{\sin^2(\varkappa_r+\varkappa_{r1})}{2}}{\sin\varkappa_r}$
	R_{tm}	$r_\beta - 0{\cdot}5\sqrt{4r_\beta^2 - s^2}$	$s\left(\dfrac{\sin 2\varkappa_r}{2} - \sin^2\varkappa_r \times \sqrt{\dfrac{2r_\beta}{s\sin\varkappa_r} - 1}\right) + 2r_\beta \sin^2\left(\dfrac{\varkappa_r}{2}\right)$	$s\dfrac{\sin\varkappa_r \sin\varkappa_{r1}}{\sin(\varkappa_r+\varkappa_{r1})} - r_\beta\left[\dfrac{\sin\varkappa_r + \sin\varkappa_{r1}}{\sin(\varkappa_r+\varkappa_{r1})} - 1\right]$

The section of the curve within the range of type I reproduction is very similar to a second-degree parabola. Therefore, instead of the formulae given in Table 6.5 for arc-shaped reproduction an approximate formula for the calculation of the theoretical mean total height of surface irregularities may be applied

$$R_{tmt1} = 125\frac{s^2}{r_\varepsilon} \quad \mu\text{m} \tag{6.8}$$

where both s and r_ε are given in mm.

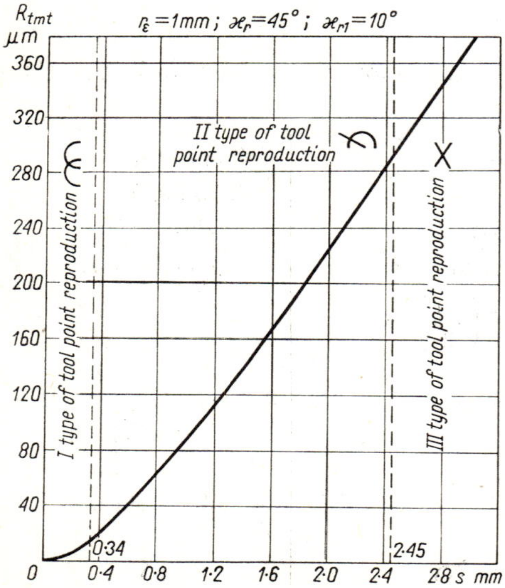

Fig. 6.9 Diagram of the dependence of theoretical mean total height of surface irregularities on feed within the ranges of I, II and III type of tool point reproduction

The section of the curve within the range of type II reproduction is similar to a curve representing a function with an exponent slightly greater than unity.

The section of the curve corresponding to type III reproduction is rectilinear.

As regards the practically applied feed ranges, it can be stated that the type I and II reproduction will predominate.

It is evident from an analysis of the formulae given in Table 6.5 that:

— when the cutting edge angles decrease, so do the theoretical mean total heights (type II and III reproduction) while limit feeds s_{12} and s_{23} increase,

— when nose radius decreases, the theoretical mean total height increases and limit feed values are reduced.

For $r_\varepsilon = 0$, i.e. when the tool point has no arc-shaped portion of cutting edge, there exists only one type of reproduction, (III). The equation for the theoretical mean total height of surface irregularities (across the lay) is then*

$$R_{tmto} = s\frac{\sin\varkappa_r \sin\varkappa_{r1}}{\sin(\varkappa_r+\varkappa_{r1})} \tag{6.9}$$

6.2.3 Increase of surface roughness by non-geometric factors

Experiments show that the real mean total height of surface irregularities is greater—to a varying degree—than the theoretical value calculated according to reproduction formulae. The difference between the real and theoretical mean total height is attributed to the influence of physical and dynamic phenomena, such as

— built-up edge,

— friction of cut surface against tool point,

— vibrations,

— embedding of particles of material being machined or of tool point material between the cut surface and tool point surfaces.

The influence of built-up edge on the tool point is very distinct in the case of machining of materials which form built-up edge within the range of practically applied machining conditions. Such materials, among others, are carbon steel and low-alloy steels.

Built-up edge formation depends mostly on cutting speed. Therefore the influence of built-up edge on surface roughness is generally represented as a function of cutting speed.

From experimental results a statistical and empirical formula for mean total height of surface irregularities resulting from the machining of BUE-forming materials has been established

$$R_{tm} = R_{tmt}\left(1+A\frac{v^{cR}}{e^{Bv}}\right) \tag{6.10}$$

where: A, B and c_R—dimensionless values determined empirically.

In most cases finishing operations are carried out under conditions of arc-shaped reproduction.

Equation (6.10) may be written in the form

$$R_{tm} \approx 125\frac{s^2}{r_\varepsilon}\left(1+A\frac{v^{cR}}{e^{Bv}}\right) \tag{6.11}$$

* This formula is a particular case of that given in Table 6.5 for rectilinear reproduction.

A typical trend of the dependence of mean total height on cutting speed in the case of BUE formation is shown in Fig. 6.10.

The cutting speed corresponding to the maximum mean total height is called the *critical cutting speed* (v_{cr}).

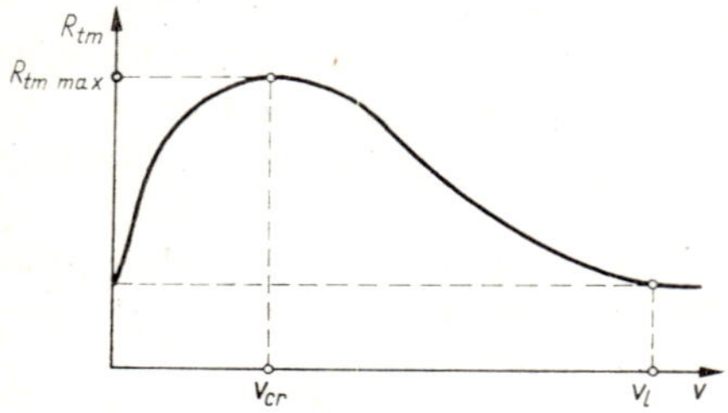

Fig. 6.10 Character of the dependence of mean total height on cutting speed with built-up edge formation

The cutting speed at which the real mean total height does not differ from the theoretical value by more than 10%, is called the *limiting cutting speed* (v_l).

It is known from research on the cutting process that the value of critical cutting speed in the machining of carbon steels is contained within the limits of 15–30 m/min.

The limiting cutting speed varies depending on the kind of material being machined, on the feed rate and on the nose radius.

When turning carbon steels with feed rates below 0·5 mm, and nose radii $r_\varepsilon \leqq 2$ mm the dependence of limiting speed on the factors listed above may be represented by the statistical and empirical equation

$$v_l \approx 200 - \frac{250}{r_\varepsilon^{0\cdot5}}\, s \tag{6.12}$$

Also in the case when a built-up edge does not form, mean total height is, within a wide range of cutting speeds, greater than its theoretical value. The cause of this is the symptoms of intense friction, viz. adhesive tacking and appearance of microcracks on the surface. Generally, with increasing cutting speed these phenomena tend to subside.

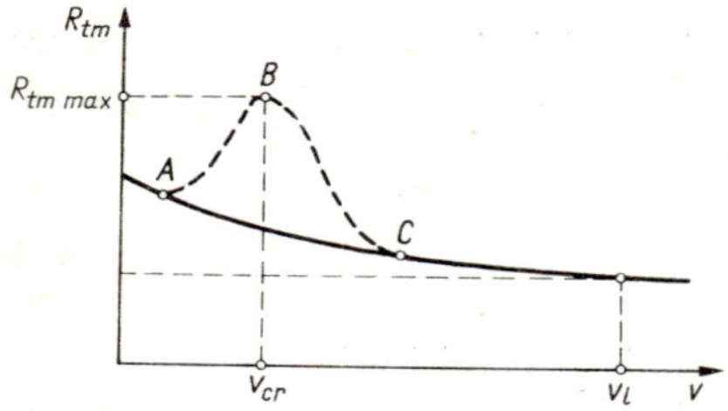

Fig. 6.11 Character of the dependence of mean total height on cutting speed in the case of absence of built-up edge, under conditions of intense friction

The form of the dependence of mean total height on cutting speed under conditions of intense friction is shown in Fig. 6.11. As long as microcracks and surface gaps do not appear, this relationship generally has a monotonic character. On the other hand, in the case when microcracks and surface gaps appear in a certain cutting speed range, mean total height may increase within this range, as shown by the curve *ABC*. Appearance of microcracks, gaps and resulting burrs is caused by the strength of material at the given temperature being exceeded in the surface layer as a result of very high friction.

In many cases of finishing, surface quality is negatively influenced by particles of workpiece material being machined and by the tool point material being abraded and chipped. These particles are present in the cutting fluid and circulate with it; entering between the cut surface and tool point they are driven into the material being machined and form the characteristic burry surface.

Experiments show that filtration of the cutting fluid can considerably improve surface finish.

Vibrations occurring during cutting and having so-called mechanical frequencies, i.e. up to 50 Hz, generally exert a stronger influence on roughness along the lay, but may also significantly increase roughness across the lay. This depends on the direction in which the vibration amplitude attains its maximum value.

On the other hand, vibrations of high frequencies and small (up to 0·05 mm) amplitude, tangential to the cutting direction, may improve surface quality, as they reduce friction, facilitate chip flow along the tool point and diminish the intensity of built-up edge formation.

The advantageous effect of high frequency vibrations is more pronounced with lower cutting speeds. At high cutting speeds the influence of such vibrations on surface roughness is practically unnoticeable.

The influence of the various factors discussed above on surface roughness may become superimposed, resulting in various relationships between surface roughness and machining conditions.

6.2.4 Selection of feed and cutting speed to achieve the desired surface finish

The influence of various factors on surface roughness, as discussed above, may reach different degrees and may be exerted within various ranges. In view of this, the relationship between surface roughness and the particular machining conditions may have a complex character. As a result, there are no generalized relationships which would enable an appropriate selection of machining conditions to achieve the necessary surface finish. On the other hand, there are various tables and nomograms or diagrams for the particular machining methods within definite ranges. An example of such a nomogram is given in Fig. 6.12.

This nomogram refers to the turning of carbon steels within the tensile strength range 45 kg/mm^2 $\leqq \sigma_B \leqq$ 85 kg/mm^2, using sintered carbide tools, under conditions corresponding to the arc-shaped type reproduction of the tool point (type I).

The following example explains how to use the nomogram. Feed and cutting speed should be selected for turning by means of a tool point with nose radius $r_\varepsilon = 2$ mm, so that surface quality would correspond to roughness number 5.

On the scale marked with 2 point *A* is chosen, corresponding to $R_{tm} = 16$ μm and thus situated well below the upper limit of roughness number 5. A line parallel to the *v* axis is drawn from point *A*. This line intersects the curves designated with $r_\varepsilon = 2$: that for feed $s = 0{\cdot}3$ mm/rev at point *B* and that for feed $s = 0{\cdot}4$ mm/rev

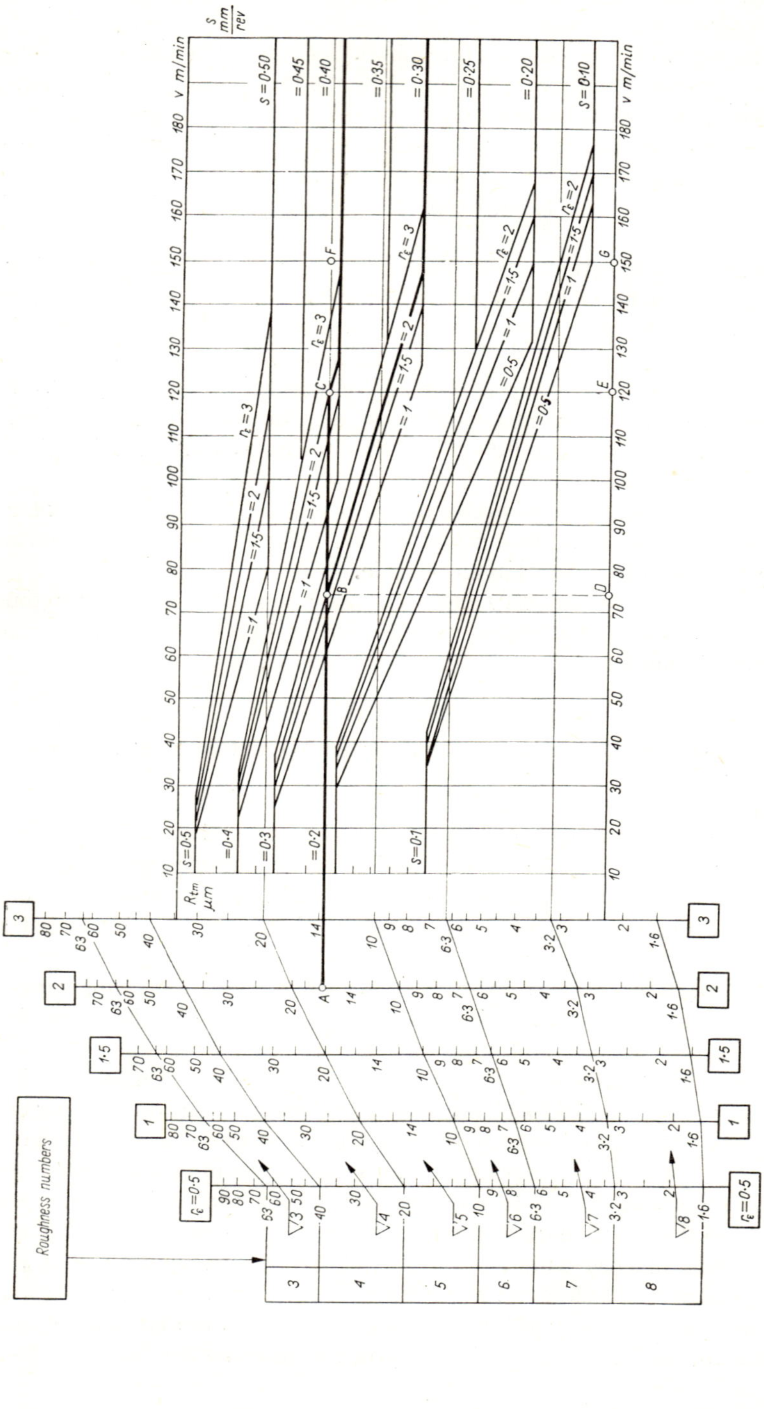

Fig. 6.12 Nomogram for an approximate selection of cutting conditions to achieve a given surface finish

at point C. To these points correspond the cutting speeds: $v = 75$ m/min (point D) and $v = 120$ m/min (point E).

Thus either the pair $s = 0{\cdot}3$ mm/rev and $v = 75$ m/min or the pair $s = 0{\cdot}4$ mm/rev and $v = 120$ m/min may be chosen. In addition, assuming that $v > v_l = 160$ m/min (point F) the feed $s = 0{\cdot}41$ mm/rev can be applied. With regard to surface finish and productivity the latter version would be most advantageous. However, in the selection of the most appropriate feed and cutting speed certain other technical and economic factors should also be taken into account.

6.3 Surface layer

6.3.1 Formation and characteristic of surface layer

6.3.1.1 *Formation of surface layer*

The field of stresses appearing in the material being machined by a cutting edge causes a definite state of deformation. These deformations are modified by the existing temperature field which changes the properties of the material in its own way.

As already discussed in Chapter 2, in the field of deformations the area of plastic deformation may be distinguished. The size and shape of this area is such (Fig. 2.6) that a part of the plastic deformations is situated below the cut surface (s–s) at a depth z.

It follows from this simplified model of surface layer formation that the thickness of this layer depends on the stress and deformation field in the material, and that the properties of this field depend, in turn, on temperature field. A conclusion may also be drawn that the thickness of surface layer is influenced by all those factors which cause changes of the stress and deformation field and the temperature field.

The picture of surface layer formation in Fig. 2.6 is simplified, as in reality the system of forces on the flank and its role in the formation of this layer should also be taken into account.

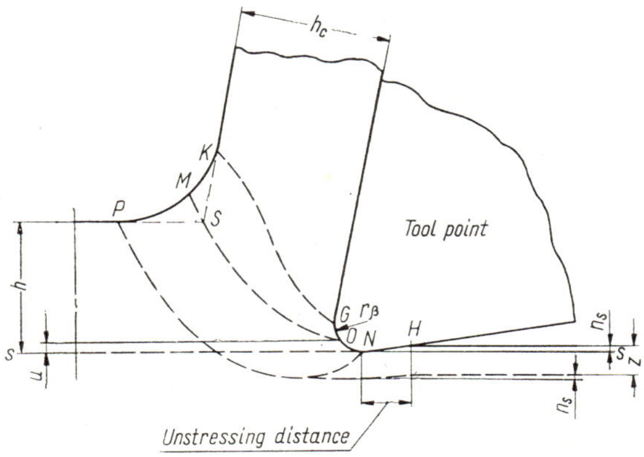

Fig. 6.13 Diagrammatic representation of the appearance of a compressed zone in the surface layer

In Fig. 6.13 the cutting edge is shown, magnified, with roundness radius r_β. This roundness depends on the sharpening method and on the amount of tool point wear. Immediately after resharpening the radius r_β is usually 20–30 μm [6/6], but it can also attain values of the order of tenths of a millimetre.

As a result of this roundness, a part of the undeformed chip, plastically strained by the action of the stress field, is pressed-in at the depth u which additionally increases the compression of the material.*

The height of point O over line s–s or depth of compression u is smaller for smaller values of edge radius.

As soon as the point passes, a part of the material which was only elastically deformed expands to the height n in relation to the cutting line s–s. This causes an increase of the length contact and friction (section NH) between tool point and machined surface. As a result a characteristic lamination of the material in the surface layer appears, called *fibrous structure*.

The frictional conditions in the zone of contact between tool point and machined surface are characterized by the coefficient of friction μ_p whose values may change within wide limits, viz. from 0·1 to 1. For dry cutting, the coefficient of friction is most often contained within the limits 0·4–0·7.

Permanent changes in the distribution of crystallites and volumetric changes caused by plastic deformations (resulting from mechanical and thermal stresses) are the cause of the appearance of machining stresses which modify the initial state of stresses in the material. Modified stresses, called the *final* or *resulting stresses*, may reach far beyond the surface layer area deep into the material, although their cause is located in the surface layer.

This explains why the removal of the surface layer in a stress-free way (electrolytic pickling) or producing lower stresses (lapping) returns the material to its initial state of stresses.

6.3.1.2 *Complex characteristic of surface layer*

It follows from the preceding description of surface layer (shortly called SL) that it should be characterized by geometric and physical features. Listing of the necessary geometric and physical features allowing an unequivocal evaluation of this layer is called the complex characteristic of the surface layer.

An example of such a characteristic is shown in Fig. 6.14.

In the upper part of the characteristic, in field 1, 2 and 3 are contained data on the geometry of the surface.

In field 1 a section of the surface roughness profilogram is given. The scale is chosen so that the section fits into field 1 while the section length fulfils the requirement that it may not be shorter than the measuring length. Mean total height is written in the upper right-hand corner and the roughness number in the upper left-hand corner.

In field 2 the curve of the load carrying capacity of the profile is presented. In order to compare it with other profiles, the diagram is plotted in percentage values. The vertical axis represents 100% of the mean total height R_{tm} and the horizontal axis represents the profile length L.

In the given case the load carrying line percentage after 50% wear of mean total height is $C_{l50} = 57\%$.

* The case considered here is that of cutting under conditions when built-up edge does not form.

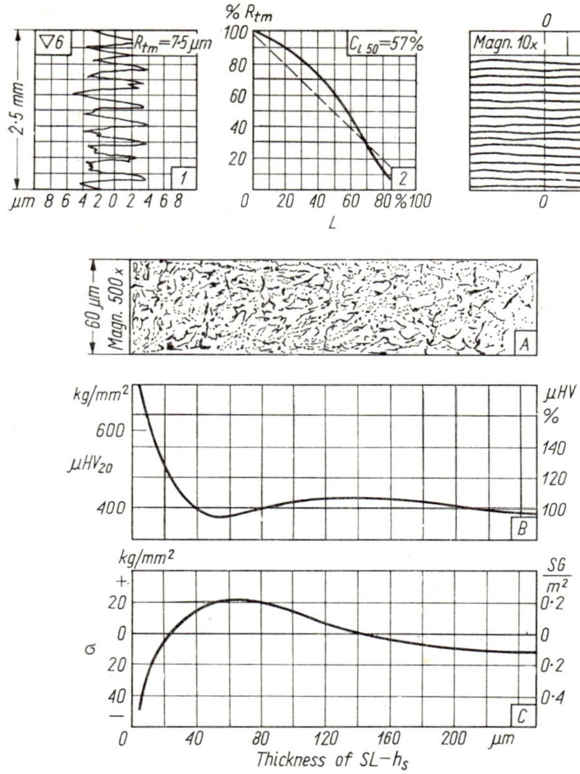

Fig. 6.14 Example of a complex characteristic of the surface layer

In field 3 is a photograph of the machined surface, magnified 10 times. In the upper right-hand corner is the symbol of tool mark lay. In the photograph the direction of surface profile measurement is given (0–0).

Fields A, B and C are used to present some features of the surface layer.

In field A part of a photograph of an etched metallographic specimen is shown. The magnification scale should be so chosen that the photograph length would be equal to SL thickness on which hardness tests* and measurements of residual stresses are made. Microstructure observation generally assists in identifying the causes of hardness and stress changes.

In field B the hardness diagram is inserted. On the left-hand side the absolute scale and on the right-hand side the percentage scale is given, where the hardness of the native material (core) is adopted as 100%. In addition, in the same field the curves of hardness change may be plotted.

Field C is used to represent the variation of residual stresses from the machined

* Hardness testing by indentation method must concern a relatively small surface, but at least comprising the area of a few crystals of various structural components. Therefore, a suitable penetrator load is selected. Hardness measured within such a load range is called macrohardness (to be distinguished from microhardness, tested under a load so selected, that the penetrator sinks on an area not larger than a single crystal).

surface inwards, all through the surface layer. In this diagram the form of the stress change gradient may also be drawn.

Prepared in this way, the complex characteristic of the state of the SL after machining assists in identifying the dependence of the various features and properties of the SL on machining conditions and the influence of the state of the SL on its useful qualities.

6.3.2 Change of structure in surface layer

For phase and structural changes to occur, sufficiently high temperatures must be present for a sufficiently long time. In the process of machining by cutting, a surface layer temperature exceeding 1000°C may be attained. This means that the cutting temperature may be high enough to cause phase and structure changes.

On the other hand, the time when a given portion of material is subjected to this temperature is related to the contact time between the tool point and the material being machined. With practical cutting speeds this time on a 1 mm section may be evaluated as 0·03 ms/mm to 2 ms/mm. This is a very short time with regard to the requirements of phase and structural changes and for this reason the occurence of these phenomena may be expected only on a very limited scale.

Experiments confirm this assumption. The structural change most often encountered when cutting steel is the pearlitic-martensitic structure, i.e. hardening or tempering. These phenomena occur under appropriate machining conditions at small depths, of the order of micrometers or hundredths of a millimetre.

Changes of surface layer properties, resulting from structural changes, are usually undesirable, mainly because they are in most cases uncontrollable and unknown. Research indicates that sometimes, for example, after machining a hardened steel by cutting, part of the surface layer undergoes tempering and rehardening. As a result, the surface layer consists of several parallel zones of different structure and properties.

The simplest way to protect the surface layer from undesirable structural changes is to select machining conditions in such a way as to suitably reduce the cutting temperature.

6.3.3 Hardening of surface layer

6.3.3.1 *Characteristic of surface layer hardening*

The phenomenon of surface layer hardening occurring as a result of permanent plastic deformation, is the more evident, the more ductile, i.e. hardenable, is the material being machined. For example, low-carbon steel, also called soft steel, hardens relatively easily and effectively.

The hardened zone is characterized by a higher yield point and ultimate tensile strength. Therefore, the hardening process is also called the process of surface strengthening.

The degree of hardening, strictly connected for the given material with the degree of compression, is not identical at the chip root and in the surface layer. An example of the topography of chip root and surface layer hardening is given in Fig. 6.15.

It is seen that the chip itself hardens considerably. In the surface layer, hardening decreases with increasing depth. The hardening role of the tool point on the flank side is evident, since immediately after the tool point passage hardening increases in relation to that appearing as a result of the deformation preceding the tool point.

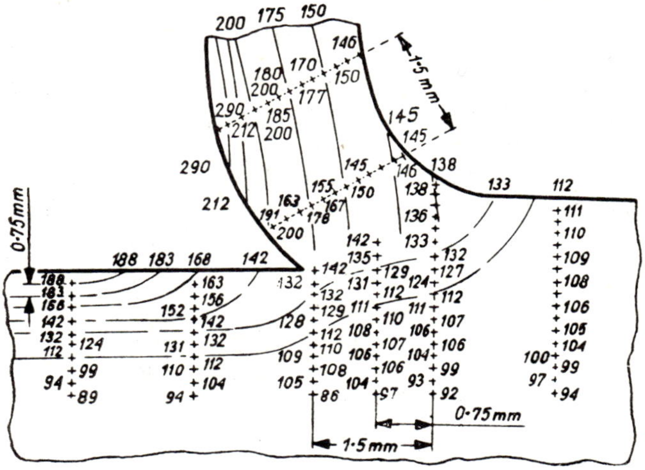

Fig. 6.15 Example of hardness topography in the chip formation zone and in part of a steel workpiece. Cutting speed $v = 10$ m/min, undeformed chip thickness $h = 1{\cdot}2$ mm (acc. to Kostetski)

The factors characterizing the compression area of the surface layer are:

(a) hardening depth, equal to the distance from machined surface to the spot where the hardness becomes identical with that of the native material of the workpiece;

(b) hardness, measured by a hardness tester on the machined surface;

(c) maximum hardness, which may appear below the machined surface. This hardness is measured perpendicularly to the metallographic specimen;

(d) maximum gradient of hardness decrease (also called the *hardening gradient*) which is calculated as

$$A_{\max} = \frac{dH_D}{dh} \qquad \frac{\text{kg/mm}^2}{\text{mm}} \tag{6.13}$$

where: H_D—macrohardness in kg/mm^2, h—distance from surface in mm. Excessively high values of the hardening gradient are disadvantageous, since high stresses usually appear which may be the cause of accelerated subsurface fatigue and even of surface layer spalling. To give an indication of hardening gradient values it can be said that, after finish turning steel, a hardening distribution may appear with a maximum gradient of some 15 000–25 000 kg/(mm^2 mm) or 15–25 kg/(mm^2 μm) at a depth of 10–15 μm.

(e) Compression intensity e_c, expressed by the ratio of the hardness increment at the given spot to the hardness of undeformed material

$$e_c = \frac{H_D - H_{Do}}{H_{Do}} = \frac{H_D}{H_{Do}} - 1 \tag{6.14}$$

where: H_D—macrohardness at the given spot, H_{Do}—macrohardness of undeformed material.

To characterize the state of hardening of the surface layer, the maximum value of e_c is usually given.

6.3.3.2 *Influence of certain machining conditions on hardening*

Of all the factors indirectly influencing the hardening of the SL changes in the deformation field, which precedes the tool point, the depth of cut, undeformed chip thickness and rake angle should be specially mentioned.

Examples of the dependance of macrohardness on the depth of cut and on rake angle γ for various distances l from the surface are given in Fig. 6.16 and 6.17. These relationships concern the turning of carbon steel with sintered carbide tools.

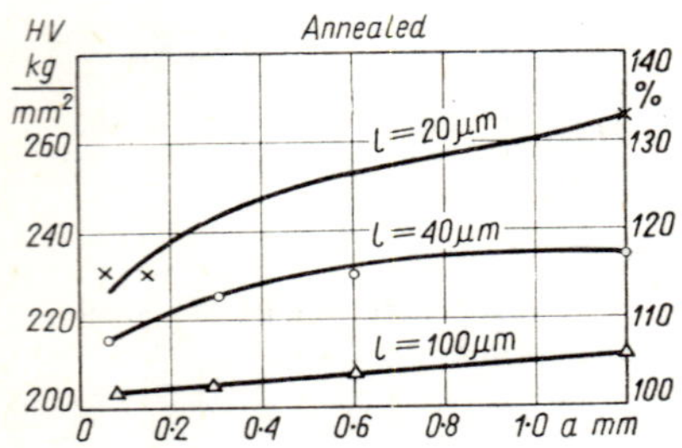

Fig. 6.16 Diagram of the dependence of macrohardness on depth of cut

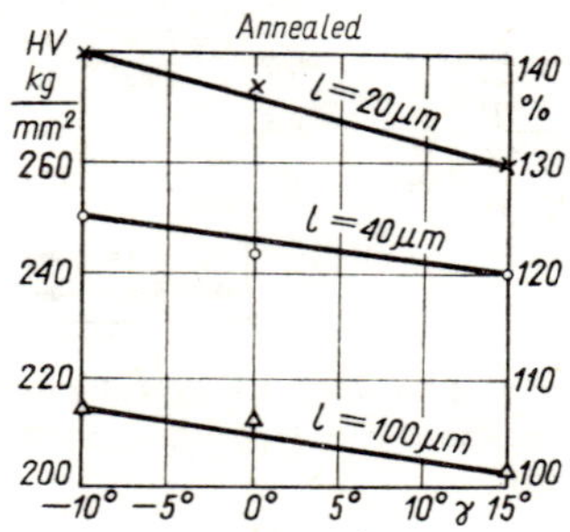

Fig. 6.17 Diagram of the dependence of macrohardness on rake angle

As is evident from the diagrams, a change of the depth of cut from 0·1 to 1·2 mm (or twelve-fold) at a distance from the surface $l = 20$ μm increases hardness by 17%.

Increases in the depth of cut also cause a less-than-proportional increment of surface layer thickness or of the thickness of plastic deformation and compression, since with increasing depth of cut the hardness grows quicker than hardening depth, compression intensity and hardening gradient increase. This, in general, is disadvantageous and, in finishing operations and cuts, the depth of cut must be suitably reduced.

Change of rake angle from $+15°$ to $-10°$, for $l = 40$ μm, causes hardness to increase by some 5%. Thus it is obvious that because of the relatively small rake angles generally encountered in practice its influence on hardening is not as great, as e.g. that of the depth of cut.

Rake angle changes also create minor variations of the hardening gradient and major ones of the hardening depth. This is connected with the fact that with rake angle reduction the plastic deformation zone reaches deeper into the material.

The influence of edge roundness radius on hardening is represented by the research results, illustrated diagrammatically in Fig. 6.18. A change of edge roundness radius from 5 to 500 μm in this case increases hardness by 40% at a distance $l = 20$ μm from the surface. The most probable range of cutting edge roundness is contained within the limits of 15–50 μm. In this range an increase of roundness radius by 10 μm causes hardness to increase by about 6 N/mm^2.

As the cutting edge roundness radius increases, so does the gradient of hardening and of the compression intensity. This means that compression increases quicker than the depth of hardened layer.

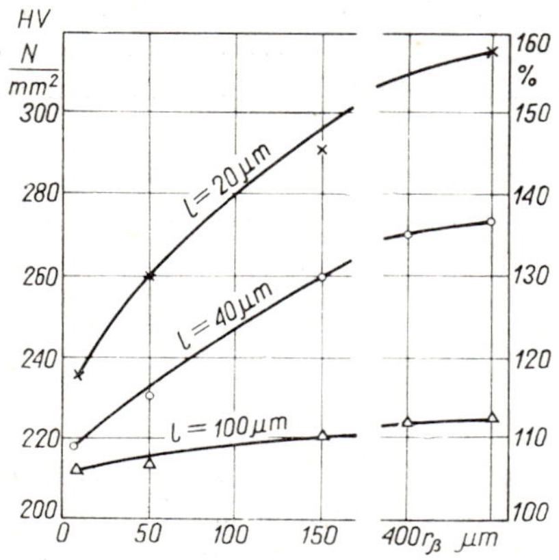

Fig. 6.18 Diagram of the dependence of macrohardness on cutting edge roundness

Cutting fluids also influence the thickness of hardened layer. Their application causes a reduction of compressed layer thickness as compared with that resulting from dry cutting. This is explained by the fact that the use of cutting fluids diminishes friction which changes the direction and value of the resultant cutting force in such a way that the depth of plastic deformation zone z (Fig. 6.13) is reduced.

The influence of cutting fluids is greater, the smaller is the undeformed chip thickness and the lower is the cutting speed. This confirms the explanation of the causes of the influence exerted by the application of cutting fluids on hardening.

The order of magnitude of this influence may be illustrated by the following example: in turning carbon steel at a cutting speed $v = 35$ m/min the hardening depth, as compared with that resulting from dry cutting, was reduced by about 20% by the use of emulsion and by about 30% by the use of neat oils.

Cutting with cutting fluids generally does not cause changes in compression intensity, but it does increase the hardening gradient.

6.3.4 Machining and final (residual) stresses

6.3.4.1 *Types of stresses*

When stresses without further definition are mentioned, so-called type I stresses are meant, i.e. stresses being in equilibrium throughout the whole volume of the given component.

As mentioned earlier, the following stresses should be distinguished:

— initial stresses, existing in the workpiece material before the given technological operation.

— machining stresses, created during the given operation,
— final (residual) stresses, appearing as a result of superposition of the field of machining stresses over the field of initial stresses.

This division indicates that the final stresses must not be ascribed exclusively to the last operation performed, since a certain state of stress may also have remained in the material after the preceding operations.

With regard to the empirical methods of research on the state of the final stress, stresses in the direction of primary motion (so-called tangential stresses), axial stresses in the direction of auxiliary motion and stresses in the direction perpendicular to the machined surface are distinguished.

In general, the greatest of these are the stresses in the direction of primary motion or of the tangential cutting force. Therefore, they are in most cases the object of interest and research.*

6.3.4.2 *Appearance of stresses*

Stresses in materials as a result of technological operations performed are caused by
— permanent disturbances of the structure created by plastic shear strain,
— volumetric changes of the structural components as a result of their phase changes, of the appearance of new components or disappearance of some of them.

Machining stresses caused by permanent deformation may be ascribed to phenomena which can be explained by one, and in some cases two, models of plastic deformation.

The first model of deformation is called the *mechanical* (or *cold*) *model* of the appearance of stresses.

As a result of the action of the tool point upon the material there appears a field of stress creating, if in this layer the yield point is exceeded, plastic deformation at a certain depth. Below the plastically extended layer are layers in a state of elastic extension (Fig. 6.19a).

As soon as the action of the field of stress ceases at point A, a contraction of the extended layers will follow. The plastically extended layer will not be able to return to its original state. Thus the lower layers, only elastically extended, will exert upon the plastic layer an elastic force which will create a state of compression in the plastically deformed layer.

Consequently, as a result of the mechanical (or cold) plastic deformation, compressive stresses will appear in the upper part of the compressed zone (Fig. 6.19b).

The second model of deformation, the so-called *thermal model*, is characterized by the appearance of stresses from a different cause.

As a consequence of the heat generated in the machining process a temperature field appears, reaching into the workpiece material. The maximum temperature appears at the surface and decreases with increasing distance from the latter.

* There are various methods of studying and calculating final stresses. In cylindrical objects stresses are mostly evaluated by Davidenkov's method (N. N. Davidenkov: Izmereniya ostatochnykh napryazhenii v trubakh—Measurements of final stresses in tubes—Zhurnal Tekhnicheskoi Fiziki, No. 9, vol. 5, 1931).

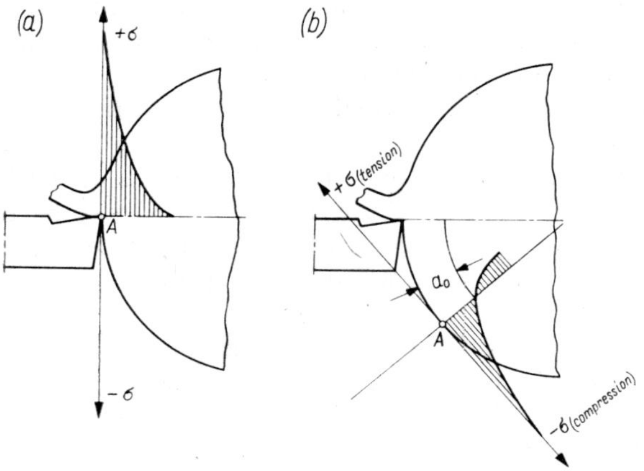

Fig. 6.19 Diagrammatic representation of the stages of appearance of stresses in the so-called cold model

When heating up, the subsurface zone can expand more than the deeper zones of lower temperature. The interconnection of these zones results in the appearance at the surface of a state of compression, while deeper the compressive stresses decrease, finally changing to tensile stresses (Fig. 6.20, position A_1).

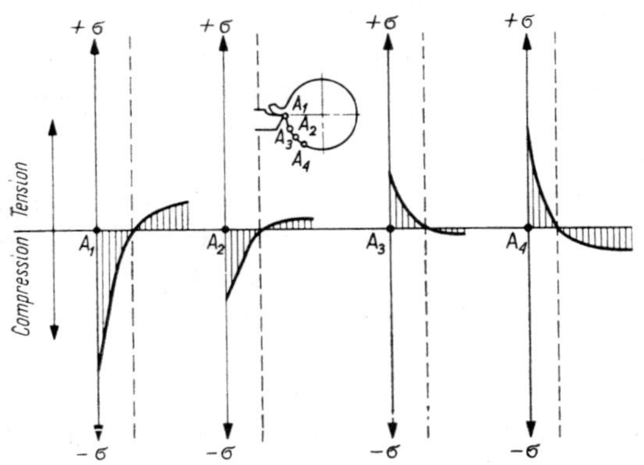

Fig. 6.20 Diagram showing the magnitude and direction of stresses in the surface layer

If the temperature reaches values at which compressive stresses will exceed the yield point, correspondingly reduced at such temperature, then a permanent deformation (by compression) will occur in the subsurface zone. Compression of the material will be the stronger, the more the compressive stresses exceed the yield point for the given temperature.

As soon as the temperature field shifts, i.e. when point A_1 acquires the position A_2, as a result of cooling, the surface layer contracts which temporarily causes a reduction of stress differences.

After some time (position A_3) elastically deformed zones will return to their normal state, while the plastically deformed subsurface zones will tend to shorten still further to reach plastic shortening.

After complete cooling (position A_4) a state of tension will appear in the upper part of the compressed (deformed) zone, whose further contraction is counteracted by deeper zones, while the latter will be in a state of compression.

It is evident from this description that the thermal model of deformation and stress yields tensile stresses in the surface layer.

Stresses which result from phase changes are connected with changes of specific volume of the particular structural components. An example of this may be the process of surface hardening. In this process martensite appears whose specific volume is larger than that of austenite. As a result, a state of compression is created in the martensite layer.

Thus structural changes connected with an increase of specific volume will create a state of compression and those connected with a reduction of specific volume a state of tension in the surface layer.

In practice, the processes of machining by cutting and by erosion create conditions under which the appearance of deformations may occur, in part, according to each of the models described. As a result, various states of stress may prevail in the surface layer. Depending on the magnitude of the stresses appearing in this layer, a more or less evident disturbance of the state of stress occurs in the whole object.

This is the reason why the depth to which the stresses reach may not be identified with the thickness of the permanently deformed zone.

6.3.4.3 *Dependence of stresses on machining conditions*

The influence of the particular machining conditions on the final stresses has not yet been studied sufficiently. The difficulties involved are connected with the measurements of stresses and with the complex character of the problems.

It can be expected, however, that those factors and machining conditions that will contribute to the cold model of deformation appearance, will also facilitate the appearance of compressive stresses in the surface layer. Such factors are undeformed chip thickness, feed rate and cutting edge roundness radius.

On the other hand, factors contributing to the existence of the thermal model of deformation will facilitate the appearance of tensile stresses in the surface layer. Such a factor is for example, under certain conditions, the cutting speed.

Of the geometrical features of the tool point the strongest influence on the appearance of stress in surface layer is exerted by rake angle. Reduction of rake angle and particularly large negative rake angles contribute to the appearance of compressive stresses in the surface layer.

It follows from studies on useful qualities of the machined component that compressive stresses in the surface layer are more advantageous than tensile stresses.

REFERENCES

6/1 Sadowski, A.: 'Gładkość powierzchni i sposoby jej pomiaru' (Surface Roughness and the Methods of Its Measurement), PWT, Warszawa 1952

6/2 Kaczmarek, J.: 'Technologiczne możliwości osiągnięcia gładkości powierzchni' (Technological possibilities of obtaining good surface finish), *Czasopismo Techniczne* 4, 1961

6/3 Szulc, S.: 'Metody technologiczne dla uzyskania warstwy powierzchniowej o określonych własnościach' (Technological methods of obtaining surface layer of definite properties), *Technika Lotnicza* 5, 1959

6/4 KACZMAREK, J.: 'Niektóre właściwości warstwy powierzchniowej obrabianej w warunkach zwiększonej wydajności skrawania' (Some properties of surface layer machined under conditions of increased rate of metal removal), *Czasopismo Techniczne* 9, 1960

6/5 KACZMAREK, J.: 'Wpływ obróbki mechanicznej na własności warstwy wierzchniej' (Influence of mechanical working on surface layer properties), *Przegląd Mechaniczny* 13, 1962

6/6 KACZMAREK, J.: 'Badania wpływu promienia zaokrąglenia krawędzi skrawającej na utwardzanie warstwy wierzchniej' (Studies on the influence of cutting edge roundness radius on surface layer hardening), *Czasopismo Techniczne* 6, 1965

CHAPTER 7

Technical and economic bases for the optimization of machining conditions

7.1 Methods of optimizing the selection of machining conditions

A rational selection of machining conditions is the result of taking into account and checking:
(a) technical specifications listed by the designer,
(b) limitations in machining,
(c) production economy.
Technical specifications listed by the designer should take into account as fully as possible the criteria of suitable quality and useful properties of the product. In practice, they are reduced to requirements concerning the accuracy of shape and dimensions, surface quality and the properties of the surface layer.

Limitations in machining result from the character of the given method (e.g. the range of roughness obtainable), of the machine tool (e.g. the range of primary and auxiliary motion speeds, motor power, strength of elements, vibrations), of tools (e.g. tool life, vibrations), and of the workpiece (rigidity, machinability).

Production economy requires a low consumption of constructional materials, tool materials and auxiliary materials, possibly high productivity (i.e. low consumption of time and labour) and small unit cost.

Optimization of the machining conditions depends on the extent to which the physical, technical and economic relationships involved in the given machining method are known.

In the case of cutting, these relationships may be approximated by continuous functions. Thus a *total* or *complex optimization* of machining conditions consists—from the mathematical point of view—in solving a system of equations describing the relationships between particular criteria and the machining conditions.

However, the present state of knowledge in the field of technology of machining by cutting and erosion does not furnish enough data for a practical application of the method of complex optimization.

Therefore, in practice, *partial optimization* is applied. It requires the assumption of certain data and in determining optimum values of some machining conditions with regard to adopted criteria.

Three types of partial optimization of machining conditions will be discussed below:

(a) one-criterion or single optimization with regard to minimum unit cost, allowing one to determine the optimum value of one machining condition (parameter), most often the cutting speed,

(b) two-criteria or double optimization with regard to minimum unit cost and the necessary surface quality, allowing one to determine the optimum values of two cutting conditions, usually cutting speed and feed rate,

(c) three-criteria or triple optimization with regard to minimum unit cost, the necessary surface quality and the necessary dimensional accuracy (or permissible dimensional error). This method allows one to determine the optimum values of three machining conditions, most often the depth of cut, feed rate and cutting speed.

Beside these types still other versions of optimization of machining conditions are possible. Thus, for example, in the case of specialized or special machine tools the optimization of machining conditions with regard to minimum unit cost and to making full use of the available motor power may be applied. This allows one to determine the optimum values of two machining conditions, usually feed rate and cutting speed.

Optimization of the machining conditions may be carried out in an analytical or graphical way. The latter method is usually used in cases when analytical solutions present difficulties because of the complicated mathematical form of the particular relationships.

To facilitate and speed up optimizing calculations, tables, diagrams and nomograms, slide rules and devices for automatic selection of machining conditions have been worked out. More and more often computers are used for the optimization of machining conditions.

7.2 One-criterion optimization of single-tool machining conditions with regard to minimum unit cost

7.2.1 Economic and productive tool life

There are unequivocal relationships between the conditions of production, and thus also of machining, and the technical and economic indices of production.

Of the latter, productivity and production quality as well as unit cost are considered as the basic ones. They are defined in the following way:

— productivity of an operation

$$W_o = \frac{1}{t_o} = \frac{Z}{t_p} \quad \text{pieces/min} \tag{7.1}$$

where: t_o—time of operation* in min, Z—number of all pieces machined in this operation during time t_p;

* This concerns single-pass operations. For multipass operations these considerations should be adapted to the particular passes.

— production quality (evaluated according to the number of rejects)

$$Q = \frac{S}{Z} \tag{7.2}$$

where S—number of acceptable pieces obtained in the given operation;
— unit cost

$$K_z = \frac{K_p}{S} \quad \text{pence/piece} \tag{7.3}$$

where K_p—total costs of the operation in pence.
A combination of eqns. (7.1), (7.2) and (7.3) allows one to establish the following relationship

$$K_z = \frac{K_p}{t_p}\frac{t_o}{Q} = \frac{k_p}{Q}t_o \quad \text{pence/piece} \tag{7.4}$$

where $k_p = K_p/t_p$—cost rate of the operation in pence/min.
The cost rate of the operation may in turn be split into costs independent of machining conditions or of tool life and costs depending on machining conditions and thus also on tool life. The situation is similar as regards the components of operation time. This may be represented by equations:

$$K_z = K_I + K_R + K_P \tag{7.5}$$

$$t_o = t_I + t_R + t_P \tag{7.6}$$

where: K_I, t_I—costs and time independent of cutting conditions, K_R and t_R—costs and time depending on cutting conditions in a regressive way, K_P and t_P—costs and time depending on cutting conditions in a progressive way.

After suitable calculations and assuming the fulfilment of condition

$$\frac{dK_z(T)}{dT} = 0 \tag{7.7}$$

and when

$$\frac{d^2K_z(T)}{dT^2} > 0 \tag{7.8}$$

the formula for tool life corresponding to minimum unit cost, or so-called *economic tool life*, is obtained in the form

$$T_e = \frac{z-1}{U_w U_c}\left(t_{ct} + \frac{K_T}{K_M}\right) \quad \text{min} \tag{7.9}$$

where: $U_c = L/l$—coefficient taking into account the difference between machine time and cutting time (L—length of mechanical travel, l—cutting length), $U_w = l+e$—coefficient taking into account the operation time component called the complementary time (e—standard complementary time), K_M—cost rate of machine tool including overheads etc., z—exponent of cutting speed from the formula for tool life, t_{ct}—tool change time in min, K_T—cost of each cutting edge (including regrinding where applicable), in pence.

When the condition

$$\frac{\mathrm{d}t_o(T)}{\mathrm{d}T} = 0 \tag{7.10}$$

is fulfilled and the inequality

$$\frac{\mathrm{d}^2 t_o(T)}{\mathrm{d}T^2} > 0 \tag{7.11}$$

established, then the formula for tool life corresponding to maximum productivity (i.e. minimum time of operation), or so-called *productive tool life* can be found in the form

$$T_{pr} = \frac{z-1}{U_c U_w} t_{ct} \quad \text{min} \tag{7.12}$$

It follows from the comparison of eqns. (7.9) and (7.11) that $T_e > T_{pr}$, since

$$\Delta T = T_e - T_{pr} = \frac{z-1}{U_c U_w} \frac{K_T}{K_M} > 0 \tag{7.13}$$

This means that machining at, simultaneously, minimum costs and maximum productivity under given conditions is impossible. On the other hand, it is possible to reduce the value of the difference ΔT.

If, e.g., $\Delta T = 1$ is to be obtained, then

$$\frac{K_T}{K_M} = \frac{U_c U_w}{s-1} \tag{7.14}$$

In the case of turning, for example, $U_c \approx 1$ and $U_w = 1{\cdot}2$. Exponent z, for sintered carbides, has an average value of 5. The ratio K_T/K_M should then be equal to 0·3. In reality, the value of this ratio is 10–60 times higher. This indicates that tool costs are an important factor impeding an economic rise of productivity.

Equation (7.9) also highlights the importance of machine tool costs. The more expensive is the machine tool and the higher are the overheads for the given workplace, the shorter becomes the economic tool life. This means that under such conditions the role of tool costs decreases and, to machine economically, it is necessary to cut at higher feed rates or cutting speeds and with the smallest possible number of passes.

In the method discussed the depth of cut and feed rate are selected with regard to other limiting conditions, such as e.g. permissible cutting forces and desired surface quality. After selecting the depth of cut and feed rate, the cutting speed is calculated from the condition of economic tool life according to the formula for the so-called *economic cutting speed* v_e

$$v_e = \frac{C_v}{a^{e_v} s^{u_v}} \sqrt[z]{\frac{U_c U_w}{(z-1)(t_{ct} + K_T/K_M)}} \tag{7.15}$$

Similarly, the formula for *productive cutting speed* v_{pr} may be written

$$v_{pr} = \frac{C_v}{a^{e_v} s^{u_v}} \sqrt[z]{\frac{U_c U_w}{(z-1) t_{ct}}} \tag{7.16}$$

where: C_v—coefficient taking into account the influence of all factors not appearing separately in the formula, a—depth of cut, s—feed rate, e_v and u_v—exponents determined in an empirical way.

7.2.2 Recommended tool life

In Fig. 7.1a is presented the dependence of unit cost and operation time on tool life and in Fig. 7.1b on cutting speed. The following conclusions may be drawn from the analysis of these diagrams:

(a) economic tool life is longer than productive tool life, while economic cutting speed is lower than productive cutting speed,

(b) application of cutting speeds higher than v_{pr} or tool lives shorter than T_{pr} is uneconomic as unit costs rises considerably and productivity decreases,

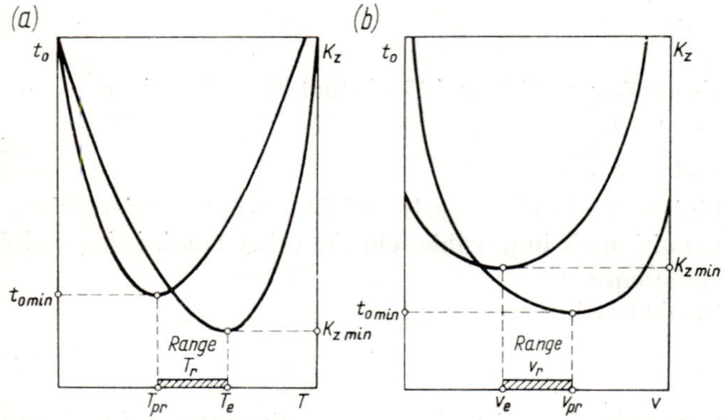

Fig. 7.1 Dependence of unit cost K_z and operation time t_o on: (a) tool life, (b) cutting speed

(c) application of cutting speeds lower than v_e or tool lives longer than T_e is uneconomic, as productivity decreases considerably and unit cost rises,

(d) if the true economic cutting speed, i.e. a speed corresponding to economic tool life, cannot be applied, then the so-called *recommended cutting speed* v_r and *recommended tool life* T_r should be adopted, fulfilling the conditions

$$\begin{aligned} v_e &\leqq v_r < v_{pr} \\ T_e &\geqq T_r > T_{pr} \end{aligned} \tag{7.17}$$

Various economic and productive tool lives may result for various conditions of production. An analysis of the variability range of these values in the whole industry yields mean recommended tool life values [7/1] and the corresponding cutting speeds which in the majority of cases fulfil the condition contained in eqn. (7.17).

Based on such assumptions, recommended standards in the form of tables or nomograms [7/2] are worked out for the selection of cutting conditions.

7.2.3 General rules and the procedure for the selection of machining conditions

On the bases of analysis of technical and economic relationships occurring in machining processes the following general rules for the selection of machining conditions have been formulated:

(a) first determine the greatest depth of cut possible under given machining conditions, i.e. aim at machining with a minimum number of passes,
(b) then choose the highest feed rate possible under given conditions,
(c) finally establish the recommended cutting speed for the adopted values of depth of cut and feed rate.

The choice of the depth of cut should be based on the principles of standardization of machining allowances [7/3].

Feed rate should be selected in a way to fulfil the conditions limiting its value for a given depth of cut. Such limitations are the forces and moments permissible with regard to the strength of machine tool elements and tools, to motor power, permissible deformations of the m–w–t system, the range of feed rate stepping and the required surface quality.

The procedure for the selection of machining conditions, based on the method of optimization with regard to minimum unit cost, should be as follows:

(a) choice of the depth of cut,
(b) choice of feed rate,
(c) determination of recommended cutting speed,
(d) calculation or determination of rotational speed of main drive, corresponding to recommended cutting speed,
(e) selection of rotational speed nearest to the calculated value (from stepped series of speed values),
(f) correction of recommended cutting speed,
(g) correction of recommended tool life according to the formula

$$T_{r2} = \left(\frac{n_1}{n_2}\right)^z T_{r1} \tag{7.18}$$

where: subscript 1 denotes the originally adopted values, and subscript 2 the corrected values,
(h) establishment of the necessary power,
(i) choice of a suitable machine tool,
(j) calculation of machine time.

7.3 One-criterion optimization of multi-tool machining conditions with regard to minimum unit cost

7.3.1 Characteristic features and classification of particular cases of multi-tool machining

The preceding considerations concerned single-tool machining. In multi-tool machining, i.e. when complete or partial working with several tools occurs, the conditions are somewhat different. Thus, for example, workpiece temperature may differ from that attained in single-tool machining, vibrations caused by the work of one tool may influence the work of other simultaneously working tool points.

Notwithstanding some differences in cutting conditions, the dependence of tool life in multi-tool machining may be described by similar formulae as in the case of single-tool machining. The differences arising have a quantitative rather than qualitative character.

The essential difference between the optimization of single- and multi-tool machining consists in the fact that for machining on automatic machine tools, and particularly on machine production lines, the calculation of costs becomes complicated. In such cases machine tool costs cannot be calculated separately, but in connection with the whole line. The correct criterion for the evaluation of optimum machining conditions must then be the unit cost of the product of the given line or of its part.

In spite of these differences the methods for determining the economic optimum machining conditions may be similar to those used in the case of single-tool machining. This is valid for the range of cutting conditions under which abrasive wear and not strength wear of the tool occurs.

In general, two cases of multi-tool machining may be distinguished.

The first case occurs when the machine time of the operation is determined by uninterrupted work of one tool. This case, let it be designated by letter *E*, is represented by Fig. 7.2. The succession and time of work of the particular tools t_{m1}, t_{m2}, t_{m3} and t_{m4} is shown there. In this case the work time t_{m4} is simultaneously the machine time of multi-tool machining t_{mo}.

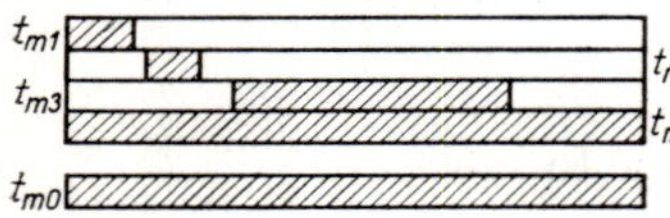

Fig. 7.2 Diagram of work succession and machine times

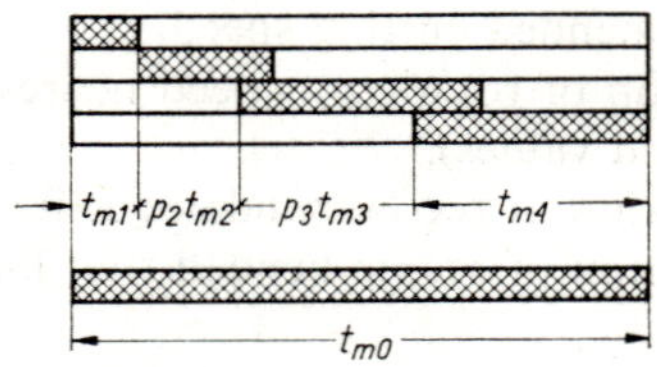

Fig. 7.3 Diagram of work succession and machine times

The second case—more frequent—occurs when the machine time of multi-tool machining is the sum of parts of machine times of particular tools. This case is illustrated in Fig. 7.3. In such a case the machine time of multi-tool machining can be generally defined by the formula

$$t_{mo} = \sum_{i=1}^{i=k} p_i t_{mi} \tag{7.19}$$

Let this case of machining be designated by the letter *S*.

From the point of view of the methods of calculating the economic tool life, two cases may also be distinguished in multi-tool machining with regard to the value of exponents *z*.

The case when working tools are of the same material and of similar type is characterized by equal exponents *z* for all tools. This case will be designated by letter *G*. On the other hand, the case when the tools are different and are made of various materials is characterized by exponents *z* differing from each other. This case will be designated by letter *V*.

As a result—from the point of view of the methods of calculating the economic tool life in multi-tool machining—four cases may be distinguished:

(a) case *GE*—with equal values of *z* and the machining time limited by one tool,

(b) case *GS*—with equal values of z and the machining time constituting a sum of parts of machine times of several tools,
(c) case *VE*—with different values of z and the machining time limited by one tool,
(d) case *VS*—with different values of z and the machining time constituting a sum of parts of machining times of several tools.

Cases *GE* and *GS* may be solved in a generalized form, in an exact way, independently of the value of exponent z.

In practice, in many cases of multi-tool machining additional requirements or limitations occur, mostly facilitating the task of optimization. Thus, for example, in multi-tool lathes rotational speeds of the workpiece are identical and so are feed rates, in single-spindle automatic lathes only the former feature appears and in milling only the latter, etc.

7.3.2 The general condition for economic selection of multi-tool machining parameters

Temchin [7/4] gives the procedure for a general case of multi-tool machining when exponent z is equal for all tools, but there is no further interdependence.

At first, the formula for variable components of unit cost of operation introduced, for the case *GE* in which rotational speed is identical for all tools

$$K_z = U_w\, t_{mx}\, K_M + \sum_{i=1}^{i=k}\left(t_{cti}\frac{t_{cut\,i}}{T_i}K_M + K_{Ti}\frac{t_{cut\,i}}{T_i}\right) \tag{7.20}$$

where t_{cut}—cutting time, subscript x denotes the tool whose machine time limits the cycle time, as e.g. in Fig. 7.2, the remaining designations are as before.

This formula is transformed in order that t_{cut} and t_m be replaced by functions of tool life. The notion of conventional, substitute tool life of a group of tools T_G is introduced here.

If under adopted cutting conditions the life of tool i is T_i, its relationship to the substitute life period T_G is

$$T_i = w_i T_G \tag{7.21}$$

After transformation, with T_G as a variable, the following formula is obtained

$$\frac{K_z}{K_M} = U_c\, T_G^m \sum_{i=1}^{i=k} C_{si}\, p_i\, U_{ci}\, w_i^m + T_G^{m-1}\sum_{i=1}^{i=g} C_{si}\, w_i^{m-1}\left(t_{cti} + \frac{K_{Ti}}{K_M}\right) \tag{7.22}$$

where $C_s = \dfrac{q}{a^{1-e_v}}\dfrac{\pi d}{s^{1-u_c}}\dfrac{1}{C_v}$, q—machining allowance, $m = 1/z$, and the remaining symbols as given before.

In turn, the first derivative $\mathrm{d}(K_z/K_M)/\mathrm{d}T_G$ is equated to zero, and the condition for economic (i.e. optimum with regard to costs) value of substitute life of a tool group is obtained

$$T_{eG} = \frac{(z-1)}{U_w}\sum_{i=1}^{i=k}\frac{t_{cti}+K_{Ti}/K_M}{p_i\, U_{ci}\, w_i} = \sum_{i=1}^{i=k}\frac{1}{p_i\, w_i}\sum_{i=1}^{i=k} T_{ei} \tag{7.23}$$

It is easy to perceive that when $k = 1$, or in the case of single-tool machining, $w = 1$ and $p = 1$ and eqn. (7.23) becomes the already known eqn. (7.9) for economic tool life under conditions of single-tool machining. On the other hand, if there are k working tools, but they are identical and identically loaded, then $p = 1$, $w = \text{const}$ and

$$T_{eG} = \frac{k}{w}\frac{(z-1)}{U_c\,U_w}\left(t_{ct}+\frac{K_T}{K_M}\right) = \frac{k}{w}T_e \tag{7.24}$$

Such a case occurs sometimes in drilling with a multi-spindle head with identical drills. Equation (7.24) is now rewritten in the form

$$w\,T_{eG} = k\,T_e = T_{ek} \tag{7.25}$$

where $T_{ek} = kT_e$ designates the economic tool life for a tool working in a set with several others, while T_e—economic tool life under conditions of single-tool machining.

It is obvious from eqn. (7.25) that the economic life of a tool working under conditions of multi-tool machining is k times longer than that in single-tool machining. Thus, for example, a drill for which a life of $T_e = 10$ min is recommended under single-tool machining conditions, should work for $T_{ek} = 60$ min in a six-spindle drill head.*

Equation (7.24) may also be written in a different form

$$1 = \sum_{i=1}^{i=k}\frac{T_{ei}}{w_i T_{eG}} = \sum_{i=1}^{i=k}\frac{T_{ei}}{T_{eki}} \tag{7.26}$$

It follows from eqn. (7.26) that always $T_{eki} > T_{ei}$ or that the economic life of a tool working under multi-tool machining conditions must always be longer than the economic tool life under single-tool machining conditions.

The guidelines given for the determination of multi-tool machining conditions indicate that the corresponding calculations are tedious (unless a suitable computer programme is available). In some cases a method of checking of adopted conditions may be applied. Equation (7.26) is helpful here.

Namely, if:

(a) $\sum_{i=1}^{i=k}\frac{T_{ei}}{T_{eki}} > 1$—this means that actually adopted cutting conditions are too difficult;

(b) $\sum_{i=1}^{i=k}\frac{T_{ei}}{T_{eki}} = 1$—then the cutting conditions selected constitute an optimum with regard to costs;

(c) $\sum_{i=1}^{i=k}\frac{T_{ei}}{T_{eki}} < 1$—cutting conditions are too easy and should be made more effective.

Presented above is the problem of economic selection of multi-tool machining parameters in cases when $z = \text{const.}$, i.e. in cases *GE* and *GS*.

Cases when $z \neq \text{const.}$, i.e. cases *VE* and *VS* cannot be solved in a mathematically

* It is assumed here that machine tool costs are equal in both cases.

exact way in a general form, when at least for one tool $z \geqq 5$. Various approximate methods are then applied, analytical or graphical,* or numerical solutions based on discrete functions.

7.4 Two-criteria optimization of single-tool machining conditions with regard to minimum unit cost and the required surface quality

The basis for the selection of machining conditions ensuring minimum costs of operation and the obtaining of a desired surface quality, is the solution of the system of equations:

$$T_e = f(s, v) \tag{7.27}$$

$$R_{tmo} = F(s, v) \tag{7.28}$$

where: T_e—economic tool life, R_{tmo}—required mean total height of surface irregularities.

Relationships (7.27) and (7.28) have a character of non-monotonic functions (see Chapters 4 and 6) and are well known only within certain ranges of monotonic character. If formulae approximating those ranges are used to solve the system of eqns. (7.27) and (7.28), this brings with it the risk that the resulting s and v values may exceed the range of validity of such formulae.

The system of eqns. (7.27) and (7.28) may be solved in its general form analytically. Such a solution is time-consuming, as it requires the calculation of the empirical constants, too. Therefore, a graphical solution is more convenient and versatile. An additional advantage of the graphical solution of the system of eqns. (7.27) and (7.28) is that, with the use of diagrams of relationships obtained for the given machine tool, the influence of dynamic action (vibrations) of this machine tool may be taken account of.

The graphical method of determining optimum values of s and v with regard to T_e and the desired R_{tm} will be explained by means of Fig. 7.4.

Diagrams $R_{tm} = F(v, s)$ and $T = f(v, s)$ are connected by the common axis of cutting speed. On $T = f(v)$, curves for the consecutive feed rates $s_1, s_2, \ldots, s_5$, points corresponding to constant values T_1, T_2 and T_3 are found.

Due to the common v axis, points corresponding to tool lives T_1, T_2 and T_3 may be found on $R_{tm} = F(v)$ curves, too, for the particular feed rates. This is carried out by drawing perpendiculars to the v axis until they intersect with the corresponding $R_{tm} = F(v)$ curves.

Points of intersection of $T =$ const. lines with $T = f(v)$ curves, determine on lines parallel to the v axis a series of feed rates fulfilling simultaneously the conditions of the assumed tool life and of the required surface quality.

Joining of points situated on $R_{tm} = F(v)$ curves for the consecutive values of s, and corresponding to the same tool life value, yields curves enabling direct determination of v and s values fulfilling both T_e and R_{tm} conditions.

This is explained by the following example. To obtain roughness number 6 it is necessary to draw a line parallel to the v axis from a point on the R_{tm} scale, designated

* E.g. by applying the so-called *Horner's diagram* (with derivatives) to solve algebraic equations or by applying the graphical method of E. Lille [7/7]

by A. This line intersects the resulting curve for T_3 at point B, that for T_2 at point C, and that for T_1 at point D. Another point of intersection with the curve for T_3 is neglected as yielding lower productivity than that corresponding to point B.

On the assumption that $T_2 = T_e$, a pair of v_c and s_c values is found which complies with the condition of simultaneous obtainment of $R_{tm} = R_{tmo}$ and $T = T_e$.

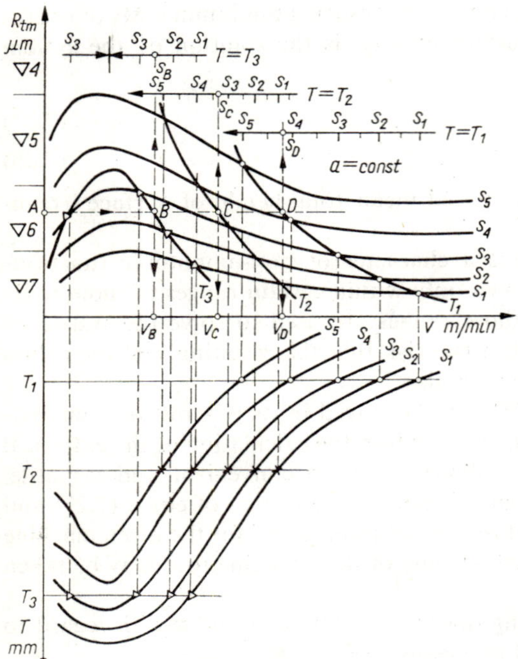

Fig. 7.4 Graphical determination of v_{TR} and s_{TR} values simultaneously fulfilling the conditions of the required mean total height of surface irregularities R_{tmo} and tool life T

Optimum values with regard to the required mean total height R_{tmo} and tool life T will be designated as follows: optimum value of feed rate S_{TR} and optimum value of cutting speed V_{TR}. In the case when the desired tool life value is the economic tool life, the designations s_{eR} and v_{eR} will be adopted, respectively.

In the example given $s_{eR} = s_C$ and $v_{eR} = v_C$. If other values are adopted, for example $T = T_1$, the pair of optimum values is constituted by $s_D = s_{TR}$ and $v_D = v_{TR}$. Similarly, for $T = T_3$, there is $s_B = s_{TR}$ and $v_B = v_{TR}$.

Figure 7.5 represents an example of a nomogram prepared by the method just explained for the selection of optimum feed rate and optimum cutting speed with regard to the necessary tool life and surface quality, in the case of turning carbon steel of σ_B = 75–80 kg/mm^2 with sintered carbide tips.

From this nomogram the $R_{tm} = F(v)$ and $T = f(v)$ curves for the given s values have been removed, leaving only the resulting curves $R_{tm} = F(v)$ for the given T values. The method of using the nomogram is illustrated by the example drawn with dashed lines. The procedure follows the alphabetical order of letters.

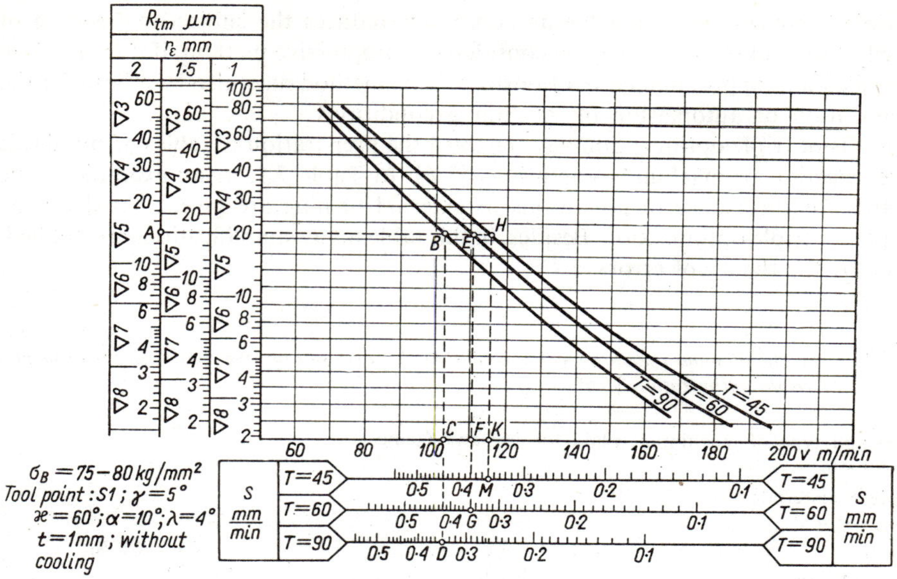

Fig. 7.5 Nomogram for the selection of optimum values of feed rate and cutting speed with regard to the required tool life and surface quality

7.5 Three-criteria optimization of single-tool machining conditions with regard to minimum costs, the required surface quality and the permissible machining error

Optimization of machining conditions with regard to minimum costs and the required surface quality may be practically applied in cases when the changes in depth of cut are slight (e.g. in finishing operations).

In those cases, when it is desirable to find optimum values of a, s and v, triple optimization of machining conditions may be applied, with regard to:

— minimum unit costs,
— required surface quality,
— permissible dimensional error or machining error.

From the mathematical point of view, the task of optimization is reduced to the solution of a system of three equations:

$$T_e = f(a, s, v) \tag{7.29}$$

$$R_{tmo} = F(a, s, v) \tag{7.30}$$

$$B_o = \varphi(a, s, v) \tag{7.31}$$

where B_o—permissible machining error.

In the case when the actual forms of functions (7.29), (7.30) and (7.31) are known, optimum values a_{TRB}, s_{TRB} and v_{TRB} may be found, simultaneously ensuring the results T_e, R_{tmo} and B_o.

The same problem may be solved by a graphical method similar to that given in Section 3.

Triple optimization of machining conditions reduces the subjective freedom of their selection and therefore may be considered a progressive method, if it is based on well studied and approximated relationships. This method may also be a base for the mechanization or automation of machining conditions.

Results of triple optimization have enabled the preparation of tables of standards for the selection of machining conditions [7/6]. In Table 7.1, as an example, some standards for the choice of conditions of external turning are given, based on the principles of triple optimization. Besides, in this table many simplifications are applied, justified by the theory of errors.

Table 7.1 *Example of some standard cutting conditions for external turning, based on the principles of triple optimization* [7/6]

Method and type of machining: plain turning and facing

Workpiece material: steel

Chemical composition	$C \leqq 0{\cdot}6\%$; Cr–Ni	$C > 0{\cdot}6\%$; Cr; Mn; Cr–Mo; Cr–Ni–Mo; Cr–Al; Cr–Al–Mo; Cr–Ni–W	Cr–Mn; Cr–Si; Si–Mn; Cr–Ni–Mn
σ_B kg/mm²	70–80	60–70	50–60

Workpiece clamping

when $l/d \leqq 8$—between centres	when $l/d \leqq 3$—in chuck

Correction coefficients of rotational speed n and motor power N_m with regard to:

Cutting edge angle $\varkappa_r$°	30	45	60	75	90
	1·15	1·0	0·9	0·85	0·8

Condition of steel	Without scale			With scale	
	normalized	softened	toughened	rolled	forged
	1·0	0·9	0·8	0·9	0·8

Roughing	Medium accurate machining	Kind of machining	Accurate machining	Very accurate machining
worse than IT 12	IT 12	Class of accuracy	IT 9	IT 7
worse than ▽3	▽3	Roughness number	▽5	▽7

Table 7.1 *continued*

Characteristics of the recommended tool

50×50 40×63	32×32 25×40	25×25 20×32	Shank cross-section	20×20 16×25	20×20 16×25
S 30	S 20	S 10	Sintered carbide	S 10	S 03
min 8 mm	min 6 mm	min 4 mm	Tip thickness	min 4 mm	min 4 mm
2 mm	1·5 mm	1·5 mm	Nose radius r	1 mm	1 mm
With land	Without land	Without land	Tool point type	Without land	Without land
18°	12°	10°	Rake angle γ	5°	−5°

Average depths of cut and necessary motor power

6	8	10	12	3	4	5	6	1·5	2	2·5	3	Depth of cut	0·75	1	1·25	1·5	0·5	0·75	1
16	22	27	32	10	14	17	20	5	6·5	8·5	10	Motor power	1·5	2	2·5	3	0·9	1·4	1·8

Recommended feed rate and rotational speed and the resulting time of machining on a length of 100 mm

s mm/rev n rev/min t_{m100} min/100 mm

s n t_{m100}	s n t_{m100}	s n t_{m100}	Diameter d mm over–to	s n t_{m100}	s n t_{m100}
		0·355 224 0·125	20–22·4	0·16 2240 0·28	0·112 3150 0·28
		0·355 2000 0·14	22·4–25	0·16 2000 0·315	0·112 2800 0·315
		0·355 1800 0·16	25–28 0·355	0·16 1800	0·112 2500 0·355
		0·355 1600 0·18	28–31·5	0·16 1600 0·4	0·112 2240 0·4
		0·45 1400 0·16	31·5–35·5	0·18 1400 0·4	0·125 2000 0·4

Table 7.1 *continued*

s mm/rev n rev/min t_{m100} min/100 mm

s	n	t_{m100}	s	n	t_{m100}	s	n	t_{m100}	Diameter d mm over-to	s	n	t_{m100}	s	n	t_{m100}
						0·45	1250	0·18	35·5–40	0·18	1250	0·45	0·125	1800	0·45
						0·45	1120	0·2	40–45	0·18	1120	0·5	0·125	1600	0·5
						0·45	1000	0·224	45–50	0·18	100	0·56	0·125	1400	0·56
			0·9	450	0·25	0·5	800	0·25	50–56	0·2	900	0·56	0·125	1250	0·63
			0·9	400	0·28	0·5	710	0·28	56–63	0·2	800	0·63	0·125	1120	0·71
			0·9	355	0·315	0·5	630	0·315	63–71	0·2	710	0·71	1·125	1000	0·8
			0·9	315	0·355	0·5	560	0·355	71–80	0·2	630	0·8	0·125	900	0·9
			1·12	250	0·355	0·5	500	0·4	80–90	0·224	560	0·8	0·125	800	1
			1·12	224	0·4	0·5	450	0·45	90–100	0·224	500	0·9	0·125	710	1·12
1·25	160	0·5	1·12	200	0·45	0·5	400	0·5	100–112	0·224	450	1	0·125	630	1·25
1·25	140	0·56	1·12	180	0·5	0·5	355	0·56	112–125	0·224	400	1·12	0·125	560	1·4
1·4	125	0·56	1·12	160	0·56	0·5	315	0·63	125–140	0·25	355	1·12	0·125	500	1·6
1·4	100	0·63	1·12	140	0·63	0·5	280	0·71	140–160	0·25	315	1·25	0·125	450	1·8
1·4	100	0·71	1·12	125	0·71	0·5	250	0·8	160–180	0·25	280	1·4	0·125	400	3
1·4	90	0·8	1·25	112	0·71	0·5	224	0·9	180–200	0·25	250	1·6	0·125	355	2·24

7.6 The principle of maximum utilization of both machine tool and cutting tool

One of the types of double optimization is the so-called *principle of maximum utilization* of both machine tool and cutting tool. It is a question of optimizing the machining conditions, most often the feed rate and cutting speed, with regard to the full use of the nominal power of the main drive motor and to minimum unit costs.

Practically, this kind of optimization of machining conditions is seldom applied in small-batch and piece production. It is more suitable for large-batch and mass production.

The principle of maximum utilization of machine tool driving power and of fulfilling the conditions of minimum unit cost should be realized in special-purpose machine tools. In such cases the problem is reduced to the choice of a suitable motor for the given machine tool.

Under average machining conditions the stepping of spindle speeds may be mentioned as the first reason for the impossibility of making full use of the machine tool driving power. When the stepping is geometrical and the common ratio is φ, the theoretical underloading will be

$$S = \left(1 - \frac{1}{\varphi}\right) 100\,\% \tag{7.32}$$

which was discussed in Chapter 5.

The second cause of underloading of the machine tool motor may be the limited range of rotational speeds.

The third and most important cause of difficulties of feed rate and cutting speed optimization with regard to N_{mn} and T_e is the relatively very small range of practically applied feed rates and depths of cut in which this aim may be attained.

Taking this into account it must be stated that the principle of maximum utilization of machine tool driving power and of the attainment of the economic tool life may be practically applied only with suitably wide tolerances.

Under actual technical and organizational conditions prevailing in the machine-building industry the tolerance of motor loading may be assumed to be as follows:

— in piece and small-batch production, not repeated periodically, motor loading may vary within the limits $\varepsilon_m = 0{\cdot}4$–$1{\cdot}0$;

— in batch production $\varepsilon_m = 0{\cdot}6$–$1{\cdot}0$;

— in large-batch and mass production $\varepsilon_m = 0{\cdot}8$–$1{\cdot}0$.

When the assumed machining conditions do not ensure motor loading within the ranges listed above, steps should be taken to transfer the production to another machine tool or to change the motor in the given machine tool.

7.7 Metal removal rate

7.7.1 Types of metal removal rate

Similar to the notion of production rate there exists the notion of *metal removal rate*. Distinguished are:

— volumetric rate of metal removal

$$Q_v = 1000 a\, s\, v \quad \text{mm}^3/\text{min} \tag{7.33}$$

when a is in mm, s in mm and v in m/min,

— surface rate of metal removal

$$Q_s = 1000 s\, v \quad \text{mm}^2/\text{min} \tag{7.34}$$

— linear rate of metal removal

$$Q_l = a \quad \text{mm/min} \tag{7.35}$$

In the case when v is replaced by:
— v_N (cutting speed ensuring a determined loading of the motor), the metal removal rate is called *machine tool productivity*,*
— v_T (tool life speed), the metal removal rate is called *tool productivity*,
— v_e (economic cutting speed), the metal removal rate is called *economic tool productivity*,
— v_{pr} (productive cutting speed), the metal removal rate is called *maximum tool productivity*.

Metal removal rate divided by cutting power is called the *specific metal removal rate*:
— volumetric

$$q_v = \frac{Q_v}{N_{me}} \quad \text{mm}^3/\text{kW min} \tag{7.36}$$

— surface

$$q_s = \frac{Q_s}{N_{me}} \quad \text{mm}^2/\text{kW min} \tag{7.37}$$

— linear

$$q_l = \frac{Q_l}{N_{me}} \quad \text{mm/kW min} \tag{7.38}$$

The specific rate of metal removal constitutes a comparative index of effectiveness with which the electric energy is used in the given machining process.

7.7.2 Relationships between machine time, production rate and metal removal rate

Metal removal rate is directly related to production rate.

If the time of operation is defined as the sum of auxiliary times (t_n) and machine time (t_m)

$$t_o = t_n + t_m \tag{7.39}$$

then the production rate will be expressed by the formula

$$W = \frac{1}{t_n + t_m} \tag{7.40}$$

* By analogy to the rate of metal removal by cutting also the rate of metal removal by erosion is determined.

Machine time—for example in turning—may be expressed by the formula

$$t_m = \frac{\pi\, d\, q\, L}{1000 a\, s\, v} = \frac{\pi\, d\, q\, L}{Q\, v} \quad \text{min} \tag{7.41}$$

where q—machining allowance and other designations are as before.

Designating $d\, q\, L = H$ and substituting eqn. (7.41) in eqn. (7.40) yields

$$W = \frac{1}{t_n + H/Q_v} \tag{7.42}$$

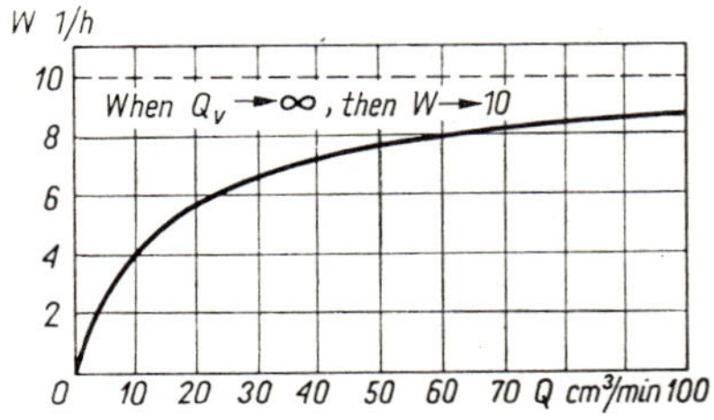

Fig. 7.6 Dependence of production rate on metal removal rate

In Fig. 7.6 the graphical interpretation of eqn. (7.42) for a certain case of machining is presented. It follows from eqn. (7.42) that—as evident in Fig. 7.6—with increasing metal removal rate production rate also rises. This rise is greatest within the range of low metal removal rates and has an asymptotic character (to asymptote $W_{max} = 1/t_n$).

REFERENCES

7/1 Normativy stoikosti rezhushchego instrumenta (Tool Life Standards), Mashgiz, Moskva 1953

7/2 Normatywy czasów obróbki skrawaniem (Time standards of machining by cutting), Vol. I–VII, CONP, Warszawa 1960–1965

7/3 ŚWIGOŃ, S.: 'Określanie naddatków materiałowych i ustalanie normatywów naddatków na obróbkę skrawaniem' (Determination of machining allowances and establishing of cutting allowance standards), Prace IOS, 1965, Seria: Materiały Instruktażowe No. 34

7/4 TEMCHIN, G. I.: 'O metodike proektirovaniya i normirovaniya mnogoinstrumentalnykh stanovochnykh operatsii' (On the methods of planning and standardizing multi-tool operations on machine tools), Avtomobilnaya i Traktornaya Promyshlennost, 10 and 12, 1953

7/5 KACZMAREK, J.: 'Metoda doboru posuwu i szybkości skrawania ze względu na żądaną trwałość ostrza i gładkość powierzchni' (A method of the selection of feed rate and cutting speed with regard to the required tool life and surface quality). Prace IOS, 1962, seria: Zeszyty Naukowe, 13

7/6 BUŁAT, A., WOLAK, B.: 'Uproszczone tablice doboru warunków skrawania'. Cz. I. Toczenie zewnętrzne stali i żeliwa. (Simplified tables for the selection of cutting conditions. Part I. External turning of steel and cast iron), Prace ISO, 1963, seria: Materiały Instruktażowe, No. 4

7/7 CHROMIŃSKI, A.: 'Matematyka stosowana' (Applied Mathematics), Kraków 1946

PART 2

Technology of chip removing processes

CHAPTER 8

Turning

8.1 Kinematic and geometric characteristic of turning

8.1.1 KINEMATICS AND TYPES OF TURNING

8.1.1.1 *Classification of turning types*

Turning is a method of machining by cutting in which the workpiece carries out a rotary motion and the tool a plane linear motion.

Depending on the orientation of the axis of workpiece rotation, *horizontal turning* (shortly called *turning* because of the commonness of its application) and *vertical turning* are distinguished.

Both the outer and inner surfaces can be turned. Turning of outer surfaces is called *external turning*. Turning of inner surfaces is called *internal turning* or *boring.*

Further division of turning into various types is carried out on the basis of the mutual situation of the basic motions.

The basic motions of turning are:

— the primary motion, which is the rotary motion of the workpiece,
— the auxiliary motion, which is the plane motion of the tool, also called the *feed motion.*

The axis of rotation constituting the primary motion is called the *turning axis.*

The following kinematic types of turning are distinguished:

— external (Fig. 8.1a) and internal (Fig. 8.1b) straight turning, when the direction of auxiliary motion is parallel to the turning axis,

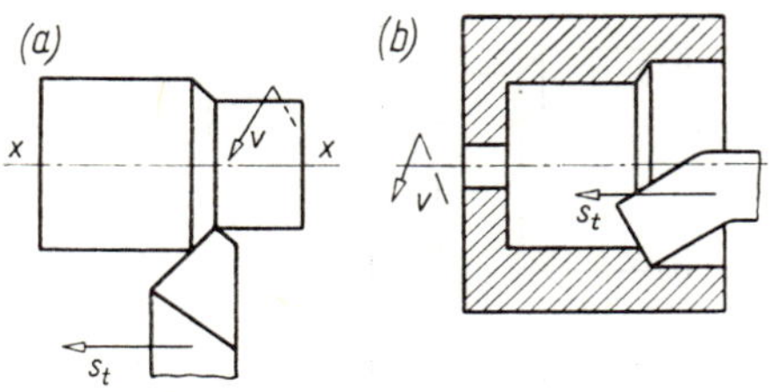

Fig. 8.1 Diagram of straight turning: (a) external, (b) internal

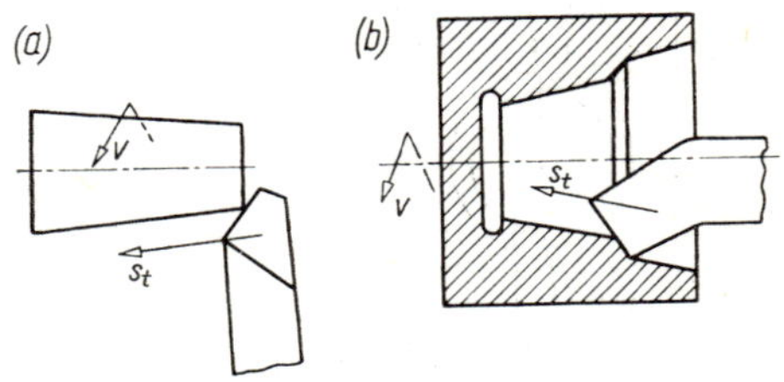

Fig. 8.2 Diagram of taper turning: (a) external, (b) internal

— external (Fig. 8.2a) and internal (Fig. 8.2b) taper turning, when the direction of auxiliary motion intersects with the turning axis,
— external and internal transverse turning when the direction of auxiliary motion is perpendicular to the turning axis. Transverse turning is further divided into facing and radial cutting.

Facing, both external and internal, may be carried out in the direction from the turning axis outwards, and is then called centrifugal facing (Fig. 8.3b), or toward the turning axis, and is then called centripetal facing (Fig. 8.3a).

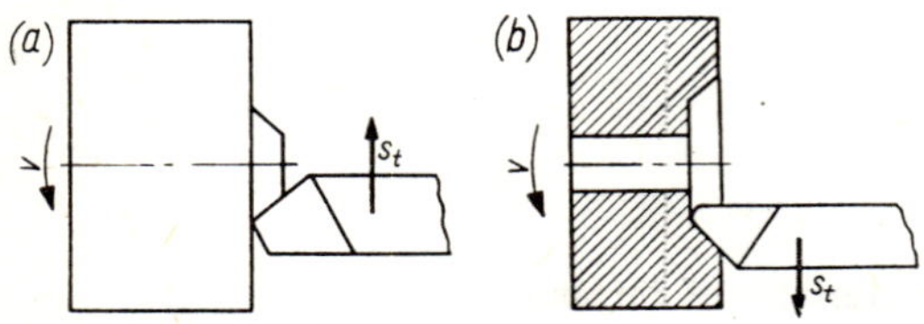

Fig. 8.3 Diagram of external facing: (a) centripetal, (b) centrifugal

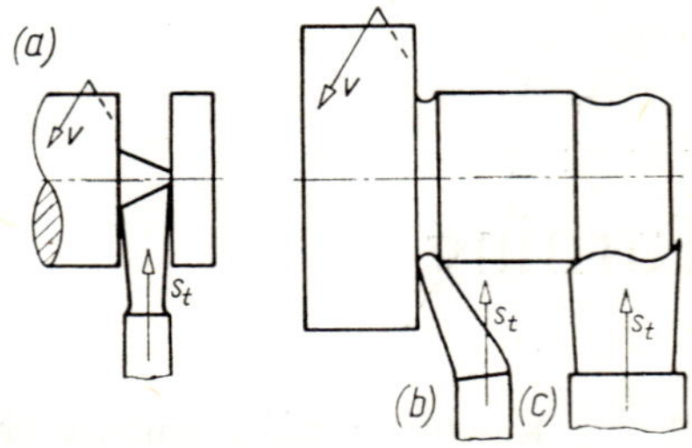

Fig. 8.4 Diagram of external radial turning: (a) parting, (b) undercutting, (c) form turning

Radial turning is divided into: parting (Fig. 8.4a), undercutting (Fig. 8.4b) and form turning (Fig. 8.4c). The latter is characterized by the cutting edge (usually curvilinear) being reproduced in the form of machined surface.

Depending on how many tools participate simultaneously in cutting, single-tool turning and multi-tool turning are distinguished.

Multi-tool turning may be performed

(a) with divided feed (Fig. 8.5a) when feed per tool

$$s_z = \frac{s}{z} \tag{8.1}$$

where z—the number of tools cutting simultaneously.

(b) with divided depth (Fig. 8.5b) when the total depth of cut

$$a = \sum_{i=1}^{i=z} a_i \tag{8.2}$$

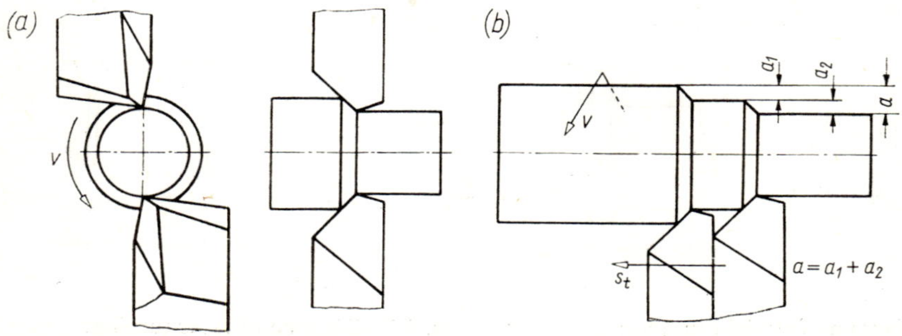

Fig. 8.5 Diagram of multi-tool turning: (a) with divided feed, (b) with divided depth of cut

8.1.1.2 *The speed of basic motions and turning time*

Basic motions of most types of turning are uniform motions. The speed of primary motion, or peripheral speed of the workpiece, is calculated for the workpiece surface being machined as

$$v_p = \pi \frac{d_w n}{1000} \quad \text{m/min} \tag{8.3}$$

where: d_m—diameter of the workpiece surface being machined in mm, n—rotational speed of the workpiece in 1/min.

Cutting speed v is conventionally assumed to be equal to the speed of primary motion, or

$$v = v_m = \pi \frac{d_w n}{1000} \tag{8.4}$$

Speed of auxiliary motion s_a or feed rate is expressed by

$$s_a = s\,n = z\,s_z n \quad \text{mm/min} \tag{8.5}$$

Turning time of all turning types is calculated from

$$t_m = \frac{l_a + l + l_o}{s_a} \quad \text{min} \tag{8.6}$$

where: l_a—length of approach in mm, l_o—length of overrun in mm, l—length of cutting in the direction of auxiliary motion.

Approach and overrun lengths are comparatively small in the case of through turning, while overrun is equal to zero in the opposite case.

If $l \geqq 200$ mm length of both approach and overrun may be neglected and then

$$t_b \approx \frac{l}{s_a} = t_c \tag{8.7}$$

where t_c—cutting time.

Equations (8.6) and (8.7) concern cutting at constant feed rate and constant cutting speed. When cutting is carried out in several passes at different values of s and n, then

$$t_m = \sum_{i=1}^{i=k} t_{mi} \tag{8.8}$$

8.1.2 Characteristics of undeformed chip and tool

8.1.2.1 *Elements of undeformed chip*

In turning, the depth of cut is calculated from

$$a = \frac{d_w - d_m}{2} \quad \text{mm} \tag{8.9}$$

where: d_w—diameter of workpiece surface being machined, d_m—diameter of the machined surface.

The remaining elements of the undeformed chip in the case of turning do not require additional explanations and definitions beside those given in Section 1.5.

The shape of the undeformed chip is characterized in the case of turning by the cutting edge angle and by the ratio of depth of cut to feed rate

$$m = \frac{a}{s} \tag{8.10}$$

In the majority of cases $m > 1$, and most frequently $5 \leqq m \leqq 10$.

Values of feed rates and depths of cut which are recommended to be applied in practice (which depend on the kind of material machined and on accuracy requirements) are given in various standard tables as in [8/1].

For rough machining and forming the feed rates encountered most often are contained within the range 0·2–0·6 mm/rev, and depths of cut within the range 0·5–5 mm.

The cutting edge angles applied most frequently, are those between 45 and 90°.

The higher the machining accuracy requirements and the harder to machine is the workpiece material, the smaller undeformed chip cross-sections and the greater cutting edge angles are generally applied.

8.1.2.2 *Characteristics of cutting tools*

Lathe tools may be monometallic or bimetallic.

Monometallic tools, made entirely of tool point material, are produced in small sizes only.

Bimetallic tools are made by means of permanent or detachable joints.

Permanently joined bimetallic tools may be pressure welded (e.g. the shank of constructional steel and the working part of high-speed steel), brazed (steel shank, sintered carbide tip) or glued (e.g. steel shank, metal oxide tip). Of these joining methods, brazing is the one most commonly used.

Detachably joined bimetallic tools are called built-up tools. With regard to material economy and other advantages, built-up tools are finding increasingly wide application.

The shank of a built-up tool is called the tool holder and the working part the tool tip.

Mechanically mounted tool points are usually in the form of exchangeable plate-shaped tips. If the latter are not regrindable, i.e. when their corners are consecutively used for cutting but are not intended for regrinding, such tool tips are called expendable or indexable.

With regard to the character of their work and to their final use, lathe tools are divided into: roughing tools, standard tools, special tools and boring tools. In Fig. 8.6 examples of the most common applications of some standard tools are presented.

Angles of a lathe tool point are measured in a system involving the holder part or shank. The basic plane of this system is the shank base plane.

Tip angles should be distinguished from point angles appearing when the tip is mounted in the shank. If tip seat angle is φ, then (according to Fig. 8.7)

$$\gamma = \gamma_t + \varphi \tag{8.11}$$

$$\alpha = \alpha_t - \varphi$$

where: γ_t—tip rake angle, α_t—tip clearance angle.

Tip rake angle γ_t is usually equal to zero and then seat angle φ is equal to point rake angle.

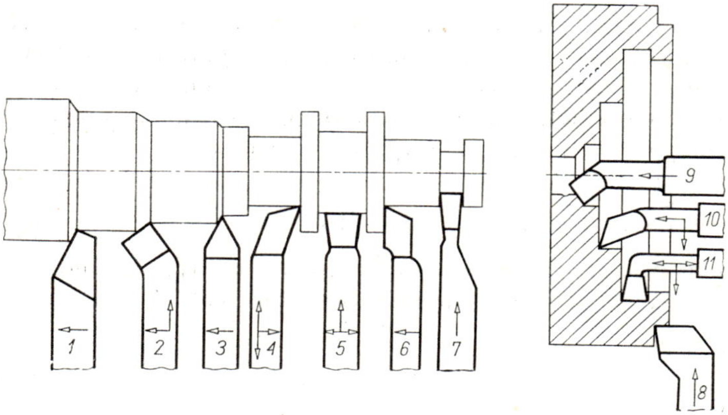

Fig. 8.6 Examples of the application of certain standard tools: 1—straight roughing tool, 2—bent roughing tool, 3—pointed finishing tool, 4—goose-neck tool, 5—square-nosed finishing tool, 6—offset tool, 7—parting tool, 8—facing tool, 9—boring tool for through-holes, 10—boring tool for blind holes, 11—recessing tool

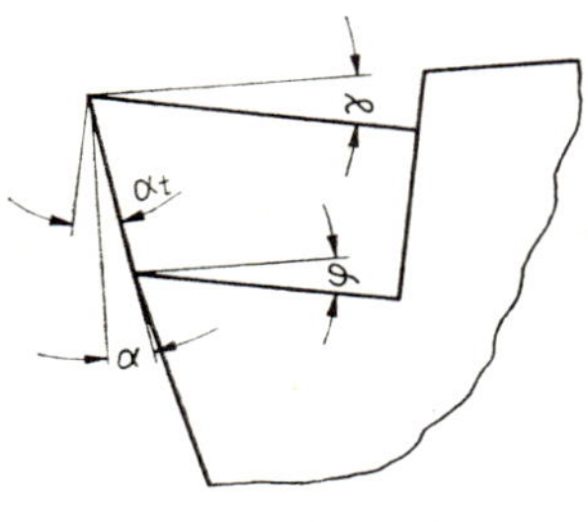

Fig. 8.7 Clearance angle and setting angle of tool tip (point)

Values of point angles depend on:

— the character and kind of work,
— machining accuracy requirements,
— machinability of workpiece material,
— mechanical and cutting properties of the point material.

The rake angles of lathe tools are contained within a wide range from -20 to $+30°$. Industrial experience indicates that conditions occurring most frequently in practice are those under which the application of rake angles of 5–20° is recommended.

Clearance angles of lathe tools are comprised within the range of 6–15°. Most often conditions occur, under which the application of clearance angles of 8–12° is recommended.

The leading cutting edge inclination angle is adopted within a narrower range of values, mostly $\lambda = 0$–5°.

Trail edge inclination angle should usually have a value of 5–10°.

Leading cutting edge angle is most often equal to 45°, 60° or 90°.

Nose radius r_ε varies from 0·5 to about 10 mm as necessary, while the values $r_\varepsilon = 1$–2 mm are most often required.

8.2 Cutting forces and resistances in turning

Physical and technological relationships in the cutting process, generally discussed in Chapters 2, 4 and 6, are mostly based on results of research on turning. Therefore the explanations and conclusions given there directly concern turning. Thus a repetition of this discussion is not necessary and it suffices to complement and stress some special features of turning in the qualitative or quantitative sense.

8.2.1 Calculation of machining resistances in turning

Between the components of cutting resistance in the tool system (F_v, F_p) and in the machine tool system (S, R) there is a relationship (acc. to Fig. 8.8)

$$\begin{aligned} R &= F_p \cos\varepsilon - F_v \sin\varepsilon \\ S &= F_p \sin\varepsilon + F_v \cos\varepsilon \\ Q &= F_f \end{aligned} \tag{8.12}$$

Equation (8.12) indicates that there exists such a value of angle ε, when $R = 0$, i.e. when

$$\frac{F_p}{F_v} = \tan\varepsilon \qquad \text{or} \qquad \varepsilon = \arctan\frac{F_p}{F_v} \tag{8.13}$$

and when $S = 0$, i.e. when

$$\frac{F_p}{F_v} = \cot\varepsilon \qquad \text{or} \qquad \varepsilon = \operatorname{arccot}\frac{F_p}{F_v} \tag{8.14}$$

For example, when $F_p/F_v = W_p = 0{\cdot}5$ then $R = 0$ for $\varepsilon = 26{\cdot}5°$ and $S = 0$ for $\varepsilon = -63{\cdot}5°$.

Equations (8.13) and (8.14) may be used when it is necessary that the vertical or transverse component should be equal or close to zero.

In some cases, for design reasons, $\varepsilon \neq 0$ is adopted, but the majority of machine tools are designed so that $\varepsilon = 0$ which means that when $R = F_p$ then also $S = F_v$.

The statistically established relationship between average (most often encountered) values of the cutting force components in the tool system is

— in turning steel:

$$\begin{aligned} F_p &= (0{\cdot}4\text{–}0{\cdot}6)F_v \\ F_f &= (0{\cdot}2\text{–}0{\cdot}3)F_v \end{aligned} \tag{8.15}$$

— in turning cast iron:

$$\begin{aligned} F_p &= (0{\cdot}3\text{–}0{\cdot}6)F_v \\ F_f &= (0{\cdot}15\text{–}0{\cdot}3)F_v \end{aligned} \tag{8.16}$$

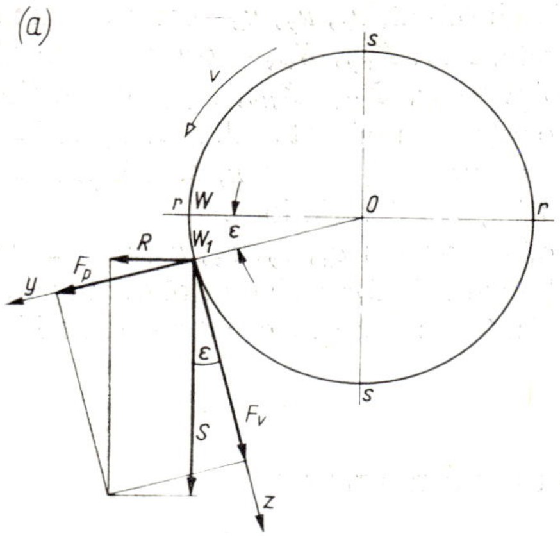

Fig. 8.8 Components of cutting resistance in the tool system (F_v, F_p) and machine tool system (S, R): (a) when $\varepsilon \neq 0$, (b) when $\varepsilon = 0$

It follows from eqns. (8.15) and (8.16) that there is no need to determine all three components accurately, it suffices to know the component F_v. Then the remaining two components can be determined approximately.

In the statistical and empirical formulae for cutting force components only the depth of cut and feed rate appear separately. The influence of the numerous other factors on cutting forces is represented in these formulae by the particular correction coefficients or contained in the so-called *constant of proportionality*.

The general form of formulae for cutting force components is as follows

$$F_v = C_z\, a^{e_z}\, s^{u_z}\, K_{Nz}\, K_{Hz}\, K_{kz}\, K_{vz}$$
$$F_p = C_y\, a^{e_y}\, s^{u_y}\, K_{Ny}\, K_{Hy}\, K_{ky}\, K_{vy} \qquad (8.17)$$
$$F_f = C_x\, a^{e_x}\, s^{u_x}\, K_{Nx}\, K_{kx}\, K_{Hx}\, K_{vx}$$

where: C_z, C_y and C_x—constants taking into account the influence of all factors and cutting conditions not appearing separately in the formulae, e_z, e_y, e_x—constant exponents representing the influence of the depth of cut, u_z, u_y, u_x—constant exponents representing the influence of the feed rate, K_{wz}, K_{wy}, K_{wx}—correction coefficients taking into account the degree of point wear, K_{Hz}, K_{Hy}, K_{Hx}—correction coefficients taking into account the hardness or strength of material, K_{kz}, K_{ky}, K_{kx}—correction coefficients taking into account the cutting edge angle $\varkappa$, K_{vz}, K_{vy}, K_{vx}—correction coefficients taking into account cutting speed variations.

Values of constants C_z, C_y and C_x and of exponents e_z, e_y, e_x, u_z, u_y, u_x are given in Table 8.1. They have been determined separately for each type of dry turning, for high-speed steel and sintered carbide tools. Tests were repeated for each kind of material of a particular hardness.

8.2.2 Influence of various factors on cutting resistance

The following factors have an influence on cutting resistance:
(a) factors connected with the cutting point,
(b) factors connected with the workpiece material,
(c) factors connected with the cutting conditions.
These are external factors whose influence on cutting resistance is of an indirect character; they act on basic physical parameters such as cutting heat and temperature, external and internal friction and the magnitude, direction and speed of any resulting deformation.

8.2.2.1 *Factors connected with the cutting point*

The following factors connected with the cutting point, influence the cutting resistance: tool point material, rake angle, face shape, clearance angle, nose radius, and cutting edge shape, dulling (wear) and edge roundness.

In the light of research conducted so far, it can be assumed that the influence of tool point material on cutting resistance is caused by:
(a) various forces of adhesion and cohesion between particles of material closely adhering to each other,
(b) various shapes and sizes of crystals mechanically influencing friction (by protruding parts),
(c) resistance of the surface layer to pressure; this feature is decisive for the actual contact area between chip and tool face.
In Fig. 8.9 diagrams based on Gritsaenko's research [8/2] are given. This research concerned the turning of 0·45 carbon steel by means of tools with tips of sintered carbides T60K6, T30K4, T5K10, WK8 and WK2.* Tool point parameters were as follows: $\gamma = -3°$, $\alpha = \alpha_1 = 13°$, $\lambda_s = 0°$, $r_\beta = 0{\cdot}8$ mm, $v = 200$ m/min, $a = 2$ mm and $s = 0{\cdot}2$ mm/rev. It is evident from the diagrams that the differences in component values vary within limits of 5–40% depending on the type of sintered carbide.

* These are symbols used in Soviet carbide production: numbers following letter T denote the percentage of titanium carbide, those following letter K the percentage of cobalt, the remaining component in all cases is tungsten (W) carbide.

Table 8.1 *Formulae for calculating the cutting resistance in turning* ($\varkappa_1 > 0$; $r = 2$ mm)

Tool point material	Turning type	Workpiece material	σ_B kg/mm²	HB kg/mm²	Depth of cut and feed rate ranges at $a/s > 1$	$F_v = C_z a^{e_z} s^{u_z}$	$F_p = C_y a^{e} y s^{u_y}$	$F_f = C_x a^{e_s} s^{u_x}$
High speed steel SW9 (see Table 4.5)	external and internal straight and transverse turning	carbon steel	75	215	$a = 1–1·2$ $s = 0·1–2·0$	$200a\,s^{0·75}$	$124·8a^{0·9}s^{0·75}$	$66·8a^{1·2}s^{0·65}$
		grey cast iron		190	$a = 1–1·5$ $s = 0·1–3·0$	$114a\,s^{0·75}$	$119·2a^{0·5}s^{0·75}$	$51·4a^{1·2}s^{0·65}$
		nodular cast iron		190	$a = 1–8·0$ $s = 0·1–2·0$	$153a^{1·03}s^{0·94}$		
		bronze, medium hard		120	$a = 1–8$ $s = 0·1–1·4$	$55a\,s^{0·65}$		
		malleable cast iron		150	$a = 1–8$ $s = 0·1–1·4$	$100a\,s^{0·75}$	$87·6a^{0·9}s^{0·75}$	$39·6a^{1·2}s^{0·65}$
	parting	carbon steel	75	215	$s = 0·04–3·5$	$247a\,s$		
		grey cast iron		190	$s = 0·04–3·5$	$158a\,s$		
		malleable cast iron		150	$s = 0·04–3·5$	$138·7a\,s$		
Sintered carbides	external and internal straight and transverse turning	carbon steel	75	215	$a = 1–1·2$ $s = 0·1–2·0$	$191a\,s^{0·75}$		
		hardened steel	150		$a = 1–2$ $s = 0·1–1·5$	$300a^{0·81}s^{0·81}$		
		grey cast iron		190	$a = 1–1·5$ $s = 0·1–3$	$92a\,s^{0·75}$		
		nodular cast iron		190	$a = 1–8$ $s = 0·1–2·0$	$145a^{0·92}s^{0·82}$		
	parting	carbon steel			$s = 0·04–3·5$	$191a\,s^{0·75}$		
		grey cast iron			$s = 0·04–3·5$	$92a^{1·1}s^{0·9}$		

The rake angle exerts a great influence on the cutting process and on the cutting force. As the value of rake angle is increased, the tangential cutting force decreases. Experiments indicate that the influence of rake angle is greater the softer and more ductile is the workpiece material. In the cutting of cast iron, bronze and other brittle metals the influence of angle γ is reduced. The same effect is achieved by an increase

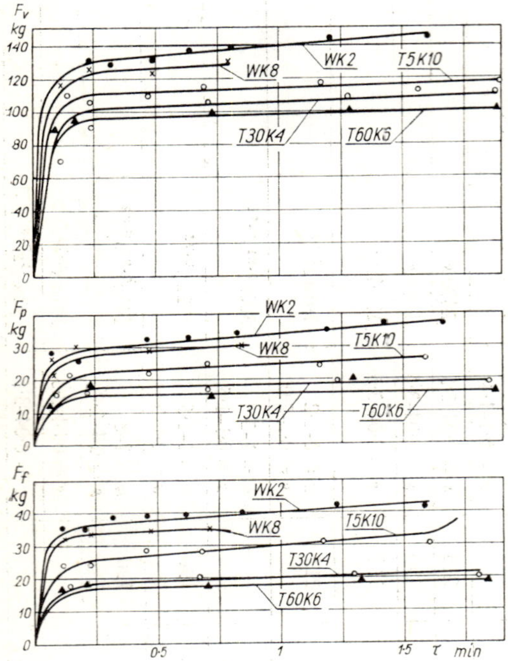

Fig. 8.9 Influence of the type of tool point material on the cutting resistance of a 0·45 carbon steel [8/2]

of cutting speed. Therefore, cutting resistance in the case of cutting at high speeds ($V > 300$ m/min) with negative rake angles increases little compared to cutting with positive angles.

A still stronger influence than on force F_v is exerted by rake angle on the components F_p and F_f. The percentage of their decrease is larger than that of the component F_v.

The influence of rake angle on cutting resistance can be explained by the fact that with increasing rake angle the magnitude of plastic deformation is reduced and so is the external friction on the tool face.

No separate correction coefficient is introduced into the calculations of cutting resistance according to eqn. (8.17) since it is assumed that to each kind of material being machined a tool point is applied with optimum rake angle with respect to tool life.

Rake angle variations within the limits of $\pm 3°$ in relation to the optimum value cause cutting resistance changes of $\pm 4\%$. Therefore, the influence of rake angle is taken account of in the values of coefficients C_z, C_y and C_x.

The land has no practical significance for the value of cutting resistance if the optimum rake angle is applied and the land width does not exceed 0·7 s.

The influence on cutting resistance of the clearance angle—within the narrow range of practically applied angle values—is very slight. As this angle decreases, the friction force rises and thus also the cutting resistance. However, this increase is quantitatively insignificant because the share of friction on the flank in the resultant cutting force under average turning conditions is small.

A similarly slight influence is also exerted upon cutting resistance by the auxiliary clearance angle α_1. For this reason in practical calculations the influence of clearance angles is neglected.

The cutting edge inclination angle λ_s exerts a considerable influence only at high values of λ_s. Considering that, in practice, the range of most frequently applied λ_s values is very narrow, from $-5°$ to $+5°$, the influence of this angle on the component F_v may be neglected. On the other hand, the influence of angle λ_s on the components F_p and F_f is more marked. It may be assumed approximately that for cutting edge angle $\varkappa = 60°$ an increase of angle λ_s by 1° causes the force F_p to increase by 2% and the force F_f to decrease also by 2%.

With regard to the influence of nose radius r_β, experiments show that machining with a curvilinear cutting edge results in greater cutting resistance, created by the relatively greater length of the edge. This causes the plastic deformation work to increase. Besides, nose roundness brings about a reduction in mean thickness of the undeformed chip and this also contributes to the increase of cutting resistance. This relationship is more evident the larger the part of undeformed chip surface is constituted by its curvilinear portion (i.e. at small depths of cut).

The greatest influence is exerted by r_β on the component F_p and it is practically non-existant as far as the component F_f is concerned. Important conclusions follow from this circumstance for the technology of machining by cutting. In cases when it is vital to reduce the component F_p (vibrations, machining accuracy) the use of points with small nose radii should be aimed at.

Correction coefficients with regard to nose radius values are given in Table 8.2.*

The influence of cutting edge shape is particularly evident in form turning. The more complicated is the shape of the cutting edge, the greater is the cutting resistance. Changes in the shape of the cutting edges of form tools may cause cutting resistance variations of the order of 20–30% as compared with ordinary tools.

Tool wear consists of a flank wearland (on a certain length of the flank the angle $\alpha = 0°$) and in the appearance of a characteristic crater on the face. In the cutting of ductile materials both these processes occur more or less simultaneously, but in the cutting of brittle materials, such as e.g. grey cast iron, the crater does not appear at all or else is very slow to do so. The appearance of the crater is accompanied by an increase of the actual rake angle and consequently a reduction of the cutting resistance should follow. On the other hand, the wearland on the flank causes an increase in friction by increasing the contact area and thus should cause the cutting resistance to increase. Experiments confirm this explanation. In the turning of steel, at first, when flank wear grows to the value $VB = 0{\cdot}2$–$0{\cdot}4$ mm, all components of cutting resistance diminish. The reason is that in this first phase of wear the influence of the crater on the rake face prevails. With further wear, the influence of growing

* In Table 8.2 and in the following tables are given the correction coefficients (factors) K for the calculation of cutting resistance, B for tool life speed and C for cutting power.

Table 8.2 *Values of correction coefficients K_{zr}, B_r and C_r, taking into account the nose radius (for $\varkappa_1 > 0$)*

Tool point material	Workpiece material	Coefficients	Type of machining	r_β mm: 1	2	3	5
				K_{zr}, B_r and C_r			
High-speed steel and sintered carbides	steel	K_{zr}	rough	0·93	1·0	1·04	1·10
			medium accurate, accurate	0·93	1·0	1·04	1·10
		B_r	rough	0·94	1·0	1·03	—
			medium accurate, accurate	0·86	1·0	1·09	1·21
		C_r	rough	0·87	1·0	1·07	—
			medium accurate, accurate	0·80	1·0	1·13	1·33
	cast iron	K_{zr}	rough	0·95	1·0	1·03	1·07
			medium accurate, accurate	0·95	1·0	1·03	1·07
		B_r	rough	0·9	1·0	1·06	—
			medium accurate, accurate	0·84	1·0	1·03	1·07
		C_r	rough	0·85	1·0	1·09	—
			medium accurate, accurate	0·8	1·0	1·06	1·15

friction on the flank becomes increasingly significant. In the turning of cast iron a minimum of cutting resistance is not observed and the growth of the latter follows from the very beginning of tool point wear.

Values of correction factors taking into account the degree of tool point wear are given in Table 8.3.

8.2.2.2 *Influence of workpiece material*

Important properties of the workpiece material are, first of all, its chemical composition and structure. These features are the result of material preparation, i.e. the production engineering applied to obtain the raw material and the semi-finished product.

The features now discussed have a fundamental influence on the strength and mechanical properties of the material, including hardness and ultimate tensile strength.

It became customary to connect the characteristic of material, as regards its cutting resistance, to its hardness or UTS. However, more accurate research indicates that there is no unequivocal, generalized connection between the cutting force and those two properties established under conditions of normal temperature and static deformations.

An important factor, influencing the cutting resistance to a high degree, is the material structure. Experiments show that the highest cutting resistance occurs in cutting steel of sorbitic structure and of spheroidal carbide. In the case of lamellar pearlite the cutting resistance is lower.

Table 8.3 *Values of correction coefficients K_w, B_w and C_w, taking into account the degree of tool point wear, measured by flank wearland* (for $\varkappa_1 > 0$)

Tool point material	Workpiece material	Coeffi-cients	Tool types	*VB* mm 0·5	1·0	1·5	2·0	2·5	3·0	4·0
High-speed steel	non-hardened steel	K_{zw}	straight and transverse turning tools	0·93	0·95	—	1·0	—	—	—
			parting tools	0·98	1·0	—	—	—	—	—
		K_{yw}	all types of tools	0·5	0·63	—	1·0	—	—	—
		K_{xw}		0·55	0·67	—	1·0	—	—	—
		B_w	straight and transverse turning tools	0·93	0·95	—	1·0	—	—	—
			parting tools	0·85	1·0	—	—	—	—	—
		C_w	straight and transverse turning tools	0·86	0·9	—	1·0	—	—	—
			parting tools	0·83	1·0	—	—	—	—	—
	cast iron	K_{zw}	straight and transverse turning tools	—	0·82	—	0·83	—	—	1·0
			parting tools	—	0·99	—	1·0	—	—	—
		K_{yw}	all types of tools	0·34	0·38	—	0·47	—	—	1·0
		K_{xw}		0·41	0·52	—	0·69	—	—	1·0
		B_w	straight and transverse turning tools	—	—	—	0·93	—	—	1·0
			parting tools	—	0·9	—	1·0	—	—	—
		C_w	straight and transverse turning tools	—	—	—	0·77	—	—	1·0
			parting tools	—	0·89	—	1·0	—	—	—
Sintered carbides	non-hardened steel	K_{zw}	all types of tools	—	1·0	—	1·05	—	—	—
		B_w		—	1·0	—	1·33	—	—	—
		C_w		—	1·0	—	1·4	—	—	—
	cast iron	K_{zw}	all types of tools	—	1·0	—	1·05	—	—	—
		B_w		—	1·0	—	1·33	—	—	—
		C_w		—	1·0	—	1·4	—	—	—

It follows that UTS and hardness may constitute accurate indices of cutting resistance of the given material only within a single type of structure.

Differences in cutting resistance in relation to the strength of material at a normal temperature are also influenced by the fact that the material shows different strength properties in the zone of plastic deformation occurring in the cutting operation.

One of the principal reasons for this is the cutting temperature. Tests of UTS and of the yield point at high temperatures allow one to have an indication of the changes of these properties in the cutting temperature field. Examples of the results of such tests are shown in Fig. 8.10.

As far as carbon steels and cast iron of lamellar pearlitic structure are concerned, quantitative relationships between the cutting force on the one hand and the hardness and UTS on the other hand have been established (Fig. 8.11).

The magnitude of the influence of material structure changes has not yet been accurately determined. However, this influence is partially taken account of in cutting force calculations in the form of correction coefficients depending on the conditions of heat treatment and the method of preparation of the semi-finished product.

Values of correction coefficients with regard to hardness or UTS of the material are listed in Table 8.4.

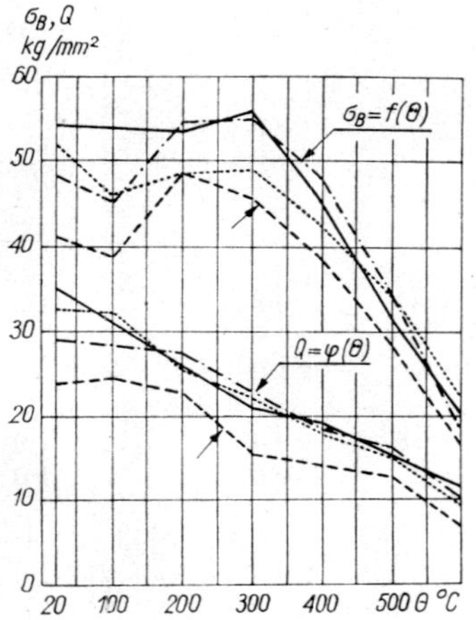

Fig. 8.10 Influence of temperature on the strength properties of some steels

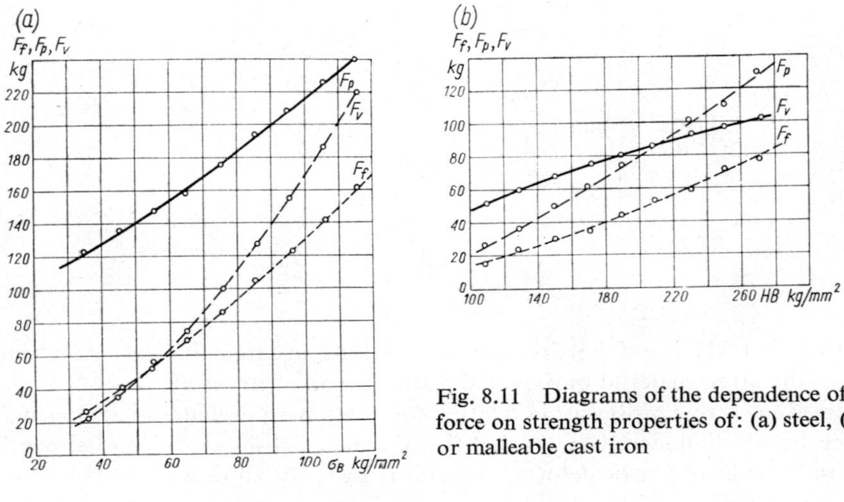

Fig. 8.11 Diagrams of the dependence of cutting force on strength properties of: (a) steel, (b) grey or malleable cast iron

8.2.2.3 *Influence of cutting conditions*

The following cutting conditions influence the cutting resistance: dimensions and shape of undeformed chip, cutting speed, cutting edge angle and the method of tool point cooling.

The dimensions and shape of undeformed chip (if its curvilinear part is neglected) may be characterized by the depth of cut and feed rate, width and thickness of the undeformed chip or its nominal surface area A and the *shape coefficient* $m = a/s$.

In practice, the influence of the undeformed chip on cutting resistance is characterized by the influence of the depth of cut and feed rate, as the primary parameters.

Research results indicate that the influence of the depth of cut is greater than that of feed rate. Thus, for example, the increase of cutting force when turning steel and cast iron is proportional to the increase of the depth of cut. On the other hand, the increase of cutting force is less than proportional to the increase of feed rate. This may be explained by the fact that stresses in the undeformed chip are not uniformly distributed in the direction of its thickness. The further from the cutting edge, the smaller become these stresses and thus also the deformations. Since feed rate increase directly causes the thickening of undeformed chip, the increase in cutting force is less than proportional. On the other hand, increase in the depth of cut only widens the undeformed chip while its thickness and conditions of deformation remain constant. Thus an almost proportional influence of this factor occurs in the majority of cases.

At very small cross-section areas of undeformed chip, the relationships described above do not necessarily hold because of the increased percentage share of the curvilinear portion of the undeformed chip and because of the influence of the rounding radius φ of the cutting edge itself.

The latter influence is the result of the actual rake angle γ_k acquiring negative values which causes an increase in deformation and cutting resistance. When the value of h_1 (Fig. 8.12) is relatively high as compared with the undeformed chip height h, then the influence of the curvature becomes important.

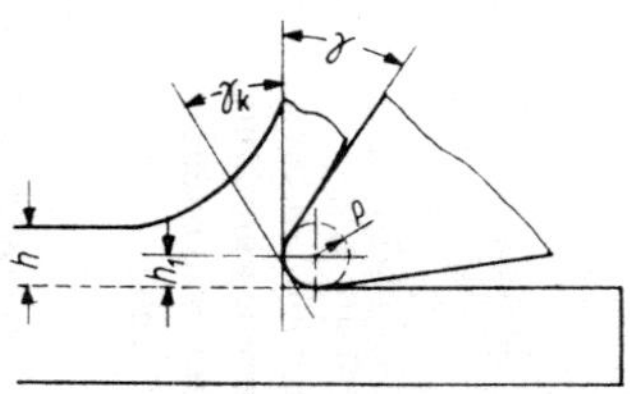

Fig. 8.12 Influence of cutting edge roundness on cutting resistance

The relationships described above between the cutting force and the depth of cut and feed rate occur when the condition $a \geqq s$ is fulfilled. In the case when $a < s$ and the shape of the undeformed chip undergoes a basic change, the influences of the depth of cut and feed rate also change.

Cutting edge angle $\varkappa$ has no direct influence on the value of the cutting force. If the geometric parameters of the tool point remain constant, and the cutting edge angle is changed by suitable grinding* and simultaneously the feed rate is varied so that the undeformed chip thickness would stay unchanged, then for various values

* Grinding of the rake face and flank face.

Table 8.4 *Values of correction coefficients K_H, B_H and C_H taking into account the strength or hardness of the workpiece material* (for $\varkappa_1 > 0$)

Tool point material	Workpiece material	σ_B kg/mm²	40–50	50–60	60–70	70–80	80–90	90–100	100–110	110–120
High-speed steels	carbon steels for constructions and tools, chromium steels	K_{zH}	0·76	0·82	0·89	1·0	1·1	1·18	1·28	1·36
		K_{yH}	0·36	0·54	0·76	1·0	1·29	1·57	1·89	2·24
		K_{xH}	0·47	0·63	0·81	1·0	1·22	1·4	1·61	1·84
		B_H	2·2	1·69	1·29	1·0	0·81	0·66	0·55	—
		C_H	1·67	1·39	1·15	1·0	0·89	0·78	0·7	—
	alloy steels: manganese, nickel, chromium-nickel, chromium-manganese, chromium-silicon, silico-manganese, chromium-silico-manganese, chromium-nickel-tungstene etc.	K_{zH}	0·76	0·82	0·89	1·0	1·1	1·8	1·28	1·36
		K_{yH}	0·36	0·54	0·76	1·0	1·29	1·57	1·89	2·24
		K_{xH}	0·47	0·63	0·81	1·0	1·22	1·4	1·61	1·84
		B_H	2·1	1·56	1·2	1·0	0·8	0·7	0·6	0·5
		C_H	1·6	1·28	1·07	1·0	0·88	0·83	0·77	0·68
	alloy steels: chromium-molybdenum, chromium-nickel-molybdenum, chrome-aluminium, chrome-aluminium-molybdenum, high-speed (tool steels)	K_{zH}	0·76	0·82	0·89	1·0	1·1	1·18	1·28	1·36
		K_{yH}	0·36	0·54	0·76	1·0	1·29	1·57	1·89	2.24
		K_{xH}	0·47	0·63	0·81	1·0	1·22	1·4	1·61	1·84
		B_H	—	—	1·18	1·0	0·85	0·73	0·65	0·55
		C_H	—	—	1·05	1·0	0·93	0·86	0·83	0·75

		HB kg/mm²	100–120	120–140	140–160	160–180	180–200	200–220	220–240	240–260
High-speed steels	grey malleable and nodular cast iron	K_{zH}	0·74	0·81	0·88	0·94	1·0	1·06	1·12	1·18
		K_{yH}	0·49	0·61	0·74	0·86	1·0	1·13	1·28	1·4
		K_{xH}	0·55	0·65	0·77	0·88	1·0	1·12	1·24	1·36
		B_H	2·67	1·94	1·52	1·21	1·0	0·85	0·72	0·64
		C_H	1·98	1·57	1·34	1·14	1·0	0·9	0·78	0·76
		σ_B kg/mm²	40–50	50–60	60–70	70–80	80–90	90–100	100–110	110–120
Sintered carbides	non-hardened steel	K_{zH}	0·76	0·82	0·89	1·0	1·1	1·18	1·28	1·36
		K_{yH}	0·36	0·54	0·76	1·0	1·29	1·57	1·89	2·24
		K_{xH}	0·47	0·63	0·81	1·0	1·22	1·3	1·61	1·84
		B_H	2·16	1·6	1·25	1·0	0·84	0·73	0·62	—
		C_H	1·64	1·31	1·11	1·0	0·92	0·86	0·79	—
		σ_B kg/mm²	110–120	120–130	130–140	140–150	150–160	160–170	170–180	180–190
	hardened steel	K_{zH}	0·84	0·90	0·95	1·0	1·05	1·09	1·15	1·19
		B_H	1·67	1·32	1·14	1·0	0·86	0·76	0·66	0·61
		C_H	1·32	1·19	1·08	1·0	0·9	0·76	0·76	073
		HB kg/mm²	100–120	120–140	140–160	160–180	180–200	200–220	220–240	240–260
	grey, malleable and nodular cast iron	K_{zH}	0·74	0·81	0·88	0·94	1·0	1·06	1·12	1·18
		K_{yH}	0·49	0·61	0·74	0·86	1·0	1·13	1·28	1·4
		K_{xH}	0·55	0·65	0·77	0·88	1·0	1·12	1·24	1·36
		B_H	2·67	1·94	1·52	1·21	1·0	0·85	0·7	0·64
		C_H	1·98	1·57	1·34	1·14	1·0	0·9	0·78	0·77

of this angle ($\varkappa = 30$–$90°$) the forces will be almost identical. Force increase depends in this case only on the variations in undeformed chip thickness.

However, cutting edge angle may also be altered by rotation of the tool in its holder or together with the holder. In such a case—for tool points with $\gamma > 0$ and $\lambda_s > 0$—with increasing $\varkappa$ the angle λ_s remains constant, rake angle γ increases and so does the undeformed chip thickness h. The increase of these parameters causes a reduction of the cutting force. On the other hand, the increase of $\varkappa$ by tool rotation reduces the trail edge angle $\varkappa'$ and raises the percentage share of the curvilinear portion of undeformed chip. Both these changes result in an increase in the cutting force.

Under such conditions a dependence of the cutting force on angle $\varkappa$ may be observed, as represented graphically in Fig. 8.13. The F_v curve plotted against $\varkappa$ has its minimum at the value $\varkappa = 60°$. This can be explained by the fact that within the range $\varkappa = 30$–$60°$ the influence of the increase of rake angle and of undeformed

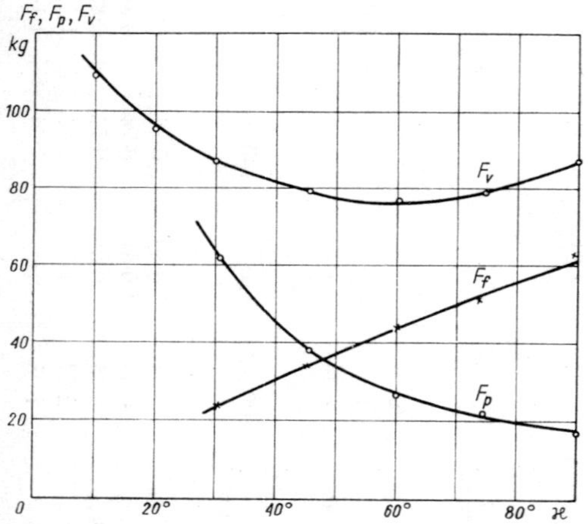

Fig. 8.13 Influence of cutting edge angle $\varkappa$ on the variation of cutting force components [I/11]

chip thickness prevails and therefore the cutting force diminishes. However, for $\varkappa > 60°$ the negative influence of rising percentage share of the curvilinear portion of undeformed chip and of trail edge angle $\varkappa$ already gains the upper hand. This increases the friction energy and makes the conditions of chip flow more difficult thus causing greater deformation and also an increase in cutting resistance.

A similar dependence of F_v on $\varkappa$ is obtained when the change of $\varkappa$ is effected by grinding the flank face only. In such a case, the change of $\varkappa$ is accompanied by variation of leading cutting edge angle λ_s while angle $\varkappa'$ remains constant.

Since in workshop practice the alteration of angle $\varkappa$ is most frequently carried out by grinding the flank face or by proper positioning of the tool, the relationship represented in Fig. 8.13 and showing a minimum is considered to be typical.

The influence of cutting edge angle $\varkappa$ on the components F_p and F_f is greater than in the case of F_v. With increasing angle $\varkappa$ the component F_p diminishes and component F_f increases (Fig. 8.13).

In the calculations of cutting resistance the influence of cutting edge angle is taken account of by the use of correction coefficients given in Table 8.5.

Table 8.5 *Values of correction coefficients $K_{\varkappa}$, $B_{\varkappa}$ and $C_{\varkappa}$ taking into account the influence of cutting edge angle $\varkappa$* (for $\varkappa_1 > 0$)

Tool point material	Workpiece material	Application range	Coefficient	Angle $\varkappa$				
				30°	45°	60°	75°	90°
High-speed steel	non-hardened steel		K_z	1·08	1·0	0·98	1·0	1·08
			K_y	1·63	1·0	0·71	0·54	0·44
		$s \geqq 0{\cdot}25$	K_x	0·7	1·0	1·27	1·51	1·82
			B	1·26	1·0	1·84	0·76	0·66
			C	1·36	1·0	0·82	0·76	0·71
	cast iron		K_z	1·05	1·0	0·96	0·94	0·92
			K_y	1·23	1·0	0·87	0·77	0·7
		$s \geqq 0{\cdot}25$	K_x	0·63	1·0	1·11	1·20	1·28
			B	1·21	1·0	0·88	0·83	0·73
			C	1·26	1·0	0·85	0·78	0·87
Sintered carbides	non-hardened and hardened steel		K_z	1·08	1·0	0·98	1·0	1·08
			K_y	1·63	1·0	0·71	0·54	0·44
		$s \geqq 0{\cdot}2$	K_x	0·7	1·0	1·27	1·51	1·82
			B	1·13	1·0	0·92	0·86	0·81
			C	1·22	1·0	0·9	0·86	0·88
	cast iron		K_z	1·05	1·0	0·96	0·94	0·92
			K_y	1·23	1·0	0·87	0·77	0·7
		$s \geqq 0{\cdot}2$	K_x	0·63	1·0	1·11	1·20	1·28
			B	1·2	1·0	0·88	0·83	0·72
			C	1·26	1·0	0·85	0·78	0·67

Cutting speed influences the cutting processes in two ways. On the one hand—directly, on the speed of deformation, on the other hand—indirectly, influencing the cutting temperature.

Within the range of cutting speeds practically applied the indirect (thermal) influence prevails.

Therefore the relationship between cutting resistance of steel and the cutting speed may be explained, according to Rozenberg and Eremin, by the typical curves shown in Fig. 8.14.

Within the range of cutting speeds v_1–v_2 occurs a reduction of cutting resistance because of the increase of actual rake angle (as a result of built-up edge) and consequently the magnitude of deformation (chip area compression) is reduced. Within the range v_2–v_3 the cutting resistance rises as a result of the disappearance of BUE and of the reduction of actual rake angle to the value of the tool point rake angle. Further increases in cutting resistance in the speed range v_3–v_4 and the corresponding temperature range Θ_3–Θ_4 are caused by a rise in the coefficient of external friction attaining its maximum at the temperature Θ_4. Beginning with cutting speed v_4, the cutting resistance decreases because of the diminishing coefficient of friction under the influence of rising temperature. Cutting speed v_5 is critical and above it the cutting resistance becomes stabilized. The cause of it is a thin liquid film of chip flowing along the face and having a friction resistance almost independent of the temperature. This phenomenon results in a smooth, mirror-like surface of chip flow along the face.

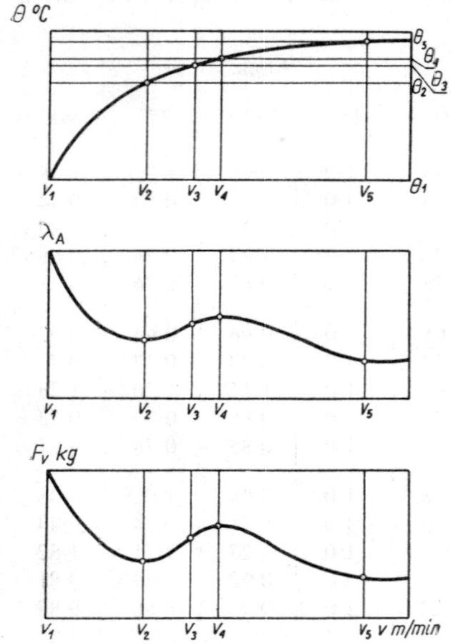

Fig. 8.14 Dependence of temperature θ, chip area compression λ_A and the cutting force F_v on cutting speed (acc. to Eremin [2/11])

These relationships have been confirmed by numerous studies.

Actually applied cutting speeds using sintered carbide tools usually exceed 50 m/min. Since speeds of this order are in most cases higher than v_4 which corresponds to $F_{v\,\max}$, it should be remembered in practice that cutting resistance decreases with increasing cutting speed.

Correction coefficients for the calculation of cutting resistance taking into account the cutting speed are listed in Table 8.6.

Table 8.6 *Values of cutting force correction coefficients taking into account the influence of cutting speed* (for $\varkappa_1 > 0$)

Workpiece material	Cutting speed v m/min	50–75	75–100	100–125	125–150	150–200	200–250	250–300	300–400	400–500
Non-hardened steel	K_{zv}	1·00	0·98	0·96	0·94	0·91	0·87	0·85	0·82	0·80
Grey cast iron	K_{zv}	1·00	0·99	0·98	0·97	0·96	0·93	0·91	0·89	0·86

Application of a coolant causes a reduction of cutting resistance as a result of decreased external friction and of a specific 'wedging' action, sometimes called the *coolant performance.*

Coolant action reducing the cutting resistance is stronger, the thinner is the undeformed chip and the lower is the cutting speed. As the latter rises, it finally reaches for every undeformed chip cross-section and for every material such a limit value beyond which the coolant influence almost vanishes. The illustration of this phenomenon is constituted by the diagram originating from research by W. A. Arshinov [8/3] and shown in Fig. 8.15. At a cutting speed of 3·7 m/min the reduction of cutting force oscillates within limits of 15–40% depending on the undeformed chip thickness. Increase of cutting speed to 60–90 m/min causes—as evident from the diagram—almost complete disappearance of coolant influence; the reduction of cutting resistance amounts to 2–6%.

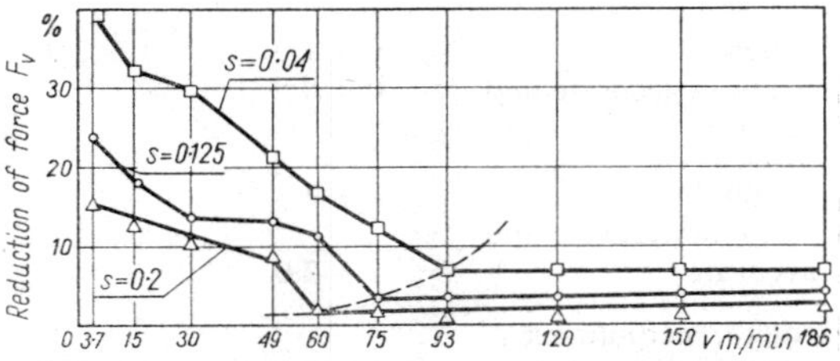

Fig. 8.15 Reduction of the cutting force in relation to such force appearing in dry cutting by the application of a coolant with surface-active agent (acc. to Arshinov [8/3])

8.2.3 Cutting power in turning

Turning power is calculated according to eqn. (2.87) in which the power of radial force and feed force* is assumed to be of a lower order of magnitude. Consequently, the cutting power in the case of turning is calculated from

$$N_e = \frac{F_v v}{6120} \quad \text{kW} \tag{8.18}$$

where: F_v—cutting force in kg, v—cutting speed in m/min.

8.3 Tool life speed and the properties of the surface layer in turning

8.3.1 Tool life speed

A discussion of the dependence of tool life speed on the life of a lathe tool completely agrees with the description contained in Section 4.3.

In order to make practical use of the formulae given there, it is necessary to know the numerical values of the criteria of lathe tool wear, since tool life must correspond to a certain defined wear of the tool.

Average values of the permissible lathe tool wear, adopted in Soviet standards for the engineering industry (prepared by the Soviet Ministry of Machine-Tool Industry) are given in Table 8.7.

For the calculation of tool life period and speed in turning, the eqns. (4.35) and (4.36) are used, respectively.

Numerical values needed for the calculation according to these equations are:

— values of C_v, e_v and u_v in Table 8.8,

— values of the correction factor $B_T = 1/T^m$ taking into account the assumed tool life period, in Table 8.9,

* The approximate formula for feed power, used in the calculation of feed mechanism is as follows: $N_{ex} \approx 6{\cdot}12\times10^6\ [F_f+\mu(F_v+F_p)]s\,n$, where μ—coefficient of friction.

Table 8.7 *Average permissible values of lathe tool wear* (acc. to Soviet Ministry of Machine-Tool Industry)

Tool point material	Tool type	Workpiece material	Type of turning	Permissible values of *VB* in mm	
				in dry turning	in turning with coolant
High-speed steel	for external straight turning	steel and malleable cast iron	rough medium accurate and accurate	0·5–1·0 0·3–0·5	1·5–2·0
		grey cast iron	rough medium accurate and accurate	3–4 1·5–2·0	—
	boring and undercutting tools	steel and malleable cast iron	rough, medium accurate and accurate	0·3–0·5	1·5–2·0
		grey cast iron	rough, medium accurate and accurate	1·5–2·0	—
	cut-off and parting tools	steel and malleable cast iron	rough, medium accurate and accurate	0·3–0·5	0·8–1·0
		grey cast iron	rough, medium accurate and accurate	1·5–2·0	
	form tools	steel	rough, medium accurate and accurate	—	0·4–0·5
	threading tools	steel	rough finish	— —	2·0 0·3
Sintered carbides	all types of tools	steel	rough, medium accurate and accurate	0·8–1·0	—
		grey cast iron	rough and medium accurate ($s > 0{\cdot}3$) accurate ($s < 0{\cdot}3$)	0·8–1·0 1·4–1·7	— —
Sintered metal oxides	all types of tools	steel and cast iron	medium accurate and accurate	0·8	—

— values of the correction coefficient B_M taking into account the kind of workpiece material, in Table 4.1,
— values of the correction coefficient B_p taking into account the state of the surface and structure of the workpiece material, in Table 4.3.
— values of the correction coefficient B_H taking into account the hardness or strength of the workpiece material, in Table 8.4,
— values of the correction coefficient B_N taking into account the kind of tool point material (if this factor is not taken care of by the constant C_v), in Table 4.4,
— values of the correction coefficient B_r taking into account the influence of nose radius, in Table 8.2.

— values of the correction coefficient B_w taking into account the degree of tool point wear, in Table 8.3,
— values of the correction coefficient B_f taking into account the cross-section area of tool shank, in Table 8.10,
— values of the correction coefficient B_k taking into account the shape of the tool face, in Table 8.11,
— values of the correction coefficient B_k taking into account the influence of leading cutting edge angle, in Table 8.5,
— values of the correction coefficient B_{k1}, taking into account the influence of trail edge angle, in Table 8.12.

8.3.2 MACHINED SURFACE AND SURFACE LAYER AFTER TURNING

8.3.2.1 *Surface quality*

Factors influencing the surface quality after turning may be classified into groups connected with workpiece material, the tool, cutting conditions and machine tool.

Of the group of factors connected with the workpiece material, the most important is material structure. Finely grained materials of a high degree of homogeneity allow one to obtain a better quality of machined surface. Thus, for example, the finer is the pearlitic structure of the steel, the smoother surface results. In the case of cast iron a higher surface quality is always achieved—under identical cutting conditions—in machining nodular cast iron with uniformly distributed fine graphite than in machining ordinary pearlitic type.

Of the group connected with the tool, the nose radius is the most important, as it participates in the kinematic and geometric reproduction of the tool point on the workpiece material. The dependence of mean total height R_{tm} on nose radius r_β is illustrated by research results obtained at various cutting speeds and feed rates, and represented in Fig. 8.16. The character of the curves agrees with equations given in Chapter 6, according to which the relationship being discussed is hyperbolic within the range of practically applied values of r_β.

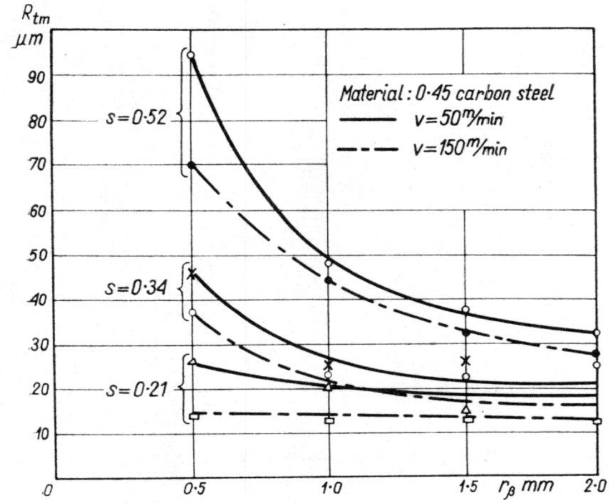

Fig. 8.16 Diagram of the influence of nose radius on mean total height of surface irregularities R_{tm} [8/4]

Table 8.8 *Values of C_v, e_v and u_v for turning*

Tool point material	Type of tool point	Type of turning	Workpiece material	σ_B kg/mm²	HB kg/mm²	T min	$a/s > 1$	Dry cutting C_v	Dry cutting e_v	Dry cutting u_v	Cutting with the use of coolant C_v	Cutting with the use of coolant e_v	Cutting with the use of coolant u_v
High-speed steel SW9 (see Table 4.5)	Tool point without land, with trail edge angle $\varkappa_1 > 0$ ($r_\beta = 2$ mm)	external straight turning (plain turning)	carbon steel	75	215	60	$a = 1–30$ $s = 0·15–4$	24	0·25	0·54	32	0·25	0·54
			grey cast iron		190	60	$a = 1–30$ $s = 0·1–4$	22·5	0·15	0·35			
			inoculated cast iron		190	60	$a = 1–30$ $s = 0·1–4$	23·4	0·2	0·35			
			malleable cast iron		150	60	$a = 1–8$ $s = 0·1–2$				45·8	0·2	0·4
			bronze, medium hard		120	60	$a = 1–12$ $s = 0·1–2$				58·7	0·12	0·4
		internal straight turning (boring)	carbon steel	75	215	60	$a = 1–12$ $s = 0·1–1·4$	21·4	0·25	0·54	28·80	0·25	0·54
			grey cast iron		190	60	$a = 1–12$ $s = 0·1–1·4$	20·2	0·15	0·35			
			inoculated cast iron		190	60	$a = 1–12$ $s = 0·1–1·4$	21	0·2	0·35			
			malleable cast iron		150	60	$a = 1–8$ $s = 0·1–1·4$				41·2	0·2	0·4
			bronze, medium hard		120	60	$a = 1–12$ $s = 0·1–1·4$				52·8	0·12	0·4
		transverse turning (facing)	carbon steel	75	215	60	$a = 1–12$ $s = 0·1–3$	25	0·25	0·54	33·3	0·25	0·54
			grey cast iron		190	60	$a = 1–12$	23·4	0·15	0·35			
			inoculated cast iron		190	60	$a = 1–12$ $s = 0·1–3$	24·4	0·2	0·35			
			malleable cast iron		160	60	$a = 1–8$ $s = 0·1–3$				47·6	0·2	0·4
			bronze, medium hard		120	60	$a = 1–12$				61	0·12	0·4

High-speed steel SW9 (see Table 4.5)		parting	carbon steel	75	215	120	$s = 0{\cdot}04$–35				67·7	—	0·66
			grey cast iron		190	120	$s = 0{\cdot}04$–35	9·78	—	0·4			
			malleable cast iron		150	120	$s = 0{\cdot}04$–35				13·4	—	0.5
		threading	carbon steel	75	215	120	roughing $s \leqq 2$ $s > 2$				 8·8 20·5	 0·1 0·6	 0·3 0·25
							finishing				22·5	0·45	0·3
Sintered carbides	Tool point with trail edge angle $\varkappa_1 > 0$ ($r_\beta = 2$ mm)	external straight turning (plain turning)	carbon steel	75	215	90	$a = 1$–1·2 $s = 0{\cdot}1$–2·0	117	0·18	0·3	In turning with $\varkappa_1 = 0$ simpler formulae may be used in workshop practice. In turning of HB = 180 kg/mm² steel, when $a > s$ $v_T = \frac{304}{T^{0\cdot18}a^{0\cdot15}s^{0\cdot3}}$ when $a \leqq s$ $v_T = \frac{304}{T^{0\cdot18}a^{9\cdot3}s^{0\cdot15}}$ In turning of HB = 200 kg/mm² cast iron when $a > s$ $v_T = \frac{304}{T^{0\cdot28}a^{0\cdot2}s^{0\cdot4}}$ when $a \leqq s$ $v_T = \frac{304}{T^{0\cdot28}a^{0\cdot4}s^{0\cdot2}}$		
			hardened steel	150		90	$a = 1$–2 $s = 0{\cdot}1$–1·5	36·8	0·18	0·3			
			grey and malleable cast iron		190	90	$a = 1$–1·5 $s = 0{\cdot}1$–3·0	62·4	0·18	0·35			
			inoculated cast iron		190	90	$a = 1$–8·0 $s = 0{\cdot}1$–2·0	73	0·19	0·34			
		internal straight turning (boring)	carbon steel	75	190	90	$a = 1$–12 $s = 0{\cdot}1$–1·4	105	0·18	0·3			
			grey and malleable cast iron		190	90	$a = 1$–15 $s = 0{\cdot}1$–2·0	74	0·18	0·35			
			inoculated cast iron		190	90	$a = 1$–8 $s = 0{\cdot}1$–2·0	65·6	0·19	0·34			
		transverse turning (facing)	carbon steel	75	215	90	$a = 1$–8 $s = 0{\cdot}1$–1·4	122	0·18	0·3			
			grey and malleable cast iron		190	90	$a = 1$–15 $s = 0{\cdot}1$–3·0	86	0·18	0·35			
			inoculated cast iron		190		$a = 1$–8 $s = 1$–2·0	76	0·19	0·35			

Table 8.9 *Values of correction coefficients B_T taking into account the assumed value of tool life period*

Tool point material	High-speed steels					Sintered carbides	
Cooling	liquid			air		air	
Workpiece material	steel, malleable cast iron			grey and inoculated cast iron		non-hardened steel, cast iron	hardened steel
Kind of tool	for straight turning and facing	parting tools	form tools	for straight turning and facing	parting tools	all kinds except form tools	
Exponent $1/z = m$	0·125	0·25	0·3	0·1	0·15	0·2	0·1
Tool life, T min 20	1·15	1·31	—	1·12	1·18	1·33	1·15
30	1·09	1·19	—	1·07	1·112	1·24	1·11
45	1·04	1·08	—	1·03	1·05	1·15	1·08
60	1	1	1·23	1	1	1·08	1·04
75	0·97	0·95	1·15	0·98	0·97	1·04	1·02
90	0·95	0·9	1·09	0·96	0·94	1·0	1·0
120	0·92	0·83	1	0·93	0·9	0·94	0·97
150	0·9	0·79	0·93	0·91	0·87	0·91	0·95
180	0·87	0·76	0·86	0·86	0·85	0·87	0·93
240	0·84	0·71	0·87	0·81	0·82	0·82	0·92
360	0·8	0·64	0·72	0·84	0·76	0·76	0·88

Of the remaining tool-point-conditioned factors, cutting edge dents and tool point wear should be mentioned.

The influence of edge dents is a consequence of their direct reproduction on the machined surface. Therefore the surface roughness resulting is of the same order of magnitude. A conclusion follows concerning the great importance of careful smoothing of the cutting edge after sharpening.

Tool point wear creates changes in the machined surface mainly because of geometric action and, to a lesser degree, as a result of physical action.

The geometric action of tool wear consists in the change of nose radius and edge dents with working time. Thus the character of tool point wear influences surface quality. This influence is different in the case of high-speed steel and sintered carbide tools (Fig. 8.17).

In the case of high-speed steel (Fig. 8.17a) at first a prevailing influence of the growing nose radius is observed, as a consequence of nose wear. This results in an improvement of surface quality. After a minimum of R_{tm} has been reached, the influence of scratches and dents on the edge increases and surface quality deteriorates.

At the moment when permissible wear, i.e. the wear criterion, is reached, the surface quality is usually of the same order or even better than at the start of cutting with a freshly ground tool point.

Table 8.10 *Values of correction coefficients B_f taking into account the tool shank cross-section area*

Shape and dimensions of shank cross section	B_f coefficients, in turning	
	steel	cast iron
10×10 Ø 10 Ø 12	0·84	0·91
12×12 Ø 16	0·86	0·93
16×16 Ø 20	0·89	0·95
20×20 16×25 Ø 25	0·93	0·96
25×25 20×32 Ø 30	0·96	0·98
32×32 25×40 Ø 40	1·00	1·00
32×50	1·04	1·02
40×63	1·07	1·04
50×80	1·14	1·07

Table 8.11 *Values of correction coefficients B_k taking into account the shape of the tool face*

Workpiece material	Tool point material	Coefficients B_k				
		Rake angle	Kind of tool point			
			without land		with land	
			plane	concave	plane	concave
Steel and cast iron	high-speed steel	$\gamma \geqq 0$	1·0	1·05	1·15	1·15
	sintered carbides	$\gamma \geqq 0$	1·0	—	1·12	1·12
		$\gamma < 0$	1·18	—	—	—

Table 8.12 *Values of coefficients $B_{\varkappa 1}$ taking into account the trail edge angle*

Kind of tool point	Tool point material	Workpiece material	Angles $\varkappa_1$					
			0°	10°	15°	20°	30°	45°
$\varkappa_1 > 0$	high-speed steel and sintered carbides	steel and cast iron	—	1·0	0·97	0·94	0·91	0·87

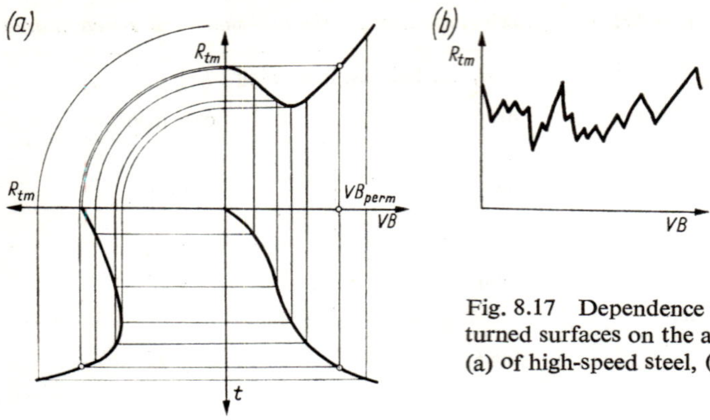

Fig. 8.17 Dependence of the roughness R_{tm} of turned surfaces on the amount of tool wear: (a) of high-speed steel, (b) of sintered carbides

In the case of tool points of sintered carbides, the relationship is mostly represented by microdents of the edge causing the wear of sintered carbide points. The more resistant to denting is the given kind of sintered carbide, the smaller are the oscillations observed in this relationship and the greater its similarity to that of high-speed steel.

The remaining magnitudes and geometric properties of the tool point have a relatively small or even negligible influence on surface quality. Only the trail edge angle exerts a substantial influence, if increased above the recommended values. In such a case the geometric and kinematic reproduction may change its character from arc-shaped to arc-shaped-and-rectilinear (see Chapter 6).

Of the group of factors connected with cutting conditions, the feed rate influences the surface quality in a stereometric and kinematic way. A typical character of mean total height changes resulting from feed rate variations is represented in Fig. 8.18.

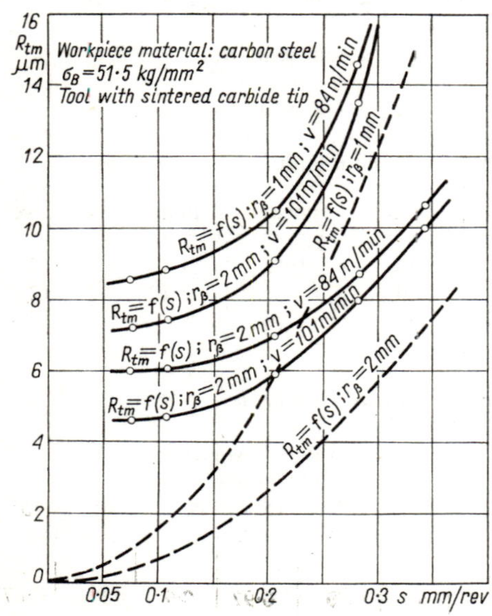

Fig. 8.18 Influence of feed rate on mean total height of surface irregularities [8/4]

It is evident from this diagram that the lower the feed rate the more the actual mean total height differs from that theoretically presumed (the theoretical values of total mean height are marked by dashed lines). Beginning with a certain feed rate value (about 0·1–0·05 mm/rev) its further reduction does not yield any improvement of surface quality and therefore is aimless.

The influence of cutting speed in turning has been discussed in Chapter 6.

The depth of cut influences the surface quality in an indirect way. Increasing depth of cut increases the cutting resistance and simultaneously increases the amplitude of vibrations. Also, cutting temperature rises and this may change the conditions of chip and BUE formation. Generally, however, the influence of the depth of cut is rather small.

Selection of turning conditions with regard to surface quality may be carried out according to the nomogram described in Chapter 6 (Fig. 6.11).

8.3.2.2 *Surface layer*

Research on surface layer properties has been conducted to date by different methods and the results are not exhaustive enough to allow a general formulation of quantitative relationships between surface layer properties and cutting conditions.

Problems concerning the surface layer, described in Chapter 6, may be complemented in the case of turning by some more detailed data.

Hardening of the surface layer depends on all those factors which influence the size and shape of the area of plastic deformation. As an example may be cited the results of research on the influence of the depth of cut, rake angle and cutting edge rounding radius [8/5], illustrated by diagrams in Fig. 8.19, 8.20 and 8.21.

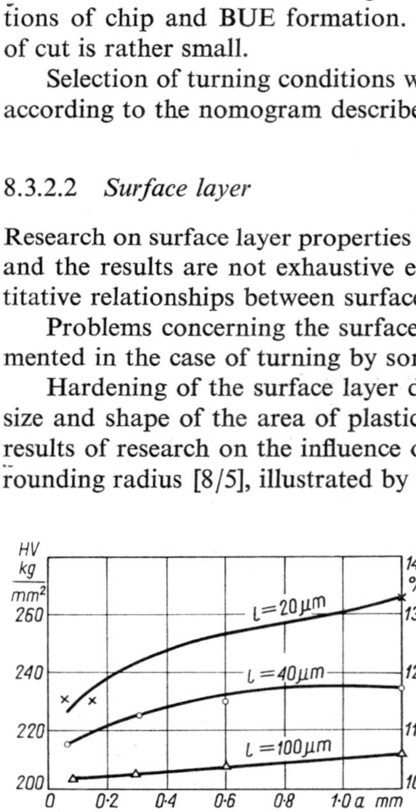

Fig. 8.19 Diagram of the dependence of macrohardness on the depth of cut; *l*— distance from the surface

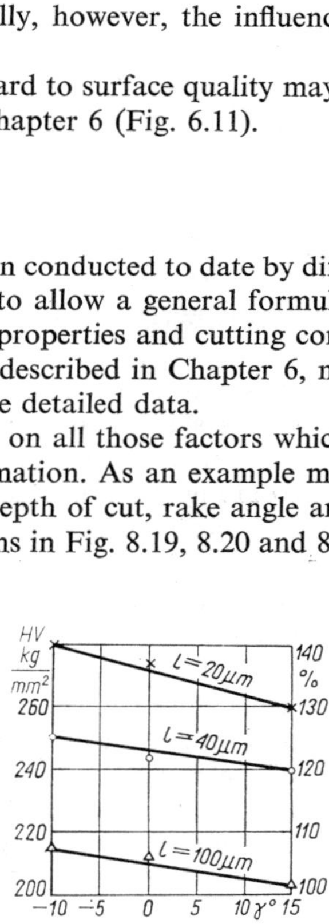

Fig. 8.20 Diagram of the dependence of macrohardness on rake angle

The workpiece material was 0·15 carbon steel in annealed state, turned by sintered carbide tools.

An increase in the depth of cut caused an increase of material compression and hardness by some 3% for each 0·5 mm of the depth of cut (Fig. 8.19).

It follows that within the range of most frequently applied depths of cut of 0·5–3 mm hardness changes of the order of 20% may occur. On the other hand, the increase in hardness may amount to 25–45% within this range.

Increasing the rake angle (Fig. 8.20) creates hardening reduction at a distance of about 20 μm from the surface (and deeper). A rake angle variation from −10° to +15° corresponds to a hardness decrease of about 10% ,i.e. from 140% to 130% of the hardness of the undeformed material. This means a hardness reduction of 2% for each 5° of rake angle increase.

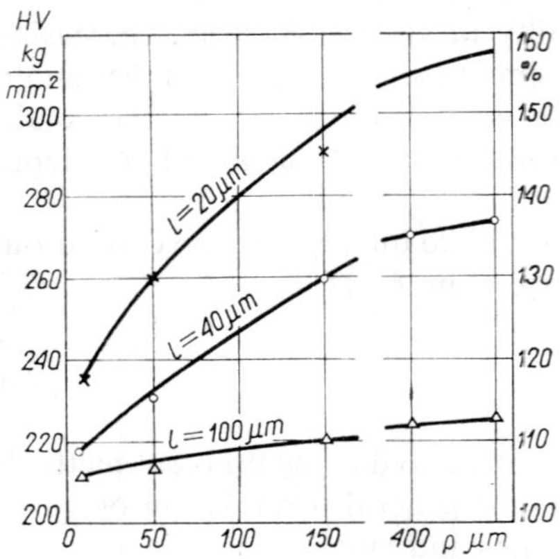

Fig. 8.21 Diagram of the dependence of macrohardness on the roundness radius of cutting edge

Roundness radius φ of a freshly ground cutting edge mostly oscillates between 20 and 30 μm. However, sharper edges occur, too (roundness radius of about 10 μm), and as a result of wear, radii of the order of tenths of a millimetre are noted.

As is evident from Fig. 8.21, the change of edge roundness radius from 10 to 400 μm causes (for l = 20 μm) an increase of hardness from 118% to 158% of that of the undeformed material. Thus, on average, a 10 μm increase of edge roundness radius results in hardness increase by 1%.

Conclusions concerning the influence of machining conditions on after-machining stresses, or so-called final (residual) stresses, are not so well founded.

According to Russian research (carried out by P. V. Timofeev [6] and others) the following conclusions may be drawn:

— the influence of the depth of cut and feed rate on final stresses is similar, i.e. increase of these factors causes an increase of maximum stress and of the depth of stress occurence and also stress changes from tensile to compressive. Absolute values of stress depend mainly on the properties of the workpiece material;

— the influence of rake angle is characterized by the fact that, as the angle varies from high positive to negative values, tensile stresses are replaced by compressive ones. Simultaneously, the depth at which the stresses occur increases, but in such a way that the stress gradient increases.

8.4 Selection of cutting conditions in turning

The principles and the process of selection of cutting conditions (as discussed in Chapter 7) fully apply to turning.

In order to optimize these conditions practically, with regard to unit cost, it is necessary to know the recommended tool life periods calculated according to formulae given in Chapter 7.

REFERENCES

8/1 Praca zbiorowa: 'Normatywy czasów obróbki skrawaniem', Tom I — Toczenie, Cz. II — Zalecane warunki skrawania przy toczeniu stali i żeliwa. Tablice i nomogramy (Collective work: Machining time standards, Vol. I. Turning, Part II—Recommended cutting conditions in turning of steel and cast iron. Tables and nomograms), CONP MPC, Warszawa 1960

8/2 GRITSAENKO, A. I.: 'K voprosu o rezhushchikh svoistvakh instrumental'nykh materialov' (To the problem of machining properties of tool materials), Vestnik Mashinostroeniya, 10, 1953

8/3 ARSHINOV, V. A., and ALEKSEEV, G. A.: 'Rezanie metallov' (Cutting of Metals), Mashgiz Moskva 1953

8/4 KACZMAREK, J.: 'Podstawy doboru warunków skrawania przy toczeniu metali' (Basis for the Selection of Cutting Conditions in the Turning of Metals) PWT, Warszawa 1953

8/5 KACZMAREK, J.: 'Badania wpływu zaokrąglenia krawędzi skrawającej na utwardzanie warstwy wierzchniej' (Studies of the influence of cutting edge rounding radius on hardening of surface layer), Czasopismo Techniczne, 6, 1965 and Annals of the CIRP, Vol. XIII, 3/1965 (in German)

CHAPTER 9

Broaching

9.1 Kinematics and types of broaching

9.1.1 GENERAL CHARACTERISTICS AND TYPES OF BROACHING

Broaching is a method of machining by cutting, in which the machining allowance is removed by a multipoint tool of such a shape, that each successive tool point projects a definite distance further than the preceding one in the direction perpendicular to that of broach length. The process of cutting by means of a broach is shown diagrammatically in Fig. 9.1.

It is seen that broaching resembles planing by means of a series of consecutive tools (1, 2, 3 in Fig. 9.1). In the example shown the workpiece remains stationary

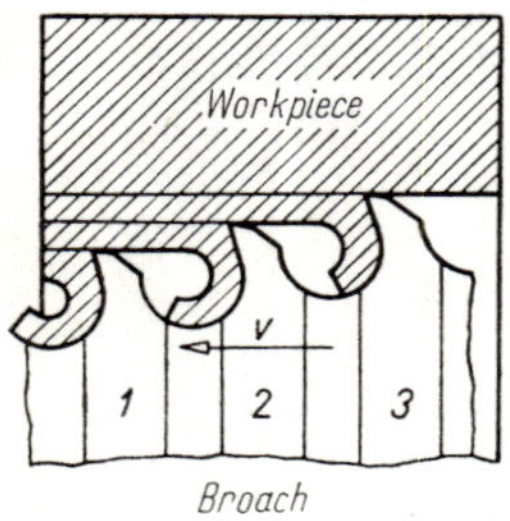

Fig. 9.1 Diagram of cutting by means of a broach

throughout the machining process. The basic motion—also the primary motion—is the rectilinear motion of the broach. In addition to this most commonly applied type, the following types of broaching are distinguished:

— the broach is stationary and the rectilinear motion is carried out by the workpiece,
— the broach carries out a rectilinear and rotary motion simultaneously and the workpiece remains stationary; such a situation occurs in the case of broaching holes with helical grooves,
— the broach is stationary and the workpiece carries out a rectilinear and a rotary motion simultaneously.

Cutting speed v in the case of broaching is the speed of progressive motion of broach or workpiece.

If an analogy to planing is drawn, the feed motion occurring stepwise at the end of each idle stroke in planing corresponds to the difference in height of the consecutive points (teeth) in broaching. Thus the feed rate is set automatically at the broach design stage.

With regard to broach position in relation to the workpiece, *external* and *internal broaching* is distinguished.

The types listed above are those of internal broaching.

In external broaching several types may be distinguished, too, as for instance:
— broaching of external grooves of various shapes,
— parting,
— broaching of planes.

In all these types the primary basic motion may be carried out by the tool or by the workpiece. Such motion may be rectilinear or curvilinear. Examples of some kinematic types are given in Fig. 9.2.

With regard to the application of the force causing the broach to move, *pull broaching* and *push broaching* are distinguished. The latter has the same characteristic features as the former, but the tool called a *push broach* is compressed and not stretched (as the *pull broach* is). Push broaching has similar types as pull broaching.

Broaches are, in general, not universal but special tools, adapted to performing a strictly defined operation. The shape of broaches used most frequently are:
(a) single groove broaches, usually for rectangular groove cross-sections (Fig. 9.3b),
(b) multiple groove broaches, e.g. with rectangular point shape (Fig. 9.3c),
(c) round broaches (Fig. 9.3a),
(d) square broaches (Fig. 9.3d),
(e) hexagonal broaches (Fig. 9.3e).

Broach teeth may be designed according to needs.

Cost of broach production is relatively high because of the required accuracy of shape and dimensions and in view of technological difficulties. However, broaching applied in mass or large-lot production is very profitable, as a result of its high output and accuracy. For these reasons broaching should be regarded as one of the modern ways of machining by cutting.

9.1.2 Broaching time

Broaching time of a single pass of a broach is calculated according to

$$t_m = \frac{L\,K}{1000v} \quad \text{min} \tag{9.1}$$

where: L—broach length in mm, K—coefficient taking into account the return stroke (if any) of the broach; usually $K = 1{\cdot}14$–$1{\cdot}5$, v—cutting speed in m/min.

By *broach travel* (*stroke*) *length* the following sum is understood (Fig. 9.4)

$$L = l + l_b + l_a + l_o \quad \text{mm} \tag{9.2}$$

where: l—length of workpiece (strictly: of the work surface) in mm, l_b—broach

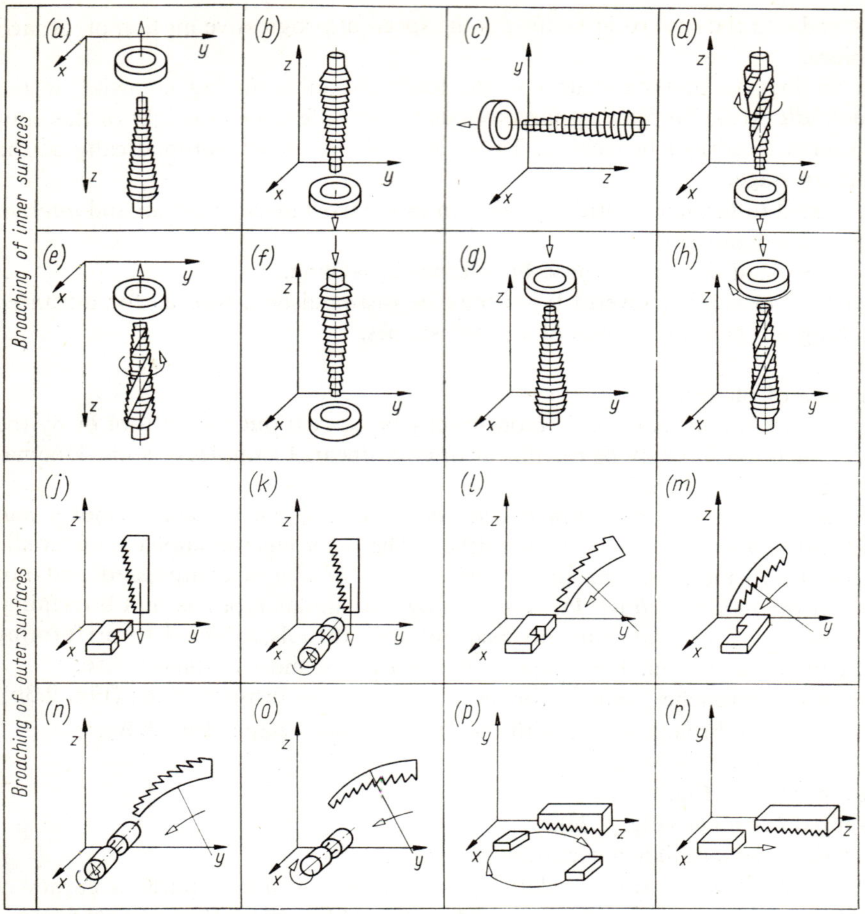

Fig. 9.2 Examples of some types of broaching

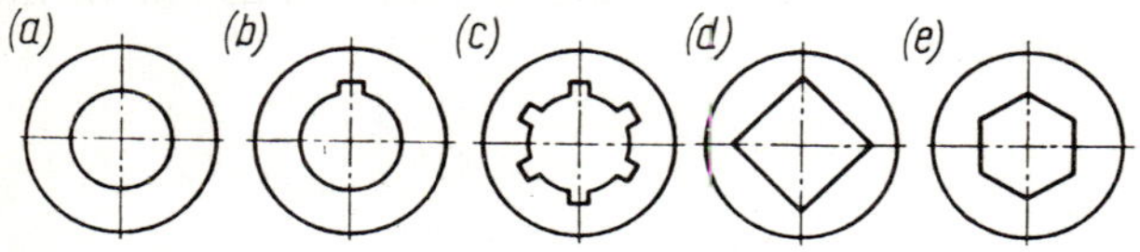

Fig. 9.3 Shapes of broached holes

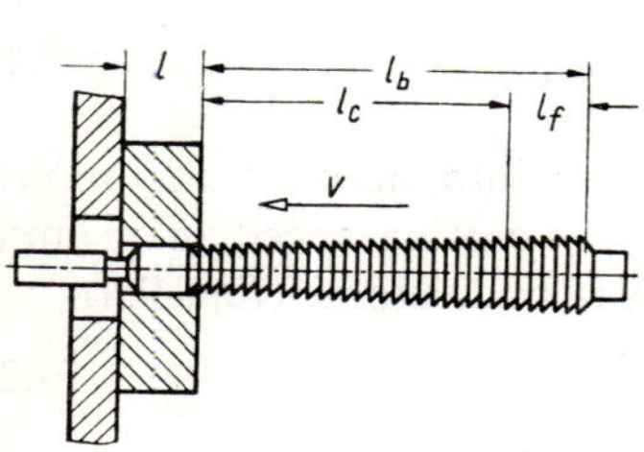

Fig. 9.4 Diagram of broaching, with designations used in the calculation of broaching time

length, l_a—length of approach, l_o—length of overrun. The latter two lengths are assumed to be

$$l_a + l_o = 10\text{–}20 \quad \text{mm}$$

Broach length is calculated as

$$l_b = l_c + l_f \quad \text{mm}$$

$$l_b = z_c p_c + z_f p_f \quad \text{mm} \tag{9.3}$$

where: l_c—length of the cutting part in mm, l_f—length of the finishing part in mm, z_c—number of cutting teeth, p_c—pitch of cutting teeth in mm, z_f—number of finishing points, p_f—pitch of finishing points in mm.

9.2 Characteristic of tool point and undeformed chip

9.2.1 Elements of broach teeth shape

Broach teeth are divided into the cutting and finishing ones. The difference between them consists in tool angles adapted to the respective working methods.

Cutting teeth are supposed to remove the machining allowance layer by layer. Finishing teeth are supposed to smooth out the work surface and to give it the final shape and dimensions within set tolerance limits.

Both the cutting and the finishing teeth resemble a turning tool and are characterized by similar parameters (Fig. 9.5).

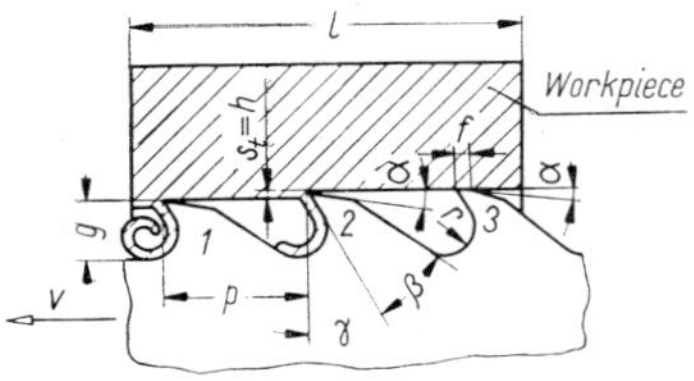

Fig. 9.5 Geometry of cutting teeth 1 and 2 and of finishing teeth 3

The reference system in which the angles of broach teeth are dimensioned is the following set of planes:

(a) a plane perpendicular to broach axis, also constituting the basic plane of the system,

(b) a plane perpendicular to the projection of the leading edge on the basic plane; this is the plane of tool point main cross-section,

(c) a plane perpendicular to both preceding ones, called the tangential plane or projection plane.

A broach tooth is characterized by the following stereometric parameters:

(a) *Clearance angle* α, in the plane of main cross-section, the angle between the tangential plane and tool point flank.

Usually, the following clearance angles are recommended:

— for external broaching

$\alpha = 10°$ for all teeth of a broach,

— for internal broaching:
$\alpha = 2°–3°30'$ for cutting teeth,
$\alpha = 30'–1°$ for finishing teeth.

Small clearance angles are aimed at in the case of internal broaches, since for greater angles the number of possible regrinds is limited. This is particularly important for finishing teeth as the accuracy of machining depends on their dimensions.

(b) *Rake angle* γ, measured in the main cross-section plane as the angle between basic plane and tool point face. The finishing teeth are usually provided with a rake angle of $\gamma = 0–5°$, while cutting teeth have various angles depending on the material to be machined. The values of those angles for steel and cast iron are given in Table 9.1.

Table 9.1 *Values of rake angle for broach cutting teeth* (acc. to N. J. Reznikov)

Workpiece material	Strength σ_B or hardness HB	Rake angle $\gamma°$
Steel	$\sigma_B < 50$	15–18
	$60 \leqq \sigma_B \leqq 100$	12–15
	$\sigma_B > 100$	8–10
Cast iron	HB < 150	8–10
	HB $\geqq$ 150	4–8

(c) *Leading cutting edge inclination angle* λ_s is measured between a plane perpendicular to the direction of broach motion and the leading cutting edge. The importance of this angle consists in that it damps the oscillations of the cutting force as consecutive teeth cut into the material. By suitable choice of λ_s it is possible to maintain a constant cutting force when broach teeth cut on the entire length of the work surface.

(d) *Land* is a chamfer on the flank side. Land width $f = 0{\cdot}05–0{\cdot}2$ mm is adopted for finishing teeth.

The relative position of teeth is characterized by their pitch and by the difference in position of tooth tips.

Tooth pitch is defined from the empirical formula

$$p = (1{\cdot}25–1{\cdot}5)\sqrt{l} \quad \text{mm}$$

where l—length of work surface in mm. Higher values of p are adopted for longer work surfaces and lower for shorter ones.

The difference between the positions of consecutive cutting teeth is positive and in the finishing portion of the broach equal to zero. This difference is called *feed per tooth* s_t or *undeformed chip thickness h*, as it plays the role of these dimensions. The cutting force and broach size depend on this value.

In addition to the parameters mentioned above, a tooth is characterized by its height g, measured from the gullet to tooth tip (Fig. 9.5) and calculated from the empirical formula

$$g = 1{\cdot}13\sqrt{k_c h l} \quad \text{mm} \tag{9.4}$$

where: k_c—coefficient of chip space filling, h—undeformed chip thickness in mm,

l—length of work surface in mm. The coefficient k_c depends on the properties of material being machined (among others: chip area factor and coil diameter) (Table 9.2).

The number of cutting teeth is calculated from

$$z_c = \frac{q}{h} + 1 \tag{9.5}$$

where q—machining allowance in mm, and the number of finishing teeth is calculated from the relation

$$z_f = \frac{q - z_a}{h_f}$$

where h_f—thickness of undeformed chip cut by a finishing tooth.

Table 9.2 *Values of broach chip space filling coefficient k_c* [9/1]

Undeformed chip thickness h mm	Workpiece material							
	Steel				Cast iron		Bronze	Brass
	C 0·35	0·45	0·40	$\sigma_B = 20$	grey	malleable		
0·100	2·10	2·45	2·25	2·60	2·00	2·48	2·15	2·08
0·090	2·12	2·47	2·27	2·62	2·05	2·52	2·19	2·08
0·080	2·15	2·52	2·31	2·66	2·12	2·60	2·24	2·08
0·070	2·20	2·57	2·36	2·72	2·20	2·72	2·31	2·12
0·060	2·27	2·63	2·43	2·80	2·30	2·88	2·38	2·15
0·050	2·38	2·75	2·52	2·92	2·45	3·08	2·50	2·22
0·040	2·50	2·92	2·63	3·10	2·68	3·33	2·65	2·35
0·030	2·70	3·17	2·82	3·37	3·13	3·67	2·92	2·57
0·020	3·20	3·70	3·10	4·00	4·50	4·50	3·90	3·00

9.2.2 Elements of undeformed chip section

9.2.2.1 *Shape of undeformed chip section*

The machining allowance may be removed in various ways, depending on the shape of consecutive teeth of the broach. Therefore the shape of the undeformed chip may vary.

There are three principal ways of removing the machining allowance by means of a broach:

(a) depth cutting,
(b) side cutting,
(c) simultaneous depth and side cutting.

Depth cutting (Fig. 9.6a) occurs when the broach teeth cut the allowance on its entire width at a time, while each tooth removes only a part of the allowance thickness.

Side cutting occurs when the broach teeth cut the allowance on its entire thickness at a time, while each tooth removes only a part of allowance width b (Fig. 9.6b).

Simultaneous depth and side cutting (Fig. 9.6c) occurs when broach teeth are staggered or arranged obliquely. In this case the width of allowance in each consecutive layer is divided among several teeth.

Of these possibilities the one applied most often is depth cutting (Fig. 9.6a). Less frequent is the application of simultaneous depth and side cutting (Fig. 9.6c) and most rarely that of side cutting (Fig. 9.6b), e.g. in the cases of broaching a rough surface.

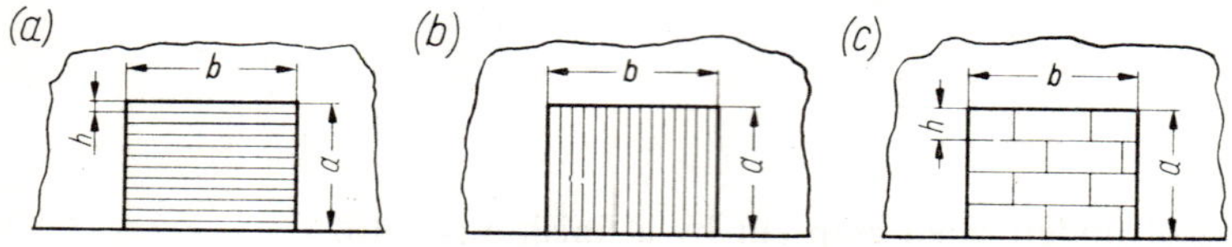

Fig. 9.6 Diagram of: (a) depth cutting, (b) side cutting, (c) simultaneous depth and side cutting

Depending on whether broach tooth edges are parallel to the final contour of the machined surface or not,
(a) parallel broaching shown in Fig. 9.7a, b, and
(b) non-parallel broaching seen in Fig. 9.7c, d
are distinguished.

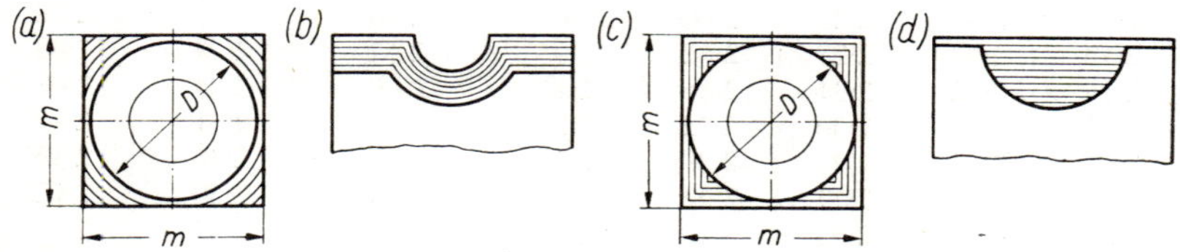

Fig. 9.7 Examples of parallel and non-parallel broaching

Broaching layer by layer non-parallel to the final contour of the machined surface has the advantage that the necessary cutting edges are easier to produce. Besides, stepwise reduction of undeformed chip area toward the finishing portion of the broach is more advantageous with regard to machining accuracy and broach wear.

9.2.2.2 *Undeformed chip dimensions*

In broaching, the characteristic dimensions of the undeformed chip are:
(a) width b,
(b) thickness $h = s_t$,
(c) length l (equal to cutting length).

By the width of undeformed chip (if the latter has a curvilinear shape) is understood the length of undeformed chip contour in transverse cross-section, i.e. perpendicular to the direction of motion. Thus, for example, when a round hole is being broached, the broaching width is the circumference of the hole cut by the given tooth, or

$$b_i = \pi d_i \tag{9.6}$$

where: b_i—width of undeformed chip to be cut by the given tooth in mm, d_i—diameter of the hole to be cut by the given tooth in mm.

In the case of square holes

$$b = 4m \quad (9.7)$$

where m—length of square side.

Undeformed chip thickness h depends on the increment of protrusion of consecutive teeth.

Knowing the machining allowance q* and undeformed chip thickness h it is possible to calculate how many cutting teeth are necessary for the required machining operation (eqn. 9.5).

The value of h is adopted from empirical formulae. For round broaches the following formula is used

$$h = s_t = C_b\, d^{x_b} \quad \text{mm} \quad (9.8)$$

where: C_b—a constant whose value is given in Table 9.3 for various broaches and broached materials, d—diameter of the broached hole in mm, x_b—exponent depending on the type of broach and material being machined (see Table 9.3).

Table 9.3 *Values of constants C_b and x_b in the formula for the calculation of undeformed chip thickness in broaching*

Workpiece material	Round broaches		Single keyway broaches		Spline broaches	
	C_b	x_b	C_b	x_b	C_b	x_b
Steel of σ_B = 45–60 kg/mm²	0·015	0·4	0·021	0·5	0·0107	0·6
Steel of σ_B = 60–80 kg/mm²	0·012	0·4	0·018	0·5	0·0096	0·6
Cast iron	0·0091	0·53	0·0234	0·5	0·0182	0·5
Bronze	0·0046	0·65	0·0225	0·6	0·0135	0·5
Babbitt	0·0058	0·65	0·0230	0·6	0·0135	0·5

In the case of broaches with rectangular teeth, the value of h is defined according to

$$h = s_t = C_b\, b^{x_b} \quad (9.9)$$

where b—tooth width in mm and other designations as in formula (9.8).

As extreme values of undeformed chip thickness $h_{min} \approx 0{\cdot}02$ mm and $h_{max} \approx 0{\cdot}2$ mm are adopted. The minimum value results from the limitation imposed by cutting edge rounding which makes it impossible to cut exceedingly thin chips and causes burnishing of the material. The maximum thickness of undeformed chip is limited, first of all, by broach strength and by the rigidity and shape of the chip produced. These characteristics determine whether the chip can be accomodated in the chip space.

* Value of the machining allowance in the case of round broaching is adopted approximately acc. to formula

$$q = 0{\cdot}005d + u\sqrt{l} \quad \text{mm}$$

where: d—diameter of the broached hole in mm, l—length of the broached hole in mm, u = 0·1–0·2 mm for less accurate broaching (after drilling) or u = 0·05–0·1 mm for more accurate broaching.

The area of undeformed chip section A is defined by

$$A = h\,b \quad \text{mm}^2 \tag{9.10}$$

Total cross-sectional area of undeformed chips to be removed by the entire broach is

$$A_c = Az_c \quad \text{mm}^2 \tag{9.11}$$

where: A—undeformed chip section per tooth, z_c—number of teeth cutting simultaneously.

The minimum number of teeth cutting simultaneously is defined by

$$z_{c\,\min} = \frac{l}{p}$$

where: p—tooth pitch, l—hole length.

9.3 Physical and technological relationships in broaching

9.3.1 Process of the chip formation

Differing in this respect from other methods of machining by cutting, broaching is characterized by very small thicknesses of undeformed chip. As a result of this, the cutting process is difficult, with high specific resistances and large plastic deformation of the chip.

In the case of broaching a carbon steel of low or medium hardness, the chip area factor is 4–5. Thus, for example, when broaching a 0·35 carbon steel at a feed rate of $s_z = 0{\cdot}03$ mm, this factor was $\lambda_A = 3{\cdot}8$–4.

A characteristic feature of broaching is that the chip produced must be completely accommodated in the chip space of the broach. As chip space dimensions are limited by the broach strength, size and shape, the chip may be accomodated there only under condition of suitable curling. Therefore the shape of the chip space must be designed correspondingly. In Fig. 9.8 the outline of the chip space is presented together with its characteristic dimensions.

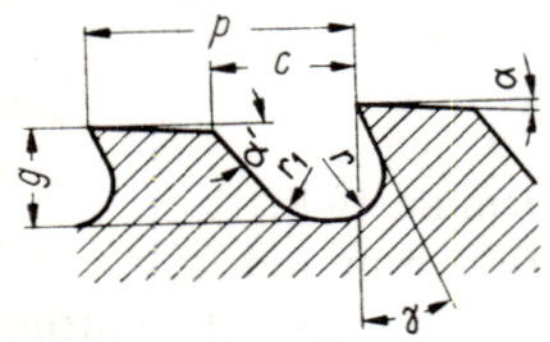

Fig. 9.8 Outline of chip space

The chip space should accomodate the chip curled into a spiral coil, as tightly as possible. To fulfil this task the chip space must:

(a) have a large volume, sufficient for cutting on the required length at the necessary undeformed chip thickness,

(b) enable tight curling of the chip coil to ensure good filling of the space.

Chip space volume is characterized by the following dimensions: tooth height g, groove length c, rake angle γ, secondary clearance angle α' and gullet radii r and r_1. The choice of those parameters may not be arbitrary as they are strictly interdependent.

If the chip is to form a tight spiral, it must be suitably bent while still in its plastic state. Thus, depending on the properties of the material and on the undeformed chip thickness, a suitable rake angle γ and gullet radius r should be adopted. Gullet rounding may not begin too far from the tooth tip as this interferes with chip curling. The radius r should be small to cause tight curling of the chip, taking into account its compression. However, if these requirements are to be fulfilled, gullet depth must be reduced and, consequently, its volume, thus limiting the cutting length.

As a result, the dimensions r and g are chosen empirically, with a view to obtaining maximum cutting length. The curvature with radius r_1 and the angle α' permit one to accomodate the chip even if it does not form an absolutely tight coil. An example of correct curling of the chip in the chip space is shown in Fig. 9.9.

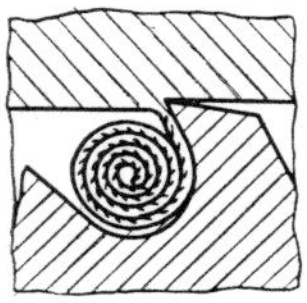

Fig. 9.9 Example of correct curling of chip in chip space

On the basis of experience gained in the cutting of steel the mutual relationships between the particular parameters of chip space have been established permitting one to optimize its dimensions:

$$c = 1{\cdot}5g \tag{9.12}$$

$$r = (0{\cdot}5\text{–}0{\cdot}7)g \tag{9.13}$$

$$r_1 = (0{\cdot}5\text{–}2)g \tag{9.14}$$

The value of angle α' is assumed to be about 60°.

Continuous ribbon-type chips have lower filling coefficients than chips consisting of short sections. It may be assumed [9/2] that chip space filling coefficient k_c is 15% (at $h \approx 0{\cdot}08$–$0{\cdot}1$ mm) to 40% (at $h \approx 0{\cdot}03$ mm) higher for discontinuous chips. The latter take up more space than chips coiled in a spiral. Values of coefficient k_c were given in Table 9.2.

9.3.2 Cutting resistance and force in broaching

Cutting resistance offered to a broach tooth may be resolved into three components (Fig. 9.10). In the case of a symmetrical tooth the components F_{f1} and F_{f2} are in balance.

Of these three components force F_v has the highest value and also the greatest importance, because it is decisive for power consumption and for broach and tooth load.

By broaching force F is understood the sum of the forces F_v acting on all teeth simultaneously, or

$$F = \sum_{i=1}^{i=z_c} F_{vi} \quad \text{kg} \tag{9.15}$$

or, when feed per tooth is identical for all teeth

$$F = z_c F_v \quad \text{kg} \tag{9.16}$$

In the case when the length of the surface being broached is not a full multiple of the pitch p, the number of teeth cutting simultaneously will vary between $z_{c\,max}$ and $(z_{c\,max} - 1)$. This is illustrated in Fig. 9.11. The length of the broached surface differs from the multiple $(z_{c\,max}+1)\,p$ by Δp or

$$\Delta p = (z_{c\,max}+1)p - l \quad \text{mm}$$

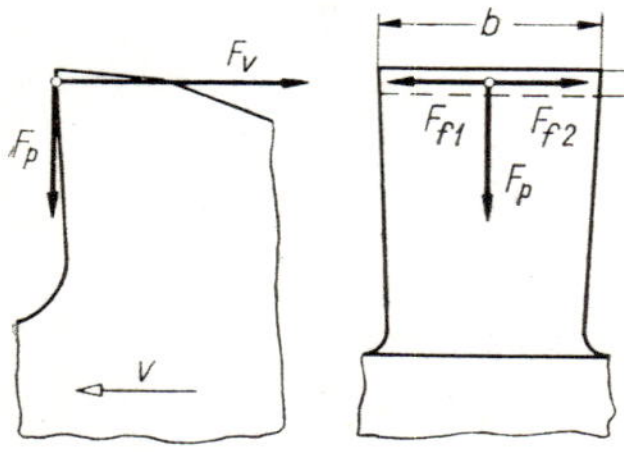

Fig. 9.10 Resolution of cutting resistance acting upon tooth

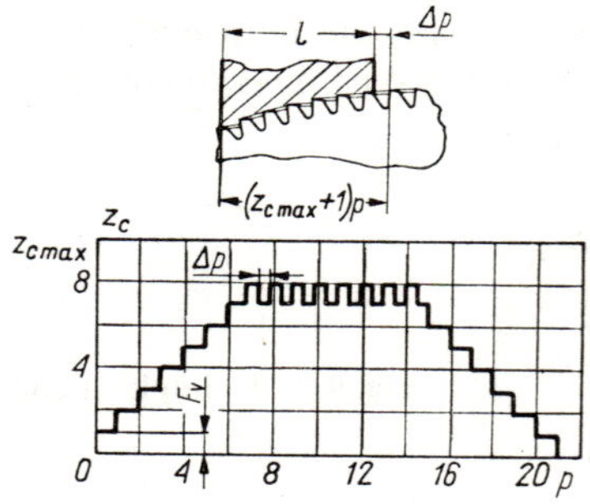

Fig. 9.11 Variations of the number of simultaneously cutting teeth in broaching

In this case the broaching force will increase to the value $F_{max} = z_{c\,max}F_v$. During the time corresponding to the travel $p - \Delta p$ broaching force F_{max} will be maintained. Then, during the time corresponding to the travel Δp, when one of the teeth has already left the material, and the succeeding has not yet reached it, the broaching force will be $F = F_{max} - F_v$.

In such a case, if the variations of cutting resistance resulting from material heterogeneity and from the gliding of undeformed chip elements are neglected, the dynamic coefficient of broaching force will be expressed by

$$K_d = \frac{z_{c\,max}}{z_{c\,max} - 1} \tag{9.17}$$

Equation (9.17) is illustrated in Fig. 9.12. It is evident there that the variations of broaching force expressed by the dynamic coefficient largely depend on the number of simultaneously cutting teeth. With three such teeth the value of K_d is 1·5, with six teeth $K_d = 1{\cdot}2$, and with eleven teeth $K_d = 1{\cdot}1$.

Variations of broaching force may be reduced or even eliminated if teeth edges are inclined sufficiently under the angle λ_s, so that at the moment when one tooth stops cutting and emerges from the broached material, another tooth will already start cutting with a portion of its cutting edge. This is illustrated by Fig. 9.13a, and the diagram of broaching force plotted against time would then have the form shown in Fig. 9.13b.

If the cutting edge inclination angle is to fulfil the condition of constant maximum broaching force, it must have the value

$$\lambda_s = \arctan\frac{p}{l} \tag{9.18}$$

Of the numerous factors, those listed below influence the cutting forces most effectively: the chemical composition, hardness and strength of material, rake and clearance angles, tooth wear (dulling) and cooling.

The influence of the chemical composition (type of material), strength and hardness of the workpiece material on the value of cutting resistance is expressed by the correction coefficient K_H. It has the same character in broaching as in turning. The

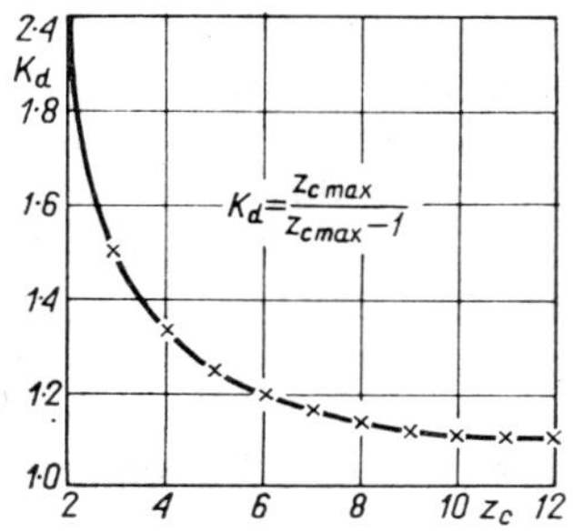

Fig. 9.12 Dependence of the dynamic coefficient on the number of teeth cutting simultaneously

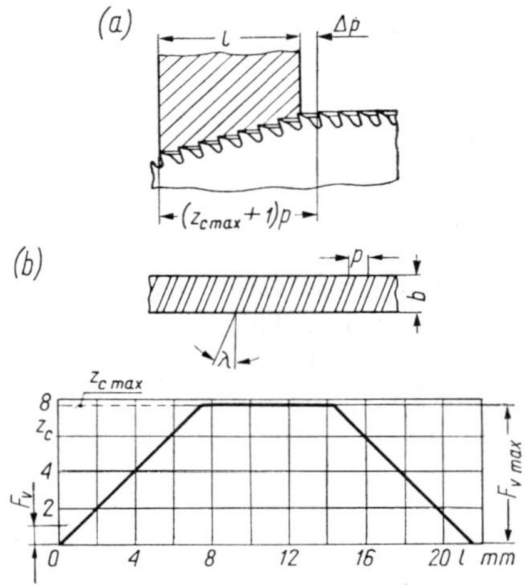

Fig. 9.13 Variations of the cutting force in broaching, when $\lambda_s = \arctan(p/b)$

changes of cutting resistance or, more accurately, of its tangential component F_v depending on the changes of mechanical strength and hardness of steel and cast iron are given in Table 9.4.

The influence of rake angle—within the range of practically applied values—on cutting resistance is considerable. This influence may be expressed by correction

Table 9.4 *Values of correction coefficients K_H appearing in the formula for tangential cutting force in broaching**

Workpiece material	HB kg/mm²	σ_B kg/mm²	Broaching of round holes	Broaching of spline holes	Broaching of tooth spaces
Structural	< 200	< 70	0·92	0·92	0·88
carbon steel	200–230	70–80	1·00	1·0	1·0
	> 230	> 80	1·10	1·24	1·24
Structural	< 200	< 70	0·91	0·81	0·81
alloy steel	200–230	70–80	1·0	1·0	1·0
	> 230	> 80	1·20	1·11	1·12
Grey cast iron	≦ 200	—	1·0	1·0	1·0
	> 200	—	1·20	1·4	1·20

* According to Soviet cutting standards

coefficient K_γ whose values are given in Table 9.5. Reduction of rake angle by 2° causes the cutting resistance to increase by 35% as compared with cutting with a rake angle $\gamma = 12°$.

The influence of clearance angle on cutting resistance is also considerable. Clearance angles applied in broaching practice are small ($\alpha \leqq 3°$) so that further

Table 9.5 *Values of correction coefficients K_γ appearing in the formula for tangential cutting force in broaching**

Rake angle $\gamma°$	12	8	2
Correction coefficient K_γ	1·0	1·13	1·35

* According to Soviet cutting standards

reduction of these values causes a great increase of friction and, consequently, of cutting resistance. This phenomenon occurs more distinctly in broaching of steel than of cast iron. Correction coefficients K_α taking into account the value of clearance angle are given in Table 9.6.

The influence of tooth wear on cutting resistance is very great. This results from the fact that the undeformed chip thickness is very small and therefore tooth friction

Table 9.6 *Values of correction coefficients K_α appearing in the formula for tangential cutting force in broaching**

Value of clearance angle $\alpha°$	Broaching of steel		Broaching of cast iron	
	2–3°	$\leqq$ 1°	2–3°	$\leqq$ 1°
Correction coefficient K_α	1·0	1·2	1·0	1·12

* According to Soviet cutting standards

Table 9.7 *Values of correction coefficient K_w appearing in the formula for tangential cutting force in broaching**

Tooth wear	Tooth point newly reground	Partly worn	Worn
	$VB = 0$ mm	$VB = 0{\cdot}3$ mm	$VB = 1$ mm
Correction coefficient K_w	1·0	1·18	1·75

* According to Soviet cutting standards

causes the resistance to change by a high percentage value. As a consequence, the cutting resistance of a sharp tooth differs by some 75% from that of teeth considered worn ($VB = 1{\cdot}0$ mm). Because of its importance, the influence of tooth wear is taken account of in the calculation of F_v through the correction coefficient K_w (Table 9.7).

The influence of cooling and lubricating on cutting resistance is very great, too, because of a relatively large share of friction energy in the cutting process. Cutting resistances in dry broaching of steel are some 33% greater than in broaching with the use of a cooling liquid. Thus the correction coefficient K_c equals 1·33 in the former case and 1·0 in the latter.

Calculation of the mean value of broaching force F is carried out according to statistical and empirical formulae.

In the case of broaching of round holes

$$F = C_z\, s_t^{u_z}\, D\, z_c\, K_H\, K_\gamma\, K_\alpha\, K_w\, K_c \quad \text{kg} \tag{9.19}$$

where: C_z and u_z—constants, depending on the type of material and its strength, s_t—feed per tool point (tooth) in mm ($s_t = h$); D—diameter of the broached hole in mm; z_c—number of simultaneously cutting teeth; K_H, K_γ, K_α, K_w, K_c—correction coefficients.

In the case of broaching of spline holes

$$F = C_z\, s_t^{u_z}\, b\, i\, z_c\, K_\gamma\, K_\alpha\, K_w\, K_c \quad \text{kg} \tag{9.20}$$

where: b—spline width in mm, i—number of splines.

In the case of broaching of holes with single grooves

$$F = C_z\, s_z^{u_z}\, b\, z_c\, K_H\, K_\gamma\, K_\alpha\, K_w\, K_c \quad \text{kg} \tag{9.21}$$

Values of constants c_z and u_z for various materials and broaching types are given in Table 9.8.

Table 9.8 *Values of constants C_z and u_z appearing in the formula for broaching force* (according to Soviet standards)

Workpiece material and cooling method	Mechanical properties		Types of broaching					
			Round		Spline		Single keyway	
	HB kg/mm²	σ_B kg/mm²	C_z	u_z	C_z	u_z	C_z	u_z
Structural carbon steel (cooling with sulphured oil)	—	75	762	0·85	230	0·85	202	0·85
Structural alloy steel (cooling with sulphured oil)	—	75	842	0·85	284	0·85	250	0·85
Grey cast iron, no cooling	190	—	300	0·73	152	0·73	115	0·73

Cutting power in broaching is calculated according to the well-known formula

$$N_e = \frac{F_v}{6120} \quad \text{kW}$$

In order to select a motor of suitable power, F_{max} is introduced into this formula. Then

$$N_e = \frac{N_{e\,max}}{\eta_e} = \frac{F_{v\,max}\, v}{\eta_e\, 6120} \quad \text{kW} \tag{9.22}$$

where the value of η_e varies between 0·75 and 0·9 depending on the design, construction and condition of the machine tool.

9.3.3 Tooth wear and tool life speed

9.3.3.1 *The process of tooth wear and permissible wear*

The wear of a broach tooth is similar to the wear process of a turning tool as described in the chapter on turning. The most distinct symptom of broach tooth wear is wear occurring on its face, on cutting edge corners—in the form of rounding—and on the flank. Wear of the latter is most intensive and is therefore adopted as the criterion of broach wear.*

Flank wear on a tooth of a spline broach (with chip dividing groove) is shown in Fig. 9.14.

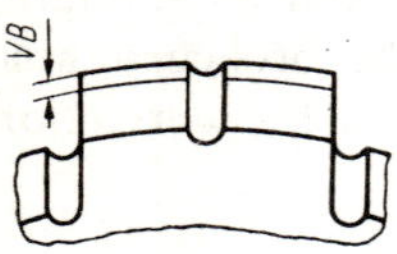

Fig. 9.14 Wear of a spline broach tooth

Permissible flank wear causing the tooth to be considered worn, has been adopted* for the broaching of steel and cast iron:
— for round broaches $VB = 0{\cdot}2\text{–}0{\cdot}3$ mm
— for single keyway broaches $VB = 0{\cdot}3\text{–}0{\cdot}4$ mm
— for spline broaches $VB = 0{\cdot}3\text{–}0{\cdot}4$ mm.

Relatively small permissible values of *VB* result from the tendency to use in a most economical way that portion of the tooth which is destined to be ground off in successive resharpenings.

Tooth wear determines the average recommended tool life. The use of a broach, i.e. the choice of machining conditions, should be such that broach tooth wear *VB* does not reach the permissible value before the average recommended tool life period (Table 9.9) has elapsed.

Table 9.9 *Average recommended broach tooth life periods* (according to Soviet standards)

Broach type	Hole diameter *D* mm	Groove width *b* mm	Workpiece material	
			Steel	Cast iron
Round	10	—	90	150
	20	—	120	180
	30	—	180	270
	40	—	270	420
	50	—	360	600
Spline	28		270	420
	32–52	—	420	600
	60–70	—	600	900
Single keyway	78–82	—	900	1300
	—	5	90	120
	—	10	120	180
	—	20	200	300

* Under the condition that dimensional tolerance has not been exceeded.

9.3.3.2 *Tool life speed*

Good surface finish may be obtained either at very low or at very high cutting speeds, if the cutting proceeds in the complete absence of vibrations. Broached surfaces are required to be very smooth. At the same time the tool construction and the dynamics of broaching make it difficult to apply high cutting speeds. Therefore the speeds applied in practice are usually contained within the range extending from a few tenths of a metre to 15 m/min.

Such speeds used in broaching have less influence on tool life than in the case of turning. This is expressed by a considerably lower value of the exponent z. Broach tooth life is determined by

$$T \approx \frac{C_T}{v^z \, s_t^{u_T}} \quad \text{min} \tag{9.23}$$

where: C_T—a constant, taking into account the influence of all factors not appearing separately in the formula (Table 9.10), v—cutting speed in m/min, s_t—feed per tooth in mm, z and u_T—exponents.

Table 9.10 *Values of constants appearing in the formula for tool life speed*

Workpiece material and cooling method			Structural carbon steel, sulphured oil	Structural alloy steel, sulphured oil	Grey cast iron, dry machining
HB kg/mm² σ_B kg/mm²			200–300 70–80	200–300 70–80	190 —
Round broaches		C_v	15·5	11·2	14
		s	1·62	1·62	2·0
		u_v	0·62	0·62	0·6
Spline broaches		C_v	14·0	10·2	17·5
		s	1·67	1·67	2·0
		u_v	0·75	0·75	0·6
Single keyway broaches	at $s_t < 0{\cdot}07$	C_v	8·8	6·3	6·2
		s	1·15	1·15	1·67
		u_v	1·4	1·4	0·95
	at $s_t > 0{\cdot}07$	C_v	7·0	5·0	6·2
		s	1·15	1·15	1·67
		u_v	1·4	1·4	0·95

In the case of broaching steel and cast iron the values of exponent $z = 1{\cdot}15$–2, for turning with high-speed steel, average z values are between 8 and 10.

Reduction of the influence of cutting speed in broaching may be explained by a different character of thermal balance of the cutting process. This is a consequence of the fact that the chip surface is much larger in relation to its volume ($b \gg h$) than in the case of chips formed by turning. Therefore, heat generated in the cutting process is absorbed much more efficiently by the coolant and the ambient air. Cutting speeds are relatively low, and thus the time of heat flow is longer which has a similar effect.

Shortening of the heat flow time as a consequence of increasing the cutting speed within limits applied in broaching is insignificant as compared with the influence of the longer undeformed chip ($b \gg h$) and the increase of heat flow surface. The resulting temperature increase is negligible and thus the tooth life is not markedly reduced.

Tool life speed v_T in broaching depends on the same factors which were discussed in the chapter on turning. Of these factors those having a particular influence are: the type and mechanical strength of workpiece material, type of tool material, coolant and the cooling method. In addition, the value of tool life speed naturally depends on the tool life period required.

Calculation of tool life speed value when the tool life period has been changed follows from the formula

$$v_{T2} = B_T v_{T1}$$

while

$$B_T = \sqrt[z]{\frac{T_1}{T_2}}$$

In the calculation of tool life speed the influence of workpiece material hardness and strength is taken account of by means of coefficients B_H given in Table 9.11.

Table 9.11 *Values of coefficient B_H appearing in the formula for the calculation of tool life speed* (acc. to Soviet metal cutting standards)

Workpiece material	HB kg/mm²	σ_B kg/mm²	Round broaching	Spline broaching	Single keyway broaching
Structural carbon steel	< 200	≦ 70	1·10	1·10	1·10
	200–230	75	1·00	1·00	1·00
	> 230	≧ 80	0·73	0·73	0·73
Structural alloy steel	< 200	≦ 70	1·37	1·37	1·37
	200–230	75	1·00	1·00	1·00
	> 230	≧ 80	0·78	0·78	0·74
Grey cast iron	≦ 200	—	1·0	1·0	1·0
	> 200	—	0·83	0·83	0·83

It may be stated that broaches with teeth made of high-speed steel allow the application of approximately 50% higher tool life speeds than broaches with teeth made of tool steel.

The type of cutting fluid is one of the important factors influencing the broaching process.

Broaching is a cutting method in which friction has a large share of total cutting energy. It follows that oil or rich oil-in-water emulsions, particularly of the so-called surface active type, are most suitable in this case. Experience [9/3] does confirm these conclusions.

In the case of broaching of grey cast iron containing graphite flakes, cooling is not applied since graphite plays to some extent the role of lubricant.

The complete formula for tool life speed in broaching, taking into account already known factors influencing tool life, is as follows

$$v_T = \frac{C_{vT}}{s_t^{u_v}} \tag{11.24}$$

where: C_{vT}, u_v are as given in Table 9.10.

Influence of the type and flow rate of the coolant is taken account of by establishing the value of C_{vT} under definite cooling conditions. In a similar way the influence of the type of tooth material and of other factors, not appearing separately in the formula, is determined.

9.3.4 Surface quality

Broaching permits one to achieve a very good surface quality. If broaching conditions are chosen in a rational way and the broach is properly designed and manufactured, then machined surfaces of roughness number 9 may be obtained. However, in some cases the quality of broached surfaces is lower. The causes of deterioration of surface quality can generally be found among the factors exerting an influence of a physical nature.

Of the factors exerting a geometrical and physical influence the smoothness of the cutting edge is important. This edge is being reproduced on the machined surface and therefore the grinding and superfinishing of the cutting edges of broach teeth is particularly important for the quality of the machined surface.

Factors which directly or indirectly influence the quality of broached surfaces are connected with the working conditions of the broach, tooth material and shape, workpiece material and machine tool.

Of the factors connected with working conditions of the broach, those most markedly influencing the surface quality are cutting speed, feed per tooth (undeformed chip thickness), cooling system and the number of teeth cutting simultaneously.

The influence of cutting speed on surface quality is similar to that appearing in other methods of machining by cutting.

Initially, increasing the cutting speed from zero to a certain value causes surface roughness to diminish [9/3]. Then roughness increases as a result of appearance of built-up edge. After the maximum of BUE formation intensity has been reached, surface roughness decreases again in a monotonic way.

In practice, cutting speeeds within the range of 1·5–15 m/min are applied. With regard to surface quality, optimum values of cutting speed should be chosen, such as e.g. 1·5–3 m/min for the machining of steel.

Surface quality deteriorates with growing undeformed chip thickness, as this increases friction and the chance of BUE appearance on the teeth.

Cooling is of considerable importance for obtaining good surface quality in broaching. The use of oils is justified for very low broaching speeds (v = 6–8 m/min). At higher speeds the effectiveness of oil cooling becomes insufficient and broaches heat up excessively which may reduce the accuracy.

At speeds over 10 m/min the use of 10–15% oil emulsions is recommended. They have quite satisfactory lubricating properties while their cooling effect is much better than that of oils.

A remarkable influence on machined surface quality is also exerted by a change of the number of teeth cutting simultaneously; the greater this number, the better is the surface finish. This influence may be explained in the following way.

As mentioned earlier, the number of cutting teeth varies from $z_{c\,max}$ to $(z_{c\,max}-1)$. If the broach has so-called straight teeth ($\lambda_s = 0$), those changes are accompanied by stepwise ones in the total cutting resistance. The greater is the number of teeth $z_{c\,max}$, the smaller—relatively—are the variations of cutting resistance created by the reduction of the number of teeth by one.

At the moment when cutting resistance increases stepwise (when the number of teeth increases to $z_{c\,max}$) so does the magnitude of thrust force F_p. In single keyway broaching and in external broaching this causes changes in elastic deflections of the broach or workpiece. In consequence, small step-like irregularities may appear.

Another phenomenon accompanying stepwise changes of cutting resistance is a momentary increase of cutting speed at the moment when the number of teeth $z_{c\,max}$ is reduced by one. On the other hand, when the number of teeth increases to $z_{c\,max}$, a momentary slowing down occurs. Those changes in cutting speed influence surface quality in such a way that more and less lustrous stripes appear on the machined surface transversely to the cutting direction. This phenomenon occurs most frequently when the machine tool deteriorates, and also depends on the broach driving system.

Of tooth shape parameters the rake angle exerts the most influence; the greater it is, the better the surface quality obtained, because the chip flow is then facilitated, while the friction resistance and the entire cutting resistance decreases.

Clearance angle plays an important role in obtaining a smooth broached surface. Of course, with regard to maximum possible number of resharpenings without loss of nominal broaching dimensions (taking into account tolerances), clearance angles should be small.

With regard to both the surface quality and broach life it can be established experimentally that optimum values of cutting teeth clearance angles should be 1–2° and those for finishing teeths 1–1·5°.

Gradual wear of the cutting edge negatively influences surface quality. Only in the case when the broach is sharpened without superfinishing of teeth, in the first phase of work (grinding in, initial wear) an improvement of surface quality may be observed. Further tooth wear, however, always results in a deterioration of surface finish.

Workpiece material is most important for the obtainment of good surface quality, the latter being influenced in a decisive way by the chemical composition, microstructure and the resulting machinability of the material.

With regard to machinability by broaching, materials may be divided into three groups.

In the first group are steels of hardness HB > 160 kg/mm². Such steels are considered to be easily machinable as far as the obtainment of good surface quality is concerned.

In the second group are steels of hardness HB < 160 kg/mm². They distinguish themselves by great plasticity and ductility and are not easily machinable in the sense stated above.

In the third group are brittle materials, yielding discountinuous chip (cast iron, bronze, etc.). In most cases the surface quality of those materials after machining is quite satisfactory.

In addition to the chemical composition and structural components, many kinds of heterogeneity of the workpiece material structure have an important influence on surface quality: the latter deteriorates as a result of non-metallic contaminations, hard spots or heterogeneity of crystal size.

The influence of the machine tool on surface quality is considerable in broaching and in other machining methods. A machine tool in good condition, only slightly used, rigid and vibration resistant permits one to obtain a superior surface finish. In broaching machines, vibrations are particularly easily generated by friction. They impair the surface quality and rapidly destroy broach teeth.

9.4 Selection of cutting conditions and increasing broaching efficiency

The general rules for the selection of cutting conditions as discussed in the chapter on turning of metals, also apply in the case of broaching. However, the selection of cutting conditions in broaching differs from that used in turning. The differences result from the use of special broaching tools for which certain cutting conditions, viz. undeformed chip width and thickness, are determined at the design stage.

Free parameters are cutting speed and cooling method.

It follows that cutting conditions are selected by both the tool designer and the production engineer.

The choice of cutting speed is made according to rules given above, i.e. based on recommended tool life periods. In broaching, the requirements concerning surface quality attain particular importance. At the same time, surface quality may be changed by means of various combinations of cutting conditions, especially cutting speed and cooling. Thus the practice has been adopted that in the case of more exacting surface quality requirements low cutting speeds and intensive oil cooling should be applied. On the other hand, in the case of less exacting requirements concerning surface finish, rather higher cutting speeds and cooling by emulsion are recommended.

REFERENCES

9/1 EREMIN, B. F.: 'Protyagivanie' (Broaching), Mashgiz, Moskva 1950

9/2 MIRACKI, J.: 'Przeciąganie' (Broaching), PWT, Warszawa 1950

9/3 STOLARSKI, B.: 'Untersuchungen über Schnittkraft und Oberflächengüte beim Innenräumen von Stahl', Paper read at XVth Congress of CIRP, Belgium 1965

9/4 PAWŁOWSKI, Z.: 'Zależności trwałości przeciągaczy do otworów wielorowkowych od niektórych parametrów eksploatacyjnych i technologicznych narzędzia', (Dependence of life periods of spline broaches on certain operational and technological parameters of the tool), A doctor's thesis at the Department of Mechanics and Production Engineering of the Warsaw Institute of Technology, 1965

9/5 KARLIC, T.: 'Badania wpływu warunków obróbczych na strukturę geometryczną powierzchni wytworzonych na żeliwie szarym w przeciąganiu zewnętrznym' (Research on the influence of machining conditions on the geometric structure of surfaces generated in grey cast iron by external broaching), A doctor's thesis at the Department of Mechanics of Warsaw Institute of Technology, 1965

CHAPTER 10

Drilling, reaming and boring

10.1 Drilling and counterboring

10.1.1 KINEMATICS OF DRILLING

10.1.1.1 *Definition and types of drilling*

Drilling is a method of machining by cutting, which consists in making holes by means of single- or double-edged tools called *drills*.

It is characteristic for drilling that this cutting process involves two basic motions, i.e. the primary, which is a rotary motion, and the auxiliary, a rectilinear one.

There are four kinematic possibilities of drilling:

(a) Drilling by means of a tool which performs simultaneously both the rotary and the rectilinear motion, while the workpiece remains stationary. This type, called *vertical drilling*, is most commonly applied and is carried out on a specially designed machine tool, the drilling machine.

(b) Drilling by means of a tool performing only the rectilinear motion, parallel to the workpiece axis of rotation, while rotary motion is performed by the workpiece. This is called *horizontal drilling* and is applied in the case of long holes, as it permits one to obtain a greater accuracy of machining. It is also used on automatic, turret and centre lathes.

(c) Drilling by means of a tool performing only the rotary motion, while the workpiece carries out a rectilinear motion, parallel to the axis of drill rotation. This type of drilling can be performed on a vertical milling machine, but is rarely applied in practice.

(d) Drilling by means of a stationary tool, both the rotary and rectilinear motion being performed simultaneously by the workpiece. This type is not used in practice.

Of the four kinematic types presented above the first one, called vertical drilling, has the most importance because of its widespread use. In this chapter the relationships and characteristics of vertical drilling will be discussed.

Another criterion for the division of drilling into technological types is the shape of the drilled hole. Thus *through hole drilling* is distinguished, in which the hole passes through the workpiece, and *blind hole drilling* when the hole reaches only a certain depth in the workpiece material.

With regard to the circumstance whether the entire length or only parts of the cutting edge are working,

(a) drilling from the solid material, shortly called *drilling* (Fig. 10.1a), and
(b) enlarging of a drilled hole, also called *secondary drilling* (Fig. 10.1b) can be distinguished.

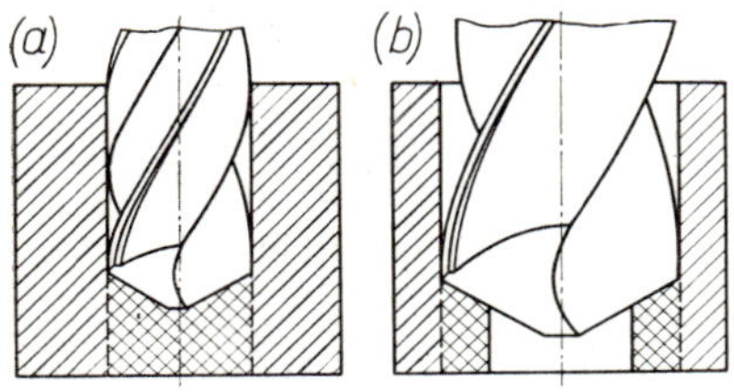

Fig. 10.1 Varieties of drilling: (a) drilling from solid material, (b) enlarging of a drilled hole (secondary drilling)

10.1.1.2 *Cutting speed and the undeformed chip*

Cutting speed v is the peripheral speed measured on the outer diameter of a drill

$$v = \frac{\pi d n}{1000} \quad \text{m/min} \tag{10.1}$$

where: d—nominal outer diameter of drill in mm, n—rotational speed in rev/min

To each point on the cutting edge, depending on the distance from the axis of rotation, corresponds a different peripheral speed. Thus the actual peripheral speed in drilling varies along the cutting edge of the tool point from zero to a maximum value measured on the drill periphery and defined by the term *cutting speed.*

The speed of rectilinear motion of the drill is called *feed rate* and designated by s. It is very low compared with the speed of rotary motion.

In drilling, the undeformed chip is characterized by the cross-section plane passing through the cutting edge parallel to the drilled hole axis and called the transverse cross-section of the undeformed chip.

The characteristic feature of the undeformed chip shape in drilling is its simultaneous generation by three cutting edges: two primary ones called *lips* and one transverse, called the *chisel edge* (Fig. 10.2a). As the length of the latter is very small in relation to the former ones, its part in the change of undeformed chip shape will be neglected.

The work of a drill may be considered as simultaneous work of two lathe tool points engaged in internal straight turning. Thus the undeformed chip to be removed by each of them will be characterized by:

(a) cutting depth a, in the case of drilling $a = d/2$, where d is the diameter of drilled hole in mm,
(b) feed rate s, i.e. the distance by which every point of the cutting edge will move in the direction of the drilling axis during one revolution; as two rigidly joined edges take part in the cutting, each of them, in its travel corresponding to feed s_z, removes an undeformed chip whose dimension in the direction of the drilling axis is s_1 and s_2, respectively.

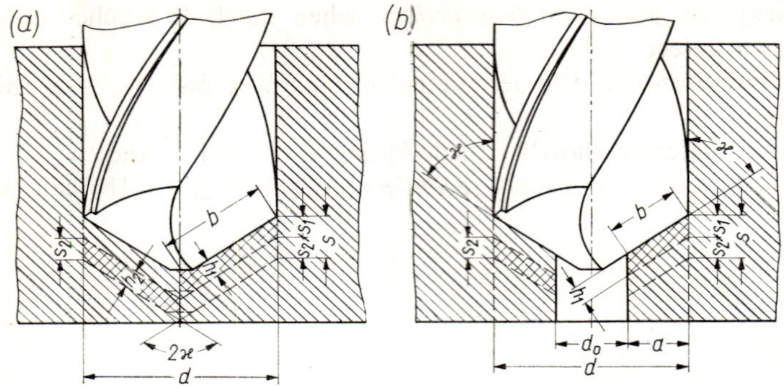

Fig. 10.2 Geometrical elements of the undeformed chip: (a) in plain drilling, (b) in secondary drilling

Since normally a drill is symmetrical,

$$s = s_1 = s_2 = \frac{s}{2} \quad \text{mm} \tag{10.2}$$

(c) cutting edge angle $\varkappa$ which in the case of drilling is constant and depends on the particular application;
(d) thickness of the undeformed chip h_z to be removed by each lip; it can be determined from the relationship

$$h_z = s_z \sin\varkappa \quad \text{mm} \tag{10.3}$$

and if the drill is symmetrical, then $h_z = h_1 = h_2$ (Fig. 10.2);
(e) undeformed chip thickness b, determined from the relationship

$$b = a\frac{1}{\sin\varkappa} = \frac{d}{2\sin\varkappa} \quad \text{mm} \tag{10.4}$$

It follows from these definitions that the transverse cross-sectional area of the undeformed chip falling to each lip will be described by

$$A_z = \frac{s_z\, d}{2} = \frac{s\, d}{4} = h_z b \quad \text{mm}^2 \tag{10.5}$$

The total transverse cross-sectional area of the undeformed chip in drilling

$$A = 2A_z = s_z\, d = \frac{s\, d}{2} = 2h_z\, b \quad \text{mm}^2 \tag{10.6}$$

In the case of secondary drilling, the undeformed chip elements are the same as in drilling, except for the depth of cut which is adopted as

$$a = \frac{d-d_0}{2} \quad \text{mm} \tag{10.7}$$

where: d—diameter of final hole in mm; d_0—diameter of primary hole in mm.

The total cross-sectional area of the undeformed chip in secondary drilling is

$$A = 2A_z = S_z(d-d_0) = \frac{s(d-d_0)}{2} = 2h_z b \quad \text{mm}^2 \tag{10.8}$$

10.1.1.3 *Drilling time*

Drilling time is determined by

$$t_m = \frac{L}{s\,n} \quad \text{min} \tag{10.9}$$

where: L—total drill travel necessary for making the given hole in mm, s—feed rate in mm/rev, n—rotational speed in rev/min.

The length of drill travel may be expressed by

$$L = l_a + l + l_0 \quad \text{mm} \tag{10.10}$$

where: $l_a = d/2 \cot \varkappa$—minimum length of approach in mm (Fig. 10.3),* l—length of drilled hole (cutting length) in mm, l_0—overrun, usually 2–3 mm.

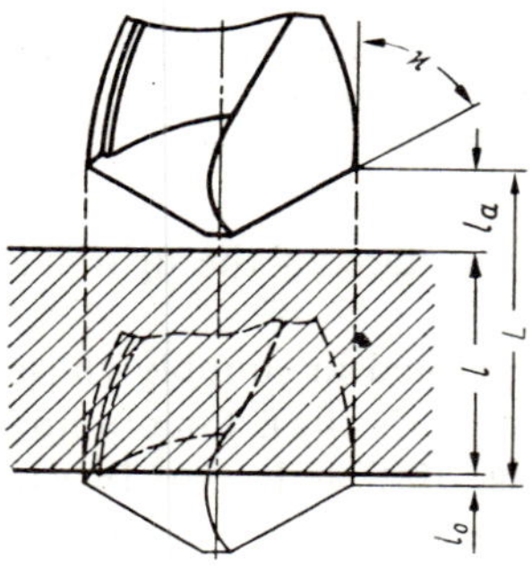

Fig. 10.3 Tool approach, overrun and cutting length

10.1.2 Drill points

10.1.2.1 *Stereometric parameters and point types*

Of several different types of drills (flat drills, twist drills, drills with straight flutes, deep-hole drills, special and combined drills), *twist drills* are the most commonly used.

Therefore, the description of point design and parameters will be given for the example of a twist drill point. In Fig. 10.4 a machine-type twist drill is shown. In a drill, as in any other cutting tool, the working part and the cutting part, or point, may be distinguished.

The drill point is characterized by:

(a) two lips 1, usually rectilinear, resulting from the intersection of face and flank, and the transverse cutting edge called the chisel edge;
(b) lip angle $\varkappa$ and double lip angle, also known as point angle $\varepsilon = 2\varkappa$;
(c) length of transverse cutting edge l_{ch} and chisel edge inclination angle λ_{ch};
(d) distance between lips $2k$, which is also the diameter of an imaginary taper called the drill web (the drill web diameter increases toward the shank at the rate of 1·4–1·8 mm per 100 mm length);
(e) land width f; the land plays the role of the trail edge in the portion adjoining the lip;

* In this formula the influence of transverse cutting edge (chisel edge) has been neglected.

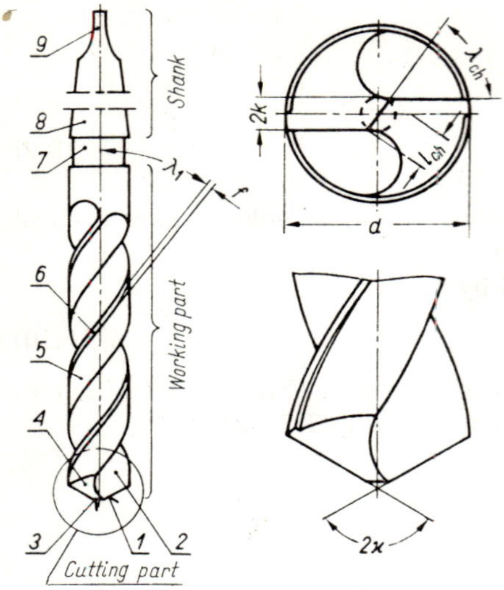

Fig. 10.4 Twist drill and its point: 1—lip, 2—face, 3—chisel edge, 4—flank, 5—flute, 6—land, 7—mark recess, 8—shank taper, 9—tang

(f) land inclination angle λ_1, measured on the peripheral surface of the drill (diameter d); λ_1 is also the inclination angle of flute helix.

In addition, the drill point is characterized by angles determined in suitably chosen drill cross-sections. The latter are: normal (perpendicular to the lip), longitudinal, transversal and cylindrical cross-sections, as shown in Fig. 10.5 and 10.6. In each of them the rake angle γ, wedge angle β and clearance angle α with suitable subscripts are distinguished (in the normal cross-section, angles are given without subscripts).

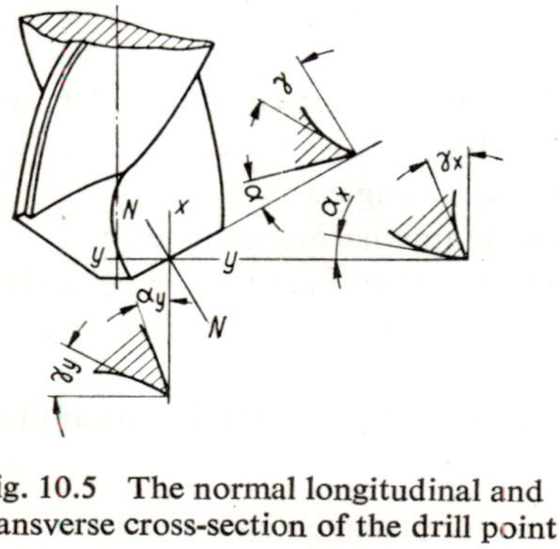

Fig. 10.5 The normal longitudinal and transverse cross-section of the drill point

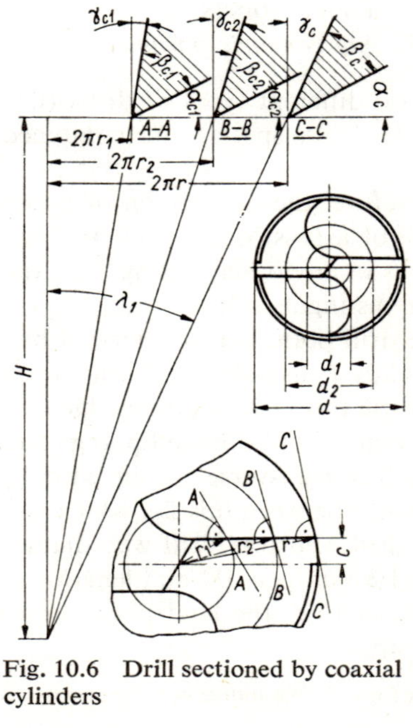

Fig. 10.6 Drill sectioned by coaxial cylinders

Of the characteristic drill point cross-sections listed above, the coaxial cylindrical cross-section has the greatest importance. Because of drill rotation the feed rate changes the point angles in the cylindrical surface (similarly in straight turning changes of angles occur in the transverse cross-sectional plane).

It follows from Fig. 10.6 in which three cylindrical cross-sections of a point are presented, characterized by radii r_1, r_2 and r, that angles vary in the cylindrical cross-section. Values of angles α_c, β_c and γ_c depend on the construction of the drill and of its point.

10.1.2.2 *Types of drill points and recommended values of point parameters*

In addition to drills with straight lips, called *single-cone drills** (Fig. 10.4), there are drills with broken-line lips, called *double-cone drills*.

The lips of these drills (Fig. 10.7) have two lip angles. On the length l_0 from the corner the lip angle is $\varkappa_0$ and on the remaining length it is $\varkappa$, as in single-cone drills.

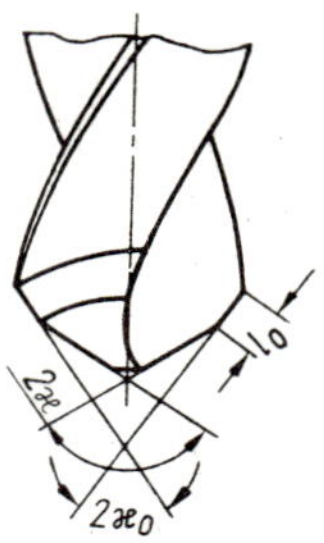

Fig. 10.7 Double-cone drill

In the case of double-cone drills $2\varkappa_0 = 70$–$75°$ and $l_0 = 0{\cdot}2d$ mm. The remaining parameters of the point are the same as in single-cone drills, and thus $l_{ch} = (1/6$–$1/7)d$ for $d \geqq 10$ mm and $l_{ch} = (1/3$–$1/4)d$ for $d < 10$ mm. Chisel edge inclination angle depends on the method of flank grinding. Values of lip angle $\varkappa$ and helix angle (trail edge inclination angle) λ, are given in Table 10.1. A small land width is adopted because the land, besides guiding the drill in the hole also causes additional friction against the wall. For medium size drills $f = (0{\cdot}05$–$0{\cdot}1)d$ is adopted and the larger the drill diameter the more f approaches $0{\cdot}05d$.

To reduce land friction, the external surface of the drill is slightly tapered towards the shank (0·04–0·1 mm on 100 mm length).

The chisel edge is often reduced or has its form changed, both in single-cone and double-cone drill points.

In Fig. 10.8 two examples of chisel edge correction are shown. Chisel edge shortening in the case of Fig. 10.8a causes a change of rake angle (reduction of γ) on a certain width of the face. Chisel edge reduction according to the case of Fig. 10.8b does not influence the rake angle.

A different shape is given to points of bimetallic drills, i.e. those with sintered carbide tips. Dimensions and shapes of such tips are shown in Fig. 10.9 and the

* These names are not accurate, as drill points may have non-conical flanks. However, these names are commonly used and therefore convenient.

Table 10.1 *Recommended values of lip angle $\varkappa$ and helix angle λ_1 of twist drills*

Workp.ece material	$\varkappa°$	$\lambda_1°$
Constructional carbon and alloy steels of $\sigma_B < 140$ kg/mm², stainless steels, sheet steel	59	25–30
Cast iron: soft and chilled	45	25–30
Armour plate, steels of $\sigma_B > 140$ kg/mm²	59	10–15
Steel for fittings (soft steels)	70	—
Brass to be machined on automatic lathes, copper, aluminium and light alloys	65	35–45
Fibre, bakelite, marbre, coal	45	10–15
Plastics (ebonite, triolite)	35	10–15

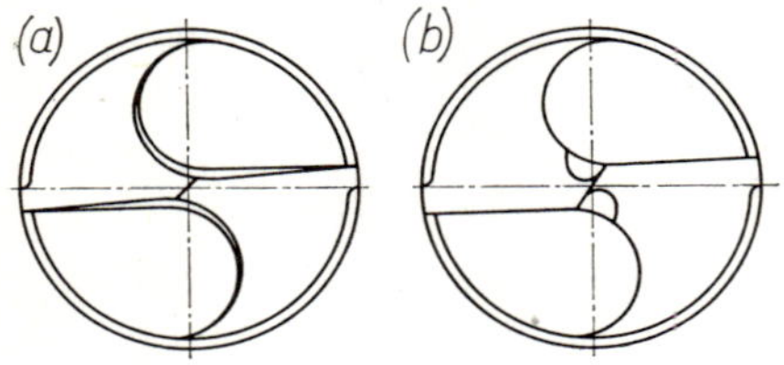

Fig. 10.8 Examples of chisel edge correction

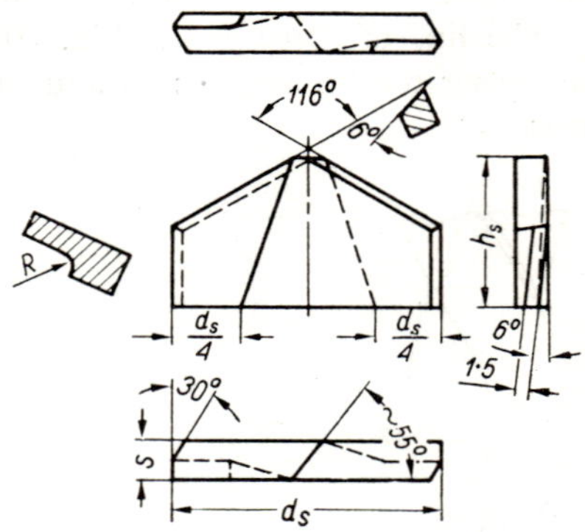

Fig. 10.9 Sintered carbide tip designed by WNII for drill points

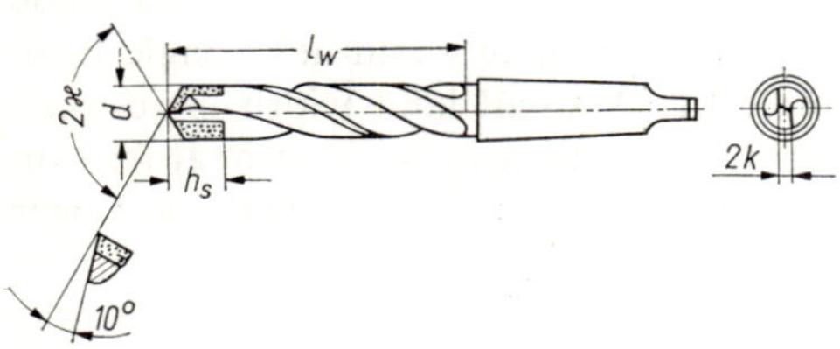

Fig. 10.10 The method of soldering a drill point of sintered carbides

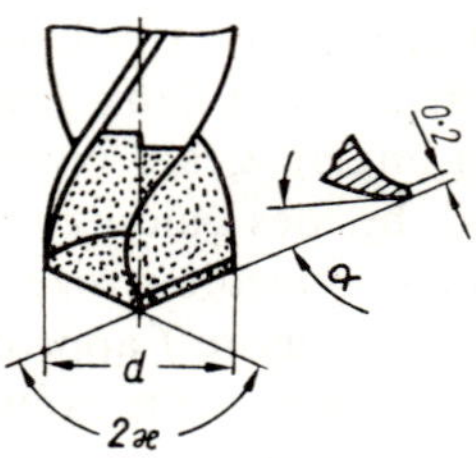

Fig. 10.11 Drill point of sintered carbides, as proposed by N.V. Zharlikov

method of soldering in Fig. 10.10. Drills with sintered carbide tips shaped exactly like extensions of the working part of the drill (Fig. 10.11) have not found wide application because of the technological difficulties of their production.

Sintered carbide drill points are shaped in such a way that rake angle and clearance angle in the normal longitudinal and transverse cross-sections are constant on the whole length of the lip. Recommended values for the point angles of a sintered carbide tip, for various workpiece materials, are given in Table 10.2.

Table 10.2 *Drill point angles for sintered carbides* (acc. to IOS*)

Workpiece material	$\alpha°$	$\gamma°$	$2\varkappa°$	$2\varkappa_0°$	Drill
Grey cast iron HB < 200 kg/mm^2	10	+7	118	70	with helical flutes
Nodular cast iron HB = 300–400 kg/mm^2		0	140		
Malleable cast iron		+4	118		
Stainless steel, chrome-nickel steel σ_B = 140 kg/mm^2	8	0	118		
Manganese tool steel, chilled casting		−2	140		
Hardened steel		−5	140		
Phosphor bronze		+4	130		
Hardened steel	7–9	8–10	130		with straight flutes
Cast iron, non-ferrous metals and their alloys		0–7	130		

* Institute of Machining by Cutting

10.1.2.3 *Drill point during cutting*

Reference lines, used in the determination of clearance angle and rake angle, change during the work of a drill point. As a consequence of feed motion there occurs the phenomenon already described in the discussion on angles in the so-called *motion system.*

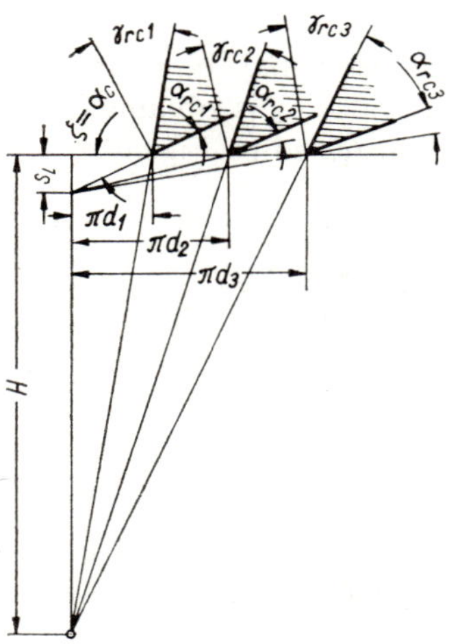

Fig. 10.12 Graphical determination of limit feed

In the drilling process, the characteristic cross-section in which drill point angles are determined is the cylindrical cross-section with its axis coinciding with that of the drill. If ζ denotes the angle through which the reference lines turn for the determination of angles γ_c and α_c (Fig. 10.12), then the values of angles γ_{rc} and α_{rc} are expressed by

$$\gamma_{rc} = \gamma_c + \zeta \tag{10.11}$$

$$\alpha_{rc} = \alpha_c - \zeta \tag{10.12}$$

and the value of ζ may be calculated from

$$\tan\zeta = \frac{s}{2\pi r_i} \tag{10.13}$$

where: s—drill feed per revolution in mm/rev, r_i—radius of the cylindrical cross-section in mm.

It is evident from eqn. (10.13) that the angle of drill motion helix ζ will be greater, the nearer the drill web it is measured. Equation (10.13) may be presented in the form

$$\tan\zeta = \frac{e_s}{2\pi\, e_r} \tag{10.14}$$

where $e_s = s/r = 2s/d$—the ratio of feed rate to outer radius of the drill (or double feed rate to outer diameter of the drill).

Values of ζ_{max} are of the order from 1°, when $s/d = 0{\cdot}005$, to 13° when $s/d = 0{\cdot}1$.

With regard to correct working of the drill point, clearance angles should be suitably greater at the chisel edge and may diminish toward the periphery of the drill, so that the inequality

$$\alpha_{rc} \geqq 0 \qquad \text{or} \qquad \alpha_c \geqq \zeta$$

is maintained. As the value of ζ decreases more rapidly towards the periphery than does the clearance angle, point seizure may occur earlier at the chisel edge (with the exception of drills sharpened by the coaxial method, i.e. when the axis of rotation in grinding coincides with the drill axis).

Thus for every grinding method a permissible limit feed may be determined, the exceeding of which causes drill seizure. Limit feed value s_l is obtained when angle ζ at the chisel edge equals α_c. This is graphically illustrated by Fig. 10.12.

10.1.3 Cutting resistance in plain and secondary drilling

10.1.3.1 *Cutting resistance and its components*

Cutting resistance in drilling consists of:
(a) resistance appearing at the lips,
(b) resistance appearing at the chisel edge,
(c) resistance appearing at the land.

The first two of these components are the resistance of the undeformed chip to plastic strain, and the resistance due to external friction. The resistance at the land mostly results from friction against the machined surface.

It is usually assumed that in a symmetrically loaded drill the cutting resistance is decomposed into two components; their origins are conventionally assumed to be situated at a distance of $0{\cdot}5r$ from the drill axis.*

Each of the resultants of cutting resistance—F' acting at one lip and F'' at another—is decomposed in three directions (Fig. 10.13):

(a) axial components F_{f1} and F_{f2}, in the direction parallel to the drill axis and, simultaneously, to feed direction,

* Results of some research indicate that this assumption is not quite accurate [10/7].

(b) circumferential components F_{v1} and F_{v2}, in the direction perpendicular to the projection of the lips on the plane normal to the drill axis,
(c) thrust components F_{p1} and F_{p2} in the direction parallel to the projections of the lips on the plane normal to the drill axis.

It is easy to perceive that such a system of forces may be reduced to a single resultant force, called the *axial force F*

$$F_f = F_{f1} + F_{f2} \tag{10.15}$$

and a resultant moment, called the *torque M*

$$M = M_v - M_p \tag{10.16}$$

while

$$M_v = F_{v1} r = F_{v2} r \tag{10.17}$$

$$M_p = 2F_{p2} c = 2F_{p3} c \tag{10.18}$$

on the assumption that $F_{v1} = F_{v2}$ and $F_{p1} = F_{p2}$.

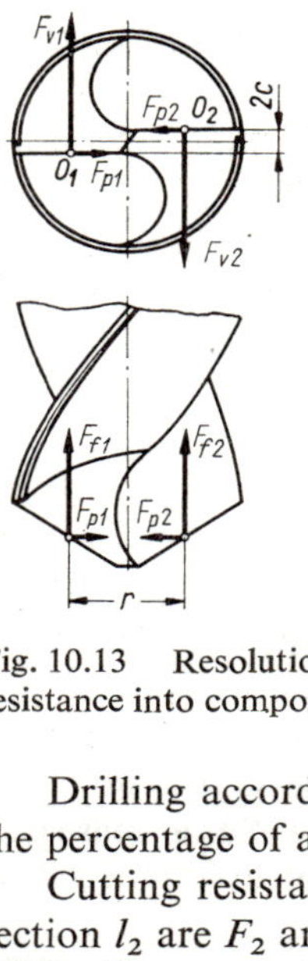

Fig. 10.13 Resolution of the cutting resistance into components

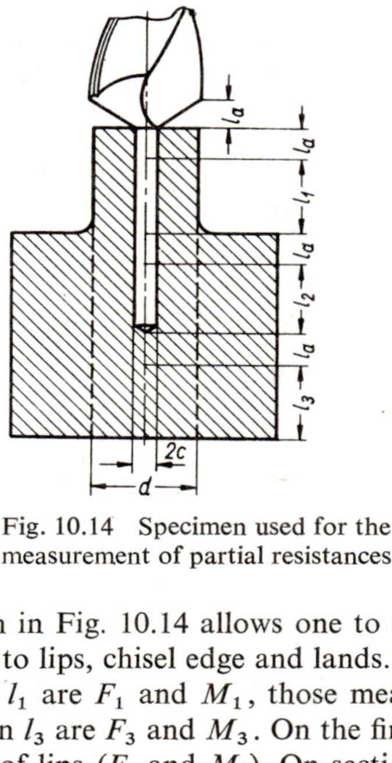

Fig. 10.14 Specimen used for the measurement of partial resistances

Drilling according to the diagram shown in Fig. 10.14 allows one to determine the percentage of axial force and torque due to lips, chisel edge and lands.

Cutting resistances measured on section l_1 are F_1 and M_1, those measured on section l_2 are F_2 and M_2 and those on section l_3 are F_3 and M_3. On the first section the resistance resulted only from the action of lips (F_k and M_k). On section l_2 land resistances F_l and M_l were added and on section l_3 chisel edge resistances F_{ch} and M_{ch}. Thus it may be determined that:

$$\begin{aligned} F_k &= F_1 &\quad\text{and}\quad M_k &= M_1 \\ F_l &= F_2 - F_1 &\quad\text{and}\quad M_l &= M_2 - M_1 \\ F_{ch} &= F_3 - F_2 &\quad\text{and}\quad M_{ch} &= M_3 - M_2 \end{aligned} \tag{10.19}$$

Experiments show that F_k, F_l and F_{ch} as well as M_k, M_l and M_{ch} vary depending on feed rate and drill diameter. Their average values may be assumed to be (within a feed range of 0·2–0·4 mm/rev and for d = 15–35 mm):

$$\left.\begin{aligned} F_k &= 50\% \\ F_l &= 10\% \\ F_{ch} &= 40\% \end{aligned}\right\} \text{of the entire axial force } F \qquad \left.\begin{aligned} M_k &\approx 80\% \\ M_l &\approx 12\% \\ M_{ch} &\approx 8\% \end{aligned}\right\} \text{of the entire torque } M$$

These data are valid for drilling to a depth not exceeding 2·5d.

10.1.3.2 *Factors influencing the cutting resistance in drilling*

Factors influencing the cutting resistance in drilling may be divided—as in the case of turning—into factors connected with:
(a) workpiece material
(b) the tool
(c) cutting conditions.
The influence of the workpiece material, i.e. of its chemical composition and structure, on cutting forces is very important. In practice, when a single kind of workpiece material is involved, the cutting resistance depends directly on the hardness (HB) or ultimate tensile strength (σ_B) of this material.

Thus, for example, in the machining of carbon steel with high-speed steel drills or sintered carbide points the following empirical formulae are valid

$$\begin{aligned} F &= C_F \sigma_B^{0\cdot 75} \\ M &= C_M \sigma_B^{0\cdot 75} \end{aligned} \qquad (10.20)$$

Similarly, the formulae for the axial force and torque in drilling of cast iron are

$$\begin{aligned} F &= C_F \, \mathrm{HB}^{0\cdot 6} \\ M &= C_M \, \mathrm{HB}^{0\cdot 6} \end{aligned} \qquad (10.21)$$

Of the drill elements the following exert a considerable influence on the cutting resistance: lip angle $\varkappa$, drill helix inclination angle λ_1 and the shape and size of the chisel edge.

It follows from research conducted so far [10/2] and [10/3] that an increase in lip angle causes an increase in axial force F and a reduction of torque M (this influence is the more distinct, the smaller is the helix inclination angle λ_1). Some reasearch concerning the drilling of plastics [10/2] indicates the existence of optimum values of lip angle.

In most research works carried out in the past, the influence of the sharpening method upon the angles α and γ along the lips was neglected. However, in each sharpening method the variation of lip angle causes a change in clearance angles. Therefore, in order to compare research results identical methods of sharpening must be used i.e. of shaping the flanks.

The influence of flute helix inclination angle is easier to analyse as there is an unequivocal relationship between the rake angle and λ_1. Increasing the angle λ_1 increases the rake angle and reduces cutting resistance. As a result, both the axial force and torque decrease. Experiments confirm this influence of angle λ_1.

The chisel edge and the form of point tip also influence the cutting resistance. An increase of chisel edge length increases both the axial force and torque.

An advantageous decrease in cutting resistance may be obtained by the application of various methods of chisel edge reduction or the creation of such point shape that the chisel edge cuts instead of scraping or burnishing the material.

Drill point wear also has a distinct influence on cutting resistance. The torque and axial force both increase as the drill becomes dull. The increase in resistance is quite insignificant at first, but then resistance rises rapidly.

Of the cutting conditions, the following exert a significant influence on cutting resistance: feed rate, depth of cut, cutting speed, drilling depth and length, cooling system and intensity.

In drilling, as in all other machining methods, increasing the feed rate causes an increase of cutting resistance, although the specific cutting energy actually decreases. Thus the increment of total resistance, i.e. of axial force and torque, is less than proportional to feed rate increase.

Depth of cut, in the case of primary drilling equal to half the diameter, and in the case of secondary drilling to a half of the difference of hole diameters, has a similar influence to the feed rate, but to a higher degree. This influence is particularly marked with regard to torque increase, since with an increase in depth of cut the circumferential cutting forces rise and their point of origin moves away from the axis and thus their arm of action increases.

Research on the influence of cutting speed on cutting resistance in drilling has shown that this relationship is of a complicated character. The influence of feed rate and cutting speed are interconnected.

Cutting speed influence increases as feed rate is reduced. For higher feed values ($s = 0{\cdot}63$ mm/rev) Barbashov's research [10/16] did not show any visible influence of cutting speed.

With increasing drilled length the working conditions of the drill deteriorate. Land friction against the machined surface becomes more intense and the lubrication deteriorates because of inadequate flow of the liquid toward the point. As a result, cutting resistance increases as the drill penetrates deeper into the material.

Cooling has a similar influence on cutting resistance in drilling as in other machining methods. The drill works under conditions of difficult cutting heat flow and therefore the use of a cutting fluid is very advantageous. Thus, for example, in the drilling of carbon steel the reduction of axial force and torque as a consequence of the application of cutting fluids amounts to 10–30%, in the drilling of grey cast iron to 10–18%, and in the drilling of light alloys even to 30–40%.

10.1.3.3 *Calculation of cutting forces in plain and secondary drilling*

Research on drilling is not as much advanced as that on turning. Therefore in formulae for the axial force and torque in drilling a lesser number of factors appear separately than in formulae for cutting forces in turning.

Such factors as the type of point material, cooling method, point angles and many others are taken account of by determining the values of the constant C_F or C_M for certain values of those parameters.

The general form of the formulae for axial force and torque in plain drilling are as follows:

$$F = C_F\, d^{b_F}\, s^{u_F}\, K_H \quad \text{kg} \tag{10.22}$$

$$M = C_M\, d^{b_M}\, s^{u_M}\, K_H \quad \text{kg mm} \tag{10.23}$$

where: C_F and C_M —constants taking into account the influence of all factors not appearing separately in the formulae (Table 10.3), d—nominal diameter of drill in mm, b_F and b_M—exponents characterizing the influence of drill diameter, s—feed rate in mm/rev, u_F and u_M—exponents characterizing the influence of feed rate, K_H—correction coefficient taking into account the mechanical properties of the workpiece material.

In the case of secondary drilling the corresponding formulae are:

$$F = C_F\, d^{b_F}\, a^{e_F}\, s^{u_F}\, K_H \quad \text{kg} \tag{10.24}$$

$$M = C_M\, d^{b_M}\, a^{e_M}\, s^{u_M}\, K_H \quad \text{kg mm} \tag{10.25}$$

where: a—depth of cut in mm, e_F and e_M—exponents indicating the influence of the depth of cut.

Values of the constants and exponents appearing in eqns. (10.22) to (10.25) are listed in Table 10.3. They are valid for machining with drills under permissible amounts of wear.

Formulae for axial force and torque as given above concern so-called average values. Although drilling is considered to be a machining operation of the continuous type, there still occur oscillations of the cutting resistance during the work of a drill. These oscillations are caused mainly by vibrations resulting from insufficient rigidity of the machine tool and of the drill mounting. Axial force and torque decrease continuously as the drill is being withdrawn from the workpiece material and the decline of axial force begins earlier than that of the torque. The reduction of cutting resistance is explained by the fact that before the drill leaves the workpiece material (i.e. when the chisel edge is still in the material) a plastic deformation of the wall occurs. Depending on the ductility of the material, the wall bends and bulges which causes characteristic burrs to appear after the passage of the drill.

10.1.3.4 *Cutting power in drilling*

Cutting power in drilling can be expressed as

$$N_{me} = \frac{M\,n}{1{\cdot}36 \times 716\,200} + \frac{F\,n\,s}{1{\cdot}36 \times 4500 \times 1000} \quad \text{kW} \tag{10.26}$$

where: M—torque in kg mm, n—rotational speed in rev/min, F—axial force in kg, s—feed rate in mm/rev.

Calculations show that within the range of practically applied feed rates the second component, i.e. the feed power, constitutes a very small percentage of the total power, its share amounting to 0·5–1·5%. As the inaccuracy of cutting power measurements considerably exceeds these values, feed power may be neglected,

Table 10.3 *Average values of constants and exponents contained in formulae for axial force and torque in plain and secondary drilling*

Kind of drill (material and diameter)	Workpiece material	Coolant	Values of constants and exponents													
			in plain drilling						in secondary drilling							
			C_F	b_F	u_F	C_M	b_M	u_M	C_F	b_F	e_F	u_F	C_M	b_M	e_M	u_M
High-speed steel points of diameter d = 10–60 mm	constructional carbon steels σ_B = 75 kg/mm²	emulsion	84·7	1·0	0·7	33·8	1·9	0·8	—	—	—	—	—	—	—	—
	grey cast iron HB = 190 kg/mm²	air	60·5	1·0	0·8	23·3	1·9	0·8	—	—	—	—	—	—	—	—
	malleable cast iron HB = 190 kg/mm²	emulsion	52·5	1·0	0·8	20·3	1·9	0·8	—	—	—	—	— —		—	—
	bronze medium hard HB = 100–140 kg/mm²	air	31·5	1·0	0·8	12·2	1·9	0·8	—	—			—	—		
Sintered carbide points of diameter d = 10–30 mm	Constructional carbon steel σ_B = 75 kg/mm²	emulsion	51·0	1·4	0·8	61·6	2·0	1·0	—		—	—	30	0·75	0·8	0·95
	grey and malleable cast iron HB = 190 kg/mm²	air	167	0·9	0·1	33·2	2·0	1·0	—				7·3	0·85	0·8	0·7

and therefore the cutting power in drilling may be calculated according to the approximate formula

$$N_{me} \frac{M\,n}{1{\cdot}36 \times 716\,200} \quad \text{kW} \tag{10.27}$$

The necessary motor power may be calculated from the well-known formula

$$N_m = \frac{N_{me}}{\eta_e}$$

where η_e—effective efficiency.

10.1.4 Drill point wear and life

10.1.4.1 *Drill wear symptoms and wear criteria*

Drills wear gradually with time, until they completely loose their ability to cut.

The most distinct symptoms of progressing wear are the increase of cutting resistance, cutting power and point wear. These symptoms are closely interconnected.

The cutting resistance may increase in different ways. Depending on the kind of point wear either the axial force may increase more quickly than the torque or vice-versa, or else their increase may be proportional.

Observations indicate that a more distinct increase of axial force accompanies chisel edge wear, while land wear has a more evident effect on the increase of torque. On the other hand, drill wear at the corners causes an almost proportional rise of both the torque and axial force.

In Fig. 10.15 examples of torque and axial force changes resulting from drill point wear at the corners (Fig. 10.15a), at the lands (Fig. 10.15b) and at the chisel edge (Fig. 10.15c) are presented. These diagrams concern the drilling of carbon

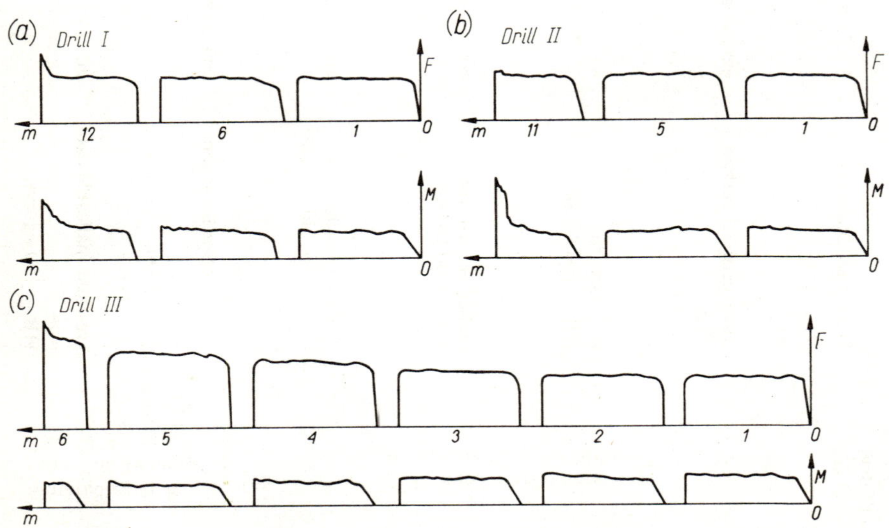

Fig. 10.15 Diagrams of the increase of axial force and torque depending on drilled length

steel with high-speed steel drills of diameter $d = 23$ mm. The moment when the cutting ability is lost is marked on the diagrams by a distinct rise of cutting resistance.

Observation of cutting resistance variations requires special measuring devices which change the drill working conditions to a certain degree. This disadvantage is eliminated in the method of indirect observation of point wear, i.e. the observation of power consumption by the motor from the mains, conducted by means of a recording wattmeter. The latter shows a distinct increase of power only in the final phase of wear, when the point looses its cutting ability.

In workshop practice drill point wear is evaluated by observing point abrasion. A drill point may be abraded at the chisel edge, at the corners and along the lip on the flank side, at the lands in their portion taking part in cutting and—less often—at the face (appearance of a groove).

Characteristic examples of the wear of single-cone drill points are given in Fig. 10.16: (a) and (e) chisel edge wear, (b) land wear, (c) wear of land and partly of the corner, (d) corner wear. A case of considerable wear at the corners, so that small chips are seen to be fused-in on the flank, is given in Fig. 10.17. This case was observed in the drilling of grey cast iron of medium hardness.

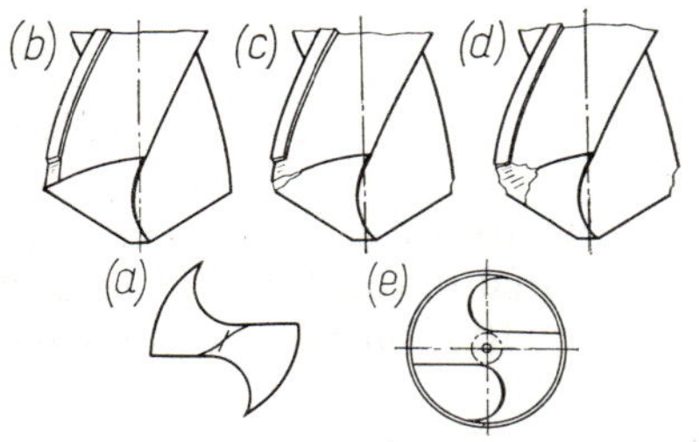

Fig. 10.16 Examples of the wear of single-cone drills

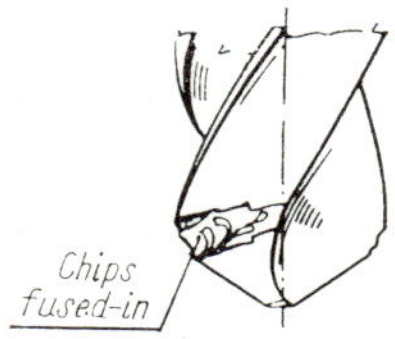

Fig. 10.17 Wear at the corners, with chips fused-in

Double-cone drills wear in a similar way to the single-cone ones. The most frequent types of double-cone wear, observed in the drilling of cast iron, is given in Fig. 10.18a and b. Very strong abrasion sometimes leads to breaking off of the drill point (Fig. 10.18c).

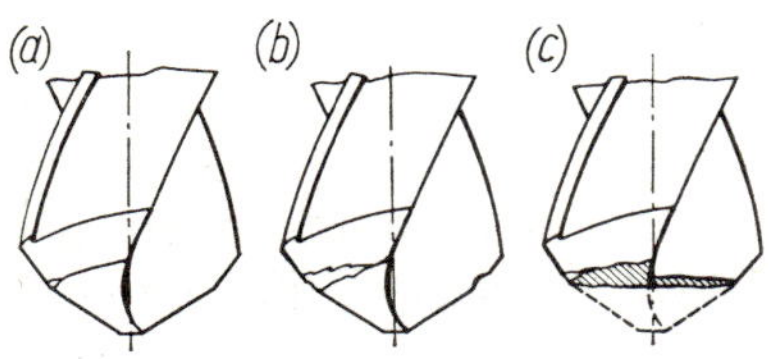

Fig. 10.18 Consecutive phases of the wear of double-cone drills

The cause of non-uniform point wear may be the heterogeneity of material and, most frequently, the point shape resulting from sharpening.

Thus, drill point wear at the corners is caused by wedge angle β diminishing toward the drill periphery, where cutting speed and thermal load are at a maximum. As a result of this, the temperature of the corner is higher than near the web and the wear process is faster.

Wear or seizure at the chisel edge is usually caused by exceeding the limit feed rate permissible with regard to the dynamic clearance angle at the chisel edge. The value of this angle may be close to zero.

When the drill is sharpened correctly and cutting conditions are suitably selected, as normal symptoms of drill wear may be regarded

(a) when drilling steel—wear of the flank along the lip, growing towards drill periphery (Fig. 10.16d);

(b) when drilling cast iron—wear of the corners with lands (Fig. 10.16c) or of the lands only.

Those symptoms have been adopted as criteria for the evaluation of drill wear, and the permissible wear is that resulting from the condition of longest working time of the drill (drill service life). This coincides with the condition of obtaining the maximum value of the product in terms of the number of sharpenings and the drill life period. The number of feasible sharpenings is:

— for a high-speed steel drill

$$i = \frac{0{\cdot}7l_w}{e+0{\cdot}2} \tag{10.28a}$$

— for a drill with sintered carbide point

$$i = \frac{h_s-0{\cdot}4d}{e+0{\cdot}2} \tag{10.28b}$$

where: l_w—length of the working part of the drill (Fig. 10.10) in mm, e—layer of drill point material ground off in one sharpening, measured in the direction of drill length, in mm, h_s—height of sintered carbide tip (Fig. 10.9) in mm, d—nominal diameter of drill in mm.

The value of e depends on the permissible wear VB of the drill point. The mean recommended values of VB are:

(a) for high-speed steel drills, when drilling steel $VB = 1$–$1{\cdot}2$ mm and when drilling cast iron $VB = 0{\cdot}5$–$1{\cdot}2$ mm

(b) for drills with sintered carbide points the values of permissible wear VB—both when drilling steels (non-hardened) and cast iron, varies with drill diameter for $d = 10$–12 mm $VB = 0{\cdot}4$ mm, for $d = 13$–18 mm $VB = 0{\cdot}6$ mm, for $d = 19$–25 mm $VB = 1{\cdot}0$ mm and for $d = 26$–30 mm $VB = 1{\cdot}3$ mm.

10.1.4.2 *Drill life and life speed*

Drill point life corresponds to a certain total length of drilling work. Therefore in some cases the total length of all drilled holes is used—instead of life period—to characterize the drill.

The tool point life period for a given drill and given conditions of its work is determined with regard to economic assumptions. That is why the cutting ability of a drill point is most often characterized by its life speed. The relationship between the life period T and cutting speed v_T is determined by the well-known formula

$$v_T = \frac{C_v}{T^m}$$

where: C_v—a constant depending on a number of factors not appearing separately in the formula, $m = 1/z$—exponent depending mainly on the kind of tool material and workpiece material.

Values of recommended drill point life periods are calculated from the formula for economic tool point life in single-tool machining.

Calculation of life speed for a changed value of recommended tool point life is carried out according to formula

$$v_{T2} = B_T v_{T1} \qquad \text{while} \qquad B_T = \sqrt[z]{\frac{T_1}{T_2}}$$

where: T_1—the previous tool point life in min, T_2—the newly adopted tool point life in min.

For the calculation of B_T a knowledge of exponent z is necessary. Average values of z are:

(a) for high-speed steel drills
- — when drilling steel $z = 5$,
- — when drilling cast iron and bronze $z = 8$,

(b) for drills with sintered carbide points
- — when drilling cast iron $z = 2{\cdot}5$,
- — when drilling steel $z = 4$.

10.1.4.3 *Calculation of tool life speed in drilling*

For the calculation of tool life speed in plain drilling the following statistical and empirical formula is used

$$v_T = \frac{C_v\, d^{b_v}}{T^m\, s^{u_v}} B_M\, B_s\, B_H\, B_l\, B_w \quad \text{m/min} \tag{10.29}$$

where: C_v—a constant taking into account the influence of factors not appearing separately in the formula (Table 10.4), m, b_v and u_v—exponents depending on workpiece material and other factors, d—drill diameter in mm (Table 10.7), T—assumed tool point in mm, s—feed per revolution in mm/rev, B_M—correction coefficient taking into account the type of material, B_s—correction coefficient taking into account the state of the material (for steel only), B_H—correction coefficient taking into account the mechanical properties of the workpiece material, B_l—correction coefficient taking into account the drilled depth, B_w—correction coefficient taking into account the adopted value of permissible wear VB (Table 10.5).

Tool life speed in secondary drilling is calculated according to

$$v_T = C_v \frac{d^{b_v}}{T^m\, s^{u_v}\, a^{e_v}} B_M\, B_s\, B_H\, B_l\, B_w \quad \text{m/min} \tag{10.30}$$

where: a—depth of cut in mm, e_v—exponent depending on workpiece material and drill point material. The remaining denotations are as in eqn. (10.29).

Values of constants contained in eqn. (10.30) are given in Table 10.4; they are valid when all correction coefficients—B_M, B_s, B_H, B_l and B_w—are equal to unity.

Table 10.4 *Values of constants contained in formulae for tool life speed in plain and secondary drilling*

Drill point material	Workpiece material	Coolant	Machining variety	Feed range	Values of constants C_v	m	b_v	u_v	e_v
High-speed steel with 9% of tungsten	constructional carbon steel σ_B = 75 kg/mm²	emulsion	plain drilling	s > 0·2	5·0	0·2	0·4	0·7	—
				s < 0·2	7·0	0·2	0·4	0·5	—
			secondary drilling	—	11·6	0·2	0·4	0·5	0·2
	grey cast iron HB = 190 kg/mm²	air	plain drilling	s > 0·3	10·5	0·125	0·25	0·55	—
				s < 0·3	12·5	0·125	0·25	0·4	—
			secondary drilling	—	16·7	0·125	0·25	0·4	0·1
	malleable cast iron HB = 150 kg/mm²	emulsion	plain drilling	s > 0·3	11·4	0·125	0·25	0·55	—
				s < 0·3	13·2	0·125	0·25	0·4	—
			secondary drilling	—	18·1	0·125	0·25	0·4	0·1
	bronze, medium hard HB = 100–140 kg/mm²	air	plain drilling	s > 0·3	23·4	0·125	0·25	0·55	—
				s < 0·3	27·2	0·125	0·25	0.4	—
			secondary drilling	—	37·2	0·125	0·25	0·4	0·1
Sintered carbides	constructional steel σ_B = 75 kg/mm²	emulsion	plain drilling	s > 0·12	15·4	0·25	0·6	0·3	—
				s < 0·12	10·2	0·25	0·6	0·5	—
			secondary drilling	—	13·9	0·25	0·6	0·3	0·2
	grey and malleable cast iron HB = 190 kg/mm²	air	plain drilling	—	17·4	0·4	0·5	0·5	—
			secondary drilling	—	22·65	0·4	0·5	0·45	0·15

Table 10.5 *Values of correction coefficients B_w for the calculation of tool life speed depending on the permissible drill point wear adopted*

Workpiece material	Drill diameter in mm: 10–12			13–18				19–25				25–30			
	Wear VB in mm														
	0·2	0·3	0·4	0·25	0·35	0·45	0·6	0·3	0·5	0·7	1·0	0·4	0·6	0·9	1·3
Non-hardened steel	0·81	0·92	1·0	0·78	0·85	0·92	1·0	0·7	0·81	0·92	1·0	0·7	0·8	0·9	1·0
Cast iron	0·76	0·90	1·0	0·70	0·80	0·90	1·0	0·62	0·76	0·9	1·0	0·62	0·73	0·87	1·0

10.1.5 DRILLING ACCURACY

In the case of drilling the dimensional accuracy and shape is much more important than surface quality, since drilling is seldom the hole finishing operation. Drilled holes are often enlarged by secondary drilling or reamed. Thus the drilled surface quality is frequently only a temporary one, while in cases when drilled holes do not undergo further machining, the quality of their surface is usually of no particular importance.

10.1.5.1 *Surface quality of drilled holes*

Working conditions of a drill are more difficult than e.g. those of a turning tool. As chips are removed along flutes, the machined surface becomes scratched and thus its quality deteriorates. Drill lands also frequently scrape the hole surface because chip particles become welded to them. An influence is exerted by drill feed rate and taper toward the shank, too, similar to that of turning tool feed rate and cutting edge angle. With increasing feed rate microirregularities of the surface increase.

A considerable influence is exerted on the quality of drilled hole surfaces by cutting fluids. Cooling markedly reduces surface irregularities as compared to those resulting from dry cutting.

On the basis of experiments conducted in laboratories and in factories [10/5] it can be stated that drilling allows one to obtain roughness numbers 2, 3 and 4 for steel and cast iron surfaces (which means that R_{tm} varies from 20 to 125 μm depending on the cutting conditions). Drilling with the use of a coolant, carried out on rigid drilling machines permits one to obtain roughness number 5, i.e. $R_{tm} < 20$ μm.

10.1.5.2 *Dimensional accuracy and shape*

Machining accuracy is the result of functioning of the machine tool–workpiece–tool system.

Drilling accuracy concerns diameter and ovality of the drilled hole. Quality is defined as follows

$$\Delta D = D_{max} - D_{min} \tag{10.31}$$

In Fig. 10.19 are shown the accuracy of holes drilled in grey cast iron of HB = 190–200 kg/mm² with carbon steel twist drills having single-cone and double-cone points.*

Research conducted on hole diameter (for d = 18–35 mm) shows that when drilling grey cast iron within the range studied the following conclusions can be drawn

(a) for smaller diameters, with increasing feed rate the increase in diameter diminishes,
(b) increasing the cutting speed generally causes a reduction of diameter increase,
(c) when single-cone drills are used, hole diameter increases more than in the case of double-cone drills.

* Measurements were carried out in IOS by R. Śniechowski, M.Sc. (Eng).

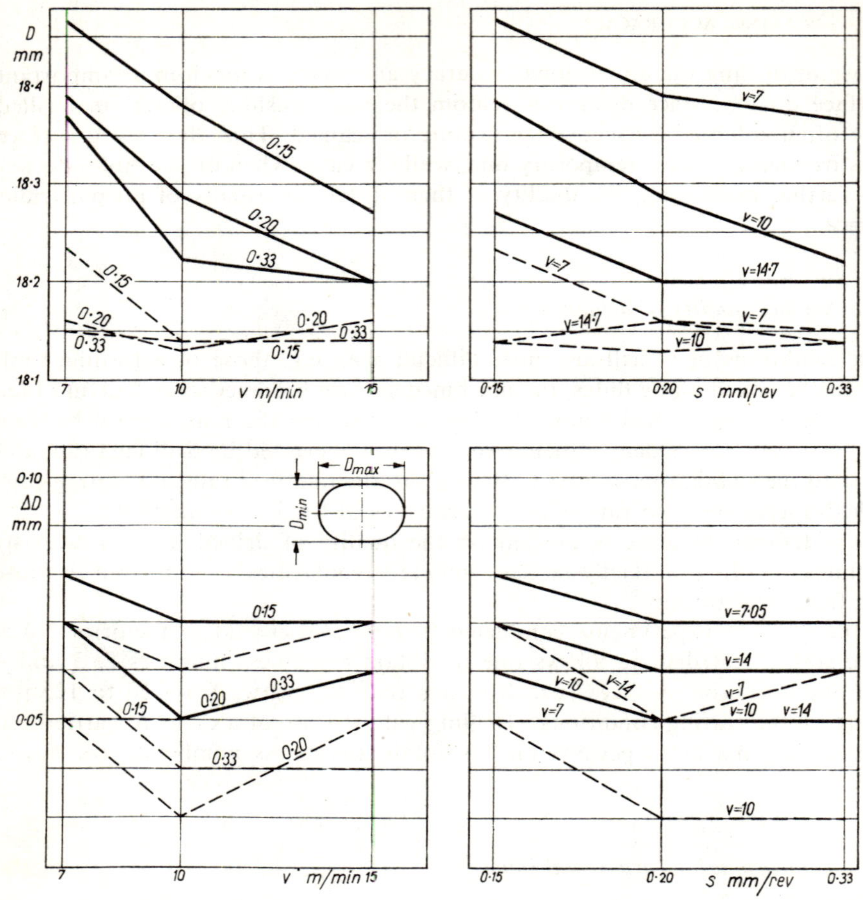

Fig. 10.19 Diameter increase and ovality of holes drilled with single-cone and double-cone drills: continuous lines denote single-cone drill with $2\varkappa = 118°$, dashed lines—double-cone drill with $2\varkappa = 118°$, $2\varkappa_0 = 75°$, $d = 18$ mm; dry cutting

The same conclusions apply to the ovality of the hole, in this case however the influence of feed rate is less marked.

One of the main causes of diameter increase is lip asymmetry which may be due to inaccurate sharpening. In this case the shape and dimensions of undeformed chip cut by the particular lips will be as shown in Fig. 10.20.

If asymmetry is e and lip angle $\varkappa$, then the original feed rate per lip $s_z = 0{\cdot}5\ s$ will change and become:

$$\begin{aligned} s_1 &= 0{\cdot}5s+k \\ s_2 &= 0{\cdot}5s-k \end{aligned} \tag{10.32}$$

While the feed increment per lip k is determined by

$$k = e\cot\varkappa \tag{10.33}$$

then

$$s_1 = 0{\cdot}5s + e\cot\varkappa \qquad s_2 = 0{\cdot}5s - e\cot\varkappa \tag{10.34}$$

From eqn. (10.34) a limit value of e can be found for which $s_1 = s$ and $s_2 = 0$, i.e. when only one lip will be cutting, i.e.

$$e_{\text{lim}} = \frac{s}{2\cot\varkappa} \tag{10.35}$$

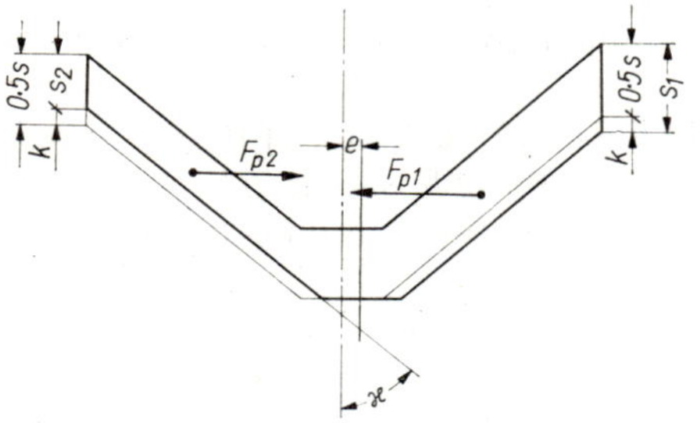

Fig. 10.20 Change of undeformed chip cross-section as a result of drill point asymmetry

In the case of drills used for steel and cast iron, when $\varkappa = 59°$

$$e_{\text{lim}} = 0{\cdot}833s \quad \text{mm} \tag{10.36}$$

As a result of a different undeformed chip thickness at each lip, thrust forces F_{p1} and F_{p2} become unequal, viz. $F_{p1} > F_{p2}$. Bending of the drill results and—as a consequence of its simultaneous rotation—hole diameter increases.

Average statistical increments of drilled hole diameter, i.e. $D_{\text{max}} - D_{\text{nom}}$, are given in Table 10.6. The accuracy of holes drilled with the use of drill guides is greater than in drilling without such guides.

Table 10.6 *Average diameter increases of drilled holes*

Nominal hole diameter D_{nom} mm	Hole increase values $D_{\text{max}} - D_{\text{nom}}$	
	Drilling without drill guide	Drilling with drill guide
less than 6	0·12	0·07
6–10	0·16	0·09
10–18	0·21	0·12
18–30	0·27	0·16
30–50	0·34	0·21

In the case of drilled holes deeper than $3d$ the increase in diameter increases by some 15% for $3d < l \leqq 6d$ and by some 25% for $\lambda > 6d$ [10/5].

10.1.6 Choice of cutting conditions and increasing the metal removal rate in drilling

10.1.6.1 *General rules for the choice of cutting conditions in drilling*

General rules for the choice of cutting conditions in drilling are the same as in turning.

Of the technical requirements involved in plain and secondary drilling, those concerning the dimensional accuracy and shape are the most important.

The depth of cut is established beforehand, so that the choice of cutting conditions is limited to the determination of s and v. The starting point for the selection of most frequently varying cutting conditions s and v is—similarly as in turning—the economic or recommended tool life.

The choice of cutting conditions in drilling is reduced to the following steps:

(a) determination of the number of passes with regard to accuracy requirements,
(b) listing of machining limitations with regard to feed rate and calculation of the permissible feed rate,
(c) correction of the calculated feed rate taking into account the available feed steps of the given machine tool,
(d) selection or calculation of the suitable drill life and calculation of the life speed,
(e) calculation of rotational speed corresponding to tool life speed,
(f) correction of the calculated rotational speed taking into account the available speeds of the given drilling machine,
(g) calculation of cutting power and checking of the motor loading,
(h) calculation of drilling time.

A specific feature of drilling is that the average loading coefficient of the machine tool motor is lower than in the case of turning. This is so, because every drilling machine is intended for work within a wide range of diameters. As both the feed rate and cutting speed increase with drill diameter, there is a very wide span of power demand as various drills within the range characteristic for the given drilling machine are used.

Thus, for example, the power demand of a drilling machine intended for drilling within a diameter range of 10–60 mm is 0·7–0·9 kW for drilling steel of $\sigma_B = 75$ kg/mm^2 with the use of coolant. It follows that for drilling the requirements concerning the coefficient of motor loading must be reduced.

Limitations of the choice of maximum feed rate differ somewhat from those applying to turning and will therefore be discussed separately.

10.1.6.2 *Feed rate limitations*

The choice of feed rate for drilling must be made taking account of limitations created by:

(a) drill strength,
(b) machine tool strength with regard to loading by axial force and torque,
(c) kinematics of cutting,
(d) stepping of feeds of the drilling machine.

In the case of drilling the surface quality does not constitute a limitation on the choice of the feed rate. This choice is reduced to the calculation of permissible feed rates with regard to the limiting factors listed above. The lowest feed rate is then selected and it fulfils all limiting conditions.

Drill strength [10/6] under correct working conditions, when the possibility of buckling may be eliminated, should be calculated for simultaneous compression by the axial force and torque resulting from the cutting forces. It can be stated that axial force does not exert a significant influence upon drill strength. Therefore, drill strength may be determined by the torque only.

For high-speed steel twist drills within the diameter range of 7·5–24 mm, the destructive torque may be expressed by the formula* [10/6]

$$M_d = c\,d^{2\cdot65} \quad \text{kg cm} \tag{10.37}$$

where: $c = 1$, d—drill diameter in mm.

Thus the condition limiting feed rate with regard to drill strength can be written as

$$M \leqq M_d \tag{10.38}$$

Hence the permissible rate with regard to tool strength s_s is

$$s_s \leqq \sqrt[u_M]{\frac{d^{(2\cdot65-b_M)}}{c_M}} \quad \text{mm/rev} \tag{10.39}$$

When numerical values of the exponents for high-speed steel drills with sintered carbide points are taken into consideration, eqn. (10.39) may be written

— for high-speed steel drills as:

$$s_s \leqq \frac{d^{0\cdot94}}{c_M^{1\cdot25}} \quad \text{mm/rev} \tag{10.40}$$

— for sintered carbide drills

$$s_s \leqq \frac{d^{0\cdot65}}{c_M} \quad \text{mm/rev} \tag{10.41}$$

Machine tool strength is taken into consideration in the choice of feed rate by fulfilling the conditions

$$F < F_{\text{perm}} \quad \text{and} \quad M < M_{\text{perm}}$$

where: F_{perm}—permissible axial force, M_{perm}—permissible torque, with regard to machine tool strength.

Thus, substituting the expressions for axial force and torque:

$$C_F d^{b_F} s^{u_F} < F_{\text{perm}} \tag{10.42}$$

$$C_M d^{b_M} s^{u_M} < M_{\text{perm}}$$

Permissible values of feed rate can be derived with regard to axial force

$$s_F < \sqrt[u_F]{\frac{F_{\text{perm}}}{c_F d^{b_F}}} \quad \text{mm/rev} \tag{10.43}$$

and with regard to torque

$$s_M < \sqrt[u_M]{\frac{M_{\text{perm}}}{c_M d^{b_M}}} \quad \text{mm/rev} \tag{10.44}$$

F_{perm} and M_{perm} values appearing in those formulae should be taken from technical data of the given drilling machine. They are determined by the designer.

* A similar formula is given by Stoewer [10/2], with $c = 1\cdot3$ and the exponent equalling 2·7.

Feed rate limitation with regard to the kinematics of cutting is caused by the changes in dynamic clearance angle values depending on the s/d ratio. Thus the necessary condition is that at every point of the lip the dynamic clearance angle α_{dc} must be positive. Its maximum value may be $\alpha_{dc} = \zeta$.

Taking into consideration that drill seizure may occur when the angle α_{dc} is still positive but very small (as elastic deformations of the workpiece and tool point should also be taken into account), limit values s_l may vary in practice from 0·023d to 0·042d in mm/rev, depending on the sharpening methods used, while the working angle α_{dc} on a diameter equalling 0·14d is between 0°54′ and 1°13′.

The feed rate chosen must also fulfil the condition

$$s_{min} \leqq s \leqq s_{max} \tag{10.45}$$

where: s_{min}—the lowest feed rate, s_{max}—the highest feed rate.

After a comparison of feed rate values s_s, s_F, s_M and s_l, the lowest of them is chosen. If this feed fulfils the condition of eqn. (10.45), then the nearest lower step of feed rate of the given machine tool is adopted.

10.2 Preliminary reaming

10.2.1 General characteristic of preliminary reaming

Reaming is a method of machining holes with multi-point tools, with an accuracy higher than that of drilling. The kinematics of reaming is similar to that of drilling.

From the point of view of machining accuracy and the tools applied, reaming is divided into preliminary and finish reaming.

Roughing and medium accurate reaming is called *preliminary reaming.*

Accurate and highly accurate reaming is called *finish reaming.*

Differences between preliminary reaming and secondary drilling consist in the application of tools of a different design and in that machining allowances in the case of reaming are much smaller and machining accuracy higher.

In the machining of holes by plain and secondary drilling, accuracy classes 14 to 10 and roughness numbers 3 to 6 are obtained. Preliminary reaming permits one to obtain accuracy classes 12 to 9 and roughness number 6 ($R_{tm} \leqq 10$ μm). This is possible since the reamer has more cutting points than the drill, i.e. more than two. Therefore, the distribution of cutting forces is more uniform and tool deformations are reduced, thus producing a more accurate hole.

Preliminary and finish reaming are cuts often applied after hole drilling, in order to obtain a higher hole accuracy. Average machining allowances in preliminary reaming are given in Table 10.7.

Table 10.7 *Average machining allowance in preliminary reaming* (after drilling)

Reamer diameter mm	up to 20	21–35	36–45	46–50	50–60	61–70	71–80
Allowance (each side) mm	0·5	0·75	1	1·25	1·5	1·75	2·0

10.2.2 Tool point shape and cutting conditions
10.2.2.1 *Design of a roughing reamer*

Design of a roughing reamer (Fig. 10.21) is similar to that of a drill used for secondary drilling. The essential difference is that a reamer has no chisel edge and has a larger number of points (cutting edges).

The points of roughing reamers are characterized by the following parameters:
(a) in the main cross-section *A–A*: the clearance angle α and rake angle γ,
(b) leading cutting edge angle, or lip angle $\varkappa$,
(c) taper angle, or trail edge angle or land angle $\varkappa_1$,
(d) flute helix angle, or trail edge inclination angle λ_1.
In addition to parameters characterizing a roughing reamer, the nominal diameter d and land width f are also important. Land clearance angle α_1 is usually equal to zero.

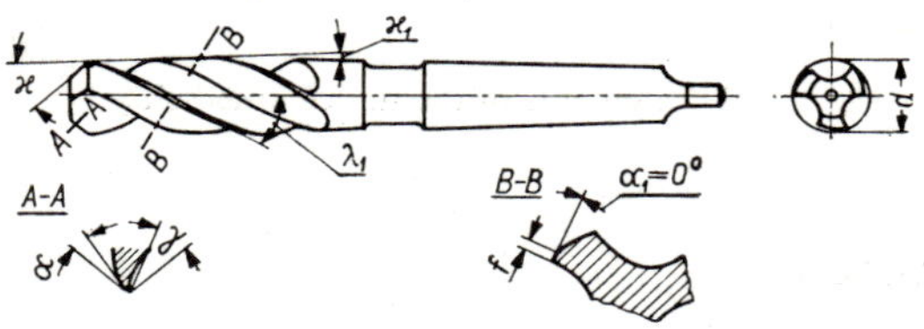

Fig. 10.21 Design of a roughing reamer

With regard to the work of a reamer, as in the case of a drill, angles α and γ in the cylindrical cross-section are of great importance. The clearance angle in the cylindrical cross-section depend on the sharpening method used and those of rake angle on the shape of the flute. In the case of counterbores with straight flutes, e.g. with sintered carbide points, the rake angle is constant along the lip and is usually $\gamma_k = -5°$ when machining cast iron, $\gamma_k = 0°$ when machining steel with $\sigma_B < 90$ kg/mm^2 and $\gamma_k = -5°$ when machining steel with $\sigma_B \geqq 90$ kg/mm^2.

Lip angles $\varkappa$ are between 30° and 60°.

The taper, determined by angle $\varkappa_1$, yields a diameter difference of 0·05–0·08 mm on a length of 100 mm for sintered carbide tools and 0·04–0·15 for high-speed steel tools.

Flute helix angle is:
(a) when machining soft steels ($\sigma_B < 60$ kg/mm^2), brass and light alloys $\lambda_1 = 25$–40°,
(b) when machining hard steels ($\sigma_B > 60$ kg/mm^2), carbon and alloy steels, cast iron (HB < 220 kg/mm^2), brittle bronze or brass $\lambda_1 = 15$–20°,
(c) when machining brittle and hard cast iron (HB > 220 kg/mm^2) $\lambda_1 = 10°$.

10.2.2.2 *Cutting conditions*

Geometric parameters of the undeformed chip are defined in the cross-section passing through the axis of the drilled hole.

As in the case of secondary drilling, the undeformed chip is characterized (Fig. 10.22) by:
(a) depth of cut $a = (d - d_0)/2$ mm,

(b) feed rate per tool point $s_z = s/z$ mm/point (where: s—feed rate per revolution in mm/rev, z—number of points),
(c) undeformed chip thickness $h = s_z \sin \varkappa$ mm,
(d) undeformed chip width $b = a/\sin \varkappa$ mm,
(e) cross-sectional area per tool point, $A = a\,s = b\,h$ mm^2.

Reaming time is calculated according to a formula similar to that derived for secondary drilling.

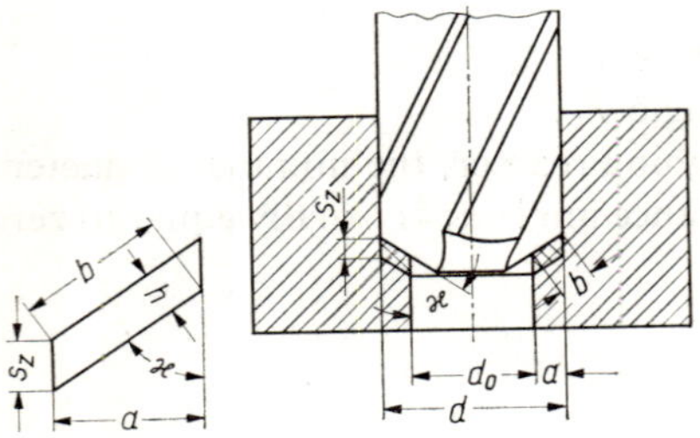

Fig. 10.22 Elements of the undeformed chip

10.2.3 CUTTING RESISTANCE AND TOOL LIFE

10.2.3.1 *Cutting resistance*

In preliminary reaming, the system of forces acting on the tool may be reduced to the resultant axial force F and torque M. The value of axial force and torque depends on the same factors as in the case of secondary drilling. Differences concern only the constants appearing in formulae for axial force and torque. Values of C_F, C_M and the exponents are listed in Table 10.8.

Table 10.8 *Values of constants and exponents appearing in formulae for axial force and torque in preliminary reaming*

Tool material	Workpiece material	Axial force				Torque			
		C_F	b_F	e_F	u_F	C_M	b_M	e_M	u_M
High-speed steel	steel	3·5	0·5	0·95	0·7	105	1	0·74	0·75
	cast iron	2·0	0·65	1·0	0·55	72	1	0·74	0·7
Sintered carbides*	steel	—	—	—	—	950	0·75	0·8	0·95
	cast iron	—	—	—	—	195	0·85	0·8	0·7

* acc. to NIBTN MSS data

10.2.3.2 *Wear, tool life and life speed*

In rough reaming, the process of chip formation and the appearance of the machined surface have the same characteristic features as in secondary drilling. Thus the wear of a roughing reamer point is similar to that of the point of a drill used for secondary drilling.

The most frequent symptoms of tool wear are:
(a) wear of corners and flanks (of high-speed steel tool points),
(b) wear of cutting edge flanks (of sintered carbide tool points).
Wear criteria adopted are:
(a) in the case of a preliminary operation or cut—the optimum magnitude of tool point wear,
(b) in the case when rough reaming is also a finishing operation or cut—the wear which causes surface quality or dimensional accuracy tolerances to be exceeded.

Table 10.9 *Formulae for life speed in preliminary reaming*

Tool material	Workpiece material	Mandrel reamers	
		Coolant	Formula
High-speed steel	Constructional steel $\sigma_B = 75$ kg/mm²	emulsion	$V_T = \dfrac{16{\cdot}3\, d^{0{\cdot}3}}{T^{0{\cdot}3} s^{0{\cdot}5} a^{0{\cdot}2}}$
	grey cast iron HB = 190 kg/mm²	air	$V_T = \dfrac{18{\cdot}8\, d^{0{\cdot}2}}{T^{0{\cdot}125} s^{0{\cdot}4} a^{0{\cdot}1}}$
	malleable cast iron HB = 190 kg/mm²	emulsion	$V_T = \dfrac{10{\cdot}4\, d^{0{\cdot}2}}{T^{0{\cdot}125} s^{0{\cdot}4} a^{0{\cdot}1}}$
	bronze, medium hard HB = 100–140 kg/mm²	air	$V_T = \dfrac{41{\cdot}8\, d^{0{\cdot}2}}{T^{0{\cdot}125} s^{0{\cdot}4} a^{0{\cdot}1}}$
Sintered carbides	chrome and chrome–nickel carbon steels (non-hardened) $\sigma_B = 75$ kg/mm²	emulsion	$V_T = \dfrac{194\, d^{0{\cdot}6}}{T^{0{\cdot}25} s^{0{\cdot}3} a^{0{\cdot}1}}$
	hardened steels $\sigma_B = 170$ kg/mm² HRC = 49–54 kg/mm²	emulsion	$V_T = \dfrac{10\, d^{0{\cdot}6}}{T^{0{\cdot}45} s^{0{\cdot}6} a^{0{\cdot}3}}$
	grey and malleable cast iron HB = 190 kg/mm²	air	$V_T = \dfrac{123\, d^{0{\cdot}4}}{T^{0{\cdot}4} s^{0{\cdot}45} a^{0{\cdot}15}}$
		emulsion	$V_T = \dfrac{261\, d^{0{\cdot}4}}{T^{0{\cdot}4} s^{0{\cdot}45} a^{0{\cdot}15}}$

Statistical and empirical formulae for life speed in preliminary reaming, depending on tool point and workpiece material are given in Table 10.9. These formulae concern maximum permissible values of flank wear *VB*. In the case when, for technological reasons (dimensional accuracy, surface quality), a smaller wear is required, a correction coefficient B_w whose values are given in Table 10.10 should be applied in the calculation of cutting speed.

Recommended life periods for the points of reamers used in preliminary reaming are 30–90 min when machining steel and cast iron (for hole diameters between 20 and 80 mm).

Table 10.10 *Correction coefficients B_w taking into account the degree of wear of reamer points and appearing in formulae for life speed*

Work-piece material	Tool diameter *d*, mm: up to 20				21–40				41–60				61–80			
	Abrasion *VB*, mm															
	0·25	0·5	0·75	1·0	0·3	0·6	0·9	1·2	0·35	0·7	1·0	1·4	0·4	0·8	1·2	1·6
Non-hardened steel	0·66	0·81	0·9	1·0	0·66	0·81	0·9	1·0	0·66	0·81	0·9	1·0	0·66	0·81	0·9	1·0
Cast iron	0·38	0·62	0·82	1·0	0·38	0·62	0·82	1·0	0·38	0·62	0·82	1·0	0·38	0·62	0·82	1·0

10.2.4 Selection of cutting conditions for preliminary reaming

The principles for the selection of cutting conditions are the same as in drilling.

Since, in view of the small depths of cut, the cutting forces are considerably smaller than in drilling, the essential limitation in the choice of feed rate is kinematic. It results from the size of the tool point clearance angle at a diameter corresponding to that of the hole being reamed. This angle, as in the case of drills, depends on the tool sharpening method. Feed rate limitation with regard to surface quality is not essential because of the smoothing portion of the reamer.

On the basis of experiments the feed range has been established, which yields the best results. Such ranges are presented graphically in Fig. 10.23 for high-speed steel reamers and in Fig. 10.24 for sintered carbide reamers.

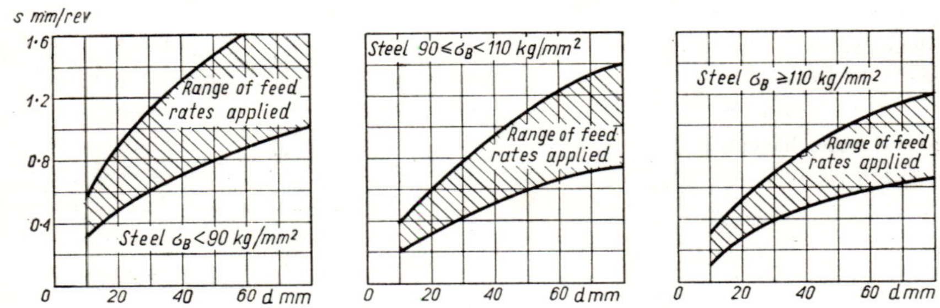

Fig. 10.23 Range of feed rates applied in preliminary reaming with high-speed steel reamers

After the feed rate has been chosen and adapted to the actual feed steps of the machine tool, and the depth of cut is known and tool point life has been adopted, the life cutting speed is calculated. The rotational speed *n* is then determined and the nearest actual speed of the given machine tool selected. A suitable correction of cutting speed is then made and the reaming time calculated.

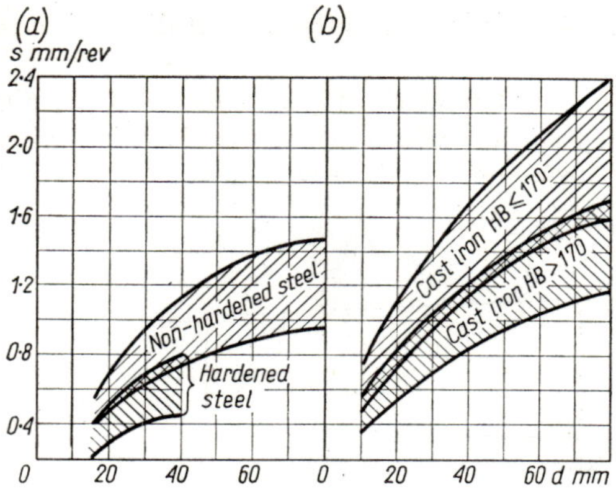

Fig. 10.24 Range of feed rates applied in preliminary reaming with sintered carbide reamers, for (a) steel, (b) cast iron

10.3 Finish reaming

10.3.1 General characteristic of finish reaming

Finish reaming usually constitutes the last cut of the hole machining cycle, i.e. it follows drilling and preliminary reaming. Reaming allows one to obtain a hole accuracy of class 8–5 and roughness number 8–10. The degree of reaming accuracy depends on the application of single, double or triple reaming. In the case of very accurate holes, particularly if their diameters are large, three passes are usually applied. Reaming may be of the machine or hand type. The latter is possible because of the small cutting resistance.

Very high accuracy is obtainable by finish reaming, since the allowances are very small and the number of points larger than in rough reamers. The number of reamer points may be from 6 to 12. Average allowances in finish reaming are given in Table 10.11.

Table 10.11 *Average allowances in finish reaming*

Reamer diameter, mm	up to 5	6–11	11–15	16–30	13–50	51–60	61–80
Allowance (each side), mm	0·07	0·1	0·125	0·15	0·15	0·2	0·25

10.3.2 Description of finish reamers and cutting conditions

10.3.2.1 *Description of finish reamers*

Finish reamers are divided into mandrel type and shell reamers. In addition, reamers may have a fixed or adjustable nominal diameter. They may be cylindrical or conical (for conical holes).

Reamer points are made either of tool steels or of sintered carbides.

A reamer (Fig. 10.25) has a working part and shank. The working part (Fig. 10.25a) is made up of

(a) centring taper *A*, present only in through-hole reamers,
(b) cutting part *B*, intended for carrying out the main cutting work,

(c) smoothing and finishing part *C*, whose task it is to give the correct dimension to the hole and to smooth its walls,
(d) exit taper *D*, which reduces friction during the work of the reamer and permits one to avoid scratching the machined surface when the reamer is being withdrawn from the hole.

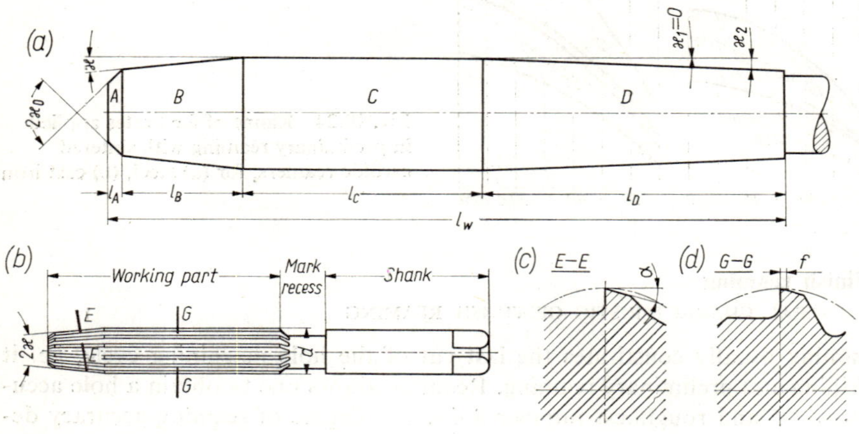

Fig. 10.25 Stereometric elements of a reamer

The shank consists of the centring portion (taper or cylinder) and driver in the form of a tang or a square prism.

Centring taper *A* is characterized by angle $2\varkappa_o$ and length l_A. Usually $\varkappa_o = 45$, length l_A equals 1–3 mm depending on reamer size.

Cutting part *B* is characterized by lip angle $\varkappa$ and length l_B. Lip angles of machine and hand reamers are different, they also depend on tool point material. In tool steel hand reamers $\varkappa = 0{\cdot}5–1{\cdot}5°$ is used while machine reamers of the same material have $\varkappa = 15°$ when reaming through-holes in steel and $\varkappa = 5°$ when reaming through-holes in cast iron. When reaming blind holes or when reaming through-holes of inferior accuracy a lip angle $\varkappa = 45°$ is used.

The smaller the lip angle, the smaller is the undeformed chip thickness. Therefore, in the case of ductile metals, below a certain minimum thickness of undeformed chip, burnishing occurs instead of cutting. To prevent this, when machining steel and other ductile metals the lip angle is increased and also the undeformed chip thickness.

In the case of reamers with sintered carbide points lip angles $\varkappa = 30–45°$ are applied.

Points in the cutting part, performing the basic cutting work, are characterized by angles in the cross-section *E–E* perpendicular to the cutting edge. In this cross-section (Fig. 10.25 b, c) clearance angles α and rake angle γ are distinguished. Rake angles are usually within the limits $\gamma = 0–10°$ for tool steel reamers and $\gamma = -5–0°$ for sintered carbide reamers. Smaller rake angles are applied when reaming brittle metals and greater angles in the case of ductile metals.

Clearance angles α are usually between 6 and 15° and the smaller the reamer diameter the greater is the angle α.

Tool points in the finishing part are characterized by angles in the cross-section perpendicular to the reamer axis. Rake angle is usually equal or close to zero and clearance angle α as in the cutting part. The characteristic feature of the points of the finishing part are lands on the flanks (Fig. 10.25d) of width $f = 0{\cdot}08$–$0{\cdot}5$ mm for tool steel reamers and $f = 0{\cdot}15$–$0{\cdot}25$ mm for sintered carbide reamers.

The length of the finishing part should be chosen such that, on the one hand, friction does not increase excessively during cutting and, on the other hand, that the reamer is well guided in the hole, this increases accuracy. The length of the finishing part is given in corresponding tool standards.

The exit taper is an extension of the guiding and finishing part. It converges towards the shank ($\varkappa_2$) usually 0·005–0·008 mm per 100 mm in hand reamers and 0·04–0·08 mm per 100 mm in machine reamers.

The working part of the reamer consists of points divided by flutes. The number of points is approximately adopted according to

$$z = 1{\cdot}5\sqrt{d} + (2\text{–}4)$$

where d—nominal diameter of the reamer in mm.

The larger the number of tool points, the higher is dimensional accuracy and roughness number obtainable and the greater is the metal removal rate of the reamer. However, the number of reamer points may not be increased excessively, since the flutes are then diminished and chip removal becomes difficult.

Flutes may be straight or helical. Straight flutes are applied when machining brittle materials and helical ones when machining ductile materials. The hand of the flute helix is usually opposite to the cutting direction. The flute helix angle on the periphery is $\lambda = 7$–$8°$ when machining hard steel and cast iron, $\lambda = 12$–$20°$ for soft steel and malleable cast iron and $\lambda = 35$–$45°$ when machining light alloys.

10.3.2.2 *Cutting conditions*

Differences in the shape of the undeformed chip in preliminary and finish reaming are created by considerably lower values of the depth of cut and clearance angle in the case of finish reaming.

Otherwise, formulae and data contained in the chapter on drilling and preliminary reaming may be applied to finish reaming.

10.3.3 CUTTING RESISTANCE AND TOOL LIFE

10.3.3.1 *Cutting resistance and cutting power*

The axial force and torque is low because of the small cross-sectional area of the undeformed chip. Thus, in practice, the cutting resistance does not play such an important role in reaming, as for example in drilling. Nevertheless, the magnitude of cutting resistance, and particularly of the torque, constitutes one of the indices of machinability in reaming.

In the calculation of axial force and torque in reaming, empirical and statistical formulae are used, similar to those applying to rough reaming.

When reaming carbon steel ($\sigma_B = 75$ kg/mm^2), when all correction coefficients are equal to unity, axial force and torque may be calculated from formulae [I/11]

$$F = 8d^{0\cdot7}\, s^{0\cdot72}\, a \qquad \text{kg} \tag{10.46}$$

$$M = 88d^{1\cdot16}\, s^{0\cdot85}\, a^{0\cdot64} \qquad \text{kg mm} \tag{10.47}$$

Under the same conditions, similar formulae when reaming cast iron (HB = 190 kg/mm^2) are as follows:

$$F = 10d^{0\cdot61}\, s^{0\cdot4}\, a^{0\cdot88} \qquad \text{kg} \tag{10.48}$$

$$M = 23d^{1\cdot42}\, s^{0\cdot79}\, a^{0\cdot74} \qquad \text{kg mm} \tag{10.49}$$

In the case when the cutting conditions differ from those assumed in these formulae, correction coefficients used in preliminary reaming may be applied in approximate calculations.

The cutting power in reaming is calculated according to a formula similar to that applicable for drilling.

10.3.3.2 *Tool life and life speed*

The wear of a finish reamer resembles that of a roughing reamer. Wear symptoms appear on the flank, corners and lands.

Maximum permissible values of wear *VB* for the cutting part of tool steel reamers are 0·6–0·8 mm. For sintered carbide reamers *VB* = 0·4–0·7 mm is adopted when machining non-hardened steels and cast iron, and *VB* = 0·3–0·35 when machining hardened steels.

In practice, a reamer is usually withdrawn from use before its wear reaches the permissible value *VB*. This occurs when such a reamer does not ensure machining results complying with the required dimensional tolerance and surface quality. In this case the technological criterion of accuracy is applied.

In finish reaming it is very important that reamer points should not be worn excessively, since unequal wear of the particular points results in a non-uniform distribution of the cutting forces among the points. Thus the reamer is being bent and hole inaccuracy follows. At the same time the surface quality deteriorates.

On the basis of experiments optimum values of tool life speed from the point of view of production engineering have been established and they are presented in Fig. 10.26.

The cutting speed corresponding to the adopted life speed is calculated according to formulae similar to those for drilling, with a suitable change of the constants (Table 10.12).

10.3.4 Choice of cutting conditions for finish reaming

The choice of cutting conditions for reaming consists in the selection and correction of feed rate, adopting a suitable cutting speed, and correcting it with regard to machine tool kinematics and reaming time.

The selection of feed rate for reaming may be carried out according to approximate empirical formulae:

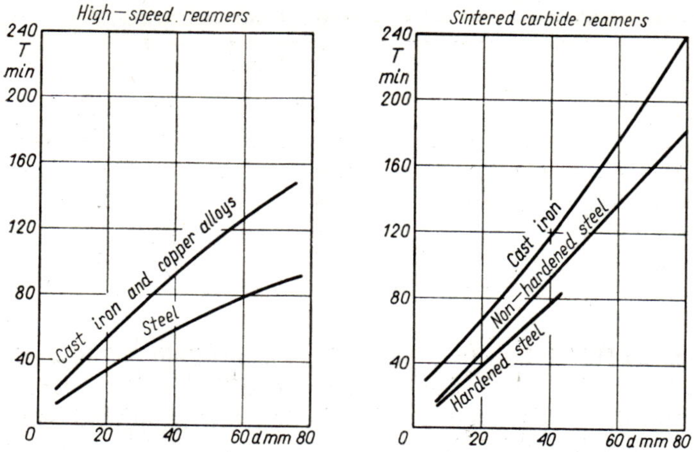

Fig. 10.26 Average values of recommended lives of reamer points

Table 10.12 *Values of constants for life speed formula*

Reamer point material	Workpiece material	Values of constants				
		C_v	m	b_v	u_v	e_v
High-speed steel	Constructional steel $\sigma_B = 76$ kg/mm^2	10·5	0·4	0·3	0·65	0·2
	grey cast iron HB = 190 kg/mm^2	15·6	0·3	0·2	0·5	0·1
	malleable cast iron HB = 190 kg/mm^2	16·9	0·3	0·2	0·5	0·1
	bronze HB = 100–140 kg/mm^2	34·8	0·3	0·2	0·5	0·1
Sintered carbides	non hardened steel $\sigma_B = 75$ kg/mm^2	100	0·7	0·3	0·65	0
	hardened steel $\sigma_B = 160$–180 kg/mm^2	14	0·85	0·4	1·05	0·75
	grey cast iron HB = 190 kg/mm^2	110	0·45	0·2	0·5	0

for tool steel reamers (diameter $10 \leqq d \leqq 80$ mm)

— when machining steel with $\sigma_B < 90$ kg/mm^2: $s = (0{\cdot}07–0{\cdot}21)d^{0{\cdot}63}$
— when machining steel with $\sigma_B \geqq 90$ kg/mm^2: $s = (0{\cdot}05–0{\cdot}15)d^{0{\cdot}63}$
— when machining cast iron with HB < 170 kg/mm, light alloys, brass: $s = (0{\cdot}22–0{\cdot}44)d^{0{\cdot}63}$
— when machining cast iron with HB $\geqq$ 120 kg/mm^2: $s = (0{\cdot}13–0{\cdot}25)d^{0{\cdot}63}$

for sintered carbide reamers (for $10 \leqq d \leqq 80$ mm):

— when machining non-hardened steel: $s = (0{\cdot}12–0{\cdot}18)d^{0{\cdot}42}$
— when machining hardened steel: $s = (0{\cdot}08–0{\cdot}12)d^{0{\cdot}42}$
— when machining cast iron with HB < 170 kg/mm^2: $s = (0{\cdot}30–0{\cdot}45)d^{0{\cdot}42}$
— when machining cast iron with HB $\geqq$ 170 kg/mm^2: $s = (0{\cdot}24–0{\cdot}36)d^{0{\cdot}42}$

In the determination of cutting speed the limiting factor is constituted by surface quality and directly by the phenomenon of the appearance of built-up edge. On the basis of production experience it has been established that in view of the impossibility of applying in reaming such high speed at which BUE does not appear any more, it is necessary to apply a sufficiently low speed at which BUE does not yet begin to form (Table 10.13).

Table 10.13 *Recommended cutting speeds for reaming steel and cast iron*

Workpiece material		Roughness number $\nabla 7$	$\nabla 8$
		Recommended cutting speed v, m/min	
Emulsion cooling	steel, $\sigma_B < 90$ kg/mm²	5	3
	steel, $\sigma_B \geqq 90$ kg/mm²	4	2
Dry cutting	grey cast iron	6	4

Results of some research [10/9] show that there exists the possibility of increasing the metal removal rate in reaming by considerably increasing the cutting speed. An obstacle to the development of this trend consists in too low rotational speeds available on drilling machines. Increasing these speeds would necessitate the improvement of drilling machine rigidity and resistance to vibrations.

A factor which deserves particular consideration in the selection of cutting conditions in reaming is the application of cutting fluids. Experience indicates that under the conditions of reamer work cutting fluids have a very great influence. The correct selection of a cutting fluid considerably improves surface quality, reduces friction, and eliminates the phenomenon of penetration of small parts of chips between the lands and hole wall. As a result of this, cutting resistance decreases and the life of the reamer is extended.

The following cutting fluids are applied in reaming:

— oil emulsion (containing 15–20% of emulsifying oil, some rape oil and a mixture of oils consisting of 25% of linseed oil, 15% of paraffin, 10% of turpentine and 50% of spindle oil) for machining steel and non-ferrous alloys,

— sulphured oil for machining steel,

— mineral oil (solar or spindle oil) for machining non-ferrous alloys.

REFERENCES

10/1 Mozhaev, S. S.: 'Analiticheskaya teoriya spiral'nykh sverl' (Analytic theory of twist drills), Mashgiz, Moskwa 1948

10/2 Schallbroch, H.: 'Bohrarbeit und Bohrmaschine', Hanser-Verlag, München 1951

10/3 Kronenberg, M.: 'Grundzüge der Zerspanungslehre', 2. Bd. 'Stirnfräsen, Bohren', Springer-Verlag, Berlin 1963

10/4 Kaczmarek, J.: 'Cutability of single and double-cone Weisker method ground drills in steel machining', Advances in Machine Tool Design and Research, Birmingham 1962 (edited by S. A. Tobias and F. Koenigsberger)

10/5 Eremin, V. F. Naumov, B. I., and Konyashov, V. V.: 'Tekhnologiya mekhanicheskoi obrabotki otverstii' (Technology of the machining of holes), Avtomobil'naya i Traktornaya Promyshlennost', 9, 1950

10/6 Barbashov, F. A.: 'Issledovanie dinamicheskikh zavisimostei pri sverlenii', Sbornik: 'Dinamika protsessa rezaniya metallov' (Study of Dynamic Relationships in Drilling. Series: Dynamics of the Metal Cutting Process), Editor: A. I. Kashirin, Mashgiz, Moskwa 1953

10/7 Vogel, Z.: 'Badanie procesu wiercenia małych otworów w stali kwasoodpornej 1H18N9T' (Study of the process of drilling small holes in acid-resistant steel 1H18N9T), A doctor's thesis at the Department of Mechanics and Production Engineering of the Silesian Institute of Technology, 1966

10/8 Bułat, A., and Pakosiewicz, B.: 'Tablice doboru warunków skrawania: Rozwiercanie wykończające narzędziami ze stali szybkotnącej. Stal węglowa konstrukcyjna 55' (Tables for the choice of cutting conditions: Finish reaming with high-speed steel tools. Constructional carbon steel 55), Prace IOS 1964, seria: Materiały Instruktażowe, No. 20; —'Rozwiercanie wykańczające narzędziami ze stali szybkotnącej. Brązy BA 1032, BA 1044, B 101, B 663' (Finish reaming by means of high-speed steel tools. Bronzes BA 1032, BA 1044, B101, B663), Prace IOS 1965, Seria: Materiały Instruktażowe, No. 48

10/9 Andreev, G. S.: 'Skorostnaya obrabotka otverstii w stalyakh razvertkami s rezhushchei chast'yu iz tverdogo splava' (High-speed machining of steels with reamers with cutting part of hard-facing alloy), Vestnik Mashinostroeniya, 9, 1953

CHAPTER 11

Plain milling*

11.1 General characteristics of milling

11.1.1 MILLING TYPES

Milling is a method of machining by cutting in which the tool carries out a rotary motion and the workpiece a plane motion, either rectilinear or curvilinear.

Milling is used in the machining of curvilinear or plane surfaces. It is the second most frequently applied method of machining by cutting (after turning). Multi-point tools called *milling cutters* or *mills* are used in milling operations.

Characteristic features of milling are as follows:
(a) the process of cutting by each point or tooth is periodically interrupted,
(b) the transverse cross-section of the undeformed chip is not constant.
Two principal varieties of milling are distinguished: so-called plain milling and face milling (Fig. 11.1).

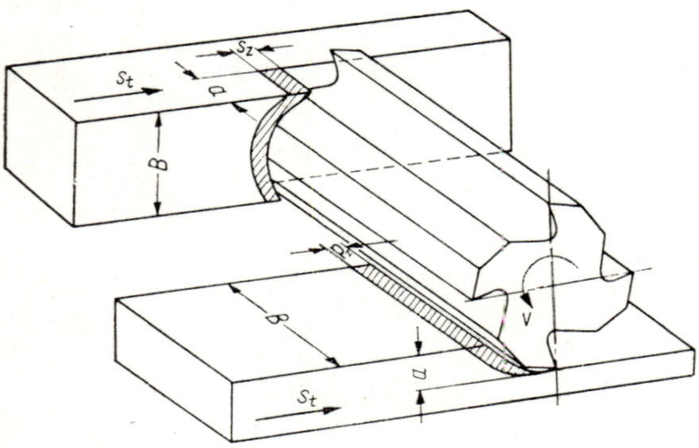

Fig. 11.1 Diagram of plain and face milling

Plain milling is defined as the type of milling in which there is a contact between the cylindrical surface of a plain milling cutter and the machined surface.

Face milling is defined as the type of milling in which there is a contact between, for the most part, the face of a face mill and the machined surface.

* Also called *slab milling*.

It follows that in plain milling the axis of cutter rotation is parallel to the machined surface and in face milling perpendicular to it.

Plain milling cutters are adapted to cutting by teeth with cutting edges situated on the surface of a cylinder which can be circumscribed on the cutter. *Face mills* are adapted to cutting by teeth with cutting edges situated on the face of the mill and partially on its cylindrical surface.

Mills which can cut with edges situated both on the face and on a large part of the cylindrical surface, are called *shell-end mills.*

Plain milling may be of the conventional or climb-cut type.

In *conventional milling* the direction of feed at the point in which the vector v is parallel to vector s_t is opposed to the direction of mill rotation (Fig. 11.2a), while in *climb-cut milling* it agrees with the latter (Fig. 12.1b). This difference is essential for the milling process, and the problem is discussed in detail in the section on plain milling.

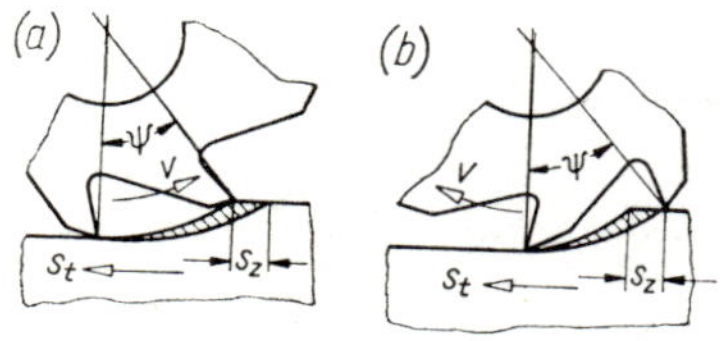

Fig. 11.2 Diagram of: (a) conventional milling, (b) climb-cut milling

In addition to those mentioned above, several other types of milling may be distinguished, their names being derived from the shape of the machined surface and from the type of cutter used (e.g. *conical face milling* or *form milling*).

11.1.2 Speed of tool and workpiece motion

Both for the kinematic and technological type of milling the corresponding definitions of the speed of rotary motion of the tool and of the workpiece feed rate are identical.

The *speed of primary motion* is the peripheral speed of rotary motion v_m of the tool, determined by the formula

$$v_m = \frac{\pi d n}{1000} \quad \text{m/min} \tag{11.1}$$

where: d—outer diameter of the milling cutter in mm, n—rotational speed of the spindle in rev/min.

The auxiliary motion of the workpiece, or *feed,* may be rectilinear or curvilinear and particularly helical. The latter is used in milling gear teeth, threads, helical flutes etc.

If the feed motion is rectilinear, its speed is called the *feed rate* and denoted by s_t, in mm/min.

If, for feed rate s_t, spindle speed is n rev/min, then the *feed per revolution* is

$$s = \frac{s_t}{n} \quad \text{mm/rev} \tag{11.2}$$

If the tool has z teeth, then during one revolution of the milling cutter the *tooth travel*, or *feed per tooth*, amounts to

$$s_z = \frac{s}{z} = \frac{s_t}{n\,z} \quad \text{mm/tooth} \tag{11.3}$$

Values of feeds per tooth applied in milling are lower than those used in cutting with single-point tools (turning, planing).

The composition of the rotary motion of the milling cutter with the rectilinear feed motion of the workpiece yields the tooth path in relation to the table. The shape of this path is an elongated cycloid.

By an ordinary cycloid is understood the path of a point situated on the periphery of a circle of radius ρ when this circle rolls without sliding along a straight line. If a second circle, of radius r, is connected rigidly but coaxially with the former one, then the path of a point situated on the periphery of that circle is called an elongated cycloid.

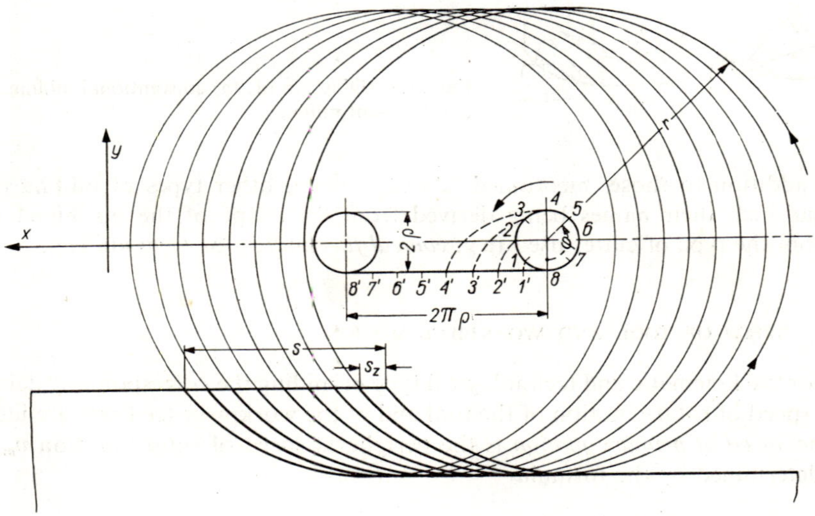

Fig. 11.3 Diagram of the motion of milling cutter teeth in relation to the workpiece (path in the shape of an elongated cycloid)

The equation of elongated cycloid parameters are (designations acc. to Fig. 11.3):

$$y = \rho - r\cos\varphi; \qquad x = \rho\varphi - r\sin\varphi \tag{11.4}$$

where: ρ—radius of the circle rolling without sliding, in mm, r—radius of the outer circle in mm, φ—angle of rotation.

It is easy to perceive that if the workpiece being milled moves by the distance s during one revolution of the milling cutter, then there exists the relationship

$$2\pi\,\rho = s \quad \text{and thence} \quad \rho = \frac{s}{2\pi} \quad \text{mm} \tag{11.5}$$

Therefore, equations of an elongated cycloid in the case of turning are as follows:

$$y = \frac{s}{2\pi} - r\cos\varphi; \qquad x = \frac{s\varphi}{2\pi} - r\sin\varphi \tag{11.6}$$

The cutting speed in milling will thus be the speed of the cycloid motion. It can be calculated (Fig. 11.4) according to the formula

$$v = \sqrt{(v_m\cos\varphi + s_t)^2 + v_m^2\sin^2\varphi} \quad \text{m/min} \tag{11.7}$$

where: v_m—peripheral speed of milling cutter in m/min, φ—angle of cutter rotation (the origin of coordinates is given in Fig. 11.3), s_t—feed rate in mm/min.

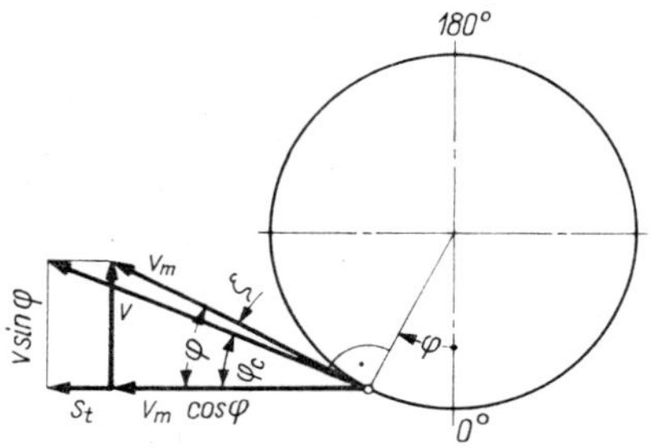

Fig. 11.4 Speed of the circular and cycloid motion

Equation (11.7) may be expanded and approximated as

$$v \approx v_m\sqrt{1 + 2\frac{s_t}{v_m}\cos\varphi} \quad \text{m/min} \tag{11.8}$$

Thus the ratio of cycloid motion speed to peripheral speed may be expressed as

$$\frac{v}{v_m} \approx \sqrt{1 \pm 2\frac{s_t}{v_m}\cos\varphi} \tag{11.9}$$

In practice, the highest values of feed rate to cutting speed ratio are 0·03. On the other hand, the angle of rotation corresponding to actual cutting time may vary in face milling from 0° to 180° and in plain milling from 0° to 45°.

It follows from these data that in extreme cases the difference between cutting speed and the resultant speed of cycloid motion may reach ±3%. In most cases the s_t/v_m ratio is less than 0·03 and, consequently, this difference is smaller.

Considering the small error involved and the convenient simplification obtainable, the peripheral speed of primary motion, i.e. of the tool, is assumed to be equal to the cutting speed in milling operations.

11.2 Plain milling

11.2.1 Description of milling cutter teeth

Plain milling cutters are made with teeth of high-speed steel or of sintered carbide.

With regard to the position of teeth in relation to cutter axis, cutters with straight teeth (parallel to the axis) and those with helical teeth (leading cutting edge forms a helix on the cylindrical surface of the cutter) are distinguished.

The main cross-section in which the teeth are characterized by the rake and clearance angles is usually the cross-section perpendicular to the cutting edge. This method of dimensioning cutter tooth angles is motivated by the way in which such teeth are made (sharpened).

With regard to cutter tooth work the tooth cross-section in a plane perpendicular to the axis of cutter rotation is of interest. In the case of milling cutters with straight teeth this cross-section coincides with the main one. In the case of cutters with helical teeth there exists a relationship between angles in the main cross-section and in that perpendicular to the axis of cutter rotation (Fig. 11.5)

$$\tan \alpha_x = \tan \alpha \cos \lambda_s \tag{11.11}$$

and

$$\cot \gamma_x = \cot \gamma \cos \lambda_s \tag{11.12}$$

where: α—clearance angle in main cross-section, α_x—clearance angle in the cross-section of plane perpendicular to the axis of cutter rotation, γ—rake angle in main cross-section, γ_x—rake angle in the cross-section of plane perpendicular to the axis of cutter rotation, λ_s—cutting edge inclination angle measured on the cylindrical surface of the milling cutter.

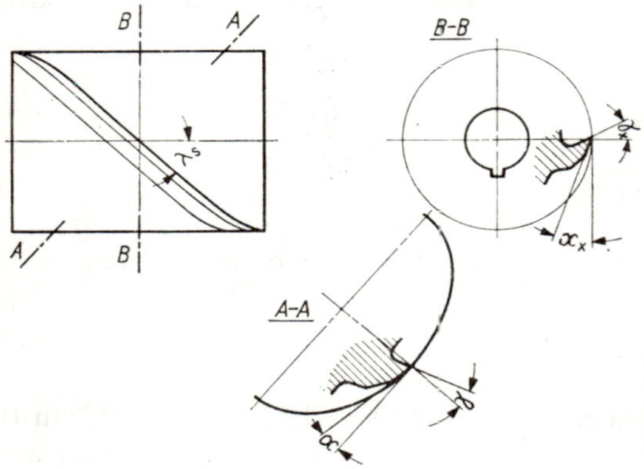

Fig. 11.5 Cutter tooth angles in main cross-section and cross-section perpendicular to the axis of rotation

It follows from eqns. (11.11) and (11.12) that clearance angles become smaller, the greater is the leading cutting edge inclination angle λ_s. It is vice-versa in the case of rake angles.

For kinematic reasons, the important angles are the angles measured in the cross-section perpendicular to the axis of milling cutter rotation and therefore they should constitute the starting point for cutter design.

Faces of milling cutter teeth may be made plane or concave, without lands. The rake angles applied in high-speed steel milling cutters are between 10° (for medium hard steel) and 20° (for soft steel).

Cutter teeth of sintered carbides usually have rake angles of 0–15°.

The milling cutter flank may have a single-land, double-land or relieved shape (Fig. 11.6a, b, c). Clearance angles vary from 10° to 20°. Teeth of plain milling cutters have a land on the flank side, made while equalizing tooth height differences to avoid so-called *run-out* during work. Land width f is of the order of tenths of a millimetre.

The helix angle of the cutting edges increases the number of teeth working simultaneously and the uniformity of cutter load. It follows from Fig. 11.7 in which a development of cutting edge helices on a plane is presented, that the number of simultaneously cutting edges depends on the axial pitch

$$p_o = \frac{H}{z} \quad \text{mm} \tag{11.15}$$

where: H—helix pitch in mm, z—number of cutter teeth, and on the milling width B.

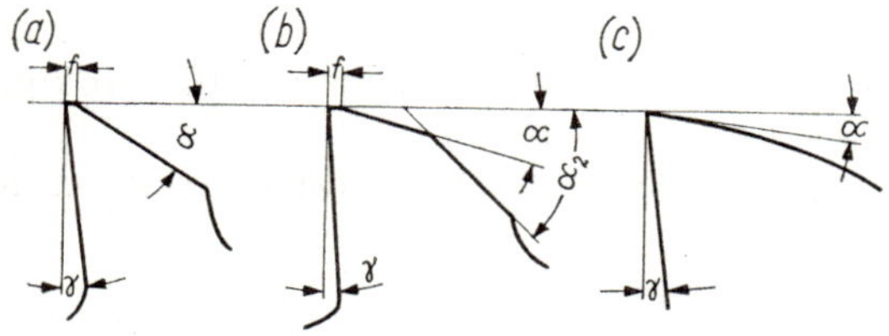

Fig. 11.6 Flank shapes

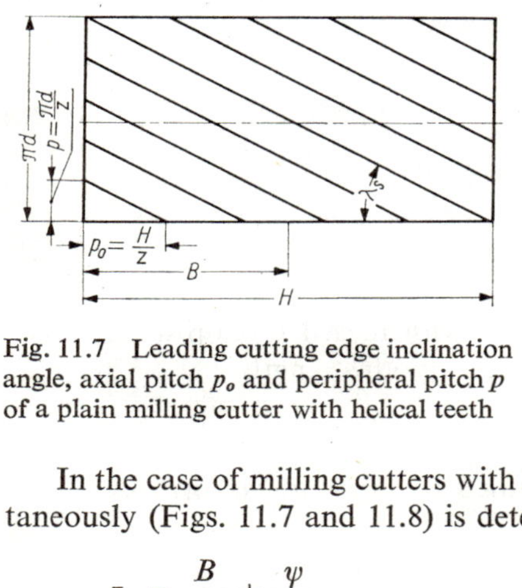

Fig. 11.7 Leading cutting edge inclination angle, axial pitch p_o and peripheral pitch p of a plain milling cutter with helical teeth

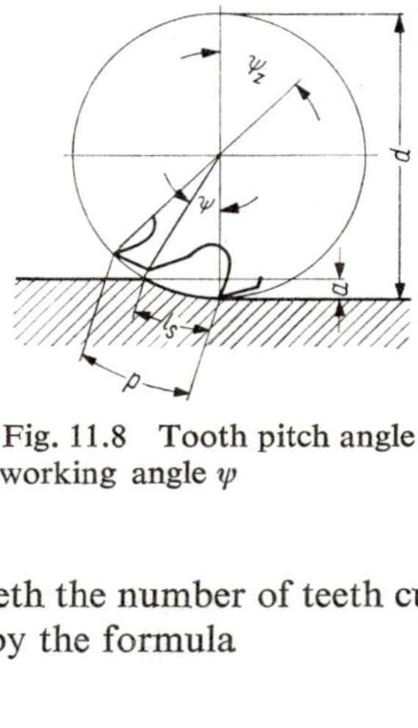

Fig. 11.8 Tooth pitch angle ψ_z, working angle ψ

In the case of milling cutters with helical teeth the number of teeth cutting simultaneously (Figs. 11.7 and 11.8) is determined by the formula

$$z_c = \frac{B}{p_o} + \frac{\psi}{\psi_z} \tag{11.16}$$

If helix pitch H is expressed by helix diameter and inclination angle λ_s, and $\psi_z = 360°/z$, then

$$z_c = \frac{Bz}{\pi d} \tan\lambda_s + \frac{z\psi}{360°} = z\frac{B\tan\lambda_s}{\pi d} + \frac{\psi}{360°} \tag{11.17}$$

Thus, the higher the B/d ratio and the greater the number of cutter teeth z and angles λ_s and ψ, the greater is the number of teeth cutting simultaneously.

If the value z_c from eqn. (11.17) is not a whole number, then the number of teeth cutting simultaneously will oscillate between the nearest lower and the nearest higher whole number. In this case the load will also vary as a result of the changing number of simultaneously cutting teeth. This number is influenced by milling width B, too, since d and λ_s are constant for the given cutter.

A disadvantage of a milling cutter with helical teeth is the appearance of an axial component of cutting force which grows with angle λ_s. Therefore, in practice inclination angles $\lambda_s < 45°$ are applied. In particular cases the negative influence of angle λ_s may be eliminated by joining two milling cutters (forming a so-called *gang type cutter*) with angles λ_s so chosen that the axial forces are balanced.

In the case of milling cutters with straight teeth the number of teeth cutting simultaneously is

$$z_c = \frac{l_s}{p_z} = \frac{l_s z}{\pi d} \tag{11.18}$$

where: $p_z = \pi d/z$, called the *peripheral tooth pitch* in mm, l_s—length of the working travel of a tooth (in contact with workpiece material) in mm.

On the assumption that the tooth path in relation to the workpiece material is circular (Fig. 11.8)

$$l_s = 0{\cdot}5d\,\psi \quad \text{mm} \tag{11.19}$$

where: d—outer diameter of the milling cutter in mm, ψ—working angle of the tooth (angle of contact with workpiece material) in radians.

Since

$$\psi = \arccos\,(1-2a/d) \tag{11.20}$$

the number of straight teeth cutting simultaneously is

$$z_c = \frac{z}{2\pi}\arccos(1-2a/d) \tag{11.21}$$

A limitation of the number of teeth is necessary with regard to chip space, which must correspond to chip volume, and with regard to tooth strength.

11.2.2 Parameters of the undeformed chip and cutting conditions

11.2.2.1 *Elements of the undeformed chip*

Depth of cut a is the dimension of the undeformed chip (removed in one pass) measured in a direction perpendicular to the machined surface (Fig. 11.1). This definition is valid for all types of plain milling.

Milling width B is the dimension of the undeformed chip measured in a direction perpendicular to the feed and parallel to the machined surface. This definition is also valid for all types of plain milling.

Each cutter tooth remains in contact with the workpiece material during the cutting time corresponding to the angle of cutter rotation ψ (Fig. 11.2), also called working angle of the tooth or the angle of contact between tooth and workpiece material.

The working angle of a milling cutter is calculated from

$$\cos\psi = 1 - \frac{2a}{d} \tag{11.22}$$

The time of tooth contact t_ψ is calculated from

$$t_\psi = \frac{\psi}{360°}\frac{60}{n} = \frac{\arccos(1-2a/d)}{6n} \quad \text{s} \tag{11.23}$$

where: ψ—working angle of the tooth, a—depth of cut in mm, d—diameter of milling cutter in mm, n—rotational speed of the cutter in rev/min.

The current angle of cutter rotation is denoted by φ. Thus the working angle ψ is a specific magnitude of the rotation angle φ, corresponding to the cutting time of a single tooth during one revolution of the cutter.

Another specific value of rotation angle φ is that corresponding to maximum thickness of the undeformed chip. This value of angle φ is denoted as φ_h (Fig. 11.9) and the maximum thickness of undeformed chip by $h_{\max}$

$$h_{\max} = s_z \sin\varphi_h = 2s_z\sqrt{a/d(1-a/d)} \quad \text{mm} \tag{11.24}$$

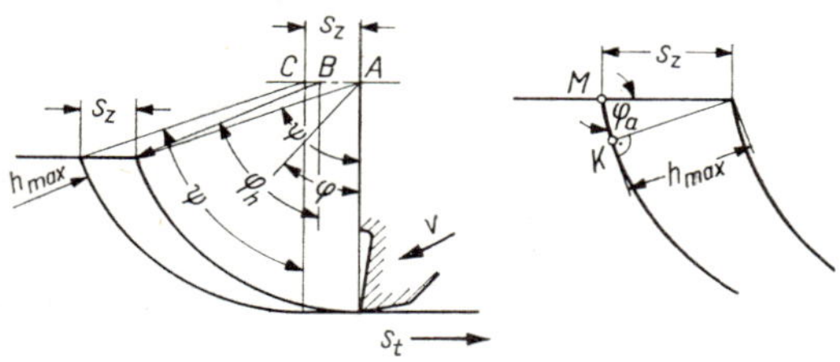

Fig. 11.9 Auxiliary drawing for the calculation of maximum thickness of the undeformed chip

The relationship presented in eqn. (11.24) may be generalized for an arbitrary angle of rotation φ

$$h = s_z \sin\varphi \quad \text{mm} \tag{11.25}$$

It follows from these formulae that the *undeformed chip thickness* in milling is the distance between two consecutive cut surfaces. This distance varies during the work of the tooth and is measured in the direction perpendicular to the preceding cut surface.

In milling by means of a plain milling cutter with straight teeth the undeformed chip thickness is uniform along each tooth and varies according to eqn. (11.25) during the time of tooth work in the machined material.

In milling by means of a plain milling cutter with helical teeth the undeformed chip thickness varies along the particular active teeth, the cause of this being the cutting edge inclination angle λ_s. Consequently, each point of the active cutting edge has a different rotation angle φ. Thus, according to eqn. (11.25) the undeformed chip thickness will vary along the cutter tooth.

It is evident from Fig. 11.10 in which the distribution of undeformed chip thickness among the particular active teeth is presented, that for angle λ_s and working angle ψ the teeth cutting simultaneously are teeth G, H, K, A and C.

Figure 11.10 concerns conventional milling. In the case of climb-cut milling, undeformed chip thickness will change in the opposite sense.

Undeformed chip width in milling is called the dimension of undeformed chip as measured along the cutting edge, in the direction perpendicular to the undeformed

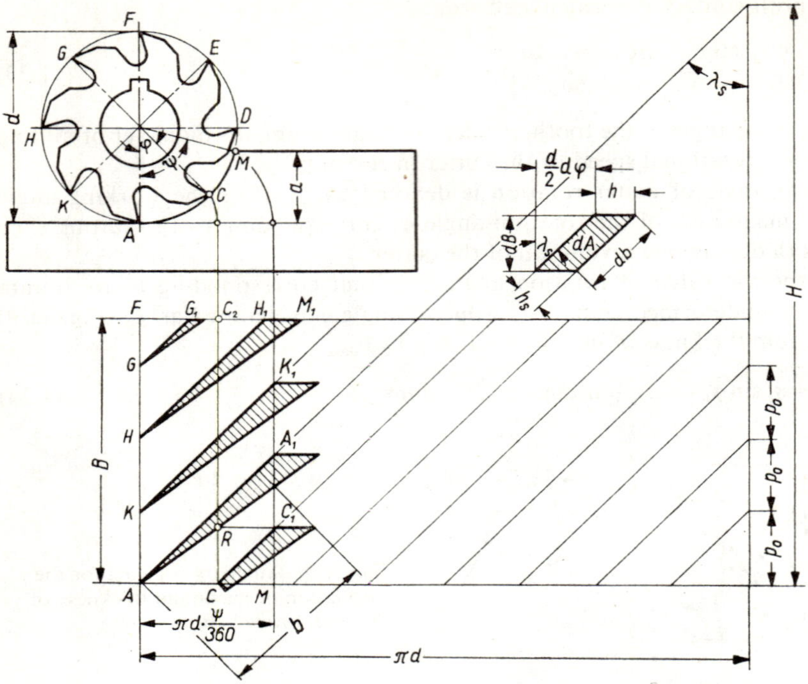

Fig. 11.10. Undeformed chip thickness in milling with a plain milling cutter having helical teeth

chip thickness. It follows from this definition that in the case of plain milling with cutters having straight teeth, the width of the undeformed chip is equal to milling width, or

$$b = B \quad \text{mm} \tag{11.26}$$

where: b—width of undeformed chip in mm, B—milling width in mm.

On the other hand, in the case of plain milling with a cutter having helical teeth, the width of the undeformed chip cut by a single tooth is variable and grows from zero to a maximum, then diminishes to zero again. Using Fig. 11.10 it is possible to determine generally the instantaneous width of the undeformed chip as

$$b = \frac{\pi d}{360^\circ} \frac{(\varphi_2 - \varphi_1)}{\sin \lambda_s} \quad \text{mm} \tag{11.27}$$

where: d—diameter of milling cutter in mm, φ_2—exit angle, i.e. the angle of rotation measured at the point where the work of the cutting edge finishes, φ_1—entry angle, i.e. the angle of rotation measured at the point where the contact between the cutting edge and workpiece material begins.

In Fig. 11.10 exit angles mean angles of rotation at points G_1, H_1, K_1 and C_1. Maximum value of the exit angle is equal to that of the working angle ($\varphi_{z\,max} = \psi$). In the same diagram entry angles mean angles of rotation at points, G, H, K, A and C.

The undeformed chip cross-section area falling to each tooth is calculated in milling by means of a cylindrical cutter with straight teeth from the formula

$$A_z = h\,b = s_z \sin\varphi b \quad \text{mm}^2 \tag{11.28}$$

Since the thickness of the undeformed chip varies from zero to h_{max}, the area of undeformed chip cross-section per tooth will vary from zero to

$$A_{z\,max} = h_{max} b = s_z \sin\varphi_h b \quad \text{mm}^2 \tag{11.29}$$

Taking into consideration the fact that $b = B$ and eqn. (11.24), then eqn. (11.29) may be rewritten in the form

$$A_{z\,max} = 2B\, s_z \sqrt{a/d(1-a/d)} \quad \text{mm}^2 \tag{11.30}$$

In the case of milling with a cutter having helical teeth the cross-sectional area of the undeformed chip per tooth A_z may be determined by integration of the element area (Fig. 11.10)

$$\mathrm{d}A = h_s \mathrm{d}b = s_z \cos\lambda_s \sin\varphi\, \mathrm{d}b \tag{11.31}$$

On the assumption that $\varphi_2 - \varphi_1 = \varphi$ the result

$$\mathrm{d}b = \frac{d}{2\sin\lambda}\mathrm{d}\varphi$$

is obtained.

Integration of the eqn. (11.31) within limits from $\varphi = \varphi_1$ to $\varphi = \varphi_2$ yields the formula for cross-section of undeformed chip per tooth

$$A_z = \frac{d}{2}\frac{s_z}{\tan\lambda_s}(\cos\varphi_1 - \cos\varphi_2) \quad \text{mm}^2 \tag{11.32}$$

Thus the maximum value of A_z occurs for $\varphi_1 = 0$ and $\varphi_2 = \psi$. The total area of undeformed chip machined by the active teeth

$$A = \sum_{i=1}^{i=z} A_z = A_{z1} + A_{z2} + \ldots + A_{zz} \quad \text{mm}^2 \tag{11.33}$$

11.2.2.2 *Plain milling time*

In plain milling, milling time is calculated according to the formula

$$t_m = \frac{l_a + l + l_o}{s_z z\, n} \quad \text{min} \tag{11.34}$$

where: l_a—length of approach in mm, l—length of the surface being milled in mm, l_o—length of overrun in mm.

Approach l_a depends on milling depth and cutter diameter

$$l_a = \sqrt{a(d-a)} = a\sqrt{d/a-1} = c\,a \quad \text{mm} \tag{11.35}$$

where: $c = \sqrt{d/a-1}$. The value of the coefficient c may be read from the diagram in Fig. 11.11.

Overrun is usually taken as l_o = 2–5 mm, depending on cutter diameter.

Thus the formula for milling time will be

$$t_m = \frac{c\,a + l + (2\text{–}5)}{s_z\, z\, n} \quad \text{min} \tag{11.36}$$

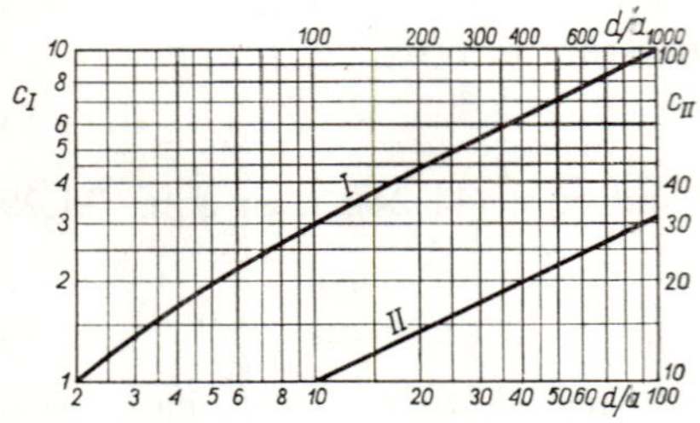

Fig. 11.11 Values of the coefficient $c = (d/a-1)^{0.5}$ depending on the d/a ratio

11.2.3 Cutting resistance and power

11.2.3.1 *Resolution of forces in plain milling*

In plain milling (as in turning) two reference systems have been adopted to determine the components of the cutting forces:

(a) the reference system connected with the machine tool,

(b) the reference system connected with each tooth of the milling cutter.

The reference system connected with the machine tool consists of:

(a) the reference line constituting the axis of cutter arbor (spindle); the projection of the cutting force on this line is called the *axial component* and denoted by Q;

(b) the reference line perpendicular to the spindle axis and situated in the vertical plane. The projection of the cutting force on this reference line is called the *vertical component* and denoted by S;

(c) the reference line perpendicular to both the preceding lines (horizontal). The projection of the cutting force on this reference line is called the *transverse component R*.

The reference system connected with the machine tool does not depend on the kinematics of cutting and is, therefore, stationary.

The reference system connected with the cutter tooth consists of:

(a) a line constituting the cutter axis; the projection of the cutting force on this direction is called the *axial component* F_f. This term is justified by the fact that, the milling cutter being correctly mounted, the cutter axis and spindle axis coincide ($Q = F_f$),

(b) a line perpendicular to the cutter axis and passing through the tip of the given tooth. The projection of the cutting force in this direction is called the *thrust component* or *radial component* and denoted by F_p,

(c) a line perpendicular to both preceding lines, tangential to the circle circumscribed on the contour of cutter cross-section. The projection of the cutting force on this line is called the *peripheral component* and denoted by F_v.

The resolution of cutting forces in milling with a cutter having straight teeth, both by the climb-cut and conventional method, may be considered approximately* as a plane system in which axial forces equal zero.

* This approximation is more accurate, the more rigid is the milling arbor and the milling machine.

Fig. 11.12 shows a system of cutting forces resulting from the action of one tooth on the workpiece material. In the diagram, designations without subscripts refer to conventional milling.

Assuming that the undeformed chip shape is identical in both conventional and climb-cut milling in each particular tooth position, the following relationship is obtained

$$|F_v| = |F'_v| \quad \text{and} \quad |F_p| = |F'_p|$$

Hence follows the equation $|W| = |W'|$.

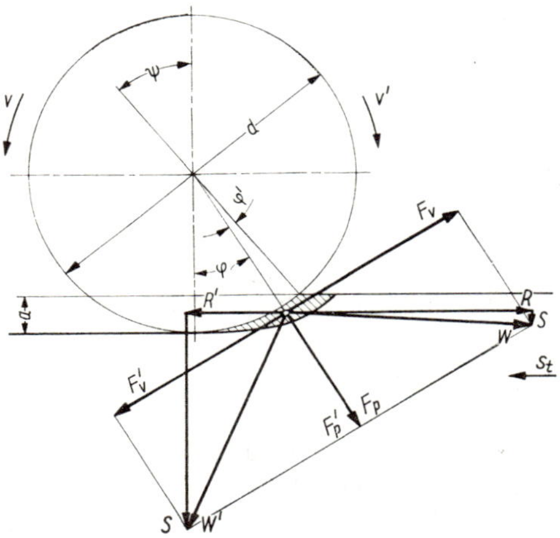

Fig. 11.12 Resolution of cutting forces in conventional milling and climb-cut milling

On the other hand, cutting forces in the machine tool system are different as regards their value and sense in conventional and climb-cut milling.

Based on Fig. 11.12 it may be stated that

$$R = F_v \cos\varphi + F_p \sin\varphi \tag{11.37}$$

$$R' = F'_v \cos\varphi - F'_p \sin\varphi \tag{11.38}$$

It has been assumed that $F_v = |F'_v|$ and $F_p = |F'_p|$ and in plain milling $0 \leqq \varphi < 90°$.

It therefore follows from eqns. (11.37) and (11.38) that the transverse force acting in the direction of table feed is always smaller in climb-cut milling $|R'| < R$. The difference between the transverse forces in conventional and climb-cut milling amounts to

$$\Delta R = R - R' = 2F_p \sin\varphi \tag{11.39}$$

Figure 11.12 allows one to derive the following formulae:

$$S = F_p \cos\varphi - F_v \sin\varphi \tag{11.40}$$

$$S' = F'_p \cos\varphi - F'_v \sin\varphi \tag{11.41}$$

When $|F'_p| = F_p$ and $|F'_v| = F_v$ and $0 \leqq \varphi < 90°$, then the vertical force $|S'|$ in climb-cut milling is always smaller than the vertical force S in conventional milling. The difference between those forces may be determined from

$$\Delta S = |S'| - S = 2F_v \sin\varphi \tag{11.42}$$

It is easily perceived that the maximum values of the difference occur for $\varphi = \varphi_h$. Then

$$|R'| = R-2F_p \sin\varphi_h \tag{11.43}$$

$$|S'| = S+2F_v \sin\varphi_h \tag{11.44}$$

As a result of this, in climb-cut milling the transverse force acting in the direction of the feed is considerably smaller than a similar force in conventional milling. Therefore, the feed screw of a climb-cut milling machine may be subject to less wear than that of a conventional milling machine. On the other hand, in the former method the pressure exerted on the table guide surface increases and so do elastic deformations of the machine tool parts and assemblies. However, since the guide surface is usually much larger than that resulting from strength calculations, this increase of vertical force has no marked influence on the rate of guide wear.

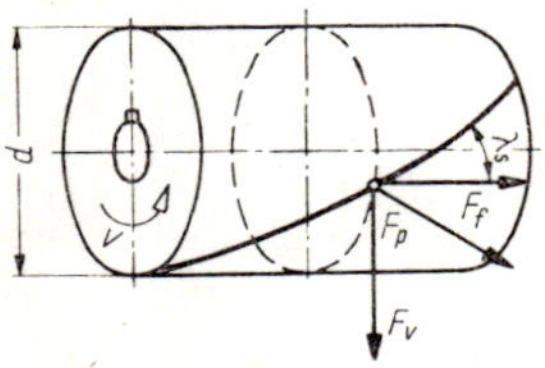

Fig. 11.13 Resolution of forces acting on a cutter with helical teeth

In milling with a cylindrical cutter having helical teeth, in addition to components F_v and F_p there is the component F_f. The resolution of these forces is shown in Fig. 11.13.

Depending on helix hand or on the mounting method of the helical cutter on its arbor and the sense of cutter rotation, axial forces may press either against the spindle bearing or against the arbor nut on the brace side. Experience indicates that more favourable machining conditions are obtained if the axial force acts towards the spindle.

In addition to the cutting forces a torque appears during milling

$$M = 0{\cdot}5F_v d \quad \text{kg mm} \tag{11.45}$$

11.2.3.2 *Non-uniformity of cutting forces in plain milling*

The thickness of the undeformed chip to be cut by each tooth increases from zero to maximum and then decreases to zero again. This indicates that the cutting force will vary in a similar way.

Within the practically applied range of feeds per tooth and a/d ratios, these variations may be expressed by formulae

$$F_v = C_z h^z \qquad F_p = C_y h^y \qquad F_f = C_x h^x$$

Since $h = s_z \sin\varphi$, the components of the cutting force exerted by the tooth upon the workpiece material will depend on the angle of rotation.

The time of contact between the tooth and workpiece material under milling conditions is, according to strictly kinematic calculations, of the order of hundredths or thousandths of a second. On the other hand, the time taken to reach maximum undeformed chip thickness may be, in climb-cut milling, of the order of thousandths or ten-thousandths of a second.

The appearance and disappearance of loads in such short time periods may under certain conditions cause impacts. In the case of plain milling high impacts generally do not occur. This results from the facts that:

(a) the applied cutting speeds are not very high,

(b) at the moment when the force increases, elastic deflections of the cutter arbor and table occur which reduce the actual speed of tooth motion in the workpiece material,

(c) the undeformed chip cross-sections falling to each tooth are relatively small.

Figure 11.14 presents diagrammatically the character of the changes of peripheral cutting force in conventional and climb-cut milling using a cutter with straight teeth. These diagrams are valid for the case when the contact angle is smaller than the angular pitch of the cutter teeth (equal to $360°/z$). Thus the number of teeth cutting simultaneously is one.

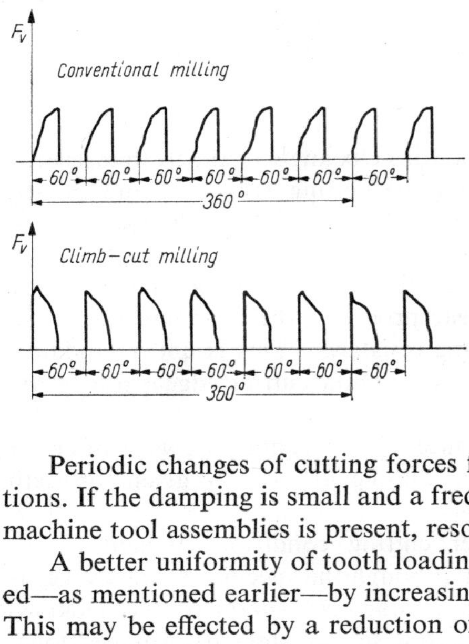

Fig. 11.14 Diagrammatic course of the changes of peripheral cutting force in milling with straight teeth

Periodic changes of cutting forces from zero to a maximum cause forced vibrations. If the damping is small and a frequency similar to that of the free vibrations of machine tool assemblies is present, resonance may occur.

A better uniformity of tooth loading and, consequently, a quieter work is obtained—as mentioned earlier—by increasing the number of teeth cutting simultaneously. This may be effected by a reduction of the peripheral pitch of the teeth and by inclining the teeth i.e. by using helical teeth.

11.2.3.3 *Factors influencing the cutting resistance in plain milling*

Factors influencing the cutting resistance in plain milling may be divided into groups according to:

(a) tooth material and shape,

(b) workpiece material,

(c) cutting conditions.

The rake angle exerts a distinct influence on the cutting work. In Fig. 11.15 is presented—according to Rozenberg's research—the dependence of cutting power on rake angle when milling steel and cast iron. The diagram was plotted on the assumption that for rake angle $\gamma = 0°$ the cutting power is 100%. The influence of rake angle on cutting resistance and power is explained by the fact that the degree of plastic deformation of the chip is lower, the greater is the rake angle.

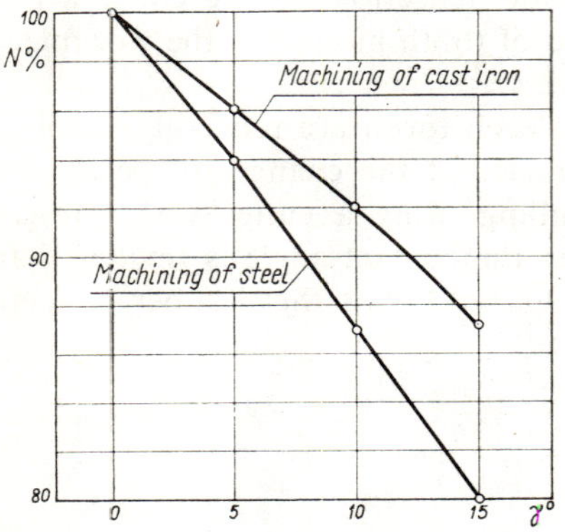

Fig. 11.15 Dependence of power consumption on rake angle in plain milling [11/1]

The leading cutting edge inclination angle λ_s (or helix angle of the teeth measured on the periphery) has no marked influence on the cutting resistance and power. Studies—as e.g. those carried out by Rozenberg—did not indicate the existence of such an influence within the range $\lambda_s = 20$–$45°$.

The radius of the cutting edge and the degree of tooth wear has a relatively large influence on the specific cutting force. As wear proceeds, the rounding radius increases and so does, simultaneously, the cutting resistance. This is understandable, since when a cylindrical cutter tooth wears the radius of the cutting edge is $\rho \approx 50$ μm (compared to some 15 μm after sharpening).

Taking into account the considerable influence of edge radius and wear of cutter teeth, all relationships between forces in plain milling are usually given for teeth worn to the permissible limit.

Increasing the cutter diameter—all other cutting conditions remaining unchanged—causes the cutting forces to diminish although the specific cutting energy is greater. This is explained by the fact that with growing cutter diameter the thickness of the undeformed chip decreases (Fig. 11.16), but so also does its cross-sectional area ($A = h\,b$) and the influence of the latter phenomenon prevails.

In the case of cutting simultaneously with more than one cutter tooth the increase of cutter diameter while preserving the same number of teeth may reduce the number of teeth cutting simultaneously and thus, of course, reduce the resultant cutting resistance, too. An increase of the number of cutter teeth, if the feed rate s_z mm/tooth remains constant, does not cause any changes in the specific cutting energy. On the other hand, it may increase the total cutting resistance if, as a result of the reduction of the angle pitch of cutter teeth, the number of teeth cutting simultaneously increases.

The influence of the undeformed chip parameters i.e. undeformed chip thickness, depth of cut and feed rate does not differ from the influence of similar parameters in other machining methods.

Increasing the depth of cut and feed rate causes an increase in thickness and therefore also of the cross-sectional area of the undeformed chip. A larger thickness

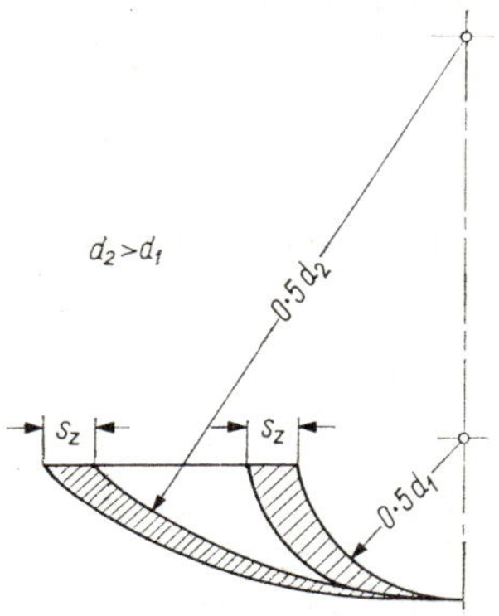

Fig. 11.16 Influence of milling cutter diameter on undeformed chip thickness

of the latter brings about a reduction of the specific force, but at the same time a growth of the undeformed chip surface correspondingly increases total resistance, while the increase of the undeformed chip cross-section has a prevailing influence.

In addition, an increased depth of cut involves a rise in the number of teeth cutting simultaneously. The influence of the depth of cut in milling using high-speed steel cutters is greater than the influence of the feed per tooth. On the other hand, in plain milling using cutters with sintered carbide teeth, research indicates the feed per tooth influence to be stronger of the two.

The increase of the width of undeformed chip only augments its cross-sectional area and thus also the cutting resistance.

11.2.3.4 *Calculation of cutting forces and power*

(1) *Plain milling using cylindrical cutters with straight teeth*

The peripheral component F_v acting on the tooth at any moment of the cutting process may be defined as the product of the specific cutting force k_s and the instantaneous cross-sectional area of the undeformed chip, or

$$F_v = k_s A_z = k_s h\,B \quad \text{kg} \tag{11.47}$$

while

$$k_s = C_z h^m \quad \text{kg/mm}^2 \tag{11.48}$$

where: C_z and m—numbers determined empirically.

Substitution of the value of h from eqn. (11.25) allows one to write the formula for the instantaneous value of *peripheral cutting force*

$$F_v = C_z B\, s_z^{1+m} \sin^{1+m}\varphi \quad \text{kg} \tag{11.49}$$

The maximum value of peripheral force per tooth occurs approximately when $\varphi = \psi$ or

$$F_{v\max} = C_z^1 B s_z^{1+m} \sin^{1+m}\psi \tag{11.50}$$

In addition to the instantaneous value of cutting force varying with tooth position, the notion of *mean peripheral force* of the tooth is introduced. By this term the quotient of work done by one tooth and of the length of arc corresponding to the working angle is understood

$$F_{vm} = \frac{E_z 360^\circ}{\pi d \psi} \quad \text{kg} \tag{11.51}$$

where: E_z—cutting energy of one tooth in kg mm, d—cutter diameter in mm, ψ—working angle of the tooth in degrees.

The cutting energy can be determined by integration

$$E_z = \int_0^\psi F_v \frac{d}{2} \mathrm{d}\varphi = 0{\cdot}5 C_z B\, d\, s_z^{1+m} \int_0^\psi (\sin\varphi)^{1+m} \mathrm{d}\varphi \tag{11.52}$$

This integral is solved by substituting $\left(\sin\frac{\varphi}{2}\right)^2 = x$. After suitable calculations and simplifications the following result is obtained

$$E_z = 2^{m+1} 0{\cdot}5 C_z B\, d\, s_z^{1+m} \int_0^{x_1} (x - x^2)^{m/2} \mathrm{d}x \tag{11.53}$$

(where x_1 corresponds to ψ).

Now the expression $(x-x^2)^{m/2}$ is resolved into a Maclaurin's series and only its first term is used in further computations. The error involved is less than 1% (within the limits $\varphi_2 - \varphi_1 \leqq 60^\circ$).

Integration and return to former denotations yield

$$E_z = \frac{2^{m+1}}{m+2} C_z B\, d\, s_z^{1+m} \left(\sin\frac{\psi}{2}\right)^{2+m} \tag{11.54}$$

Moreover, taking into consideration that

$$\sin\frac{\psi}{2} = \sqrt{0{\cdot}5(1-\cos\varphi)} \tag{11.55}$$

and that $\cos\psi = 1 - 2a/d$, it is possible—substituting these relations into eqn. (11.54)—to obtain the formula for cutting energy of one tooth as

$$E_z = \frac{2^{m+1}}{m+2} C_z B\, d^{-0{\cdot}5m} a^{0{\cdot}5m+1} s_z^{m+1} \tag{11.56}$$

Substitution of the expression for E_z from eqn. (11.56) to eqn. (11.51) gives a formula for the mean value of peripheral cutting force

$$F_{vm} = \frac{360}{\pi\psi} \frac{2^{m+1}}{m+2} C_z B\, d^{-(1+0{\cdot}5m)} a^{0{\cdot}5m+1} s_z^{m+1} \tag{11.57}$$

Making use of the notion of mean cutting force of one tooth it is also possible to introduce the notion of mean peripheral force acting during cutting. By the *mean peripheral cutting force* of a milling cutter is understood the product of the mean value of peripheral force acting on one tooth and of the number of teeth cutting simultaneously (this number is not being rounded off to the nearest whole number)

$$F_{vp} = F_{vm} z_c = F_{vm} \frac{\psi}{\psi_z} = \frac{z\,\psi}{360°} F_{vm} \tag{11.58}$$

or
$$F_{vp} = \frac{2^{m+1}}{\pi(m+2)} C_z B\, z\, d^{-(1+0{\cdot}5x)} a^{0{\cdot}5m+1} s_z^{m+1} \tag{11.59}$$

Assuming, moreover, that

$$C_{zp} = \frac{2^{m+1} C_z}{\pi(m+2)} \tag{11.60}$$

and
$$b_z = -(1+0{\cdot}5m) \qquad e_z = 0{\cdot}5m+1 \qquad u_z = m+1 \tag{11.61}$$

one can write eqn. (11.59) in the form

$$F_{vp} = C_{zp} B\, z\, d^{b_z} a^{e_z} s_z^{u_z} \tag{11.62}$$

The mean value of cutting power in plain milling with straight teeth is

$$N_{me} = \frac{F_{vp} v}{6120} = \frac{F_{vp} \pi d\, n}{6120 \times 1000} \quad \text{kW} \tag{11.63}$$

Taking into account eqn. (11.62) it may be written that

$$N_{me} = \frac{C_{zp}\pi}{6120 \times 1000} B\, z\, n\, d^{1+b_z} a^{e_z} s_z^{u_z} \quad \text{kW} \tag{11.64}$$

Substituting

$$C_N = \frac{2^{m+1} C_z}{\pi(m+2)} \frac{\pi}{61{\cdot}2} = \frac{2^{m+1}}{61{\cdot}2(m+2)} C_z \tag{11.65}$$

and
$$1+b_z = b_N \tag{11.66}$$

one can write the formula for mean value of cutting power in milling in the following form

$$N_{me} = 10^{-5} C_N B\, z\, n\, d^{b_N} a^{e_z} s_z^{u_z} \quad \text{kW} \tag{11.67}$$

It follows from eqns. (11.60) and (11.65) that

$$C_{zp} = 19{\cdot}5 C_N \tag{11.68}$$

(2) *Plain milling using cutters with helical teeth*

The instantaneous value of peripheral cutting force may be calculated in a similar way to that for milling with straight teeth, viz.

$$dF_v = k_{sz} \mathrm{d}A = C_z h^m h_s \mathrm{d}b = C_z h^m h\, \mathrm{d}B \tag{11.69}$$

while (Fig. 11.17)

$$h_s = h \cos\lambda_s = s_z \cos\lambda_s \sin\varphi \tag{11.70}$$

and
$$\mathrm{d}b = 0{\cdot}5d \frac{\mathrm{d}\varphi}{\sin\lambda_s} \tag{11.70}$$

When these relationships are taken into account

$$dF_v = 0{\cdot}5C_z d\, s_z^{m+1} \cot \lambda_s (\sin \varphi)^{m+1} d\varphi \tag{11.71}$$

and after integration within the limits φ_1 to φ_2

$$F_v = \frac{2^{m+1}}{m+2} C_z d \cot \lambda_s\, s_z^{m+1} \left[\left(\sin \frac{\varphi_2}{2} \right)^{m+2} - \left(\sin \frac{\varphi_1}{2} \right)^{m+2} \right] \tag{11.72}$$

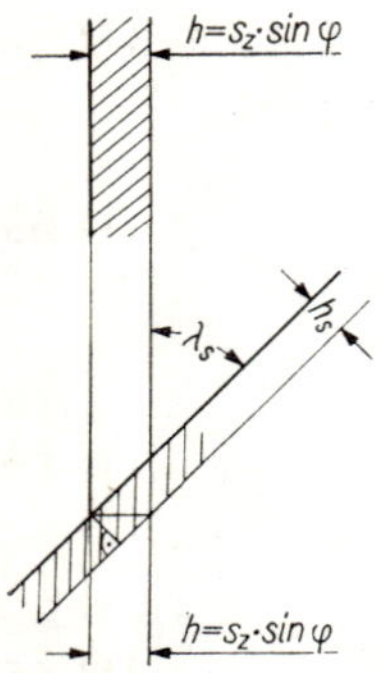

Fig. 11.17 Undeformed chip thickness in the cross-section perpendicular to the cutting edge (h—for straight cutter teeth, h_s—for helical cutter teeth)

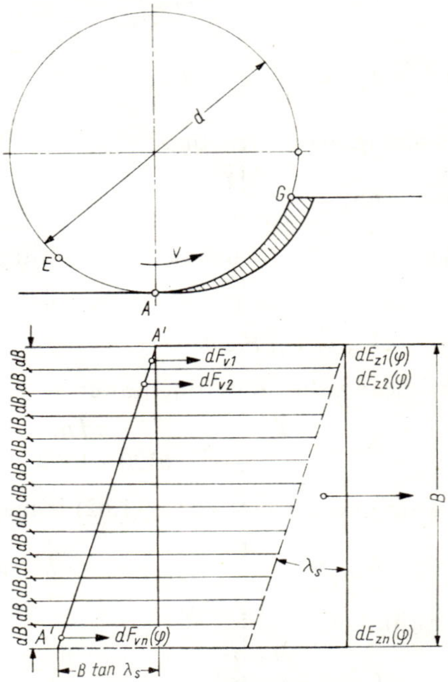

Fig. 11.18 Cutting energy of a helical tooth as the total energy of straight teeth with milling widths dB

The maximum value of peripheral cutting force occurs when $\varphi_1 = 0$ and $\varphi_2 = \psi$. Then

$$F_{v\,\max} = \frac{2^{m+1}}{m+2} \cot \lambda_s C_z d\, s_z^{m+1} \left(\sin \frac{\psi}{2} \right)^{m+2} \tag{11.73}$$

Further substitutions for $\sin \frac{\psi}{2}$ give eqn. (11.73) in the form

$$F_{v\,\max} = \frac{2^{m+1}}{m+2} C_z \cot \lambda_s s_z^{m+1} d^{-0{\cdot}5m} a^{0{\cdot}5m+1} \tag{11.74}$$

The total energy for straight teeth of widths dB, adding up to B (Fig. 11.18), may be considered as the energy of a single tooth, while each of the elementary teeth is displaced by d$B \tan\lambda_s$ under the action of force

$$dF_v = k_{sz} dA = C_z h\, dB = C_z s_z^{1+m} \sin \varphi^{1+m} dB \tag{11.75}$$

Tooth energy is determined by

$$\mathrm{d}E_z = 0{\cdot}5d\mathrm{d}\varphi\mathrm{d}F_v \tag{11.76}$$

By integration,

$$E_z = \frac{2^{m+1}}{m+2} C_z B\, d\, s_z^{1+m} \left(\sin\frac{\psi}{2}\right)^{m+2} \tag{11.77}$$

is obtained, and the use of the same substitutions as for eqn. (11.56) yields

$$E_z = \frac{2^{m+1}}{m+2} C_z B\, d^{-0{\cdot}5m} a^{0{\cdot}5m+1} s_z^{m+1} \tag{11.78}$$

The path on which the helical tooth works is

$$l_s = \frac{\psi}{360^\circ}\pi d + B\tan\lambda_s \tag{11.79}$$

As a result, the mean value of peripheral cutting force is

$$F_{vm} = \frac{360^\circ}{\pi\psi}\frac{2^{m+1}}{m+2} C_z B\, d^{b_z} a^{e_z} s_z^{u_z} \frac{1}{1 = \dfrac{360^\circ}{\psi}\dfrac{B}{d}\dfrac{\tan\lambda_s}{\pi}} \tag{11.80}$$

If the mean value of peripheral force for a milling cutter with straight teeth is denoted by F'_{vm} and for one with helical teeth by F''_{vm}, then using the relationships (11.57) and (11.80) one may write that

$$F''_{vm} = \frac{1}{1 + \dfrac{360^\circ}{\psi}\dfrac{B}{d}\dfrac{\tan\lambda_s}{\pi}} F'_{vm} = \frac{1}{K} F'_{vm} \tag{11.81}$$

It follows from this relation that the mean value of peripheral cutting force of a helical cutter tooth is $1/K$ of the mean value of peripheral force of a straight tooth.

Thus for example, for $\psi = 20^\circ$, $B/d = 1$, $\lambda = 30^\circ$, the value of $K = 4{\cdot}31$ or $F''_{vm} = 0{\cdot}232F'_{vm}$.

The mean peripheral force of cutting with a milling cutter having helical teeth F''_{vp} is the product of the mean value F''_{vm} and the number of teeth z_c cutting simultaneously

$$F''_{vp} = F''_{vm} z_c \tag{11.82}$$

The use of eqns. (11.17) and (11.81) permits one to write the equation (11.82) in the form

$$F''_{vp} = \frac{F'_{vm}}{K} z \left(\frac{B}{d}\frac{\tan\lambda_s}{\pi} + \frac{\psi}{360^\circ}\right) \tag{11.83}$$

The ratio in which F'_{vp} is determined by eqn. (11.58) is

$$\frac{F'_{vp}}{F''_{vp}} = \frac{z\psi F'_{vm} K}{360^\circ F''_{vm} z \left(\dfrac{B}{d}\dfrac{\tan\lambda_s}{\pi} + \dfrac{\psi}{360^\circ}\right)} \tag{11.84}$$

after substitution of the value for K from eqn. (11.81), and after some transformations this leads to the conclusion that

$$\frac{F'_{vp}}{F''_{vp}} = 1 \qquad \text{or} \qquad F''_{vp} = F'_{vp} \tag{11.85}$$

Cutting power in milling using cutters having both straight and helical teeth is calculated according to eqn. (11.67).

Constant values necessary for the calculation of F_v, F_{vm}, F_{vp} and N_{me} are listed in Table 11.1.

Table 11.1 *Constant values appearing in formulae for cutting force and power in plain milling* (for permissible tooth wear)

Cutter tooth material	Workpiece material	Constant values								According to
		C_z	m	C_{zp}	b_z	e_z	u_z	C_N	b_N	
High-speed steel	steel	224	−0·28	68·2	−0·86	0·86	0·72	3·5	0·14	standards of the Soviet Ministry of Machine Tool Industry
	malleable cast iron	98·5	−0·28	30	−0·83	0·83	0·72	1·54	0·17	
	grey cast iron	98·5	−0·28	30	−0·86	0·86	0·65	1·54	0·14	
	bronze	73·8	−0·28	22·5	−0·86	0·86	0·72	1·155	0·14	
Sintered carbides	constructional steel	140	−0·26	43	−0·87	0·87	0·74	2·2	0·13	Morozov and Oznobitin [11/2]
	chromium-nickel steel	175	−0·26	53·5	−0·87	0·87	0·74	2·74	0·13	

(3) *Calculation of components in the machine tool system*

With regard to machining accuracy and to machine tool loading, a knowledge of components in the Q, R, S system is necessary.

The components of the cutting resistance may be determined graphically (Fig. 11.19) or analytically.

The graphical determination (adding up of vectors) consists in projecting the components F_f, F_p and F_v of each cutter point on the axes Q, R and S.

Since the direction of the Q axis coincides with the x axis, then

$$\sum_{i=1}^{i=n} \vec{F}_{fi} = \vec{F}_{fA} + \vec{F}_{fB} + \dots + \vec{F}_{fn} = \vec{Q} \tag{11.86}$$

Components R and S are determined by geometrical addition

$$\sum_{i=1}^{i=n} \vec{F}_v = \vec{F}_{vA} + \vec{F}_{vB} + \dots + \vec{F}_{vn}$$

$$\text{and} \quad \sum_{i=1}^{i=n} \vec{F}_p = \vec{F}_{pA} + \vec{F}_{pB} + \dots + \vec{F}_{pn} \tag{11.87}$$

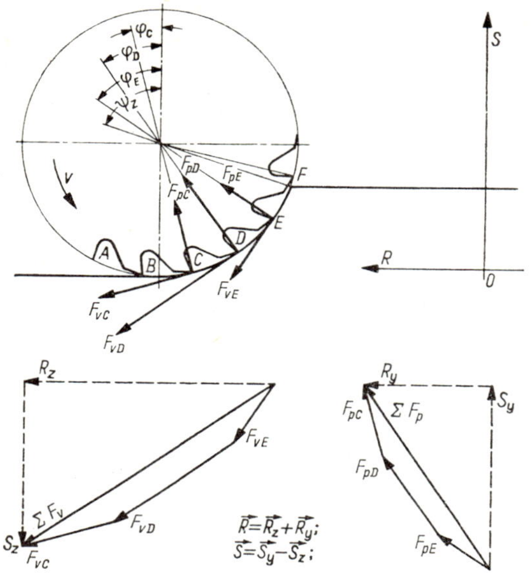

Fig. 11.19 The sum of components of the cutting resistance in the machine tool system (Q, R, S)

Projections of F_v and F_p on the axes R and S yield

$$R = R_p + R_v$$
$$S = S_p - S_v \tag{11.88}$$

Instead of geometrical addition, the analytical method may be used. If it is assumed that $F_p = w_y F_v$, then, generally,

$$S = \sum_{i=1}^{i=n} (w_y \cos\varphi_i - \sin\varphi_i) F_{vi} \quad \text{kg} \tag{11.89}$$

$$R = \sum_{i=1}^{i=n} (w_y \sin\varphi_i + \cos\varphi_i) F_{vi} \quad \text{kg} \tag{11.90}$$

Now values calculated according to eqn. (11.49) for milling with straight teeth and to formula (11.74) for milling with helical teeth are substituted for F_v.

The value of the ratio $w_y = F_p/F_v$ is taken by analogy to turning as $w_y = 0{\cdot}4$. The value of the ratio $w_x = F_f/F_v$ largely depends on the cutter tooth helix angle λ_s

$$F_f = Q = K \tan\lambda_s F_v \tag{11.91}$$

where K—a number found in an empirical way.

On the basis of several experiments it has been established that in plain milling of the conventional and climb-cut type, the most frequently encountered values of components R and S are:
for conventional milling

$$S = (0{\cdot}2\text{–}0{\cdot}3) F_v \quad \text{kg}$$
$$R = (1\text{–}1{\cdot}2) F_v \quad \text{kg} \tag{11.92}$$

for climb-cut milling

$$S = (0{\cdot}75\text{–}0{\cdot}8)F_v \quad \text{kg}$$
$$R = (0{\cdot}8\text{–}0{\cdot}9)F_v \quad \text{kg} \qquad (11.93)$$

11.2.4 TOOTH WEAR AND LIFE

11.2.4.1 *Tooth wear*

The wear of cylindrical cutter teeth occurs in a similar way to the wear of other cutting tools. The amount of wear *VB* (Fig. 11.20) measured on the flank is adopted as a wear criterion.

A different value of permissible wear *VB* is adopted for roughing and finishing operations. When the permissible wear *VB* for roughing and profiling operations is determined, the necessity of obtaining maximum tool life—i.e. economic considerations—is usually taken into account.

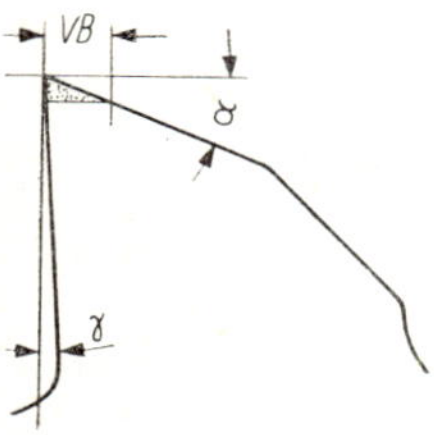

Fig. 11.20 Wear of a tooth of cylindrical milling cutter

In the roughing of steel, wear $VB = 0{\cdot}4\text{–}0{\cdot}6$ mm is permissible for cylindrical cutters and for profiling $VB = 0{\cdot}2\text{–}0{\cdot}3$ mm.

In the roughing of cast iron, $VB = 0{\cdot}5\text{–}0{\cdot}8$ mm is adopted for cylindrical cutters, and for profiling $VB = 0{\cdot}2\text{–}0{\cdot}4$ mm.

In the case of finishing with high accuracy and surface quality requirements the degree of tool wear adopted as the criterion often has to be reduced.

Several factors influence the tooth wear of cylindrical milling cutters.

An important factor influencing the uniformity of tooth wear is the accuracy of cutter mounting on its arbor and the accuracy of tooth grinding. So-called run-out of the cutter, causing non-uniform loading of the particular teeth, results in non-uniform wear. Unequal hardness of cutter teeth has a similar significance.

A characteristic influence on type and intensity of wear in plain milling is the cutting kinematics.

Experiments indicate that the tooth wear of cylindrical cutters in climb-cut milling proceeds slower than that of cutters milling conventionally (Fig. 11.21).

The cause of faster wear in conventional milling are less advantageous conditions of tooth entry into the material. As the cutting edge always has a certain roundness, in conventional cutting the edge is exposed to greater pressure and friction on the flank while entering the material. This results in increased wear of the tooth.

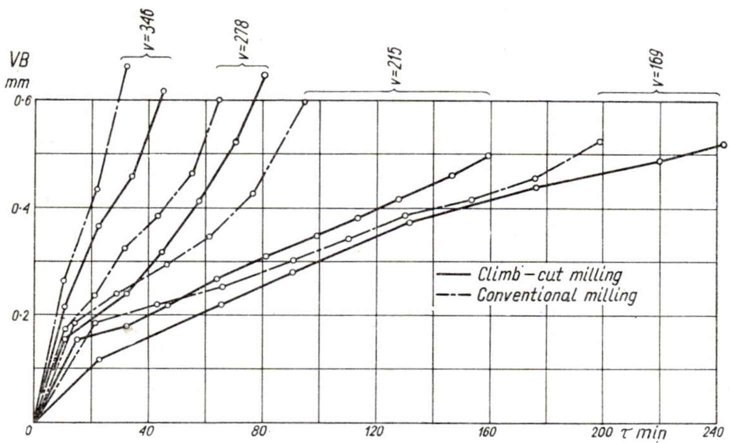

Fig. 11.21 Cutter tooth wear curves in climb and conventional plain milling. Workpiece material: alloy steel, $s_z = 0{\cdot}1$ mm/tooth, $a = 6$ mm, $B = 12$ mm, $\gamma = -5°$, $\alpha = 15°$, $\lambda_s = 10°$ [11/2]

11.2.4.2 *Influence of some cutting conditions on life speed*

Of the many factors influencing tool life speed those whose influence is particularly remarkable should be primarily considered.

As a result of cutter diameter increase the tooth travel increases, but at the same time the undeformed chip thickness decreases. This causes an increase of work time and friction, and simultaneously diminishes the resistance met by cutter tooth, while the angular pitch of teeth and the mass of the milling cutter increases. These phenomena are accompanied by an improvement of cooling conditions, better heat abstraction and lowering of tooth temperature. As a result, a larger diameter of cutter raises tool life speed. It follows from the research conducted on this subject that the dependence of life speed on cutter diameter may be expressed by an exponential function

$$v_t = C_v d^{b_v} \tag{11.94}$$

while the exponent has the value $b_v = 0{\cdot}17$–$0{\cdot}7$ depending on tooth and workpiece materials (Table 11.2).

Feed per tooth s_z directly influences the undeformed chip thickness. It is well known that thicker undeformed chips require more cutting energy and at the same time more cutting heat is generated and the cutting temperature rises. As a result the tooth life shortens.

When the depth of cut increases, so does the volume of undeformed chip per tooth and the time of tooth work in the material being machined. This results in a greater energy of plastic deformation, a rise of cutting temperature and an extension of the time of friction of the tooth against workpiece material. As a consequence of all this, doubling of the depth of cut may cause a reduction of life speed by 70–90% depending on tooth and workpiece materials.

Milling width does not influence the specific cutting energy. On the other hand, in proportion to greater milling width the cutting energy and heat rise. As, simulta-

Table 11.2 *Values of constants appearing in the formula for cutter life speed in plain milling*

Cutter tooth material	Workpiece material	Within range of			Constant values						
		B	a mm	s_z mm/point	C_v	m	b_v	e_v	u_r	r_v	n_v
High-speed steel	steel $\sigma = 75$ kg/mm²	—	—	$\leqq 0{\cdot}1$	55	0·33	0·45	0·3	0·2	0·1	0·1
				$> 0{\cdot}1$	35·4	0·33	0·45	0·3	0·4	0·1	0·1
	grey cast iron HB = 190 kg/mm²	—	—	$\leqq 0{\cdot}15$	57·7	0·25	0·7	0·5	0·2	0·3	0·3
				$> 0{\cdot}15$	27	0·25	0·7	0·5	0·6	0·3	0·3
	malleable cast iron HB = 150 kg/mm²	—	—	$\leqq 0{\cdot}1$	64·1	0·33	0·45	0·3	0·2	0·1	0·1
		—	—	$> 0{\cdot}1$	41·2	0·33	0·45	0·3	0·4	0·1	0·1
	bronze HB = 100 –140 kg/mm²	—	—	$\leqq 0{\cdot}1$	115·5	0·33	0·45	0·3	0·2	0·1	0·1
				$> 0{\cdot}1$	74·3	0·33	0·45	0·3	0·4	0·1	0·1
Sintered carbides	0·50 carbon steel	$\leqq 35$	$\leqq 2$	—	417	0·33	0·17	0·19	0·28	−0·05	0·1
			> 2	—	475	0·33	0·17	0·38	0·28	−0·05	0·1
		> 35	$\leqq 2$	—	660	0·33	0·17	0·19	0·28	0·08	0·1
			> 2	—	750	0·33	0·17	0·38	0·28	0·08	0·1
	cast iron	$\leqq 30$	2·5	$\leqq 0{\cdot}2$	892	0·42	0·38	0·13	0·19	0·23	0·14
				$> 0{\cdot}2$	568	0·42	0·38	0·13	0·47	0·23	0·14
			2·5	$\leqq 0{\cdot}2$	1145	0·42	0·38	0·4	0·19	0·23	0·14
				$> 0{\cdot}2$	725	0·42	0·38	0·4	0·47	0·23	0·14

neously, conditions of heat abstraction deteriorate, a slight rise of temperature is observed when milling width increases. This phenomenon is more pronounced with higher cutting speeds. If the influence of milling width is approximated by the exponential function $v_r = C_v B^{r_v}$, then the value of the exponent is $r_v = 0{\cdot}1$–$0{\cdot}3$ (depending on cutter and workpiece materials).

The influence of the number of cutter teeth on life speed is explained by the fact that with the growing number of teeth, all other cutting conditions remaining unchanged, there occurs:

(a) an increase of mean value of peripheral force and cutting power and therefore also of mean energy and cutting heat,

(b) a reduction of space between teeth which worsens the conditions of chip flow and heat abstraction,

(c) a reduction of volume and mass of particular teeth.

All this causes an increase of tooth temperature, and thus an acceleration of tooth wear. The overall influence of the number of cutter teeth is of the same order as that of milling width.

Influence of the helix angle of cutters with helical teeth or of the cutting edge inclination angle within practically applied limits is rather small. However, it can be observed that tool life shortens with increasing angle λ_s, because then the number of teeth cutting simultaneously increase. At the same time the general amount of cutting teeth increase and—although the cutting edges lengthen, thus improving heat abstraction—the tooth temperature rises. Considering that in cutters with helical teeth the angle $\lambda_s = 20$–$45°$, the influence of the cutting edge inclination angle may be practically neglected.

As far as the influence of tool material and shape, of workpiece material and its properties, and of the method and intensity of cooling are concerned, the explanations given in Part I of this book and chapters dealing with other machining methods are also valid here.

11.2.4.3 *Calculation of life speed*

The relationship between tool life speed and cutting conditions in plain milling has been established on the basis of numerous studies. The statistical and empirical formula describing this relationship has the general form

$$v_T = \frac{C_v d^{b_v}}{T^m a^{e_v} s_z^{u_v} B^{r_v} z^{n_v} \lambda_s^{q_v}} B_M \, B_H \, B_s \, B_p \, B_m \tag{11.95}$$

where: C_v—a constant taking into account the influence of all factors not appearing separately in the formula, d—cutter diameter in mm, T—tool life in min, a—depth of cut in mm, s_z—feed per tooth in mm, B—milling width in mm, z—number of cutter teeth, λ_s—cutting edge inclination angle in degrees, b_v, m, e_v, u_v, r_v, q_v—exponents determined empirically, B_M—correction coefficient taking into account the kind of workpiece material, B_H—correction coefficient taking into account the hardness or strength of workpiece material, B_s—correction coefficient taking into account the state of steel, B_p—correction coefficient taking into account the kind of semi-finished product and the state of its surface, B_m—correction coefficient taking into account the kind of machining (for rough milling $B_m = 1{\cdot}0$, for accurate milling $B_m = 0{\cdot}8$).

The relationship between life speed and tool life may be represented by means of a curve as in Fig. 11.22 (for conventional milling). In practice, however, formulae are used corresponding only to that part of the curve which shows a tool life decrease for rising cutting speed.

The influence of cutting speed on tool life largely depends on the value of exponent m. It follows from experiments that for high-speed steel teeth $m = 0{\cdot}15$–$0{\cdot}33$ and for sintered carbide teeth $m = 0{\cdot}2$–$0{\cdot}6$. These values indicate that cutting speed has a stronger influence on the tooth life of high-speed steel than on that of sintered carbide teeth. Average values of the exponent m are given in Table 11.2 together with values of the constant C_v and of exponents b_v, e_v, r_v and n_v for milling steel and cast iron with high-speed steel and sintered carbide teeth.

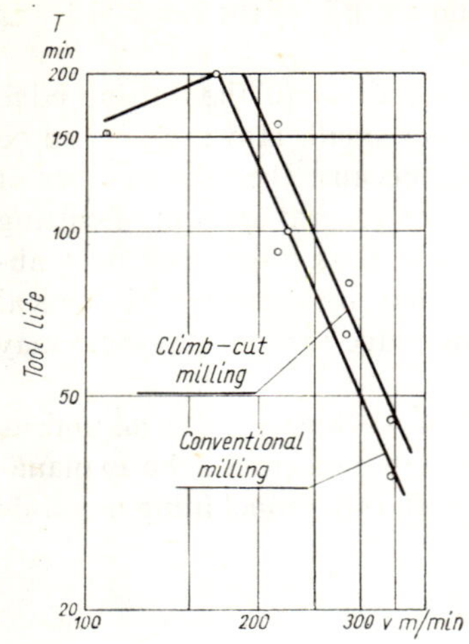

Fig. 11.22 Influence of cutting speed on cutter tooth life in conventional and climb-cut milling. Workpiece material: alloy steel, $s_z = 0{\cdot}1$ mm/tooth, $a = 6$ mm, $B = 22$ mm, $\gamma = -5°$, $\alpha = 15°$, $\lambda_s = 10°$ [11/2]

Values of the constant C_v given in Table 11.2 are valid for conventional milling. In the case of climb-cut milling the correction coefficient $B_m = 1{\cdot}05$ is used.

In practice, cutting speeds v corresponding to the recommended values of tooth life T_2 are used. As an example, to illustrate the order of those values, cutter life speeds are given in Table 11.3.

Table 11.3 *Values of life of cylindrical milling cutters of high-speed steel*

Tool	Cutter dimensions diameter × length in mm	T, min	
		in machining of steel	in machining of cast iron
Cylindrical cutters	40 × 40 50 × 50	30	45
	60 × 60	30	45
	75 × 75	55 50	90 85

11.2.5 SURFACE ROUGHNESS

11.2.5.1 *Roughness as a kinematic and geometric reproduction*

Cycloidal motion of cutter teeth in relation to the machined surface is the cause of the appearance of surface irregularities whose height—according to Fig. 11.23—may be expressed by

$$R_{tmt} = 0{\cdot}5d - \sqrt{(0{\cdot}25d^2 - 0{\cdot}25s_z^2)} \quad \text{mm} \tag{11.96}$$

where: d—cutter diameter in mm, s_z—feed per tooth in mm.

As already demonstrated in Chapter 6, eqn. (11.96) may be simplified with sufficient accuracy to

$$R_{tmt} \approx 250 \frac{s_z^2}{d} \quad \text{mm} \tag{11.97}$$

Experiments indicate that the actual height of the irregularities considerably differs from that calculated from eqn. (11.17). Thus, for example, for a milling cutter of diameter $d = 100$ mm and for a feed per tooth $s_z = 0{\cdot}2$ mm, the theoretical height of

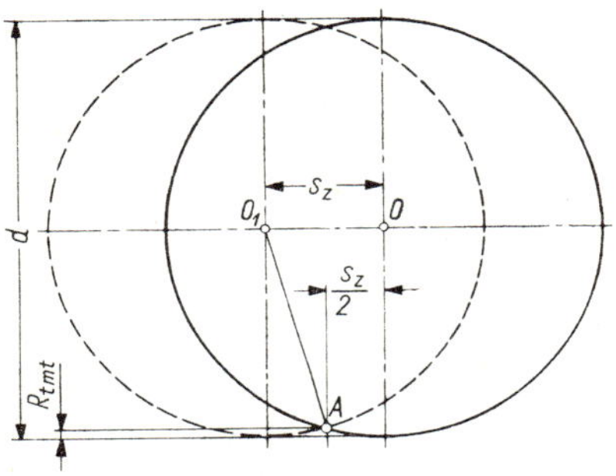

Fig. 11.23 Theoretical height of surface irregularities in plain milling

the irregularities amounts to $R_{tmt} = 125.\ 0{\cdot}04/100 = 0{\cdot}05$ mm. However, the measured height of the irregularities of a surface machined under these conditions is of the order of several micrometers. Such a divergence between the theoretical and actual values can only be explained by the run-out of cutter tooth tips, resulting from:

(a) sharpening errors (eccentric grinding of lands in relation to the axis of cutter rotation),
(b) mounting errors (eccentric mounting of the cutter on its arbor),
(c) a bent arbor,
(d) periodically varying rigidity of the machine-tool–workpiece–tool system (particularly the varying rigidity of the arbor).

As a result of these causes it may be observed in practice that there exists another relationship, much nearer to reality, viz. that between surface roughness and feed per cutter revolution $s = z\,s_z$. This relationship may be represented by means of the formula

$$R_{tmt} = \frac{s^2}{4(d+e)} \quad \text{mm} \tag{11.98}$$

where: $s = z\,s_z$—feed rate in mm/rev, d—cutter diameter in mm, e—run-out in mm.

There are two types of run-out. The first type (Fig. 11.24a, b) consists in that the axis of arbor rotation describes a closed curve, e.g. a circle of radius $r = 0{\cdot}5e$. This may be caused by curvature of the arbor or by its varying rigidity.

The second type of run-out (Fig. 11.25) consists in that the axis of rotation does not change its position, but tooth tips are at unequal distances from the axis. This may be the result of incorrect mounting of the arbor or of eccentric grinding.

In the first type of run-out two cases may occur:

(a) when $R_{tmt} < e$; then R_{tmt} is calculated from eqn. (11.98) (Fig. 11.24a),

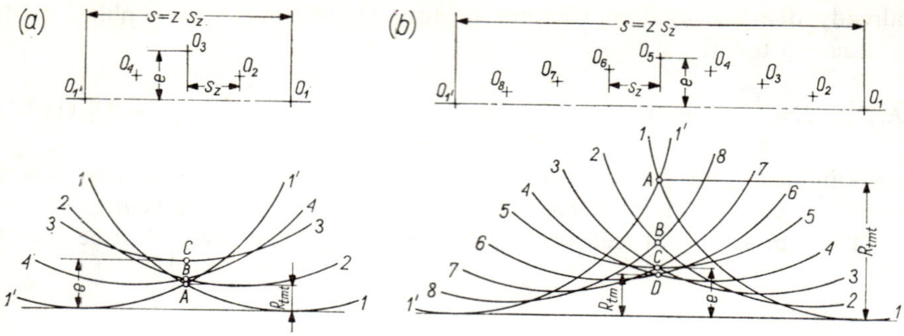

Fig. 11.24 Surface irregularities as a result of eccentricity: (a) of a cutter with four teeth, (b) of a cutter with eight teeth

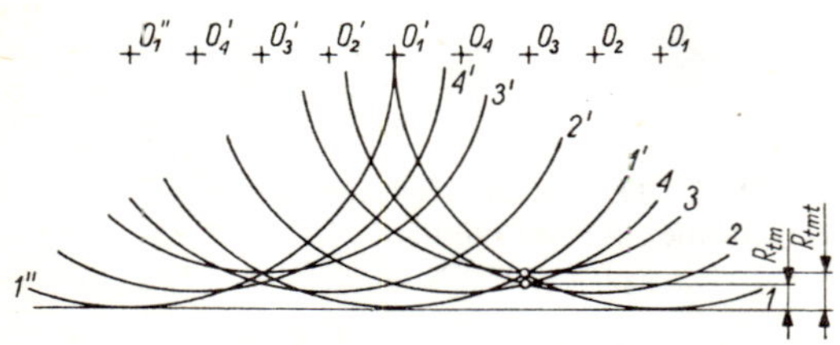

Fig. 11.25 Surface irregularities as a result of unequal height of teeth or eccentric grinding

(b) when $R_{tmt} > e$; then the actual total mean height R_{tm} is smaller than e (Fig. 11.24b). In this case the cause of the reduction of R_{tmt} to the value of R_{tm} is that the successive teeth cut the irregularities left over by the preceding teeth.

Which of these cases occurs is determined, as evident from Fig. 11.24, by the e/s ratio. For sufficiently small values of this ratio, most frequently met in practice, only the type shown in Fig. 11.24b can occur.

Thus, most frequently,

$$R_{tm} \leqq e \quad \text{mm} \tag{11.99}$$

In the second type of run-out (Fig. 11.25) also always

$$R_{tm} \leqq R_{tmt} = e \quad \text{mm} \tag{11.100}$$

where e is the difference between the largest and the smallest distance of a tooth tip from the axis (on the assumption that the cutter periphery is a circle).

Thus, generally, the maximum height of irregularities in plain milling by means of cutters with the number of teeth $z = 1$ may attain the value of the run-out, or

$$R_{\max} = e \quad \text{mm} \tag{11.101}$$

Experiments show that the greater is the feed per revolution, the more the theoretical height approaches the actual height.

Another factor (besides run-out) are irregularities of the cutting edge which are reproduced on the machined surface. Therefore, the condition for obtaining a milled surface of good quality is careful sharpening and smoothing of cutter teeth.

11.2.5.2 *Other factors influencing surface quality*

In addition to factors influencing the surface quality mainly in a geometric way, a distinct influence on the finish of surfaces milled by means of cylindrical cutters is exerted by:
(a) rounding of the cutting edge and tooth wear,
(b) built-up edge,
(c) vibrations of the machine-tool–workpiece–tool system.
The quality of milled surfaces is also influenced by the workpiece material, cutter material and tooth shape, coolant and the method of cooling. These factors have been described in the preceding chapters.

The significance of cutting edge rounding appears clearly when the quality of surfaces milled conventionally and by the climb-cut method are compared. A large radius of cutting edge rounding results in a greater compression of the material. In conventional milling, in the initial phase this compression occurs on the machined surface. The more ductile and deformable is the workpiece material, the easier it is for surface irregularities to appear caused by burrs and cracks of the subsurface layer.

In the case of climb-cut milling, the influence of compression, particularly as the tooth enters the material, has no such significance for the finished surface, since cutting into the material occurs in the undeformed chip zone which is then removed by the following teeth.

The final quality of a surface milled by means of cylindrical cutters is also influenced by the pressing-in of chips into the machined surface. This occurs particularly in conventional milling and to a greater extent if the material is ductile. The cause of this phenomenon is that some of the chips become welded to teeth or are carried away by a tooth, are 'spread' over the material and pressed into it or else leave scratches on the machined surface.

For these reasons both research and practical tests indicate that the quality of climb-cut milled surfaces is generally higher than that of conventionally milled ones.

In Fig. 11.26 the results of research on surface quality are shown as an example. Alloy steel workpiece, sintered carbide cutters and both the climb-cut and conventional milling method were used [11/2].

The phenomenon of built-up edge has similar effects and occurs in a similar way in milling as in other machining methods. Therefore a deterioration of surface quality as a result of BUE should be taken into account, particularly in milling with high-speed steel teeth characterized by cutting speeds at which BUE appears. In peripheral milling with sintered carbide teeth at the appropriate cutting speeds (over 120 m/min) surface quality is higher, since the phenomenon of BUE does not occur any more.

In milling, much more often than in any other machining method, vibrations occur lowering the surface quality and destroying the tools. The appearance of vibrations is particularly connected with tooth wear and the increase of cutting edge radius. Therefore, in milling (more than in other machining methods) attention should be paid to sufficiently early regrinding of cutter teeth and to avoiding working with worn out tools.

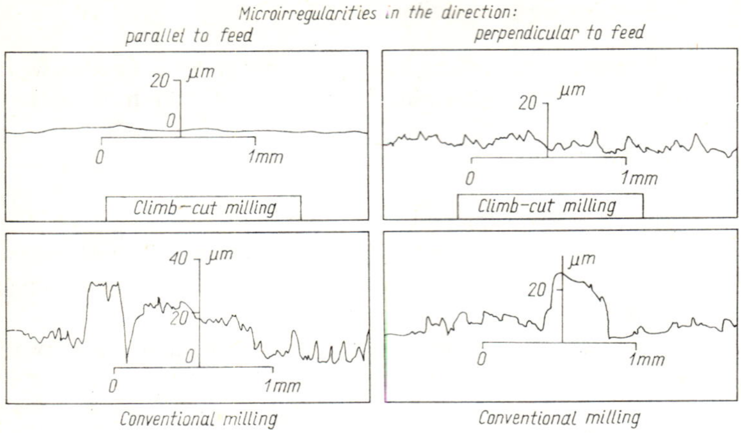

Fig. 11.26 Height of the irregularities of conventionally milled and climb-cut milled surfaces, measured in the direction perpendicular and parallel to feed motion, cutter diameter d = 208 mm, number of teeth z = 1, cutting conditions: v = 125 m/min, a = 4 mm, B = 22 mm, s_z = 0·2 mm; tooth geometry: $\alpha = 15°$, $\gamma = -5°$, $\lambda_s = 10°$ [11/2]

11.2.6 Selection of conditions for plain milling

General principles of the selection of cutting conditions discussed in Chapter 7 are valid also for the case of plain milling.

Milling width usually depends on the dimensions of the surface being machined or on cutter length. The choice is also limited as far as cutter diameter and the number of teeth are concerned. Standard cutters have a determined diameter to milling width ratio and a constant number of teeth for the given cutter diameter (a smaller number for roughing and a greater number for finishing).

The depth of cut should be as high as possible, i.e. as permitted by limitations mentioned above and by the total machining allowance. In practice the removal of the entire machining allowance in one or two passes (roughing and finishing) is aimed at.

Usually, other limitations permitting, the total machining allowance is removed in one pass when surface quality requirements correspond to roughness numbers 1 to 3. For numbers 4 to 6, and for high-speed steels, a single pass is applied if the machining allowance does not exceed 3 mm and two passes if it does. An allowance of 1–2 mm is then left for the finishing operation.

A high feed rate is also chosen, based on recommended feed values listed in standard tables. A check is then conducted of whether all limitations concerning the machine tool, workpiece and tool have been taken into account.

The check whether the selected feed rate is permissible with regard to arbor strength and rigidity is carried out according to the inequalities

$$F_v \leqq F_s \qquad \text{and} \qquad F_v \leqq F_d \tag{11.102}$$

where: F_v—mean peripheral force in kg, F_s—permissible force with regard to arbor strength in kg, F_d—permissible force with regard to arbor deflection.

Values of F_s and F_d may be approximated according to formulae:

$$F_s = \frac{0 \cdot 1 k_b d_a^3}{0 \cdot 08L + 0 \cdot 65 \sqrt{(0 \cdot 25L)^2 + (0 \cdot 5\alpha\, d)^2}} \quad \text{kg} \tag{11.103}$$

$$F_d = \frac{4Efd^4}{L^3} \quad \text{kg} \tag{11.104}$$

where: k_b—permissible bending stress in kg/mm^2, d_a—arbor diameter in mm, L—arbor length between supports in mm, $\alpha = k_b/(1 \cdot 3k_t)$—coefficient in which k_t—permissible torsional stress in kg/mm^2, d—cutter diameter in mm, E—modulus of elasticity of arbor material in kg/mm^2, f—permissible value of arbor deflection. In roughing operations $f = 0 \cdot 2$ mm, in finishing $f = 0 \cdot 05$ mm.

Cutting speed is selected according to eqn. (11.95) from tables of recommended cutting conditions or by means of other calculation aids. The chosen values are then corrected with regard to limitations resulting from the stepping and range of spindle speeds.

REFERENCES

11/1 ROZENBERG, A. M.: 'Dinamika frezerowaniya' (Dynamics of Milling), Mashgiz, Moskva 1945

11/2 MOROZOV, V. D., and OZNOBITIN, N. V.: 'Skorostnoe frezerowanie tsilindricheskimi frezami'. Sbornik: 'Novye raboty v oblasti rezaniya metallov' (High-speed Milling by Means of Cylindrical Cutters. Collection: New Works in the Field of Metal Cutting), Moskva 1952

CHAPTER 12

Face milling

12.1 Types of face milling and cutting conditions

12.1.1 TYPES OF FACE MILLING

The following types of face milling are distinguished (Fig. 12.1):

(a) *full-face milling* when milling width B is equal to mill diameter d (6 in Fig. 12.1);

(b) *part-face milling* when milling width B is smaller than mill diameter d (1–5 and 7–9 in Fig. 12.1).

Part-face milling in turn may be divided into:

(i) *unilateral*, when the mill protrudes beyond the work surface on one side only, or when $w_1 = 0$ and $0 < w_2 < d$ (1, 2, 3, 7, 8, 9, in Fig. 12.1),

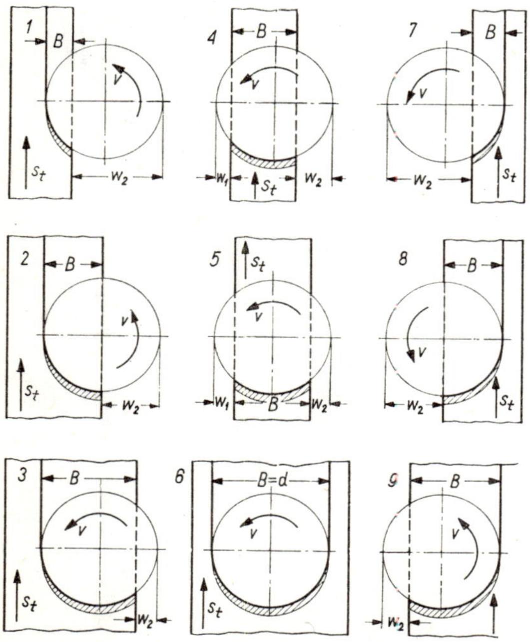

Fig. 12.1 Types of face milling

(ii) *bilateral,* when the mill protrudes beyond the work surface on both sides, or when $0 < w_1 < d$ and $0 < w_2 < d$ (4 and 5 in Fig. 12.1).

Bilateral part-face milling may be: *asymmetrical* when $w_1 \neq w_2$ (4 in Fig. 12.1) or *symmetrical* when $w_1 = w_2$ (5 in Fig. 12.1).

Thus the classification of face-milling types is based on the differences in workpiece situation relative to the mill, and on milling width.

Independent of the above is the distinction between conventional and climb-cut milling. This is based on the relative situation between the cutting speed and feed rate vectors. However, in face milling, particularly for full-face and bilateral part-face types, a tooth always works conventionally when entering the material and climb-cut mills when leaving it.

12.1.2 Undeformed chip parameters

Undeformed chip parameters in face milling are the same as in plain milling.

The only difference is that—according to the depth of cut definition given previously—in face milling this dimension of undeformed chip stretches in the direction in which, in plain milling, the milling width is measured (compare a and B in Fig. 11.1).

Denotations of undeformed chip parameters in face milling are given in Fig. 12.2. Bilateral part-face milling (Fig. 12.2a) is characterized by a tooth penetrating the material with the entering angle φ_e and leaving the material with the leaving angle φ_l.

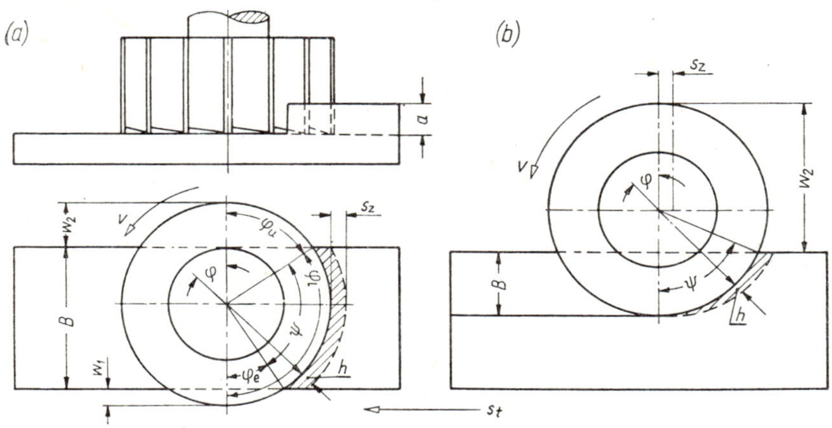

Fig. 12.2 Parameters of the undeformed chip in part-face milling: (a) bilateral, (b) unilateral

Since the undeformed chip thickness is expressed by the same formula as in plain milling, in bilateral part-face milling the tooth begins to cut a layer of thickness

$$h_e = s_z \sin \varphi_e \quad \text{mm} \tag{12.1}$$

and finishes the cutting with thickness

$$h_l = s_z \sin \varphi_l \quad \text{mm} \tag{12.2}$$

The maximum value of undeformed chip thickness in face milling occurs when

$$\varphi = \varphi_l \quad \text{if} \quad \varphi_l \leqq 90° \tag{12.3}$$

$$\varphi = 90° \quad \text{if} \quad \varphi_l > 90° \tag{12.4}$$

Thus the tooth working angle in face milling is equal to the difference between leaving angle and entering angle

$$\psi = \varphi_l - \varphi_e \tag{12.5}$$

Unilateral milling (Fig. 12.2b) is a particular case of face milling when $\varphi_e = 0$.

Leaving angle φ_l is determined from the relationship

$$\varphi_l = 180° - \varphi_u \tag{12.6}$$

while $\cos\varphi_u = 1 - \dfrac{2w_2}{d}$

and the entering angle φ_e by

$$\cos\varphi_e = 1 - \frac{2w_1}{d} \tag{12.7}$$

Thus, for example, when $w_1 = 0{\cdot}15d$ and $w_2 = 0{\cdot}25d$, then $\varphi_l = 180° - 60° = 120°$ and $\varphi_e = 45°$ or $\psi = 75°$.

Knowing the working angle ψ one can calculate the tooth work period using the appropriate formulae for plain milling.

The cross sectional area of the undeformed chip for face mill teeth is expressed by

$$A = h\,b = s_z \sin\varphi b \quad \text{mm}^2 \tag{12.8}$$

where: h—undeformed chip thickness in mm, b—undeformed chip width in mm, s_z—feed in mm/tooth, φ—mill rotation angle in degrees.

In the case of milling with a face mill having tooth angles $\varkappa > 0$ and $\lambda_s \neq 0$, the undeformed chip width is expressed by formula [12/7]

$$b = a\frac{\cos\lambda_s}{\sin\varkappa} \quad \text{mm} \tag{12.9}$$

The total area of the undeformed chips machined by all teeth cutting simultaneously is calculated in the same way as for plain milling.

12.1.3 Calculation of face milling time

The milling time in bilateral part face milling is calculated according to (Fig. 12.3a)

$$t_m = \frac{l_a + l - l_s + l_o}{s_t} = \frac{0{\cdot}5d + l - w_1\sqrt{d/w_1 - 1} + l_o}{s_z z\, n} \quad \text{min} \tag{12.10}$$

where: l_a—length of approach in mm, l—length of work surface in mm, l_s—shortening of the cutting length due to part-face milling, in mm; when $w_1 \leqq w_2$, then $l_s = w_1\sqrt{d/w_1 - 1}$; l_o—length of overrun in mm.

Milling time of full-face and part-face milling (Fig. 12.3b) for which $w_1 = 0$ is calculated according to

$$t_m = \frac{l_a + l + l_o}{s_z z n} \quad \text{min} \tag{12.11}$$

Equation (12.11) is thus a particular case of eqn. (12.10). In full-face milling $l_a = 0{\cdot}5d$ and in unilateral part-face milling

$$l_a = B\sqrt{d/B-1} \quad \text{mm} \tag{12.12}$$

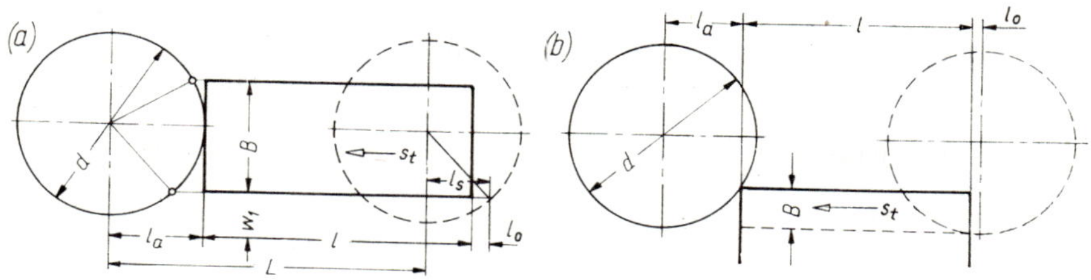

Fig. 12.3 Diagram for the calculation of machine time

12.1.4 Face mill teeth

Geometric parameters of face mill teeth are shown in Fig. 12.4. Recommended optimum values of these parameters are listed in Table 12.1 for high-speed steel, sintered carbide, and of sintered metal oxide teeth.

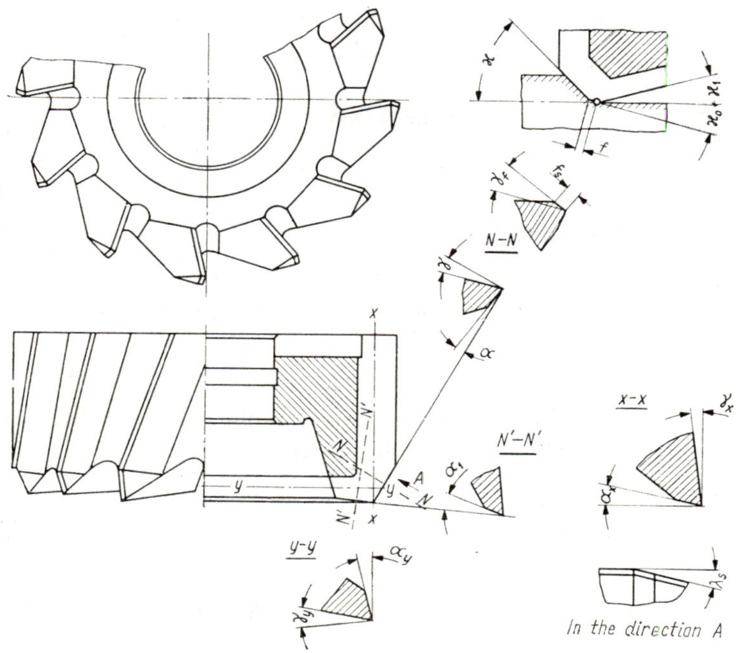

Fig. 12.4 Parameters of face mill teeth

Table 12.1 *Values of angles and other dimensions of face mills*

Mill tooth material	Workpiece material	Values of angles and other dimensions									
		$\alpha°$	$\alpha_1°$	$\gamma_f°$	$\gamma°$	f_s mm	$\lambda_s°$	$\varkappa°$	$\varkappa_0°$	$\varkappa_1°$	f mm
Sintered carbide*	steel										
	$\sigma_B \leqq 110$ kg/mm^2	15	20	5–5	10–0	1·5	15	60	30	5	1·0
	$\sigma_B > 110$ kg/mm^2	20	25	−10	−5	1·5	15	60	30	5	1·0
	grey cast iron	12	17	5	10	1·5	0–15	60	30	5	1·0
High-speed steel	carbon and alloy steel $\sigma_B < 60$ kg/mm^2 soft brass	12			20			30–45	25		2·0
	cast, carbon, alloy and tool steel $\sigma_B = 60$–100 kg/mm^2 grey and malleable cast iron HB < 150 kg/mm^2	16			15						
Sintered oxide	grey cast iron HB = 180–200 kg/mm^2	14	11	—	−10	—	9	35	10	10	2

* Acc. to 'Rezhimy skorostnogo rezaniya metallov' (Conditions of High-speed Metal Cutting), Mashgiz, Moskva 1951
† Acc. to Warunki skrawania metali narzędziami ze stali szybkotnącej (Conditions of Cutting Metals with High-speed Steel Tools), PWT, Warszawa 1953

High-speed steel teeth usually have no lands. Sintered carbide teeth are provided with lands and a greater cutting edge inclination angle, in order to strengthen the teeth when entering the material.

The relationship between tooth angles in the transverse and main cross-section is as follows

$$\tan\gamma_x = \tan\gamma \sin\varkappa + \tan\lambda_s \cos\varkappa \tag{12.13}$$

$$\cot\alpha_x = \cot\alpha \sin\varkappa + \tan\lambda_s \cos\varkappa \tag{12.14}$$

On the other hand, values of dynamic angles in cutting are determined in the same plane as angles γ_x and α_x. Angles in the dynamic system γ_{xd} and α_{xd} can be calculated as

$$\gamma_{xd} = \gamma_x + \zeta \qquad \alpha_{xd} = \alpha_x - \zeta \tag{12.15}$$

Angle ζ is computed according to the formulae for plain milling, but the depth of cut a is replaced by milling width B. In face milling maximum values of angle ζ_{max} occur when $\varphi = 90°$. Then

$$\zeta_{max} = \arctan \frac{s_z z}{\pi d} \tag{12.16}$$

Within the practical range of s_z, z and d the value of ζ may be neglected and it may be assumed that

$$\gamma_{xd} \approx \gamma_x \qquad \text{and} \qquad \alpha_{xd} \approx \alpha_x \tag{12.17}$$

For example, when $s_z z = 3{\cdot}14$ mm and $d = 100$ mm, then $s_z z/(\pi d) = 0{\cdot}01$ and $\zeta \approx 33'$.

With regard to the number of mill teeth, the observations made in discussing plain milling cutters are also valid here.

12.2 Cutting resistance and power in face milling

12.2.1 Cutting force component

The tool system in which the cutting forces are determined remains the same as in plain milling and consists, for each tooth, of the axis of mill rotation—x, the axis passing through the tooth profile top and perpendicular to the mill axis—y, and the axis passing through the tooth profile top and perpendicular to both former axes—z.

The machine tool system also remains the same as in plain milling. On the other hand, the relative position of both systems changes. The x-axis of the tool system is vertical instead of horizontal which was the case in horizontal milling machines. Consequently, in face milling the component F_f coincides with component S, while the following relationships occur between components F_p, F_v, Q and R in conventional milling:

$$Q = F_v \sin\varphi - F_p \cos\varphi \quad \text{kg} \qquad (12.18)$$
$$R = F_v \cos\varphi + F_p \sin\varphi \quad \text{kg}$$

in climb-cut milling:

$$Q' = F_v' \sin\varphi + F_p' \cos\varphi \quad \text{kg} \qquad (12.19)$$
$$R' = F_v' \cos\varphi - F_p' \sin\varphi \quad \text{kg}$$

Using eqns. (12.18) and (12.19) one should keep in mind that angle φ may change between 0° and 180°.

Calculation of instantaneous peripheral force F_v, mean peripheral force acting upon tooth F_{vm} and mean milling force F_{vo} is carried out according to the same formulae as derived for plain milling. However, suitable substitutions should be made: i.e. cutting width for the depth of cut and vice versa.

12.2.2 Statistical and empirical formulae for the calculation of peripheral cutting force and cutting power

The general form of peripheral force formulae is as follows

$$F_{vo} = C_{zo} a^{e_z} s_z^{u_z} z^{n_z} B^{r_z} d^{b_z} v^{f_z} \qquad (12.20)$$

In this formula previously explained denotations have been used. Values of C_{zo}, e_z, u_z, n_z, r_z, b_z, f_z are given in Table 12.2.

According to research results mean values of the cutting forces component most frequently occurring in the machine tool system have been established:

(a) in symmetrical bilateral part-face milling

$$Q = (0{\cdot}85\text{–}0{\cdot}95)F_v; \qquad R = (0{\cdot}3\text{–}0{\cdot}4)F_v; \qquad S = (0{\cdot}5\text{–}0{\cdot}55)F_v$$

Table 12.2 *Values of constants in the formula for the peripheral cutting force*

Mill tooth material	Workpiece material	Values of constants C_{zo}	e_z	u_z	n_z	r_z	b_z	f_z
High-speed steel	steel $\sigma_B = 75$ kg/mm²	82·4	0·95	0·8	1·0	1·1	−1·1	0
	malleable cast iron HB = 190 kg/mm²	50	0·95	0·8	1·0	1·1	−1·1	0
	grey cast iron HB = 190 kg/mm²	50	0·9	0·72	1·0	1·14	−1·14	0
	bronze HB = 100–140 kg/mm²	37·5	0·95	0·8	1·0	1·1	−1·1	0
Sintered carbide	steel $\sigma_B = 75$ kg/mm²							
	$a > 3$; $s_z \leqq 0{\cdot}12$	214·2	1·0	1·42	0·93	1·2	−1·0	−0·1
	$a > 3$; $s_z > 0{\cdot}12$	61·2	1·0	0·79	0·93	1·2	−1·0	−0·1
	grey cast iron HB = 190 kg/mm²	50	0·9	0·72	1·0	1·14	−1·14	0

(b) in unilateral part-face conventional milling

$$Q = (0{\cdot}45\text{–}0{\cdot}7)F_v; \qquad R = (0{\cdot}6\text{–}0{\cdot}9)F_v; \qquad S = (0{\cdot}5\text{–}0{\cdot}55)F_v$$

(c) in unilateral part-face climb-cut milling

$$Q = (0{\cdot}9\text{–}1{\cdot}0)F_v; \qquad R = (0{\cdot}15\text{–}0{\cdot}3)F_v; \qquad S = (0{\cdot}5\text{–}0{\cdot}55)F_v$$

For calculating the face milling power eqn. (11.63) of Chapter 11 is used.

12.2.3 Actual variations of the peripheral cutting force

The mean value of the peripheral force in milling as given by formulae of the type of eqn. (12.20) may differ considerably from actual instantaneous values of such force. Particularly large differences occur between its mean and maximum values. Those differences are greater, the smaller is the number of teeth cutting simultaneously. This is illustrated by Fig. 12.5 in which the changes of instantaneous values of the peripheral cutting force on each tooth are represented graphically, as well as the actual cutting force of teeth cutting simultaneously and the mean peripheral force of milling F_{vo}.

In the case given, the range variation is: $(0{\cdot}72\text{–}1{\cdot}4)F_{vo}$ and therefore the dynamic coefficient $k_d = 1{\cdot}95$.

Independently of the variation of resultant peripheral force of milling, the peripheral forces acting upon each tooth also show dynamic changes. At the instant when the tooth penetrates the material impact forces may occur. A necessary condition for the determination of peripheral force variations of this kind is the use of measuring apparatus of sufficiently low inertia.

In Fig. 12.6 a typical case of variations of the peripheral cutting force on the tooth in full-face milling is shown ($B = d$). Characteristically, the increase of cutting force is not continuous from the moment the tooth contacts the workpiece material. It is only after a certain length of travel (in the example given, after a rotation through

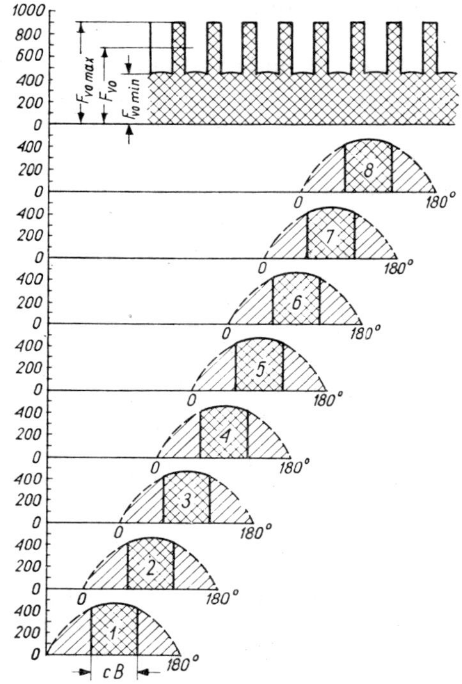

Fig. 12.5 Adding up of the peripheral force (kg) in face milling by means of mill with eight teeth

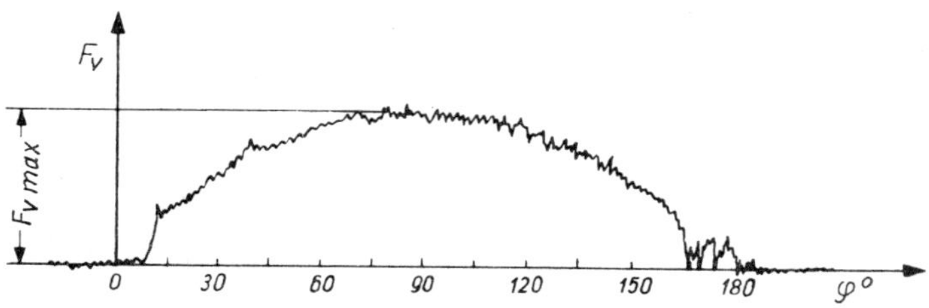

Fig. 12.6 Variations of the peripheral cutting force on one tooth in full-face milling

some 10°) that a rapid increase of peripheral force occurs. This is probably the result of initial sliding of the tooth on the material until the undeformed chip thickness builds up sufficiently for the plastic deformation and cutting stage to begin.

In the case of part-face milling, depending on several factors and first of all on the value of entering angle φ_e, on cutting speed and on undeformed chip dimensions (a and s_z), the impact at the moment of the tooth penetrating the material may increase the cutting force by 10–80%.

12.2.4 Influence of some factors on the peripheral cutting force

In addition to the factors whose influence on the peripheral cutting force has been considered in eqn. (12.20), considerable influence is also exerted by some other factors and this was taken account of by the introduction of appropriate correction coefficients.

Equation (12.20) concerns the milling of steel with ultimate tensile strength $\sigma_B = 75$ kg/mm². As in the case of other cutting methods, so in milling, too, the strength of the material strongly influences the cutting resistance. If the UTS of steel is other than 75 kg/mm², a correction coefficient should be introduced into the calculation of the peripheral cutting speed (Table 12.3).

Table 12.3 *Correction coefficients for the peripheral cutting force K_H taking into account the tensile strength of the steel*

B kg/mm²	55	65	75	85	95	105	115
K_H	0·85	0·925	1·00	1·075	1·15	1·225	1·30

Among the tooth parameters the rake angle γ has a distinct influence upon peripheral cutting speed. If the value of this force for $\gamma = 0°$ is assumed to be 100%, then with angle γ diminishing to $-10°$ the peripheral cutting force increases to 112%, while increasing the rake angle to $\gamma = 10°$ reduces the peripheral force to 89%.

Within the range of cutting edge angle $\varkappa = 15$–$90°$ studied so far, an optimum value of this angle appears, amounting to $\varkappa = 60°$ for face mills with sintered carbide teeth. For greater and smaller cutting edge angles correction coefficients should be used according to Table 12.4.

Table 12.4 *Correction coefficients for the peripheral cutting force $K_\varkappa$ taking into account the cutting edge angle $\varkappa$*

$\varkappa°$	15	30	45	60	75	90
$K_\varkappa$	1·23	1·15	1·06	1·0	1·06	1·14

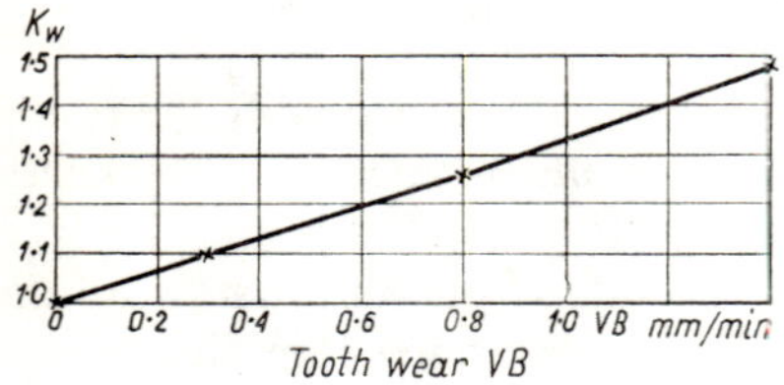

Fig. 12.7 Correction coefficient K_w with regard to tooth wear

In order to give a general notion of the influence of progressing tooth wear on the peripheral cutting force, a diagram is shown in Fig. 12.7, illustrating the growth of the correction coefficient K_w with tooth wear.

A greater cutting resistance results in face or end milling than in plain milling since mill length exceeds cutting width. This may be explained by the fact that in the case of end milling the work of the face milling teeth is added to that of teeth situated on the cylindrical surface. Thus an analogy to orthogonal and non-orthogonal cutting may be drawn, as explained by Fig. 12.8.

The auxiliary cutting energy amounts to 4–8% and grows with increasing s_z.

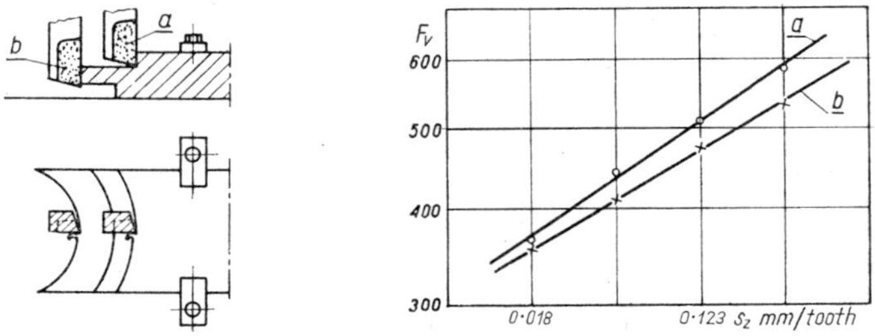

Fig. 12.8 Peripheral cutting force in milling (in kg); *a*—non-orthogonal cutting, *b*—orthogonal cutting

12.3 Tooth wear and life

12.3.1 The wear process and dependence of tool life on some cutting conditions

The wear of face mills, both of the solid and inserted tooth type, has the same character as the wear of plain cutters.

Tooth material does not influence the wear character, the differences being only quantitative and not qualitative.

Among the tooth parameters clearance angle α, rake angle and cutting edge inclination angle λ_s have a distinct influence on tooth life.

The influence of clearance angle is characterized, in the case of increasing angle, by diminishing friction, but also by simultaneous weakening of the tooth corner. The latter circumstance is conducive to microdents and faster wear. It may thus be supposed that a certain optimum value of angle α should exist for which tooth life would attain its maximum.

Confirmation of this assumption has been found by experiment. As an example, in Fig. 12.9 a diagram of tooth wear and life is represented showing their dependence on clearance angle when milling grey cast iron with HB = 180–200 kg/mm^2 with sintered carbide teeth WK8 [12/1] (see note on page 182). Under the conditions described in the caption it has been established that the optimum value of clearance angle is $\alpha_{opt} = 15°$.

A considerable influence is exerted on the optimum value of clearance angle by the mechanical properties of the workpiece material and by the undeformed chip thickness (or—when the conditions of tooth penetration into the material and the working angle ψ remain constant—the influence is exerted by the feed per tooth). Figure 12.10 illustrates the dependence of the optimum clearance angle on feed per tooth in face milling steel with sintered carbide teeth (upper curve) and with high-speed steel teeth (lower curve) [12/2].

The higher values of optimum angles α for sintered carbides are explained by better abrasion and heat resistance which has a greater influence than the lower bending strength.

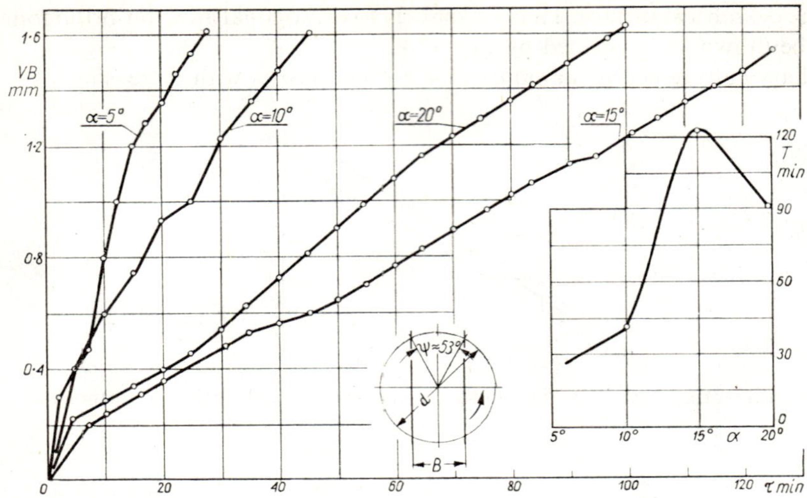

Fig. 12.9 Dependence of tooth wear *VB* on cutting time for various clearance angles and the influence of angle α on tooth life *T*. Conditions under which the experiments were conducted: $B = 105$ mm, $a = 4$ mm, $v = 147{\cdot}3$ m/min, $d = 235$ mm, $z = 2$, $\gamma = 5°$, $\alpha = \alpha_1 = 15°$, $\varkappa = 60°$, $\varkappa_0 = 30°$, $\varkappa_1 = 5°$, $f = 1$ mm, $s_z = 0{\cdot}12$ mm [12/1]

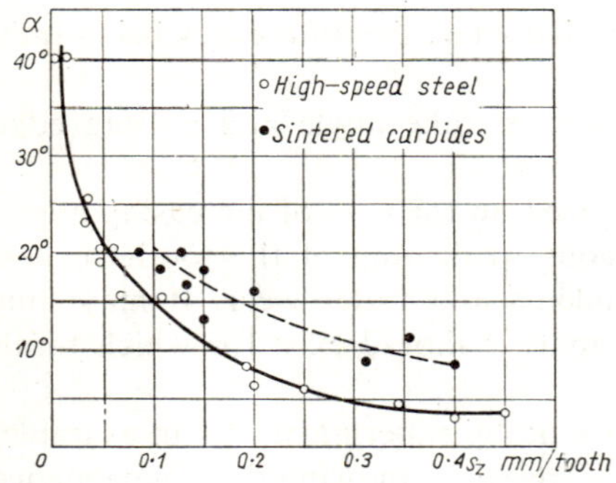

Fig. 12.10 Influence of the feed per tooth s_z on the optimum value of clearance angle α when milling steel [12/2]

The rake angle has a similar influence on tooth life, the optimum value being between $\gamma = -10°$ and $0°$. The existence of an optimum value is due to the rake angle having opposing effects.

On the one hand a reduction of rake angle increases both the mechanical strength and wear resistance of the tooth, but on the other hand, as rake angle decreases, the energy of plastic deformation increases and so does the amount of heat generated. In the case of small rake angles the latter influence prevails and tooth life decreases.

Experiments indicate that the optimum value of rake angle for face mill teeth depends, first of all, on the type of material and its properties. Thus e.g. when face milling various cast-irons of different hardnesses a certain relationship between the optimum value of rake angle and material hardness has been established (Fig. 12.11).

The influence of leading cutting edge inclination angle on tool life has a similar character to that of clearance and rake angle. However, in this case also other indirect consequences of angle changes should be taken into account. These consequences will be discussed when analysing the influence of mill position in relation to the workpiece.

Angles of the leading cutting edge, nose cutting edge and trail edge, like the angles α and γ discussed above, exert two types of influence opposed to each other.

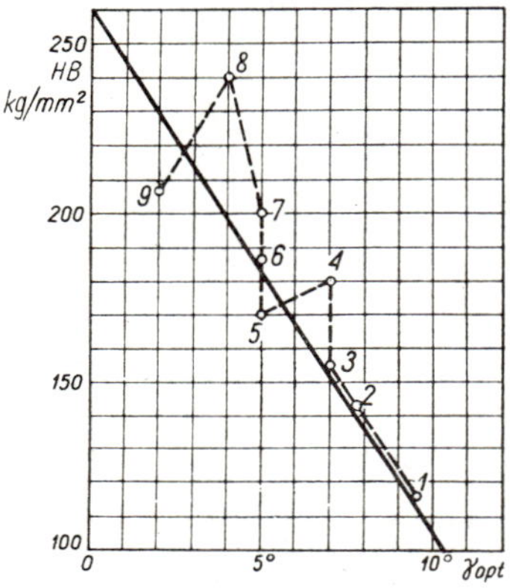

Fig. 12.11 Dependence of the optimum rake angle γ_{opt} on cast iron hardness HB in face milling: 1—grey cast iron, 2—grey cast iron, 3—malleable cast iron, 4–7—grey cast iron, 8—chromium–nickel cast iron, 9—grey cast iron [12/1]

The reduction of cutting edge angle causes a growth of tooth mass and thus an improvement of its strength and heat resistance. On the other hand, small cutting edge angles result in rising friction and unfavourably influence the resolution of cutting forces by increasing the component F_p, which in turn may cause the appearance of microdents in the tooth. Therefore, as in the case of angles α and γ, also for angles $\varkappa$ and $\varkappa_1$ optimum values exist (for the given material and other determined cutting conditions).

The influence of the depth of cut, milling width, feed per tooth and cutting speed on tooth life, within the range normally applied in milling using sintered carbide teeth, is unidirectional (when the values of these parameters increase, tooth life decreases). The strongest influence is that of cutting speed. Only for the case of sintered metal oxide teeth milling cast iron within $s_z = 0{\cdot}15$–$0{\cdot}3$ mm/tooth is the influence of feed rate greater. Otherwise, in machining with sintered carbide or high-speed steel teeth the influence of feed rate is smaller than that of cutting speed, but greater than that of the remaining parameters. The milling width and the depth of cut generally have an influence of the same order of magnitude (the influence of milling width being slightly greater). Because of frequent changes in the cutting conditions (v, a, s and B) their influence on tool life is expressed in the form of mathematical formulae.

Attempts to apply cooling in face milling using sintered carbide teeth, with a view to extending tooth life, were unsuccessful, as the intermittent work of the tooth results in rapid periodical heating and cooling. This in turn causes or accelerates the cracking of inserted teeth and therefore liquid cooling is not applied in face milling.

Experiments indicate that in bilateral part-face milling an important influence on tooth life is exerted by the position of the workpiece relative to the mill.

Tool face π_γ contacts the undeformed chip by a parallelogram surface (while the influence of trial edge angle is considered negligible): $V_0 U_0 T_0 S_0$ (Fig. 12.12). The first contact of this parallelogram with the workpiece material plane π_M (Fig. 12.13) may occur at one of the points V_0, U_0, T_0 and S_0 or along the edges, $V_0 U_0$, $U_0 T_0$, $T_0 S_0$ and $S_0 V_0$ or on the entire surface $V_0 U_0 T_0 S_0$.

These possibilities are effected by the tooth angles γ, λ_s and $\varkappa$ and by the workpiece position relative to the axis of mill rotation, as characterized by the so-called angle of entry ε_e (Fig. 12.13).

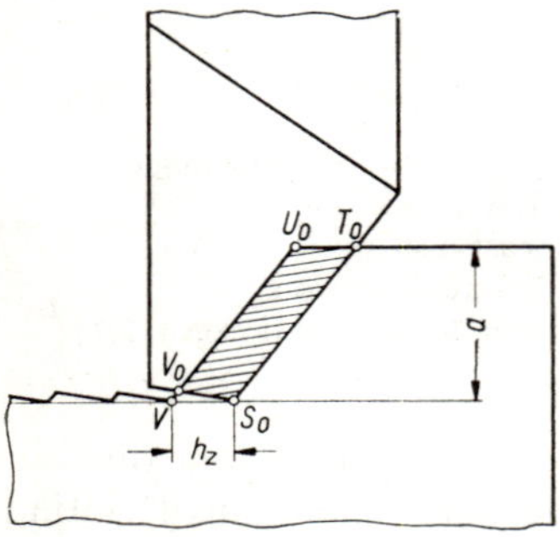

Fig. 12.12 Determination of the contact surface between the tool face and undeformed chip

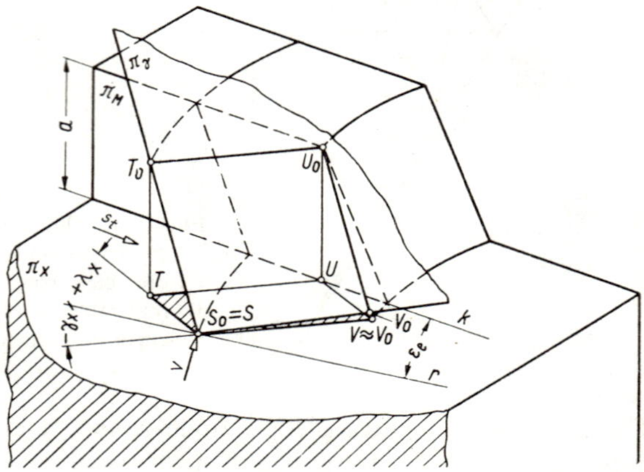

Fig. 12.13 Shape and position of the parallelogram *VUTS* under conditions of first contact between mill tooth and workpiece material [12/4]

By projecting the parallelogram $V_0 U_0 T_0 S_0$ on the plane π_x which is perpendicular to π_m (they intersect along the line k), the parallelogram *VUTS* is obtained. Taking advantage of the relationship occurring in the projection on plane π_x it is found that three types of tooth entry into the workpiece material are possible:

(a) *VUT* type, when $\gamma_y < 0$ or $\lambda_x > \gamma_x$ or $\tan\varkappa \tan\lambda_s > \tan\gamma$. In this type of entry the first contact may be established at *V*, *VU*, *U*, *UT* or *T* (which means five contact possibilities);

(b) *VFT* type, when $\gamma_y = 0$ or $\lambda_x = \gamma_x$ or $\tan\varkappa \tan\lambda_s = \tan\gamma$. In this type of entry the first contact may be established at *V*, *VUTS* (the entire surface) or *T* (which means three contact possibilities);

(c) *VST* type, when $\gamma_y > 0$ or $\lambda_x < \gamma_x$ or $\tan\varkappa \tan\lambda_s < \gamma$. In this type of entry the first contact may be established at *V*, *VS*, *S*, *ST* or *T* (which means five contact possibilities).

The location of first contact determines the entry time, i.e. the time from the instant of first contact to the establishment of contact with the undeformed chip over the entire parallelogram $V_0 U_0 T_0 S_0$. The stresses occurring in the tooth are also the result of the first contact position.

It has been concluded that of the three types of entry, *VFT* and *VST* are to be considered disadvantageous (to exclude the possibility of first contact at *VUTS* or *S*). In practice, therefore, *VUT* type entry should be aimed at, yielding five contact possibilities (*V*, *VU*, *U*, *UT*, or *T*) depending on the entry angle ε_e in relation to γ_x and λ_x [12/3].

Under conditions when this principle is not adhered to, wide variations in tooth life may result.

It has been observed while machining 0.45 carbon steel [12/5] that depending on the various conditions of tooth entry, tooth life may vary within the limits $T_{max}/T_{min} = 4/1$ and in milling of alloy steels this ratio is 6·5–8 [12/6].

The causes of this are seen in the following phenomena:

(a) At the instant the work begins, a local impact occurs which may be accompanied by a local exceeding of the strength of the material. This is followed by the appearance of a microdent and by tooth wear. In view of this the least advantageous are the first contacts along the cutting edge, particularly at S_0.

(b) The phenomenon of local impact and local exceeding of the strength of the material is more harmful, the quicker the contact area grows from zero to *A*. This phenomenon is considered to be of paramount importance.

(c) In a similar way to tooth loading at tooth entry, a violent unloading occurs at tooth exit. At the same time the possibility arises of a change from a compressive to a tensile stress as part of the undeformed chip is torn from the material. These phenomena are of considerable importance when the undeformed chip thickness is large at the tooth exit. This factor may have a decisive importance for tooth life and therefore a minimum cross-sectional area of the undeformed chip is desirable at tooth exit.

Industrial experience indicates that the correct positioning of the workpiece relative to the mill in part-face milling can increase tooth life by 50–60%.

A favourable influence on tooth life in face milling is also obtained by the use of a flywheel. Its advantages are particularly evident in symmetrical and unilateral milling. Due to the flywheel the uniformity of rotary motion is increased, torsional vibrations are reduced and thus the working conditions of the teeth improve. According to Larin's research [12/1] the use of a flywheel when face milling cast iron extended tool life by 20–30%.

12.3.2 Calculation of life speed

For the calculation of life speed in face milling a formula of identical structure as that given for plain milling is used. Differences consist only in the numerical values of the constants C_v and the exponents, as given in Table 12.5. In the case when the use of correction coefficients is necessary, values of B_M, B_p, B_s and B_H may be used as in plain milling.

12.4 Surface quality in face milling

All remarks concerning the surface quality in turning and plain milling are also valid for the quality of face-milled surfaces. Certain specific features of face milling cause only quantitative differences in the relationship between the quality of face-milled surface and cutting conditions as compared with other machining methods.

In the case of milling using face mills with sintered carbide teeth, cutting speeds

Table 12.5 *Values of constants C_v and exponents in the formula for life speed in face milling*

Mill tooth material	Workpiece material	Other parameters	Within the range of: a mm	Within the range of: s_z mm	Constant and exponent values: C_v	m	b_v	e_v	u_v	r_v	f_v
High-speed steel	carbon steel $\sigma_B = 75$ kg/mm² (with cooling)	—	—	0·1 0·1	41 64·7	0·2 0·2	0·25 0·25	0·1 0·1	0·4 0·2	0·15 0·15	0·1 0·1
	grey cast iron HB = 190 kg/mm²	—	—	—	42	0·15	0·2	0·1	0·4	0·1	0·1
	malleable cast iron HB = 150 kg/mm²	—	—	0·1 0·1	57·4 90·5	0·20 0·20	0·25 0·25	0·1 0·1	0·4 0·2	0·15 0·15	0·1 0·1
	bronze, medium hard HB = 100–140 kg/mm²	—	—	0·1 0·1	86·2 136	0·2 0·2	0·25 0·25	0·1 0·1	0·4 0·2	0·15 0·15	0·1 0·1
Sintered carbide	carbon steel $\sigma_B = 75$ kg/mm²	—	—	0·04–0·08 0·08–0·2	714 334	0·2 0·2	0·2 0·2	0·1 0·1	0·20 0·4	0·20 0·20	— —
	grey cast iron HB = 190 kg/m²	—	—	—	165	0·2	0·2	0·15	0·4	0·20	—
Sintered oxide	grey cast iron HB = 180–200 kg/mm²	$d = 250$ mm	1–2	0·14 0·15–0·3	2880 8·4	−0·74 −0·74	— —	0·35 0·35	−0·51 −0·45	— —	— —
			—	0·14 0·15–0·3	4600 15·4	−0·74 −0·74	— —	−0·48 −0·48	−0·51 −3·45	— —	— —

exceed the range of built-up edge formation. Therefore, an increase of cutting speed improves surface quality.

The influence of feed per tooth and feed per revolution is similarly explained in both face and plain milling.

Experiments indicate that an increase of the number of teeth in a mill favourably influences surface quality, but to a lesser degree than a cutting speed increase.

A very distinct influence on the quality of face-milled surfaces is exerted by the trail edge angle.

In Fig. 12.14 results of research [12/1] on surface quality when milling grey cast iron are presented. It follows from this diagram that an increase of angle $\varkappa_1$ from 1° to 10° is accompanied by surface quality deterioration from roughness number 7 to 4.

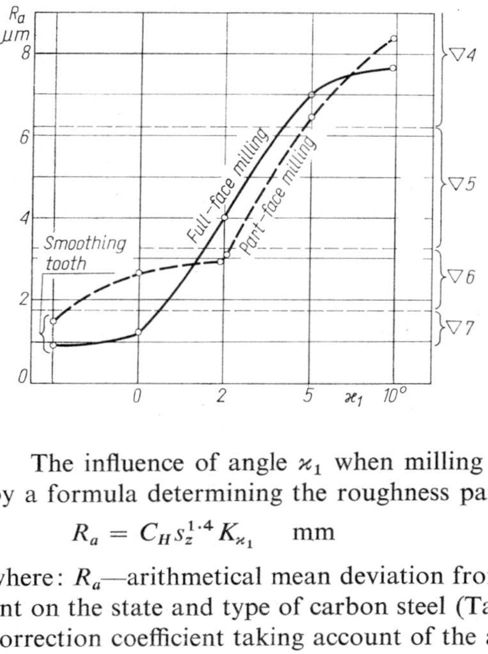

Fig. 12.14 Influence of trail edge angle $\varkappa_1$ on surface quality defined by parameter R_a. Measuring conditions: workpiece material—grey cast iron HB = 190 kg/mm², mill tooth material—sintered carbides WK8 (see note on page 182), $d = 235$ mm, $z = 4$, $B = 105$ mm, $a = 1$ mm, $s_z = 0{\cdot}094$ mm/tooth, $\alpha = \alpha_1 = 15°$, $\gamma = 5°$, $\varkappa = 60°$, $\varkappa_0 = 30°$, $f = 1$ mm, $v = 143{\cdot}5$ m/min [12/1]

The influence of angle $\varkappa_1$ when milling carbon steels has been expressed [12/4] by a formula determining the roughness parameter

$$R_a = C_H s_z^{1\cdot4} K_{\varkappa_1} \quad \text{mm} \tag{12.21}$$

where: R_a—arithmetical mean deviation from the mean line, C_H—a constant dependent on the state and type of carbon steel (Table 12.6), s_z—feed in mm/tooth, $K_{\varkappa_1}$—correction coefficient taking account of the angle $\varkappa_1$ (Table 12.6).

Table 12.6 *Constant C_H and coefficient $K_{\varkappa_1}$ used in the calculation of roughness parameter R_a*

Sort and state of steel	Hardness HB kg/mm²	Value of C_H	Value of angle $\varkappa_1^\circ$: 1	2	3	4
			Value of coefficient $K_{\varkappa_1}$			
0.25 carbon steel non-hardened	163–170	1·5				
0.50 carbon steel non-hardened	207–217	11·5	1·0	1·2	2·0	3·3
Hardened steels	364–387	6·75				

Note: data valid for $\gamma = -10°$, $\alpha = 15°$, $\varkappa = 60°$, $\varkappa_0 = 45°$, $f = 1$ mm

Table 12.7 *Table for the selection of feed rate s_z with regard to the desired surface quality*

Workpiece material	$\varkappa_1^0$	Tool material	∇4	∇5	∇6	∇7	∇8
	R_{tm}		40	20	10	6,3	3,2
	Roughness number		Accurate milling	Accurate milling	Very accurate milling	Very accurate milling	Very accurate milling
			s_z mm				
Carbon steels	0	sintered carbide		0·3	0·15 0·03		0·01 0·08 0·005
		high-speed steel	0·3 0·2	0·15	0·05	0·08	0·008
	2	sintered carbide		0·16 0·08	0·01	0·008	0·005
		high-speed steel	0·18 0·12 0·08 0·04		0·008 0·005		
	5	sintered carbide	0·15 0·11	0·08	0·01	0·008 0·005	
		high-speed steel	0·04	0·008	0·005		
Cast iron	0	sintered carbide			0·3 0·15	0·08	0·010 0·008 0·005
		high-speed steel		0·3	0·15	0·05 0·008	0·005
	2	sintered carbide	0·3 0·2	0·15 0·1	0·08 0·04	0·008	0·005
		high-speed steel	0·3 0·2	0·1 0·04	0·08	0·05	
	5	sintered carbide	0·2 0·15 0·1	0·07	0·04	0·08	
		high-speed steel	0·4 0·3 0·2 0·07	0·04	0·008 0·005		
Bronze	0						
	2	high-speed steel			0·3	0·1 0·08 0·03	0·008 0·005
	5						

——— high-speed steel
– – – – sintered carbide

Equation (12.21) has been worked out for a tooth having: $\gamma = -10°$, $\alpha = 15°$, $\varkappa = 60°$, $\varkappa_0 = 45°$, $f = 1$ mm.

Progressing wear of mill teeth causes—after an initial improvement of surface quality—a steady growth in surface roughness (Fig. 12.15).

In view of the intermittent character of milling, increasing the rigidity of the *m–w–t* system and the uniformity of rotary motion contributes very much to the

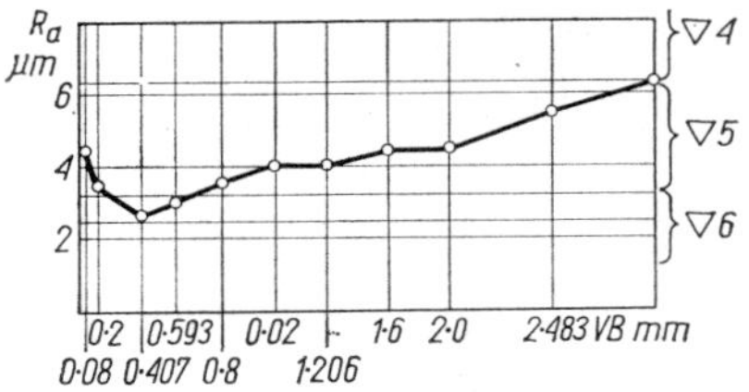

Fig. 12.15 Influence of tooth wear *VB* on surface quality defined by parameter R_a. Measuring conditions: $v = 162$ m/min, $s_z = 0{\cdot}084$ mm, $\varkappa_1 = 5°$, other conditions as in the caption to Fig. 12.14 [12/1]

improvement of surface quality. And vice-versa, vibrations may cause (given resonance) very bad surface roughness. Consequently, the application of a flywheel on a milling machine spindle gives a possibility (due to increased uniformity of motion) to reduce surface roughness, usually by 25–40%.

Generally, within the range of practically applied feed rates $s = 0{\cdot}05$–$0{\cdot}4$ mm/tooth it is possible to obtain surfaces of roughness numbers 4–8 when face milling steel, cast iron and bronze. Table 12.7 illustrates diagrammatically the possibility of attaining the desired surface quality depending on the feed per tooth and the trail edge angle used.

REFERENCES

12/1 LARIN, M. N., and IGNATOV, B. A.: 'Skorostnoe frezerovanie chugunnykh zagotovok'. Sbornik: Novoe v tekhnologii gidromashinostroeniya (High-speed Milling of Cast Iron Workpieces. Collection: New Developments in the Technology of Hydromechanical Engineering), Trudy UIGM, N. XIII, Mashgiz Moskva 1951

12/2 ROZHDESTVENSKII, L. A.: 'Skorostnoe frezerovanie tortsevokonicheckimi frezami'. Sbornik: Rezanie metallov (High-speed Milling by Means of Conical End Mills. Collection: Metal Cutting), Mosk. VTU im. Baumana, Mashgiz, Moskva 1951

12/3 KACZMAREK, J.: 'Die Abhängigkeit des Einschneidtyps beim Stirnfräsen von der Gestaltung der Werkzeugschneide', CIRP-Annalen, Band X

12/4 'Tochnost mekhanicheskoi obrabotki metallov' (red. A. P. SOKOLOVSKII) (Accuracy of Metal Machining, edited by A. P. SOKOLOVSKII), Mashgiz, Moskva 1953

12/5 ETIN, A. O.: 'Vliyanie uslovii vrezaniya tortsevoi frezy na ee stoikost'. Sbornik ENIMS: Dinamika protsessa rezaniya metallov (Influence of the Conditions of Face Mill Entry on Its Life. Collection of ENIMS: Dynamics of the Process of Metal Cutting), Mashgiz, Moskva 1953

12/6 ZOREV, N. N.: 'Vliyanie ustanovki zagotovki na stoikost' frez pri tortsevom frezerovanii' (Influence of the Setting of the Workpiece on Mill Life in Face Milling), Vestnik Mashinostroeniya, 1952 (8)

12/7 KIEPUSZEWSKI, B.: 'Podstawy obróbki metali skrawaniem', cz. II (Elements of Metal Machining by Cutting, Part II) Wyd. Uczelniane Politechniki Poznańskiej, Poznań 1965, p. 75

PART 3

Theoretical and technical elements of abrasive processes

CHAPTER 13

Cutting process and theory of abrasion cycle

13.1 Specific features of abrasive processes

Cutting methods described so far were characterized by the use of single-point or multi-point tools or—generally—tools with a definite number of points. In addition, each of these points had determined, in most cases purposefully selected, geometric features.

Beside these methods of machining by cutting, there are the so-called *abrasive processes* having several specific features different from those of chip removal processes, but at the same time having one common feature, which is fundamental for machining. This feature is the removal of a machining allowance mainly in the form of chips.

Thus abrasive processes is the name used for those methods of machining by cutting in which the machining process is carried out using abrasive tools or lose abrasives, i.e. by tools of an indefinite number and shape of cutting points which transform the undeformed chip layer into actual chips (or dust) of a size mostly invisible to the naked eye and partly oxidized.

The following characteristic features of abrasive processes should be noted:

(a) as a result of different point forms situated in random positions, only a part of them carries out the cutting work, another part plastically scratches the work surface, whilst still other part rubs against this surface causing only elastic deformations,

(b) the size of the undeformed chip cut by a single point is very small (of the order of micrometres) compared with chip removing processes, and varies from point to point,

(c) maximum cutting speeds may be more than ten times higher (from 4000 m/min) than those used in chip removing processes.

These features [and particularly feature (a)] justify the name of abrasive processes.

In considering abrasive processes some laws of probability and statistics should be taken into account to a higher degree than in chip removing processes.

The small size of the undeformed chip cut by a single point and small machining allowances for abrasive processes result in a relatively greater influence of machine tool properties and features in this case than in that of chip removing processes. Of particular importance here are machine tool rigidity and vibrations and in some machining methods thermal deformation of the machine tool. Therefore, under identical machining conditions, but on various machine tools of different quality and in different conditions, a different machining accuracy and surface finish is obtained.

13.2 Classification of abrasive processes

There are several abrasive processes and types, and also various bases for their classification. In the first place, however, the type of tool used is accepted as a classification criterion.

From this point of view abrasive processes using abrasive tools and those using loose abrasives are distinguished (Table 13.1).

Abrasive processes with using tools (AT) are also called the *bonded abrasive processes*. The tools used are of various shapes and sizes and are called grinding wheels, stones and bands. Their common characteristic is that abrasive grains are imbedded in a bonding material (so-called *bond*) and during the cutting process do not change their position until worn by abrasion or chipping.

Abrasive processes with the use of loose abrasives (AL) consist in cutting using abrasive grains which may change their position during the process. Two methods are discerned here.

Table 13.1 *Classification of abrasive processes*

A Abrasive processes

AT Using abrasive tools	AL Using loose abrasives
ATG Grinding by means of grinding wheels	ALL Lapping by means of laps
ATB Grinding by means of abrasive bands	ALP Abrasive polishing
ATH Honing	ALI Impact grinding
ATS Superfinishing	ALB Barrel finishing
	ALV Vibratory finishing
	ALA Abrasive-blast treatment

In the first, the so-called *pressure method*, abrasive grains are temporarily supported by or imbedded in the soft or heterogeneous material of a tool called a *lap*. This is accompanied by suitable pressure and relative motions.

In the second, the so-called *kinetic method*, grains are accelerated to such a speed that their kinetic energy suffices for the carrying out of cutting work when grains impinge upon the workpiece material.

Within the range of abrasive processes using tools, there are distinguished:

(a) *grinding by means of grinding wheels* (ATG) the most common method of abrasive machining, having many types,

(b) *grinding by means of abrasive bands* (ATB) on paper or cloth base, in the form of sheets or belts,

(c) *honing* (ATH) or *long-stroke smoothing*, carried out by means of abrasive tools called *hones*,

(d) *superfinishing* (ATS) or *short-stroke smoothing*, also carried out by means of hones, but with different kinematics of the working motions.

Within the range of abrasive processes using loose abrasives, there are distinguished:

(a) *lapping* (ALL) by means of laps or matching parts,

(b) *polishing* (ALP) carried out by means of soft cloth impregnated with abrasive paste,

(c) *abrasive blast-treatment* (ALA) in which abrasive grit is accelerated using fluids (gas or liquid) and formed into a stream impinging upon the work material,

(d) *vibratory finishing* (ALV) consisting in the use of kinetic energy of grit in a vibrating liquid. This type is often joined with the preceding one into a *vibro-barrelling process*,

(e) *barrel finishing* (ALB) consisting in the use of kinetic energy of grit due to rotation of the container in which abrasive, liquid and workpiece are placed, while the latter press against the grains,

(f) *impact grinding* (ALI) in which cutting work is performed by grit subjected to periodical, very rapidly changing (over 16 Hz) loads applied by a beater.

13.3 Cutting and friction in abrasive processes

As mentioned before, the size and volume of the undeformed chip cut by a single abrasive grain point is very small compared with chips cut in other chip removing processes. Thus, for example, the volume of chip cut by a single grain is about 1/400 000 of that cut by a single tooth of a face mill.

Such large quantitative differences in undeformed chip size result in cutting process features which are characteristic of the abrasive processes.

As abrasive grains penetrate the material to a depth of the order of micrometers or hundredths of a millimetre, a considerable number of grains may not cut at all, i.e. may not produce chips.

Since grain corners are rounded off at the top, a certain minimum penetration of the grain into the material is necessary for the cutting process to begin. This minimum value depends on the properties of the workpiece material, too.

In Table 13.2 minimum tool penetration depths $h_{z\,min}$ are given, for which the cutting process still occurs.

Table 13.2 *Values of minimum tool penetration into workpiece material for which the cutting process occurs* [13/1]

Workpiece material	Grain corner radius r µm	$h_{z\,min}$ µm	$k = \frac{h_{z\,min}}{r}$
0·40 carbon steel, hardened	6	1·5	0·25
Bronze	6	2	0·33
Cast iron	6	3	0·50
0·40 carbon steel, annealed	6	4	0·66
Silumin	6	5	0·83
Copper	6	5	0·83
Brass	6	7	1·16

Research results indicate that the more ductile is the material, the higher is $h_{z\,min}$. The capability of the material to form chips is characterized by the ratio

$$k = \frac{h_{z\,min}}{r} \tag{13.1}$$

The lower is k, the more is the material capable to form chips.

These remarks are valid for very low cutting speeds. At high speed the coefficient k diminishes, i.e. the chip formation capability increases and thus chips are formed at sufficiently high speeds even if grains with relatively large corner rounding radii are used.

In the case when $h_z < h_{z\,min}$ the cutting process does not occur. Depending on h_z there may be:

— friction within the limits of elastic deformations,

— friction within the limits of plastic deformations, accompanied by scratching i.e. formation of permanent traces or scratches.

Mean values of the friction coefficient of particular grains within the elastic deformation range are of the same order as in friction between solids and depend mainly on the type of medium in which friction takes place.

In cutting there also occur friction and burnishing processes. The larger is the undeformed chip thickness h_z as compared with $h_{z\,min}$ the larger becomes the share of chip plastic deformation energy and the more the percentage share of external friction energy and of outer layer burnishing energy is reduced.

The process of cutting with a single grain consists of the following phases (Fig. 13.1): elastic friction and deformation (I), burnishing (II) and cutting (III). The duration of phases I and II is practically independent of the thickness h_z. It is therefore understandable that the percentage share of cutting work will be higher, the larger is h_z.

The number of points which do not cut but only burnish or rub against the work surface is considerable, because of the non-uniformity of grit distribution in the grinding wheel, or too small rake angles resulting from stray positions of loose or bonded grains. If two successively working grains are situated very close to each other then even for a relatively large depth of cut a case may occur that $h_z < h_{z\,min}$

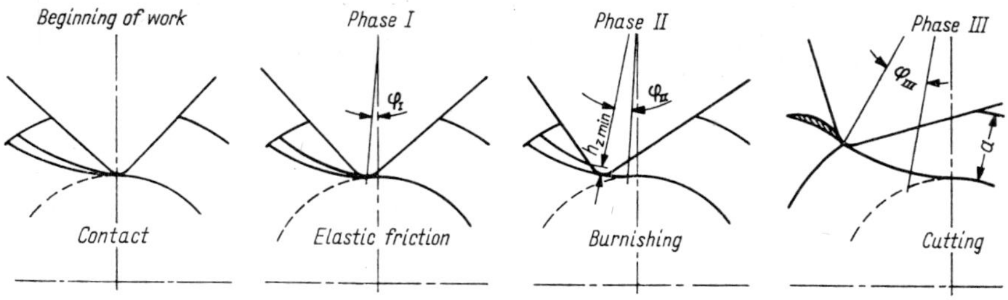

Fig. 13.1 Process of cutting with a single grain

and burnishing will occur instead of cutting. The following point will even have to cut material that is partially burnished and therefore hardened. This is one of the causes of higher specific cutting force in abrasive processes.

The shape of the formed chips varies. Under suitable magnification the chips show a structure similar to that obtained in milling: of straight ribbon or wavy ribbon type. There are also shapeless, partly fused and partly burnt chips.

The percentage ratio of chips of typical structure similar e.g. to that obtained in milling, to formless chips is an index of the uniformity of grain distribution and spacing. The wider is their spacing and the more uniformly are the grains distributed on the work surface the larger is the percentage of chips of typical structure.

A characteristic index of the share of friction and scratching energy is the ratio of the components of cutting force: thrust force F_p and tangent force F_v. In the case of chip removing processes the ratio

$$w_y = \frac{F_p}{F_v} \tag{13.2}$$

is within 0·4–0·8. On the other hand, in abrasive processes this ratio is considerably higher and amounts to 10 (depending on the amount of grain wear, undeformed chip thickness, workpiece material etc.).

13.4 Deformation speed and cutting temperature

In abrasive processes the *speed of plastic deformations* which depends on cutting speed may be very high. On the assumption that the average length of working travel of a point $L = 1$ mm and cutting speed $v = 30$ m/s, a working time $t_\varphi = 0{\cdot}000033$ s is obtained. This indicates that the cutting processes are almost instantaneous. In relation to tooth working times in high-speed face milling the point working times in abrasive processes are 1/800 to 1/500 of the former.

The speed of plastic deformations may cause two processes in the deformed material:

(a) hardening of the material and a lowering of its ductility if the process occurs without temperature increase,

(b) softening of the material and an increase in ductility if the deformation is accompanied by considerable increase in temperature.

If the change of strength properties caused by hardening is denoted by $\Delta\sigma_v$ and that caused by softening by $\Delta\sigma_\Theta$ then the resulting change will be

$$\Delta\sigma = \frac{\Delta\sigma_v}{\Delta\sigma_\Theta} \tag{13.3}$$

When $\Delta\sigma > 1{\cdot}0$, this means that the mechanical influence of cutting speed prevails, and when $\Delta\sigma < 1{\cdot}0$ then the temperature influence prevails and the yield stress is lower or ductility is enhanced.

Heat generated by cutting work in a very short time concentrates almost exclusively in the undeformed chip and actual chip. This very short time of contact between the cutting point and workpiece allows only a relatively small percentage of heat to conduct deeper into the material.

The temperature of the undeformed chip being cut by a single grain is very high and may even reach the melting temperature of the given material. For example, a spark jet* accompanying the grinding operation and constituted by chips heated to red or white heat, clearly indicates the temperature level in the undeformed chip and in the actual chips. In addition, solidified pieces of chips are sometimes found, indicating that they have been in a liquid state.

Heat generation and cutting temperature increase in abrasive processes are so fast that liquid cooling has little influence on the maximum temperature values, particularly on those appearing in the undeformed chip and even in the actual chips. This follows from the fact that even ample liquid cooling does not prevent sparks from flying.

Only a small percentage of the heat generated by the work of each grain penetrates into the material, but the total heat produced by all points of a grinding wheel raises the temperature in the work surface layer.

Therefore within this range the use of coolants is of great importance, since it considerably lowers the mean temperature as compared with that prevailing in dry abrasive processes.

The high temperature of the undeformed chip and actual chip has a favourable influence on the cutting process because it facilitates chip formation, while surface layer temperature exceeding that of the ambience is harmful. Depending on its value it may cause inaccurate machining (expansion and warping of the material) and even damage the surface layer (annealing and renewed hardening, cracks etc.).

A large temperature difference between the undeformed chip and the core of the machined material accelerates heat conduction in the material. However, due to the small thickness of the machining allowance, the absolute amount of heat penetrating the material is not great, particularly when coolant is used.

On the other hand, the large number of passes and relatively long heat conduction time causes the temperature of the work surface layer to rise with machining time.

As an example of this relationship the results of research on temperature Θ in the surface layer of sintered carbides through-ground by means of the face of a carborundum grinding wheel are shown in Fig. 13.2. This research was conducted with

* Colour and shape of particular sparks depend to a large extent on the chemical composition of the given material. This fact has been taken advantage of in the determination of alloy components and of carbon content in steel. The corresponding method is called the *spark test*.

the use of a thermo-couple placed at a depth $l = 0{\cdot}1$–$0{\cdot}8$ mm below the surface being ground.

Measurements were taken both in dry grinding and in grinding with ample emulsion cooling.

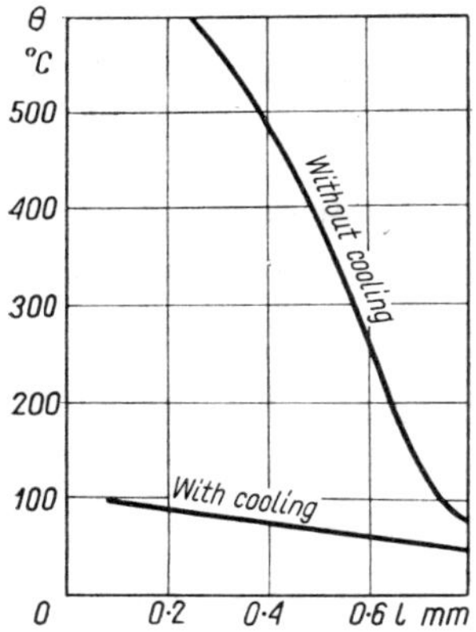

Fig. 13.2 Mean temperatures in the surface layer in through plane grinding of sintered carbides by the face of a grinding wheel (acc. to [13/2])

It is evident from the diagram in Fig. 13.2 that:

— the use of coolant considerably lowers the surface layer temperature, particularly near the surface,

— temperature gradient, in the case of dry grinding, is several times greater than in the case of grinding with the use of coolant.

Experiments indicate that the surface layer temperature rises slowly as the abrasive process time grows. This rise is slight when a coolant is applied and much greater in dry machining.

It follows that the use of a coolant in abrasive processes is of particularly great importance.

Already a relatively slight rise of surface layer and workpiece temperature may cause accuracy errors exceeding the machining tolerance.

For example in grinding stepped buffer shafts made of steel when the mean workpiece temperature rises by some 30°C, the measuring error on a length of 500 mm will amount to about 0·18 mm.

In addition to being influenced by the use of a coolant, the cutting temperature in abrasive processes also depends on all those factors which influence the cutting temperature in chip removing processes and the character of this influence is also similar.

13.5 The role of unit pressure and m-w-t system deformability in the abrasive process cycle

13.5.1 Significance of the rigidity of the m-w-t system

For all methods of machining the following relationship determining tool advance is valid

$$a_t = a+f+e$$

where: a—depth of cut, f—displacement of the m–w–t system caused by elastic deformation and play, measured in the direction of the depth of cut, e—tool wear, measured in the same direction.

In the case of abrasive processes this relationship has a particular importance since the deformability of grinding machine tools and equipment is usually greater than that of chip removing machine tools. Thus e.g. the deformability of a m–w–t system for manufacturing-type grinders is mostly within the limits of 0·5–4 μm/kg which corresponds to a stiffness j = 0·25–2 kg/μm. Within the range of grinding forces up to 10 kg deformations between 5 and 40 μm result.

Since tool advances and depths of cut are also of the order of micrometers and hundredths of a millimetre, the deformability and consequent deformations of the m–w–t system may considerably influence the cutting process.

This can be illustrated by the example of grinding a shaft.

On the assumption that the grinding wheel wears in grinding but during one wheel life its dimensions do not change, $e \approx 0$, or

$$a_t \approx a+f \tag{13.5}$$

By definition, system displacement

$$F = \frac{S}{j} \tag{13.6}$$

where: S—vertical force, perpendicular to shaft axis, j—system stiffness in the direction of action of force S.

It is evident from Fig. 13.3 that the vertical force

$$S = F_p\left(\cos\frac{\psi}{2} - \frac{1}{w_y}\sin\frac{\psi}{2}\right) \tag{13.7}$$

since the position of resultant thrust force is assumed to coincide with the bisecant of the contact angle between grinding wheel and workpiece material.

The thrust force may be defined from

$$F_p = k_{sy} Z_c A_{zm} \tag{13.8}$$

where: k_{sy}—mean specific cutting force, Z_c—number of grains cutting simultaneously, A_{zm}—mean value of the transverse cross-sectional area of the undeformed chip cut by a single grain.

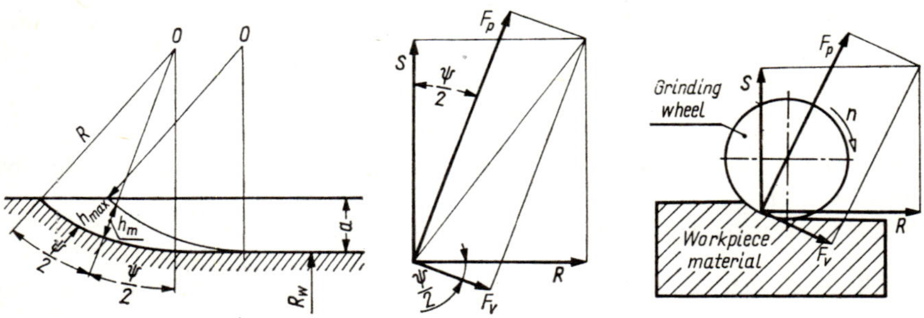

Fig. 13.3 Interdependence of force in the tool and machine tool system on the example of grinding

The angle of contact between the grinding wheel and workpiece material is very small (of the order of 1–2°). It may thus be assumed with sufficient accuracy that

$$S \approx F_p = k_{sy} Z_c A_{zm} \tag{13.9}$$

Since the number of grains cutting simultaneously is determined by

$$Z_c = \frac{B_a L_\varphi}{l_z^2} \tag{13.10}$$

where: B_a—active width of grinding wheel, l_z—average distance between grains cutting simultaneously, L_φ—length of travel in contact (accurate formulae for L_φ are derived in Chapter 15), while

$$L_\varphi \approx \sqrt{\frac{d\,d_s}{d+d_s}a} \tag{13.11}$$

where: d—grinding diameter, d_s—shaft diameter, a—depth of cut.

From eqns. (13.10) and (13.11)

$$Z_c = \frac{B_a}{l_z^2}\sqrt{\frac{d\,d_s}{d+d_s}a} \tag{13.12}$$

The mean cross-sectional area of the undeformed chip cut by a single point may be approximated by*

$$A_{zm} \approx h_{zm}^2 \tan\varepsilon\, K \tag{13.13}$$

where: h_{zm}—mean thickness of the undeformed chip cut by a single point, ε—half the grain corner angle, K—correction coefficient taking into account the influence of factors not appearing separately in the formula.

The value of h_z is calculated from**

$$h_{zm} = C\sqrt{\frac{d+d_s}{d\,d_s}a} \tag{13.14}$$

where: C—coefficient taking into account the influence of factors not shown separately in the formula.

After suitable substitutions

$$a_t = a + K\,C^2 \tan\varepsilon \frac{B_a}{l_z^2}\frac{k_{sy}}{j}\sqrt{\frac{d+d_s}{d\,d_s}}\,a^{3/2} \tag{13.15}$$

putting

$$A = K\,C^2 \tan\varepsilon \frac{B_a}{l_z^2}k_{sy}\sqrt{\frac{d+d_s}{d\,d_s}} \tag{13.16}$$

gives

$$a_t = a + \frac{A}{j}a^{3/2} \tag{13.17}$$

* Accurate calculation of A_z value is given in Chapter 15.

** This formula is derived in Chapter 15.

The relationship expressed by eqn. (13.17) is represented graphically in Fig. 13.4 for the common values of A/j from 0 (in the case of grinding with spark out) to 0·10 (without spark out).

It is evident from eqn. (13.17) that for a constant tool advance a_t and constant machining conditions the depth of cut depends on stiffness j. The more rigid is the m–w–t system, the nearer will be the value of the grinding depth to that of tool advance.

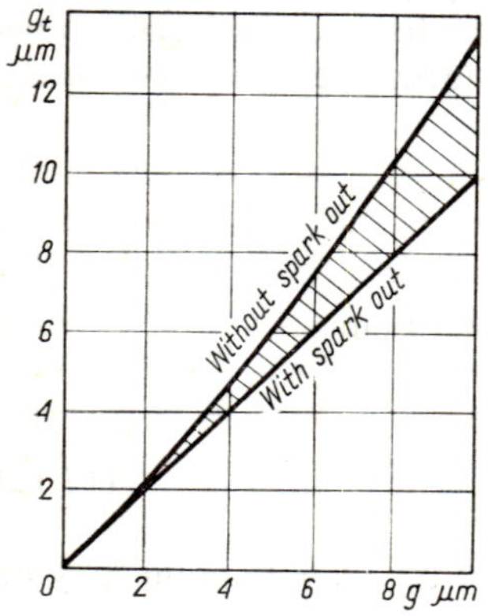

Fig. 13.4 Dependence of the depth of cut a on tool advance a_t without spark out and with spark out under determined machining conditions

13.5.2 The significance of unit pressure

The *unit pressure* is of particular importance in abrasive processes. The undeformed chip thickness and width depend on it, and consequently so do all physical and technological factors.

The pressure exerted on a single grain should be so chosen that, on the one hand, maximum output, i.e. high values of undeformed chip cut by a single point can be obtained and, on the other hand, the pressure should not be too great as this causes a deterioration of surface quality, increases cutting temperature and grain loss by chipping.

Thus optimum pressure depends on machining requirements and possibilities. For so-called *high-output grinding* Maslov [13/1] gives the optimum pressure per cutting grain as 0·5–1·0 kg.

In the case of abrasive processes using abrasive tools unit pressure is given in kg per unit of active nominal surface area (nominal area of contact e.g. between grinding wheel and workpiece material). This pressure oscillates, depending on abrasive process type and basic requirements, from a few to about 20 kg/cm^2.

Between the depth of cut and unit pressure q_y there is a relationship which may be expressed by the empirical curve shown in Fig. 13.5. This curve indicates that an increase of q_y, at small unit pressure, is accompanied by progressive growth of the depth of cut. On the other hand, beginning with a certain value of q_y this growth becomes slower and slower.

In the same diagram the curve representing the reduction e of grinding wheel dimensions is given, plotted for the most general case, i.e. that of grinding wheel work with partial self-dressing. The sum of the depth of cut and dimensional wear of the grinding wheel

$$a_r = a+e \tag{13.18}$$

is also given in Fig. 13.5.

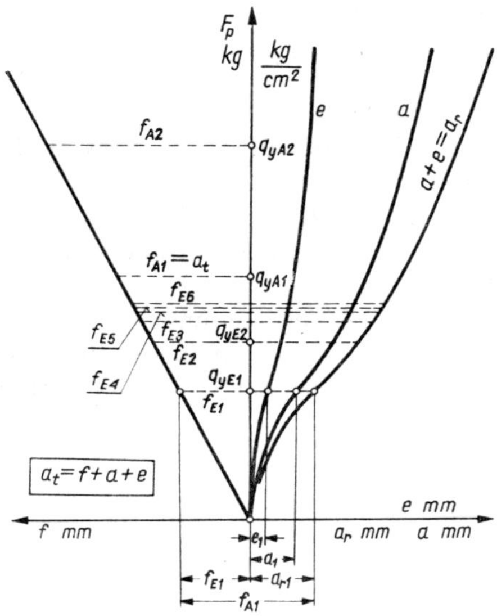

Fig. 13.5 Diagram of the dependence of elastic deformation f on thrust force F_v and of the dependence of depth of cut a, linear reduction e of grinding wheel dimensions and the sum $a_r = e+a$ on unit pressure q_y

The left hand part of Fig. 13.5 represents the deformability (rigidity) characteristics of the given grinder.

Using Fig. 13.5 it is possible to explain for the example of grinding with constant tool advance (the most commonly applied machining method) how the abrasive processes are carried out by means of abrasive tools.

In the first pass a constant tool advance a_t is established. At the instant when the grinding wheel comes into contact with the workpiece material $a_1 = 0$ and therefore $a_t = f_{A1}$. After the first pass from which a depth of cut a_1 and grinding wheel wear e_1 resulted, a deformation f_{A1} has remained.

Since a constant tool advance a_t was adopted, in the next pass the deformation will be at first $f_{A2} = f_{E1}+a_t$. As this will create a greater force F_p and a greater unit pressure, the depth of cut at the end of the second pass (a_2) will exceed that obtained in the first pass. The same applies to grinding wheel wear.

After the second pass a deformation f_{E2} will remain. However, all the time eqn. (13.3) is valid.

Repeating this cycle one may remark that final elastic deformations grow less and less while a stable balance between thrust force and the m–w–t system elasticity is being approached. This means that beginning with a certain pass the process is carried out with constant depth of grinding. Or, beginning with a certain pass $f =$ const and thus, if e is neglected (since, measured within short time intervals, its value is small in relation to a_t, a and f) then under conditions of stabilization

$$a_t \approx a \tag{13.18a}$$

Thus, beginning with the pass in which eqn. (13.18a) is fulfilled, the entire tool advance in each pass corresponds to the cutting depth.

If, on the other hand, the value of e counts as compared with a_t, then

$$a_t = a_r = a+e \tag{13.18b}$$

13.5.3 Influence of rigidity and unit pressure on machining cycle (under conditions of constant tool performance)

Variations of the m–w–t system rigidity influence the depth of cut changes at the time when the tool penetrates the workpiece material.

This is represented graphically in Fig. 13.6a. Shown there by the method described above are depth of cut changes in successive passes on two grinders of stiffnesses j_a and j_b, while $j_a > j_b$.

In both cases tool advance was identical in each pass and amounted to a_t. The number of passes i necessary to practically equalize a_r and a_t is smaller in the case of the more rigid m–w–t system, as the equation $a_{ra} = a_t$ becomes valid there after

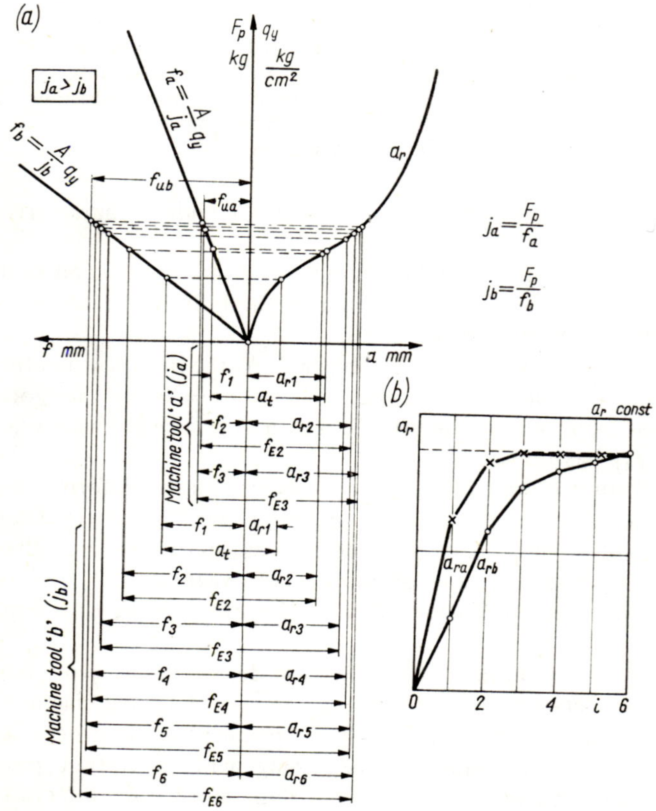

Fig. 13.6 Variations of the sum of the depth of cut and grinding wheel wear a_r for different machine tool rigidities

three passes. Instead, to obtain a similar equation ($a_{rb} = a_t$) in the case of a less rigid machine tool six passes were needed. This is shown in Fig. 13.6b where depth of cut is shown on the vertical axis and the number of passes on the horizontal axis.

A similar situation but in reverse order will repeat itself in relief grinding, i.e. when tool advance ceases. Then in consecutive spark out passes the equivalent of tool advance will be elastic displacements (in the first pass—amounting to f_{ua} and f_{ub}, respectively). When the entire record of elastic displacement is exhausted, then the relief is completed. In addition, in the case of a machine tool of greater rigidity the end of relief will occur earlier than in the case of a less rigid grinder.

Figure 13.7 illustrates graphically the full *machining cycle** on the basis of the m–w–t system rigidity influence upon depth of cut changes, as described above.

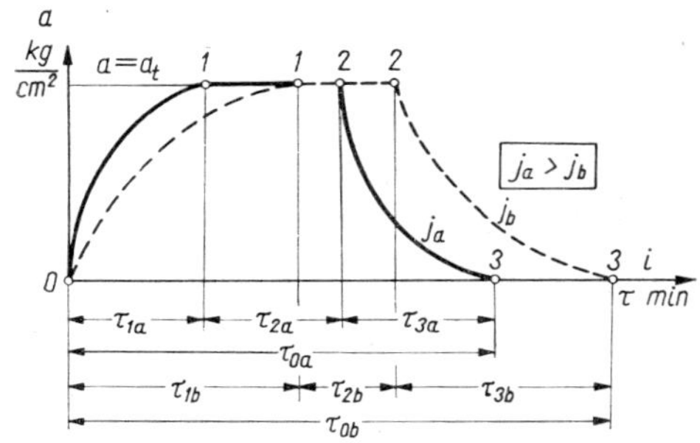

Fig. 13.7 The variations of the depth of cut, workpiece material penetration time, stabilization and relief within a single machining cycle, for different m–w–t system rigidities

A cycle of abrasive process consists of:

- penetration of the tool into the workpiece material (from 0 to 1 in Fig. 13.7) notable for the growth of the depth of cut to $a = a_t$ and of elastic deformation to the value $f = f_u$;
- stabilization period characterized by constant depth of cut $a = a_t$ (from 1 to 2 in the diagram);
- relief, notable for the lack of tool advance ($a_t = 0$) and the reduction of elastic displacement from f_u to 0 and of the depth of cut to 0, too (from 2 to 3 in the diagram). At the same time machining errors diminish or accuracy improves.

As is evident from Fig. 13.7, the machining cycle is longer in the case of lesser m–w–t system rigidity, because then the penetration and relief take more time. It follows that high rigidity of the m–w–t system is desirable in order to shorten the machining time.

The practical application of a machining cycle of this type meets with difficulties, since for a given machining allowance and an unknown reserve of elastic deformation of the m–w–t system it is impossible to determine exactly the instant when relief should start. Thus e.g. in the example shown in Fig. 13.8 the relief should begin at point O_A on the grinder of stiffness j_A and at point O_B on the grinder of stiffness j_B.

Therefore in practice the machining cycle type represented in Fig. 13.9 is preferred. Beginning with point A the penetration period lasts until B. From point B to point E,

* This term was borrowed from Lur'e [13/5].

i.e. to the removal of allowance and obtainment of the nominal dimension (with permissible tolerance), is the stabilization period. The attainment of the nominal dimension is determined by hand or by mechanical measurements (while the machining lasts). As soon as the nominal dimension is reached, i.e. from point E on, begins the so called *spark out period* or, shortly the *spark out* (section EF in the drawing).

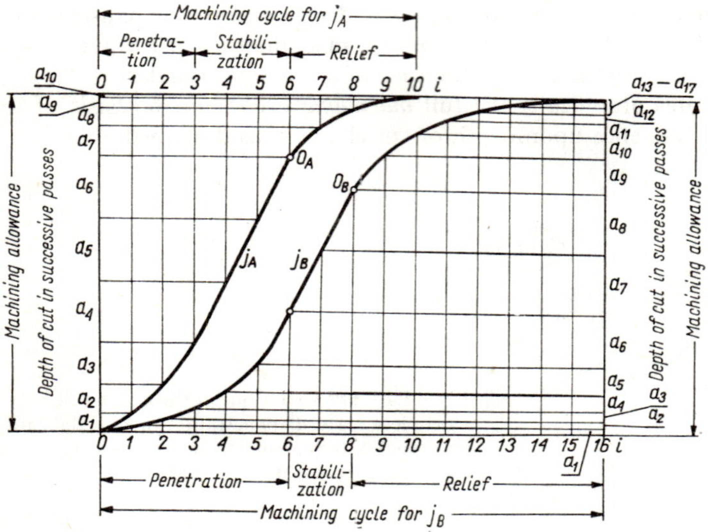

Fig. 13.8 Graphical representation of the machining cycle and its component periods, in the case of two different rigidities of the m–w–t system

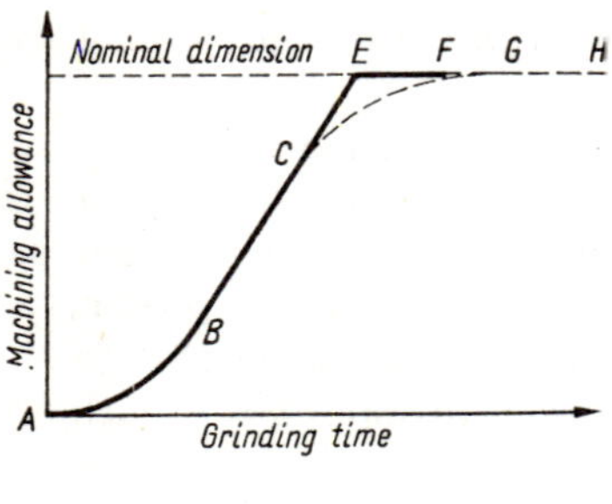

Fig. 13.9 Types of the machining cycle carried out by means of an abrasive tool

Spark out differs from relieving by the lack of elastic forces between the grinding wheel and the m–w–t system. Such a situation arises when, in order that measurements may be taken, the grinding wheel is moved away from the workpiece and, the measurement having established that the nominal dimension has been obtained with the given tolerance, is approached again only to restore contact without exerting pressure.

Spark out occurs with no tool advance and when the nominal depth of cut equals zero. However, some grinding wheel points smooth out surface irregularities appearing on the machined surface. Those irregularities result from the fact that an abrasive tool has no continuous cutting edge and protruding points will cause scratching while receding points will leave bulges. Thus the principal aim and advantage of spark out is the improvement of surface quality by a reduction of irregularities of the machined surface.

In addition to the cycle just described other types of work cycles may be used, such as e.g. *A–B–C–G–H*, i.e. penetration–stabilization–relief–spark out (Fig. 13.9). This is a full machining cycle.

13.5.4 Influence of abrasive tool wear on the variations of the depth of cut

The machining cycles described so far concerned the case when the cutting ability of the tool does not undergo any changes. This corresponds either to a short time period when grit wear has no practical importance, or to abrasive process conditions of self-dressing, when the restoration of sharpness occurs automatically—by releasing the worn grains.

The penetration speed of the abrasive tool into the workpiece material in the direction of the depth of cut v_t decreases with lengthening cutting time, if abrasive grains become worn, i.e. if cutting performance of the tool deteriorates. This relationship may be approximated by a function of the type

$$v_t = \frac{A}{B + C\,t^b} \tag{13.19}$$

where: *A*, *B*, *C* and *b* are determined in an empirical way.

In the case when the abrasive process consist of several passes then

$$t = i\,t_i \tag{13.20}$$

where: *i*—number of passes, t_i—time of a single pass.

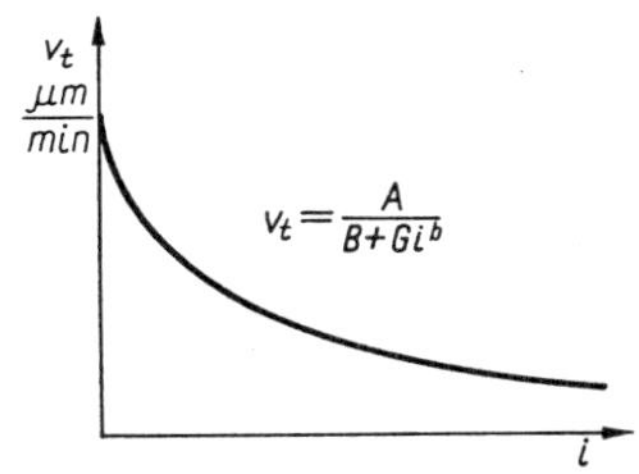

Fig. 13.10 A typical relationship between depth of cut and number of passes under conditions of abrasive tool wear

In such a case (Fig. 13.10):

$$v_t = \frac{A}{B + G\,i^b} \tag{13.21}$$

while $G = C\,t^b$.

These formulae are valid in the case when q_y = const.

When the abrasive process is carried out with constant tool advance in consecutive passes then the unit pressure undergoes variations as discussed previously.

In Fig. 13.11 a graphical representation is given of the depth of cut variations in a machining cycle, when in each successive pass the cutting ability of the tool decreases.

On the left-hand side of the diagram (Fig. 13.11b) is the deformation of the m–w–t system, or f as a function of F_p. On the right-hand side are shown the curves characterizing the variations of the depth of cut a, obtained at the end of the particular passes depending on unit pressure q_y.

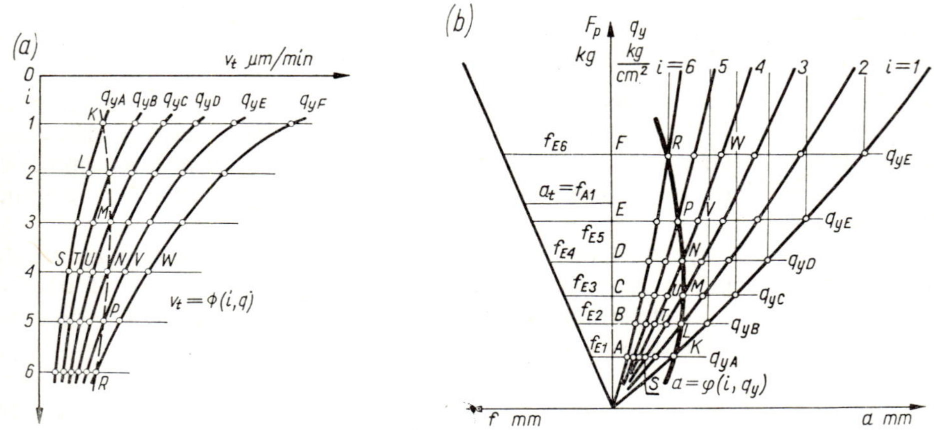

Fig. 13.11 Diagram of grinding depth variations in successive passes in the case of grinding wheel wear ($e = 0$, $F_p = \text{const} \times q_y$)

These curves have been plotted on the basis of empirically established dependence of v_t on the number of passes at various unit pressures (Fig. 13.11a). The plotting of the curves may be explained by the example of $i = 4$. Points S, T, U, N, V, W situated at the intersection of the line $i = 4$ with v_t curves for unit pressures q_{yA}, q_{yB}, q_{yC}, q_{yD}, q_{yE}, and q_{yF}, respectively, are projected on the lower diagram, on horizontal lines corresponding to the unit pressure values just mentioned. By joining these points on the lower diagram the curve $a = \varphi(q_y)$ is obtained for $i = 4$.

Once the curves are drawn, it is possible to observe how the depth of cut will change at the end of each pass, at constant tool advance a_t.

Starting with tool advance $a_t = f_{A1}$ at the beginning of the first pass, at its end the depth $a_1 = AK$ will be obtained, thus leaving elastic deformation f_{E1}. Therefore at the start of the second pass the total tool advance will be: $a_t + f_{E1}$. Consequently, the unit pressure will rise, so that the depth reached in the second pass will be $a_2 = BL$ and the remaining elastic deformation f_{E2}.

In a similar way it can be found that in the third pass there is $a_3 = CM$ and f_{E3}, in the fourth $a_4 = DN$ and f_{E4}, in the fifth $a_5 = EP$ and f_{E5}, in the sixth $a_6 = FR$ and f_{E6}.

Joining points K, L, M, N, P and R yields the curve of depth of cut variations under machining conditions of constant tool advance in successive passes, and gradual wear of the abrasive tool, i.e. curve $a = \varphi(i, q_y)$ in Fig. 13.11b.

It follows from this curve that under conditions of grinding wheel wear the depth of cut diminishes after a certain time, although elastic deformation f_E grows steadily and so does unit pressure. In an extreme case conditions may be attained when no cutting but only friction will occur.

The course of the resultant curve just described is accompanied by a deterioration of machining accuracy, temperature rise and damage to the surface layer.

The conclusion follows that the higher the accuracy and quality requirements, the more attention should be paid to keeping the abrasive tool properly dressed.

This justifies the relatively short work periods of abrasive tools (under conditions of wear) after each dressing operation.

REFERENCES

13/1 Maslov, E. N.: 'Osnovy teorii shlifovaniya metallov' (Elements of the Theory of Metal Grinding), Mashgiz, Moskva 1951. 'Mekhanizm raboty abrazivnogo zerna pri shlifovanii'. Sbornik: Osnovnye voprosy wysokoproizvoditelnogo shlifovaniya (Mechanism of the Work of Abrasive Grit in Grinding. Collection: Basic Problems of High-output Grinding), Mashgiz, Moskva 1960. 'Wysokoproizvoditelnoe shlifovanie' (High-output Grinding), Izd. AN SSSR, 1962

13/2 Eilmes, J., and Kaczmarek, J.: 'Wpływ niektórych warunków szlifowania węglików spiekanych na powstawanie pęknięć i mikropęknięć' (Influence of some conditions of grinding sintered carbides on the appearance of cracks and microcracks), Zeszyty Naukowe IOS, No. 20, 1965

13/3 Backer, W. R., and Merchant, M. E.: 'On the Basic Mechanics of the Grinding Process', ASME, pap. No. 56–A–43, Nov. 1956

13/4 Kaczmarek, J.: 'Einfluss des Einzeldrucks und der Steifigkeit des Systems *W–W–W* auf Schleifvorgang', CIRP-Annalen, No. 4/62

13/5 Lur'e, G. B.: 'Teoriya rabochego tsikla pri kruglom shlifovanii kak osnova wysokoproizvoditelnoi obrabotki'. Sbornik: Osnovnye voprosy wysokoproizvoditelnogo shlifovaniya (Theory of the Work Cycle in Cylindrical Grinding as a Basis for High-output Machining. Collection: Basic Problems of High-output Grinding), Mashgiz, Moskva 1960

CHAPTER 14

Properties and wear of abrasive materials and tools

14.1 Shape and classification of abrasive materials

14.1.1 DIMENSIONS AND SHAPE OF ABRASIVE GRAINS

The term *grains* or *grit of an abrasive material* is used to describe the latter's minute lumps (crystals or chips) which have a cutting capability, of a width not exceeding 4 mm and length to height ratio not greater than 3:1.

In practice grains of considerably smaller size are used.

Grain dimensions are called the dimensions of the smallest rectangular prism which can be circumscribed on the given grain. Height h, length l and width a of the prism are distinguished. The relationships between these dimensions are established as $l \geqq a \geqq h$. Grain width a is taken as the characteristic dimension.

Abrasive grit may have the form of monocrystals or twin crystals. The former always split along determined cleavage planes while the latter usually split along intercrystalline planes, showing no regularity.

Under ideal conditions of crystallization symmetrical crystals are formed, while under industrial conditions abrasive materials with grains of asymmetrical structure are obtained. These grains show the structure of solids with many corners, and are similar to pyramids. The cutting part of the grain consists of those corners which may be considered as minute tool points.

Abrasive grain points are characterized in two main cross-sections perpendicular to each other and so situated that the line of their intersection should be normal to the corner top. In the case of symmetrical corners a single main cross-section suffices. In most cases only the characteristic in this main cross-section is given, which shows the less advantageous features with regard to the cutting performance of the point.

Abrasive grain points are characterized by two parameters:

— included angle 2ε,

— top rounding radius r, shortly called top radius.

It has been established on the basis of statistical research [14/3] that definite relationships occur between grain size and the geometric parameters of grain points.

In Fig. 14.1 the dependence of included angle and top radius on the characteristic dimension a of the grain is shown (for aluminium oxide and corundum).

The statistically confirmed rules in this respect are:

a) top radius increases with grain size; within the range of a = 30–420 μm top radius grows almost linearily from 3 to 28 μm,

b) included angle grows with grain dimensions. For grains of width a = 20–70 μm this angle increases from 90 to 100° and for a = 70–420 μm from 100 to 110°.

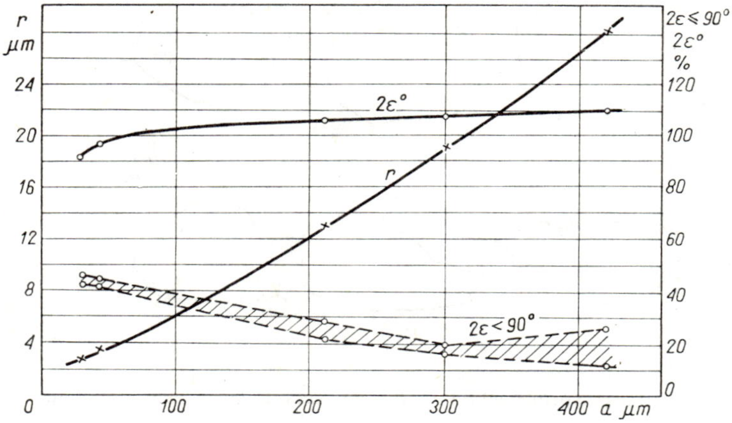

Fig. 14.1 Dependence of included angle 2ε, top radius r and percentage share of $2\varepsilon < 90°$ on grain size (acc. to data given by Vakser [14/3])

These rules are also valid for other abrasive materials although both the included angle and top radius depend also on the kind and type of such material.

Given above are average values of the geometric parameters of abrasive grains. Actually, each grain has corners with different included angles.

With regard to cutting performance it is important what percentage of corners has included angles smaller than 90°, as this indicates the probability of grains cutting with positive rake angles.

In Fig. 14.2 the results of measurements of included angles and top radii are presented [14/3]. As is evident from this figure, the included angles oscillate between 80 and 145°.

In Fig. 14.1 the results of statistical research on the percentage share of included angles $2\varepsilon < 90°$ of abrasive grains are plotted against their size. As seen in the diagram, with increasing grain size the percentage share of acute angles decreases which indicates a deterioration of cutting performance.

14.1.2 Classification of abrasive grains

For practical purposes classification rules have been established and the abrasive grains classified.

Generally, statistical grain size is called *graininess*.

By *grain size* is understood the number corresponding to the range of dimensions characteristic for the basic grit fraction appearing in the given abrasive material.

In every such material, although in principle homogeneous grain size is aimed at, there is a certain percentage of grains smaller than the basic fraction and a certain percentage of larger ones. Standards determine for each grain number the permissible

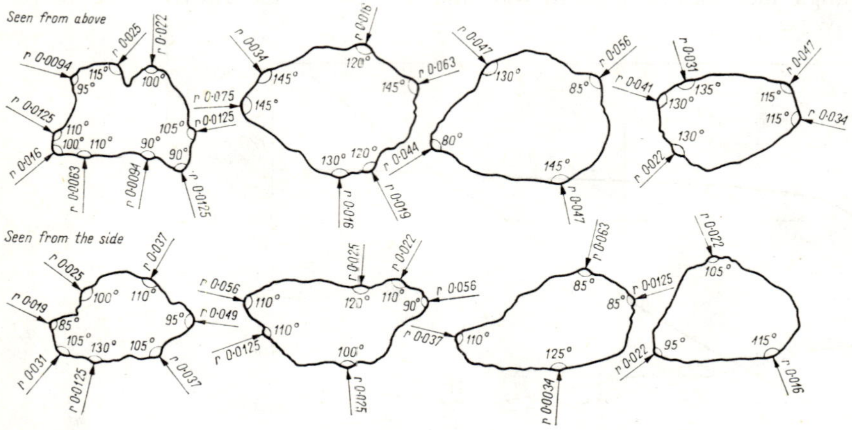

Fig. 14.2 Shape of abrasive grains as projected on main cross-section planes, values of included angles and top radii

percentages of grains differing in size from the basic fraction. It may be said generally that, on average, 40–45% of grains should belong to the basic fraction* and 65–80% jointly to the basic fraction and to that containing grit of the nearest a finer size.

There are two criteria for grit classification and denotation. The first criterion is the characteristic grain dimension a (in micrometres). Another criterion is the number of meshes on the length of one inch of the sieve (of standard thread thickness and mesh dimensions) through which the given grit was screened, called *grit size*.

Grit of the given size may not pass through the next finer sieve in the standardized series of mesh sizes.

Grain size denotation consists of two numerical symbols: the first is the smallest characteristic dimension of the grain s_{min} for the basic grain fraction, in 10 μm, the second (given in parenthesis) grit size.

Between grit and the maximum characteristic dimension there is the relationship

$$a_{max} \approx \frac{25{,}4}{\text{grit size}} \quad \text{mm} \tag{14.1}$$

Grain size or number is one of the most important parameters of abrasive materials and tools as it strongly influences the course and results of abrasive processes.

Denotations of grain sizes are given in Table 14.1.

* Fraction—in the meaning used here denotes grains belonging to the same grain number.

Table 14.1 *Classification and denotation of abrasive materials*

Material	Grain size denotation (grain number)	Characteristic dimension *a* μm from	to	Material	Grain size denotation (grain number)	Characteristic dimension *a* μm from	to
Abrasive grit	250(8)	2500	3150	Abrasive dust	12(100)	125	160
	200(10)	2000	2500		10(120)	100	125
	160(12)	1600	2000		8(150)	80	100
	125(14)	1250	1600		6(200)	63	80
	100(16)	1000	1250		5(240)	50	63
	80(20)	800	1000		4(280)	50	50
	63(24)	630	800		4(280)	40	50
	50(30)	500	630		3(320)	28	40
	40(36)	400	500				
	32(46)	315	400	Abrasive micro-dust	28	20	28
	25(60)	250	315		20	14	20
	20(70)	200	250		14	10	14
	16(80)	160	200		10	7	10
					7	5	7
					5	3,5	5

14.2 Characteristic of abrasive tools

14.2.1 Abrasive tools and their features

An *abrasive tool* is a porous solid of a determined shape and dimensions, containing in its volume abrasive grains held in a sufficiently permanent manner by a binding material called the *bond.*

Abrasive grains perform the function of cutting edges, and the bond the function of tool shank. Grains may be chipped off the abrasive tools and are then replaced by freshly uncovered grains imbedded in a deeper layer of the bond.

Among abrasive tools grinding wheels, abrasive segments, abrasive sticks and hones are distinguished.

Regardless of its type, if an abrasive tool is to be defined in an unequivocal way, the following properties should be known:
(a) type of abrasive material,
(b) dimensions of abrasive grains (grain number),
(c) type of bond,
(d) tool grade,
(e) structure, porosity and cohesion,
(f) tool shape and dimensions.

The abrasive tool features listed above influence the course and results of abrasive processes. Therefore a good knowledge of those features is the basis for optimization of abrasive process conditions.

14.2.2 Bond of abrasive tools

Bond is an abrasive tool component whose task is to bind the particular abrasive grains into a porous solid.

In addition, a bond should have the following properties:
(a) suitable strength,
(b) resistance to chemicals and humidity,
(c) capability of forming in the tool large pores which play the role of flutes.
When bond strength is insufficient, grain is prematurely released by the action of cutting forces, before it actually wears. This causes excessive wear of the tool. And vice-versa, if the bond is excessively strong, then the tool glazes over and the machined surface is burned.

The following bonds are now in use: mineral (magnesite bond, also known as cold bond), silicate, ceramic, resin, rubber and metal.

Mineral bond is characterized by brittleness and low strength. It has also little resistance to humidity and temperature changes. Grinding wheels with a mineral bond are seldom used in machining operations.

Silicate bonds constitute an intermediate group between magnesite and ceramic bonds. Silicate bond tools perform like abrasive tools of natural sandstone. Since their strength is not too high, they do not allow cutting with large undeformed chip cross-sections per point and this favours the lowering of cutting temperature.

A ceramic bond is commonly used. In manufacturing corundum tools a so-called fused (liquefied) bond is applied and in manufacturing carborundum grinding wheels a so-called sintering type.

In both cases the bond cements abrasive grains forming between them characteristic bridges constituting the mounts of particular grains.

A ceramic bond has high strength, thus allowing the application of cutting speeds of the order of $v = 35$ m/s, but it is relatively brittle, and scratching of the machined surface as a result of the high bond hardness may also occur.

A resin bond does not take part in the cutting process because of its softness and this improves surface quality. The strength of this bond is high, and cutting speeds of the order of 45 m/s can be used. Also, the cases of surface burning are less frequent since the resin-on-metal friction energy is less than that of a ceramic bond.

A rubber bond has a still greater tensile strength and therefore allows high cutting speeds (up to 65 m/s). This bond is suitable for grinding wheels used for finishing work, because sulphur-vulcanized rubber not only does not scratch the surface, but smoothes and polishes it. In addition, rubber bonds are applied in the case of cut-off wheels used to cut hard-to-machine metals, such as through-hardened steel.

Metal bonds are seldom used at present. They are encountered in the case of diamond wheels with e.g. bronze, aluminium or copper bonds, which have a very high tensile strength.

14.2.3 Grade of abrasive tools

By the grade of an abrasive tool is understood the resistance offered by a bond to the tearing of abrasive grains off the tool surface by external forces.

Thus bond grade depends on its strength properties and on the thickness of the bond layer between grains. As the latter depends, *inter alia,* on grain size, the grade of the same bond may vary for different grain sizes.

Grinding wheel grade is determined under conventional conditions by one of the following hardness testing methods:

(a) the method of eroding a groove in the tool by means of a sand jet, used in the case of grain number 200(10)–16(80) (grainy abrasives);

(b) the method of indenting by means of a ball of determined diameter and under determined pressure on a Rockwell type hardness tester, used within the range of grain numbers 12(100)–14 μm. The indication of hardness is the depth of ball penetration;

(c) the test consisting in drilling a hole of determined depth ($l = 2{\cdot}5$ mm) under a determined axial pressure. Indication of hardness is the drilling time. This test is used in the case of resin and rubber bond wheels;

(d) various cutting methods (grain chipping) using a diamond tool or workpiece material to cut the grinding wheel. The measure of hardness is the cutting (chipping) resistance.

In practice tool grade is denoted by capital letters. The scale comprises 19 grades: E, F, G—very soft, H, I, J, K—soft, L, M, N, O—medium hard, P, Q, R, S—hard, T, U, W, Z—very hard.

The choice of tool grade is one of the most important functions in the selection of machining conditions, since optimum hardness depends on numerous factors.

Thus for example a change of speed causes such symptoms as if a grinding wheel of a different grade were used. Increasing the cutting speed causes symptoms typical of a grinding wheel of a harder grade and vice-versa.

Therefore the notions of static and dynamic hardness are distinguished, the latter being conditioned by the cutting speed and—to a lesser degree—by still other factors.

14.2.4 Structure, porosity and cohesion of an abrasive tool

The total volume of an abrasive tool consists of abrasive grains, bond and pores. Abrasive grains play the role of cutting edges, the bond constitutes their mounting and pores are similar to chip flutes in multipoint tools. The percentage shares of these three components in the grinding wheel volume influence the conditions and course of the machining process.

From the denotations:

V_g—volume of abrasive grains in relation to abrasive tool volume, in per cent,
V_b—volume of bond in relation to abrasive tool volume, in per cent,
V_p—volume of pores in relation to tool volume, in per cent, while the latter $V_t = 100\%$

$$V_g + V_b + V_p = 100\%$$

The percentage share of abrasive grains in the total tool volume is called the *tool structure*. However, the adopted practice is not to determine the tool structure directly by the percentage content of abrasive grains, but by a conventional number indicating such content.

If the structure number is denoted by L_g (which can take on whole number values between 0 and 12), then the relationship between the percentage share of abrasive grains in the tool and the structure number may be expressed as

$$V_g = 62 - 2L_g\% \qquad (14.2)$$

In Table 14.2 the structure numbers and percentage shares of abrasive grains in the tool are given, and in Fig. 14.3 four principal structure types are presented.

As a rule, the more a structure is open, the larger are its pores. Therefore the use of more open structures is recommended when grinding with greater undeformed chip thicknesses and with a longer cutting travel of particular grains.

Table 14.2 *Grinding wheel structure numbers and percentage content of abrasive grains*

Structure	Dense					Medium				Open			
Structure number L_g	0	1	2	3	4	5	6	7	8	9	10	11	12
Percentage share of abrasive grains V_g	62	60	58	56	54	52	50	48	46	44	42	40	38

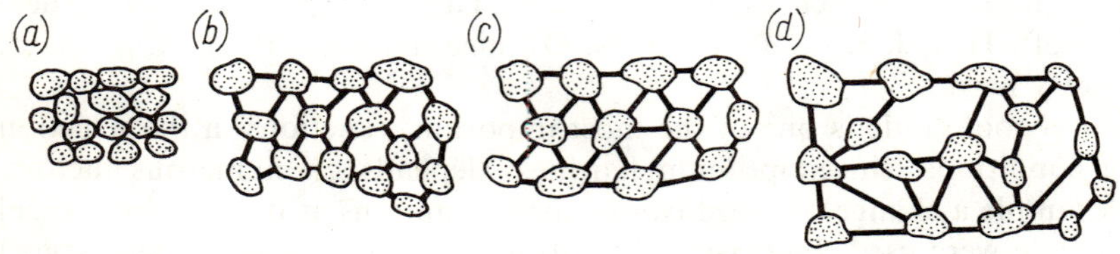

Fig. 14.3 Examples of grinding wheel structure: (a) dense, (b) medium, (c) open, (d) wide open

The percentage ratio of pore volume to the total tool volume is called *porosity*. In practice, however, porosity is not defined by the percentage share of pores, but by the corresponding porosity numbers L_p, while the relationship between these magnitudes is

$$V_p = 46{\cdot}5 - 1{\cdot}5L_p\% \tag{14.3}$$

Porosity number L_p may take on whole number values between 0 and 13.

For a determined structure and porosity of an abrasive tool, its bond content is already unequivocally established. This follows from the fact that the ratio of pore and bond percentage contents, for a given structure, cannot be arbitrary, but conforms to a strict relationship which may be represented by

$$V_l = 2L_g + 1{\cdot}5L_p - 8{\cdot}5\% \tag{14.4}$$

The percentage ratio of bond volume to the total tool volume is called *cohesion*.

Cohesion is strictly interconnected with abrasive tool grade. The greater is the cohesion, the thicker are the bonds ('bridges') between abrasive grains. Since, for a given structure, cohesion strictly depends on porosity, there is also an interdependence between tool grade and porosity.

Within the range of dense structures there exists an unequivocal relationship between hardness grades (from G to T) and porosity numbers. This relationship is shown in Table 14.3.

Among abrasive tools particularly advantageous are those with large pores or open structure, especially grinding wheels. The latter differ from grinding wheels of dense structure by the size of pores, and not by porosity which depends on the re-

Table 14.3 *Relationship between grade and porosity*

Grade	G	H	I	J	K	L	M
Porosity (and hardness) number L_p	0	1	2	3	4	5	6
Porosity V_p (%)	46·5	45·0	43·5	42·0	40·5	39·0	37·5
Grade	N	O	P	Q	R	S	T
Porosity (and hardness) number L_p	7	8	9	10	11	12	13
Porosity V_p (%)	36·0	34·5	33·0	31·5	30·0	28·5	27·0

quired hardness. Grinding wheels of open structure have, for the same grade as those of medium or dense structure, correspondingly larger pores, which means that the same percentage volume of pores consists of a smaller number of larger pores.

Experiments indicate that large-pore grinding wheels have two advantages over small-pore ones: they enable cutting with larger depths and feeds, since the increased pore size can accomodate larger chips.

Due to large pore volumes much better cooling is possible during cutting and this favourably influences the cutting temperature.

The quality of surfaces ground by large-pore grinding wheels may be very good if grains of suitably small size are used.

Generally, large-pore grinding wheels may be described as considerably more efficient than the small-pore ones.

14.2.5 Shape and dimensions of abrasive tools as exemplified by grinding wheels

Industry produces a large number of grinding wheels of various shapes and sizes. A selection is made according to manufacturers' catalogues. The choice of shape and dimension of a grinding wheel depends first of all on the character of the machining operation, the shape of the work surface and the type of machine tool.

Typical types of grinding wheel shape are shown in Fig. 14.4. In addition, there exists a great number of intermediate varieties and of grinding wheels of special profiles. Grinding wheels may be solid or of a segmental type, i.e. made up of suitably shaped parts fixed (usually by gluing) on a metal body. Examples of such wheels are shown in Fig. 14.5. Their use makes possible a considerable economy of abrasive materials.

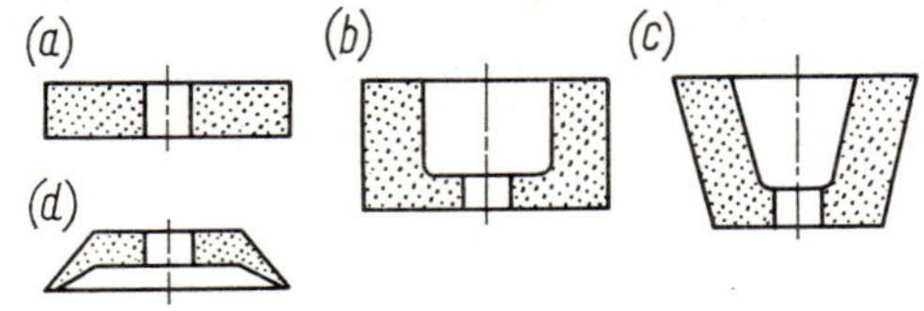

Fig. 14.4 Typical grinding wheel shapes: (a) disc type, (b) cup type, (c) flaring cup type, (d) dish type

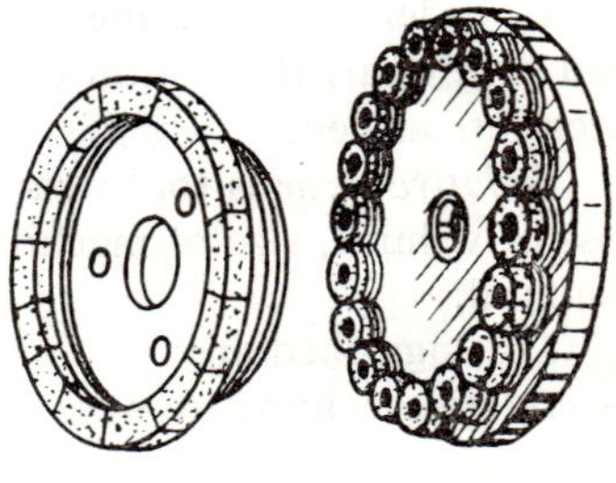

Fig. 14.5 Examples of segmental grinding wheels

A grinding wheel symbol should contain the designation of type, principal dimensions, abrasive material, abrasive grain number, wheel grade and structure, and bond type.

In addition, each grinding wheel should bear the designation of its permissible peripheral speed in m/s and rotational speed in rev/min. The manufacturer's name and factory number should also be given.

Other abrasive tools are marked in a similar way as grinding wheels.

14.3 Wear and life of abrasive tools

14.3.1 Determination of wear of abrasive grains and tools

Change of the original shape and dimensions of abrasive grains, connected with a loss of their mass (volume reduction) as a result of cutting work is called *abrasive grain wear.*

Two typical wear processes are distinguished.

The first type of wear is caused by the friction and abrasion process. Mass losses in this process correspond to volumes invisible to the naked eye. In certain cases of friction, e.g. in adhesive friction, the losses may be the size of molecules. Generally, this type of wear is called *abrasive wear of a point.*

The second type of wear consists in exceeding the local ultimate or fatigue strength, as revealed by chipping or splitting of the grains. This type is called *strength wear.*

In practice both types occur together. However, in certain cases one of them distinctly prevails.

Wear of abrasive grains could theoretically be evaluated by the measurement of superficial wear traces. The small size of grains, their random distribution, the usual lack of a stable measuring datum and other difficulties demand other methods of expressing wear.

Together with the progressive wear of abrasive grains there occurs a change in their cutting performance. One of the principal criteria for its determination is the rate of metal removal. In the case of abrasive wear this rate decreases monotonically with time. On the other hand, in the case of strength wear the rate of metal removal generally decreases with time, but there may occur cases when, within a certain time range, the rate increases. Thus, in general, the dependence of metal removal rate on time in the case of strength wear is non-monotonic.

In order to determine objectively, and to evaluate, the course of abrasive grain and tool wear, the notion of relative wear and relative rate of metal removal have been introduced.

Relative wear is tool wear related to the effect on cutting work, such as e.g. the loss of volume, size of machined surface or depth of penetration into the workpiece material.

Distinguished are:

— relative volumetric wear

$$m_v = \frac{V_T}{V_M} \tag{14.5}$$

where: V_T—volumetric loss of abrasive tool, V_M—volumetric loss of workpiece material;

— relative superficial wear

$$m_A = \frac{A_T}{A_M} \tag{14.6}$$

where: A_T—surface area of tool abrasion, A_M—machined surface area;

— relative linear wear

$$m_L = \frac{L_T}{L_M} \tag{14.7}$$

where: L_T—linear loss of tool, L_M—linear loss of material (measured in the same direction as tool loss).

Relative rate of metal removal or the effect of cutting related to tool wear is a notion reciprocal to that of relative wear.

Similarly to the types of relative wear are distinguished:

— relative volumetric rate of metal removal

$$k_V = \frac{1}{m_V} = \frac{V_M}{V_T} \tag{14.8}$$

— relative superficial rate of metal removal

$$k_A = \frac{1}{m_A} = \frac{A_M}{A_T} \tag{14.9}$$

— relative linear rate of metal removal

$$k_L = \frac{1}{m_L} = \frac{L_M}{L_T} \tag{14.10}$$

The relative volumetric rate of metal removal is sometimes referred to as the *grinding ratio.*

As in the case of single and multiple-point tools, there also occurs relative shape wear of abrasive tools. This is measured by considering the deviation of the workpiece or tool dimension to the permissible deviation. If, for example, the permissible rounding radius of a grinding wheel corner is R_T and the measured radius is equal to R_i, then the relative shape wear will be

$$m_S = \frac{R_i}{R_T}. \tag{14.11}$$

All types of relative wear or relative rate of metal removal may be expressed in abstract numbers or in per cent.

14.3.2 Wear of abrasive tools

As a result of cutting work, abrasive tool points gradually undergo microchipping (abrasion) and loose their cutting ability. The successive phases of this process are illustrated in Fig. 14.6.

Phase I, when the point is quite sharp, does not last long. From the very beginning of cutting there may repeatedly occur small losses, gradually increasing the total loss and changing the shape of the grain.

The further course of wear may be twofold. In the first case, as a consequence of point shape change, the cutting force increases so much that it exceeds the force retaining the grain in the bond and the whole grain is released together with a bond

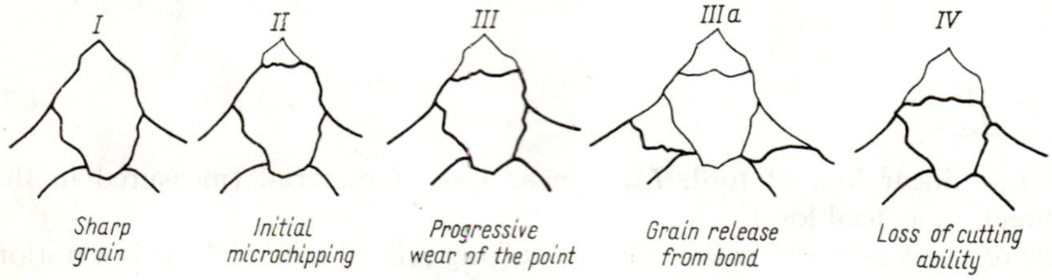

Fig. 14.6 Diagrammatic presentation of the phases of abrasive grain wear: I–II–III–IV wear, I–II–IIIa self-dressing

fragment, uncovering a new grain. Due to the pressure exerted by the grinding wheel or other abrasive tool, new grains situated in deeper bond layers now begin to work. As a result, the cutting performance of the abrasive tool practically does not change. Such a case of tool wear is called *self-dressing*.

In the second case, although the cutting force increases as a result of grain shape change, the force retaining the grain in bond is greater and the grain becomes rounder and rounder until it completely looses its cutting ability. This type of grain wear is called *grain wear* or *abrasive tool wear*.

As the wear of abrasive grains proceeds, the number of grains which do not cut but only rub and burnish the workpiece material increases more and more. Consequently, the cutting temperature rises, so-called superficial burns and vibrations may occur, surface quality deteriorates and the rate of metal removal decreases.

The self-dressing process does not require work interruption for wheel dressing, and enables higher rate of metal removal and a better surface quality to be attained while the cutting temperature remains lower.

However, with regard to technical and economic effects it does make a difference whether or not the point is released from the bond at the right moment. If this happens too early, then the relative wear of the abrasive tool increases exceedingly. If too late, the relative wear also increases, since the cutting performance is then greatly reduced (Fig. 14.7).

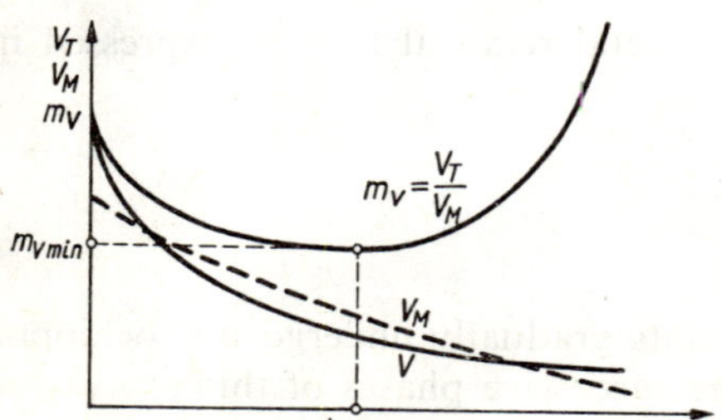

Fig. 14.7 Diagram of the dependence of relative wear m_V of an abrasive tool during self-dressing on time of cutting by grains until torn out of bond t_w

The condition under which the self-dressing process is economical is thus the occurrence of minimum relative wear $m_{V\min}$ (Fig. 14.7). This requires optimum cutting conditions to be selected, often individually for each particular operation. Such selection is time-consuming and tedious, and in some cases the self-dressing process is not feasible at all (e.g. because of the lack of grinding wheels of suitable grade).

When machining plastics there occur cases when the released grains and bond are pressed into the workpiece material in the self-dressing process, thus spoiling the surface finish.

Thus whilst the advantages obtainable by self-dressing are unquestionable, this process is applied in practice only in a limited number of cases (e.g. in the grinding of cutting tools with sintered carbide tips).

Therefore, in the majority of cases abrasive tools work under conditions of wear and consequently require periodical regeneration (dressing) or forced removal of worn grains.

14.3.3 Economical life and maximum production life of worn abrasive tools

In the case of self-dressing of an abrasive tool its life is considered to be the work period after which the permissible value of relative shape wear m_K is reached. If, on the other hand, there are no requirements concerning shape and dimensions or the tool does not show shape wear, then the notion of abrasive tool life is equivalent to that of service life, which means that the life of a tool comprises the time of its work until completely worn out.

In the case of an abrasive tool wearing, its life must be adopted conventionally. As a criterion of the life period various technical and economic indices may be adopted, such as dimensional accuracy (errors) of machining, surface roughness, rate of metal removal, unit costs etc.

Tool point life depends to various degrees on all machining conditions. The general form of a statistical and empirical formula for tool point life, taking into consideration the factors exerting the greatest influence, is as follows (for the example of cylindrical grinding)

$$w = \left(\frac{C_v v^z d^{b_v} B^{r_v} d_w^{b_{vw}} B_g B_c}{v_w a^{e_v} s^{u_v}} \right)^z \tag{14.12}$$

where: C_v—coefficient taking into account the influence of all factors not appearing separately in the formula, z, b_v, r_v, b_{vw}, e_v, u_v—exponents found in an empirical way, B_g—correction coefficient taking into account the changes of grain size, B_c—correction coefficient taking into account the influence of coolant, v_w—peripheral speed of the workpiece.

Experience indicates that for calculations made with the accuracy necessary for production purposes the following magnitudes may be eliminated from equation (14.12):

(a) grinding wheel width, as its influence is very slight (values of exponents r_v are of the order of hundredths),

(b) B_g and B_c, since the range of variations of practically used values of grain size and coolant types is small,
(c) diameter d, under condition that value of C_v will be determined for the given intervals of d.

Taking into account these simplifications one can write—using eqn. (14.12)—the formula for peripheral life speed of the workpiece

$$v_w = C_v \frac{d_w^{b_{vw}}}{T^m a^{e_v} s^{u_v}} v^z \quad \text{m/min} \tag{14.13}$$

In order to establish the recommended conditions of abrasive processes carried out using abrasive tools, the economical life period is determined which would satisfy the condition of minimizing the unit cost of machining, e.g. of a unit of surface area [14/2].

It is assumed that the cost of machining per tool life is

$$K_L = K_A + w\, t_{te} + w\, t \quad \text{pence} \tag{14.14}$$

where: K_A—cost of abrasive used, w—worker's wages per minute with overheads, t_{te}—time of exchange or regrinding of a dulled tool, t—abrasive tool life.

It is also known that in the case of tool wear the rate of metal removal is defined by

$$Q_v = \frac{E}{t^e} \quad \text{mm}^3/\text{min} \tag{14.15}$$

where: E and e—constants determined in an empirical way.

From this definition it follows that

$$\mathrm{d}V = Q_v \mathrm{d}t = \frac{E}{t^e} \mathrm{d}t \tag{14.16}$$

By integration

$$V = \frac{E}{1-e} t^{1-e} \tag{14.17}$$

is obtained, and therefore the machined surface area

$$A = \frac{E}{a(1-e)} t^{1-e} \tag{14.18}$$

Machining cost per unit surface area will thus be determined by

$$K = \frac{K_L}{A} = \frac{K_A + w\, t_{te} + w\, t}{E\, t^{1-e}} a(1-e) \tag{14.19}$$

From the condition $\mathrm{d}K/\mathrm{d}t = 0$ is determined the extreme value of t satisfying this condition, called the *economical life* T_e

$$T_e = \left(\frac{1}{e}\right)\left(\frac{K_A}{w} + t_{te}\right) \tag{14.20}$$

In order to determine tool life for maximum production, one should start with the condition for maximum surface area machined during tool service life

$$B = A\, i \tag{14.21}$$

where: A—surface area machined during 1 tool life [acc. to formula (14.18)], i—number of tool lives contained in tool service life, or

$$i = \frac{W}{t_{te}+T} \tag{14.22}$$

where W—tool service life.

Substituting eqns. (14.18) and (14.22) into eqn. (14.21), differentiating and equating the first derivative to zero yields an extreme value for tool life called *tool life for maximum production*

$$T_{pr} = \left(\frac{1}{e}\right) t_{te} \tag{14.23}$$

The actual life periods of an abrasive tool should fulfil the condition (explained in Chapter 7)

$$T_{pr} < T \leqq T_e \tag{14.24}$$

Recommended lives of worn abrasive tools are given in grinding standards and satisfy the condition (14.24).

14.4 Wear and life of abrasives

14.4.1 Wear of loose abrasives

In the case of machining by means of loose abrasives the wear process may also be twofold:

— by the wear of grain corners,

— by splitting and crushing, i.e. by breaking up of the grains.

Actually both wear processes involving abrasive and strength wear occur simultaneously. In most cases the strength type of wear distinctly prevails.

The main causes of the breaking up of abrasive grains are stresses exceeding the limits of the ultimate or fatigue strength of the grains. These stresses are created either by pressing (pressure method) or by striking (kinetic method). Therefore the process of grain wear by breaking up will principally depend on grain resistance to crushing, pressing forces or impact energy, on the quantity of abrasive in circulation and working time.

Grain resistance to crushing depends mainly on the type of abrasive material, on the size and shape of the grains.

A comparison of forces N destroying (crushing) grains of various materials is represented graphically in Fig. 14.8 [14/4]. The conclusion may be drawn from an analysis of this research that the harder is the abrasive material the greater the static force necessary to crush it.

Partly from Fig. 14.8 and also from Fig. 14.9 (graphically illustrating the results of research on grain crushing force depending on grain number) it also follows that the smaller are the grain dimensions, the smaller is the force necessary to crush the grains. It may be observed additionally that the smaller are the grains, the narrower is the scatter of the destroying force values.

It follows from research on grain shape that grains having an approximately spherical shape are more resistant to crushing (Fig. 14.10). On the other hand, oblong grains have a resistance to crushing equal to about one third of that of spherical grains.

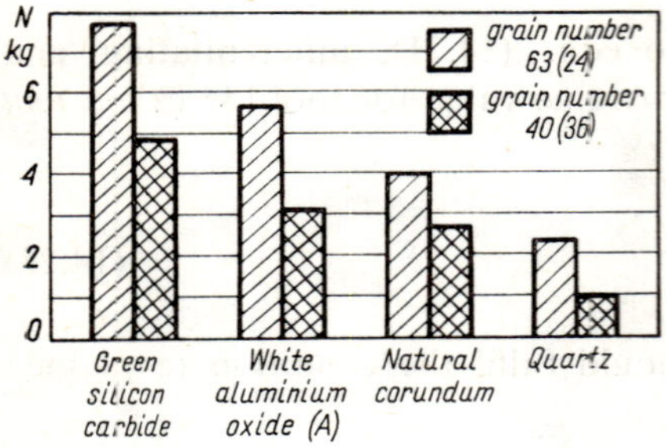

Fig. 14.8 Average value of the force crushing grains of various abrasives (acc. to data given by Zaitsev [14/4])

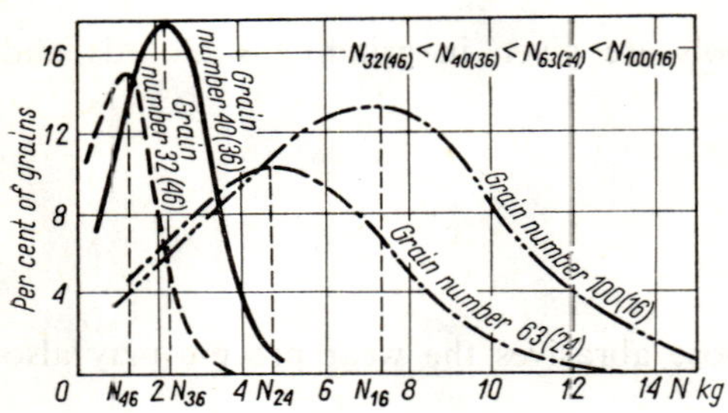

Fig. 14.9 Diagram of the dependence of force N crushing abrasive grains on grain number [14/4]

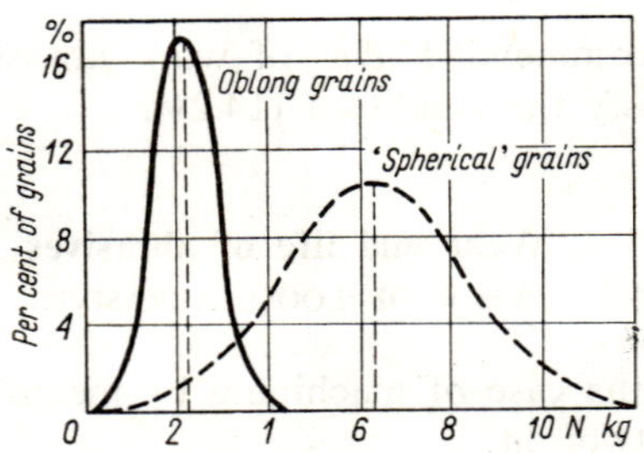

Fig. 14.10 Distribution of values of crushing force N for oblong and spherical grains of aluminium oxide [14/4]

It is true that the research results cited concern the crushing of abrasive grains by static loads, but nevertheless they give an indication of the role of abrasive strength in the breaking up of grains in machining work.

The pressing force or impact energy has a great importance in the breaking up process. For a sufficiently great pressure or impact speed the exceeding of ultimate strength occurs. The smaller is the grain load, the longer the grains work without being crushed. This indicates the fatigue character of the breaking up of abrasive grains.

This character is also confirmed by changes in the quantity of abrasive in the process. The larger is the quantity of abrasive used in the machining process, the slower the breaking up proceeds, since the probable frequency of loading of the same grain diminishes in a way inversely proportional to the amount of abrasive involved.

Photographs of aluminium oxide grains of numbers 40(36) and 10(120) in Fig. 14.11 indicate that the breaking up of grains increases with abrasive working time. It may be observed, too, that breaking up intensity is greater at the beginning of the process and then gradually decreases.

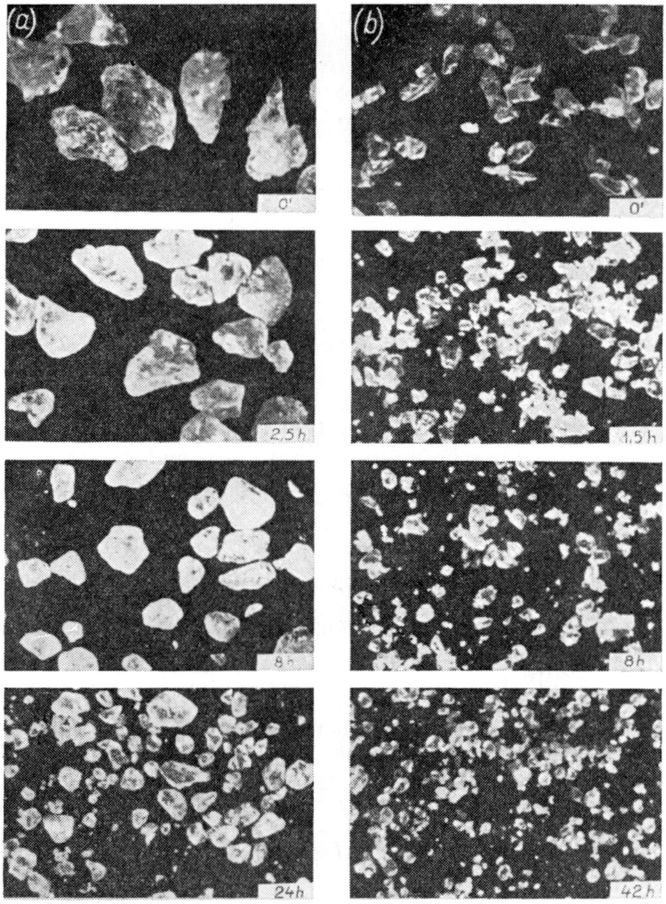

Fig. 14.11 Photographs of aluminium oxide grains broken up depending on working time: (a) grain number 40(36), (b) grain number 10(120)

14.4.2 Life of loose abrasive

To characterize the life period of abrasive material in machining by the pressure or kinetic method, the considerations of section 14.3.3 may be repeated.

This results from the fact that the breaking up of grit has a similar influence* on the rate of metal removal as grain wear. As a result, the relationship between metal removal rate Q_v and abrasive working time may be expressed by a formula of the same type as eqn. (14.15).

Research conducted on the influence of quantity of abrasive taking part in the machining process [14/2] led to the discovery of a relationship between the exponent

* Deviations from this rule are discussed in Chapter 18.

e and abrasive material quantity *M*. This is expressed by the empirical and statistical formula

$$e = \frac{u}{M^n} \tag{14.25}$$

where *u* and *n* were found in an empirical way.

Thus eqns. (14.20) and (14.23) can be complemented by the relationship contained in eqn. (14.25), or

$$T_e = \left(\frac{M^n}{u} - 1\right)\left(\frac{K_A}{w} + t_{te}\right) \tag{14.26}$$

$$T_w = \left(\frac{M^n}{u} - 1\right) t_{te} \tag{14.27}$$

It follows from these formulae that the more abrasive that takes part in the machining process the longer may be the maximum production life and economical life of this abrasive.

REFERENCES

14/1 BURNAT, L.: 'Szlifowanie i dogładzanie ścierne' (Grinding and Superfinishing), WNT, Warszawa 1962

14/2 KACZMAREK, J.: 'Ekonomicheskaya stoikost' abraziva pri ultrazvukovoi i struino-abrazivnoi obrabotke' (Economical abrasive life in ultrasonic impact grinding and abrasive-blast treatment), Vestnik Mashinostroeniya 6, 1954

14/3 VAKSER, D. B.: 'Vlianie geometrii abrazivnogo zerna na swoistva shlifoval'nogo kruga', Sbornik: 'Osnovnye voprosy vysokoproizvoditelnogo shlifovaniya' (Influence of abrasive grain geometry on grinding wheel properties, Collection: Basic Problems of High-output Grinding), Edited by MASLOV, E. N., Mashgiz, Moskva 1960

14/4 ZAITSEV, A. K.: 'Opredelenie prochnosti abrazivnykh zeren' (Determination of abrasive grain strength), Abrazivy, *1*, 16, 1956

CHAPTER 15

Grinding

15.1 Kinematic characteristics of grinding

15.1.1 ESSENTIALS OF THE CLASSIFICATION OF GRINDING TYPES

Grinding is an abrasive process in which the tool—called *grinding wheel*—carries out the primary basic motion, always a rotary one, while the auxiliary basic motion may be carried out either by the workpiece alone or by the grinding wheel alone or by both of them simultaneously.

Classification of grinding types and of their characteristic features is based on various criteria.

The criteria applied most frequently and the names of the corresponding types are listed below.

With regard to the shape of the work surfaces are distinguished:

— grinding of surfaces of revolution (shortly: rotary grinding) such as cylindrical, conical and curvilinear surfaces of revolution, either external (shafts) or internal (holes),
— grinding of other surfaces which may be plane or curvilinear. For technical reasons mainly external surfaces are considered here.

With regard to the method of clamping and driving the workpiece, within the range of rotary grinding are distinguished:

— cylindrical grinding, when the workpiece is held between the centres of spindle-head and tailstock or in a grinding fixture, and the drive of rotary motion is transmitted by means of a dog,
— centreless grinding, when the workpiece rests on or moves along a guide and the auxiliary motion, rotary or translatory is imparted by a control wheel.

With regard to the positioning of the feed direction in relation to the machined surface are distinguished:

— grinding with longitudinal feed, when at least one direction of a rectilinear component of feed is parallel to the machined surface,
— grinding with transverse feed, when the direction of rectilinear component of feed is perpendicular (or normal) to the machined surface.

When both the workpiece and grinding wheel participate in auxiliary motion, or just one of them, the following are distinguished:
— single feed when carried out by the grinding wheel only or by the workpiece only,
— composite or joint feed, when carried out simultaneously by the grinding wheel and workpiece.
With regard to the position of the grinding wheel in relation to the workpiece after the pass is completed, are distinguished:
— through-feed grinding, when the grinding wheel works with overrun,
— end-feed grinding, when the grinding wheel has no overrun.
With regard to the type of working surface of the grinding wheel are distinguished (similarly as in the case of milling):
— plain grinding, when the working surface consists exclusively or almost entirely of the cylindrical surface or periphery of the grinding wheel,
— face grinding, when the working surface consists exclusively or almost entirely of the grinding wheel face or side,
— plain-and-face grinding when the working surface of the grinding wheel consists of both the cylindrical surface and face. A particular case of this type of grinding may be the so-called deep-feed grinding, i.e. grinding at both great cutting depth and high feed rate.
With regard to the application of grinding are distinguished:
— machine part grinding, or so-called production grinding in which are distinguished additionally:
  — fettling, for the purpose of removing the damaged layer in preceding operations such as casting crust, mill or forge scale, welds etc.,
  — dimensional and form grinding for the purpose of imparting the appropriate shape and dimensions within given tolerances,
  — sharpening or grinding of the working surface of cutting tools. It has a number of specific properties differing from those of production grinding.
Most criteria listed above are not of an interdependent character. Thus, from the theoretical point of view, many types of grinding may result from combining the particular features.

In workshop practice not all possible combinations are applied. The most frequently used are those contained in Table 15.1.

Some of these types of grinding are illustrated by diagrams in Fig. 15.1 to 15.14. Besides the abbreviated name, each caption contains a classification symbol indicating all features listed in Table 15.1.

For example symbol AGR/ICXN denotes: grinding (AG) of surfaces of revolution (R), internal (I), between centres (C), with longitudinal (X) end (N) feed.

15.1.2 Characteristics of some types of grinding

15.1.2.1 *Cylindrical longitudinal grinding of external surfaces of revolution*

In the case of *cylindrical grinding* of external surfaces (Fig. 15.1) the auxiliary motion is composed of:
— the rotary auxiliary motion of the workpiece whose rotational speed is denoted by n_w and peripheral speed by v_w;

Table 15.1 *Classification of practically applied types and methods of grinding*

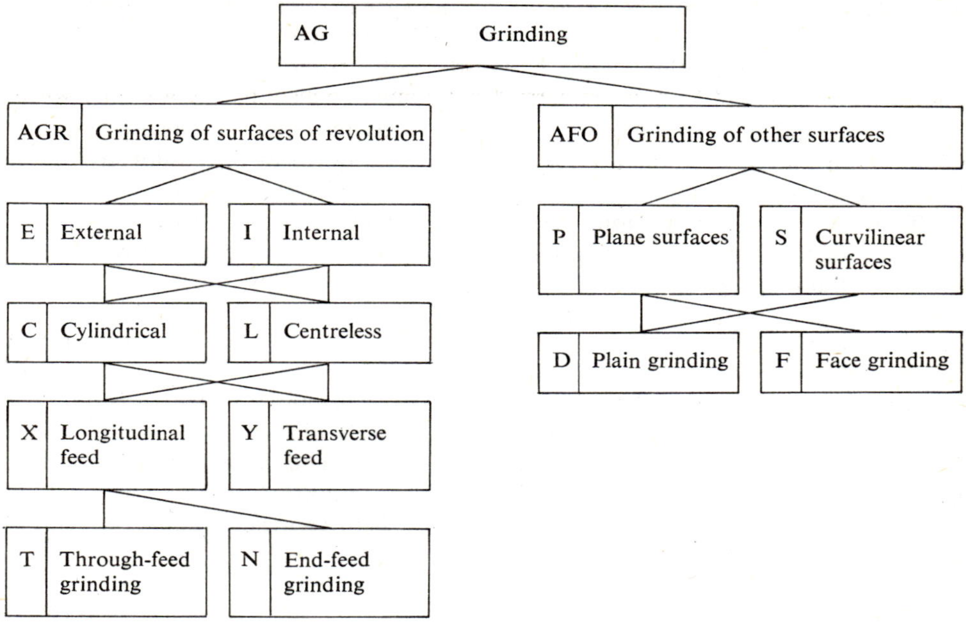

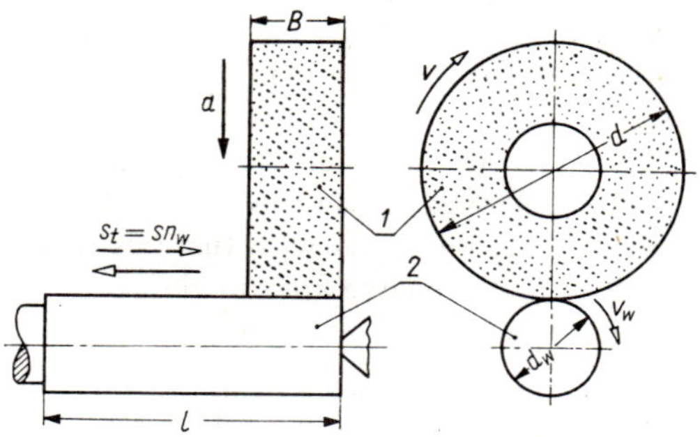

Fig. 15.1 Grinding of a cylinder between centres (AGR/ECXT)

— the rectilinear auxiliary motion which may be carried out by the workpiece (more frequently) or by the grinding wheel. This motion is parallel to the machined surface and is called the longitudinal motion.

The longitudinal motion is measured:

— by workpiece travel per revolution, in mm/rev, or feed per revolution (feed s),

— by feed rate (feed per minute) in mm/min, determined in the direction of motion (feed s_t).

The setting of the depth of cut is called *tool advance* a_t. It is also called inaccurately—in the technical slang—transverse feed. The tool advance always exceeds the depth of cut by a certain value Δa depending on the rigidity of m–w–t system, or

$$a = a_t - \Delta a \tag{15.1}$$

This setting of the depth of cut or tool advance is carried out before each pass.

Primary motion speed, or cutting speed, is given in m/s, as its values in m/min would be relatively high. The range of cutting speeds applied in grinding extends from 700 to some 4800 m/min or from about 12 to about 80 m/s.

Rotary auxiliary motion speeds within the range from a few to several tens of metres per minute are applied.

Longitudinal auxiliary motion speeds are determined depending on grinding wheel width and surface quality requirements. The range of feed rates applied is between 0·2 and 0·9 of grinding wheel width per workpiece revolution.

Transverse auxiliary motion speeds are analogous to the depths of cut, i.e. of the order of micrometers or hundredths of a millimetre per workpiece revolution.

Longitudinal grinding time between centres is expressed by

$$t_m = i\frac{l_a+l+l_o}{s_t} \tag{15.2}$$

where: i—number of passes, l_a—length of approach in mm (this length is adopted within limits of 2–10 mm; for wider grinding wheels higher l_a values are taken), l_o—length of overrun in mm (in end-feed grinding $l_o = 0$ and in through-feed grinding $l_o = (1{\cdot}1\text{–}1{\cdot}5)B$, where B—grinding wheel width in mm).

The number of passes (a pass is understood here to mean a single travel, i.e. a stroke comprises two passes) is determined by

$$i = \frac{q}{a}+i_o \tag{15.3}$$

where: q—grinding allowance in mm, a—grinding depth in mm, i_o—number of spark-out passes which even up surface irregularities and compensate elastic deformations of the workpiece and m–w–t system.

The grinding of external cones (Fig. 15.2) has the same features and properties as longitudinal grinding of cylinders, the only difference being that the direction of longitudinal feed, parallel to the machined surface, intersects with the axis of cone rotation under an angle whose value is equal to half the cone apex angle.

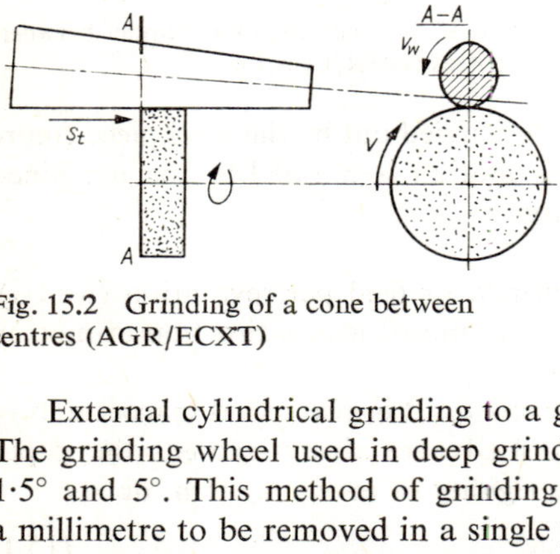

Fig. 15.2 Grinding of a cone between centres (AGR/ECXT)

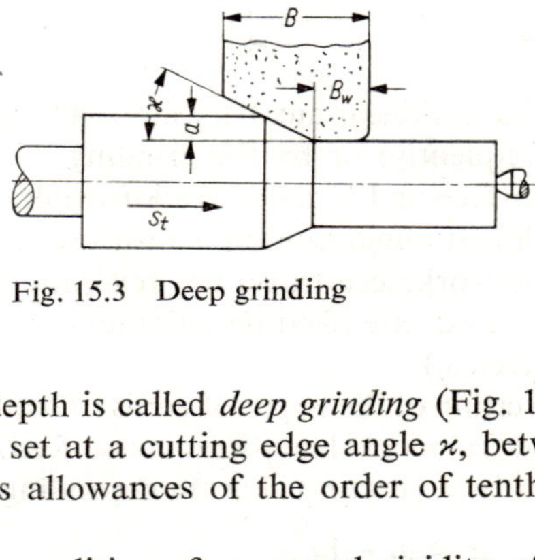

Fig. 15.3 Deep grinding

External cylindrical grinding to a great depth is called *deep grinding* (Fig. 15.3). The grinding wheel used in deep grinding is set at a cutting edge angle $\varkappa$, between 1·5° and 5°. This method of grinding allows allowances of the order of tenths of a millimetre to be removed in a single pass.

Deep grinding may be applied under the condition of very good rigidity of the m–w–t system, as cutting forces are correspondingly greater in this case. In practice, deep grinding is used in machining materials having a low cutting resistance (brass, bronze, cast iron) to average accuracy and surface quality requirements.

15.1.2.2 *Cylindrical transverse grinding of external surfaces of revolution*

In the case of *transverse grinding*, also called *plunge grinding*, the rectilinear feed is directed perpendicularly to the machined surface (Fig. 15.4). Feed per revolution (s) is in this case equal to tool advance a_t per revolution.

Within the field of transverse grinding are distinguished

— grinding of protruding surfaces (as in Fig. 15.4); the grinding wheel width B must then be greater than the length of machined surface, i.e. $B > l$;

— grooving and parting, when $B = l$.

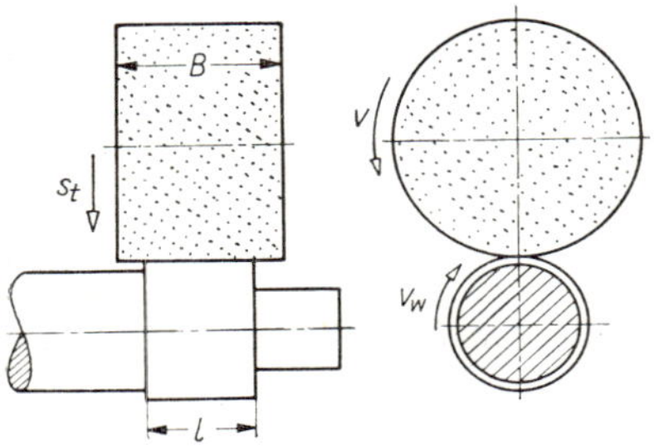

Fig. 15.4 Cylindrical transverse (plunge) grinding of a cylinder (AGR/ECY)

Transverse grinding time is calculated from

$$t_m = \frac{l_a+q}{s_t} = \frac{l_a+q}{n_w s} \quad \text{min} \tag{15.4}$$

where n_w—rotational speed of the workpiece in rev/min.

The allowance q in the case of parting equals half the diameter of the workpiece being parted. Grinding wheels of large diameter and small width are used for this purpose.

15.1.2.3 *Centreless longitudinal grinding of external surfaces*

Centreless longitudinal grinding is depicted in Fig. 15.5. The workpiece, supported by guide 1, is friction-driven by the driving disc 3. On the opposite side the workpiece is in contact with the grinding wheel 2 which performs the grinding work.

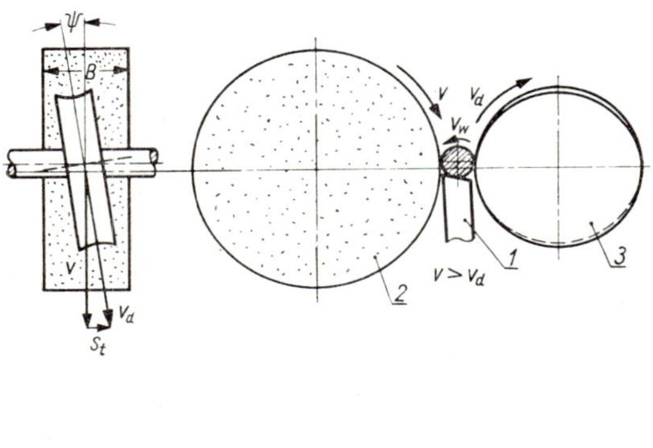

Fig. 15.5 Centreless longitudinal grinding of a cylinder (AGR/ELXT)

In the case when the driving wheel is set under a certain angle ψ in relation to the plane of grinding wheel rotation and not perpendicular to the axis of workpiece rotation, then at the point of contact with the workpiece appear: peripheral component v_{wt} perpendicular to the axis of workpiece rotation and component s_{tt} parallel to this axis.

When slipping on the friction surface is neglected, it may be assumed that the peripheral component is equal to the peripheral speed of the workpiece, or

$$v_{wt} \approx v_w \tag{15.5}$$

and, analogously,

$$s_{tt} \approx s_t \tag{15.6}$$

In order to increase the friction force and reduce the possibility of slip, the driving disc is specially formed so that the friction surface may contact the workpiece along its entire width.

The rate of longitudinal feed s_t is calculated—taking the slip into consideration—from

$$s_f = e_f v_d \sin\psi \quad \text{mm/min} \tag{15.7}$$

where: v_d—peripheral speed of the driving wheel (disc) in mm/min, e_f—correction coefficient taking slip into account.

Coefficient e_f is determined by

$$e_f = 1 - \frac{f}{100} \tag{15.8}$$

where f—slippage in per cent.

The peripheral speed of the workpiece is calculated—still considering the slippage—according to

$$v_w = e_f v_d \cos\psi \quad \text{mm/min} \tag{15.9}$$

The value of coefficient e_f changes depending on driving wheel material, the shape of the friction surface and the accuracy of its contact with the workpiece, diameter ratio, workpiece size, guide type and material etc. It also depends on the driving wheel inclination angle in a way illustrated graphically in Fig. 15.6. It is seen there that feed rates are lower by several per cent than those calculated without taking slip into consideration.

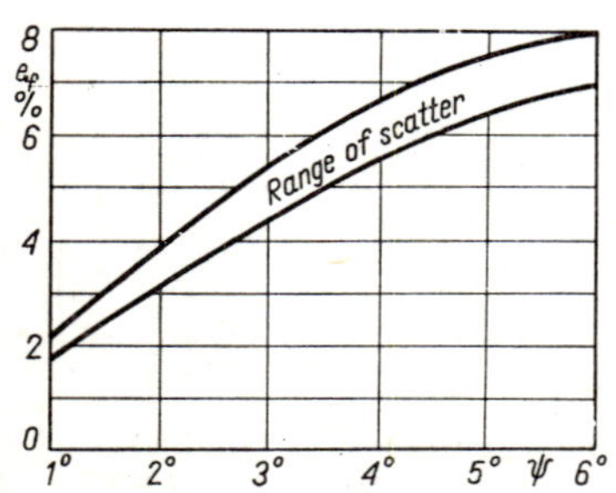

Fig. 15.6 Dependence of slippage value in percent on driving wheel inclination angle ψ (acc. to data from [15/1]

Equations (15.7) and (15.9) are valid if there is full slippage between the grinding wheel and work, i.e. when the grinding wheel performs the cutting work with a moment not exceeding the total of the moments of friction between workpiece and guide and between workpiece and driving wheel, as determined by the inequality

$$F\frac{d}{2} \leqq f_g N_g \frac{d_w}{2} + f_d N_d \frac{d_d}{2} \tag{15.10}$$

where: F—peripheral grinding force in kg, d—grinding wheel diameter in mm, f_g—coefficient of friction between guide and workpiece, N_g—load normal to guide surface (as a result of wheel pressure and workpiece weight) in kg, d_w—workpiece diameter, f_d—coefficient of friction between driving wheel and workpiece, N_d—load normal to the surface of contact between workpiece and driving wheel in kg, d_d—driving wheel diameter in mm.

Thus it follows from eqn. (15.10) that—for the workpiece not to be driven by the grinding wheel and for the eqns. (15.7) and (15.9) to be valid—the peripheral grinding force should be determined by the relationship

$$F \leqq \frac{d_w}{d} f_g N_g + \frac{d_d}{d} f_d N_d \tag{15.11}$$

Considering the possibilities available to increase the right hand side of this inequality the following statements can be made:

— the value of d_w is an independent factor as it results from design requirements and cannot be used to increase forces counterbalancing the peripheral grinding force,
— increasing f_g and N_g is not recommended, as then the resistance to longitudinal feed rises proportionally,
— increasing of diameter d_d is limited by the condition not to exceed the appropriate longitudinal feed rates and the condition that cutting speed should be higher than the peripheral speed of driving wheel,
— increasing N_d is purposeless, since it causes an increase of the thrust force and depth of cut and, consequently, of the peripheral cutting force F,
— on the other hand, increasing of the coefficient of friction f_d is advisable. The higher this coefficient, the greater peripheral cutting force F is permissible.

Centreless longitudinal through-feed grinding time is calculated according to

$$t_n = \frac{l_a + l + l_o}{e_f v_d \sin\psi} \tag{15.12}$$

Centreless longitudinal through-feed grinding finds economical application in batch production and quantity production, in grinding of surfaces of revolution without any protrusions, such as stepless shafts, bushes, rings etc. This type of grinding is also useful in machining out-of-round workpieces if only their peripheries have common tangents with peripheries of the driving and grinding wheels in all angular positions (for a given distance between the axes of the wheels).

Centreless longitudinal grinding is also used in end-feed operations (performed to a stop), in the machining of stepped shafts, neck bushes etc.

Cutting conditions applied in practice are as follows:

— depth of cut from 0·002 to 0·2 mm,

— longitudinal feed rate s_t from 150 mm/min (for large workpiece diameters) to about 10 000 mm/min (for very small diameters),
— feed rate within the same range as in the case of cylindrical grinding.

15.1.2.4 *Centreless transverse grinding of external surfaces*

The condition of *centreless transverse grinding* is that $B > l$ (as in the case of centre-type grinding).

In this case the driving wheel serves only to impart the peripheral speed v_w and thus its inclination angle equals zero. Grinding of a cylindrical surface is depicted in Fig. 15.7.

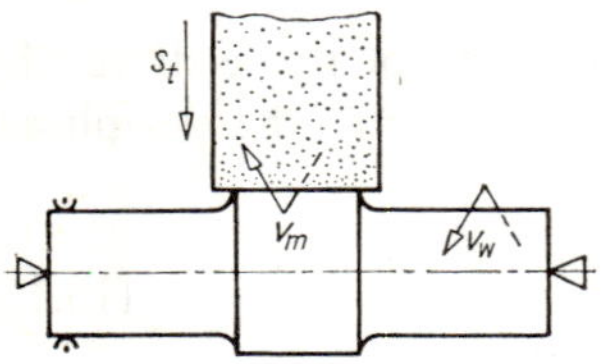

Fig. 15.7 Centreless transverse grinding (plunge grinding) of a cylindrical shaft

Centreless transverse grinding is often applied in the machining of formed surfaces of revolution.

Cutting conditions in this type of grinding are similar to those of centreless grinding with longitudinal feed. The only difference is that transverse feed rate, equalling tool advance, does not exceed 0·05 mm/rev.

Grinding time is calculated according to eqn. (15.4).

15.1.2.5 *Internal grinding*

Internal grinding may be carried out according to principles similar to those applied in cylindrical and centreless external grinding with longitudinal or transverse feed.

Internal grinding with longitudinal helical feed is depicted in Fig. 15.8a.

In this method of internal grinding the workpiece rotates at a peripheral speed v_w, and the grinding wheel performs the primary motion and longitudinal feed motion at the rate s_t. The depth of cut is set by advancing the wheel perpendicularly towards the work axis.

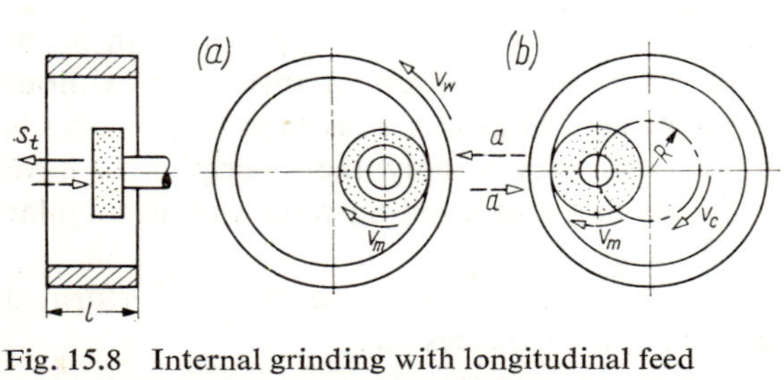

Fig. 15.8 Internal grinding with longitudinal feed (AGR/ICXT): (a) helical, (b) planetary

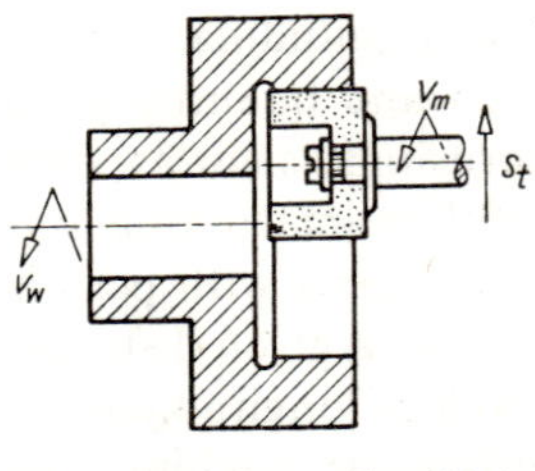

Fig. 15.9 Example of internal grinding with transverse feed (AGR/ICY)

The auxiliary feed motion results here from rotary motion (at a peripheral speed v_w) and rectilinear motion (at a speed s_t), as in the case of external grinding.

Another method of internal grinding is shown in Fig. 15.8b, viz. so-called *planetary grinding*. In this case the workpiece motion is rectilinear and the grinding wheel performs two rotary motions: about its own axis (primary motion) at a peripheral speed v_m and about the axis of the hole being machined, on a circle of radius R, at a peripheral speed of the grinding wheel centre v_c.

On account of kinematic similarity, the grinding time for the grinding methods shown in Fig. 15.8 is calculated according to eqn. (15.2) or in the same way as for longitudinal external grinding.

Figure 15.9 shows one of the cases of internal grinding with transverse feed. The workpiece carries out rotary auxiliary motion and the grinding wheel the primary motion and rectilinear auxiliary motion, perpendicular to machined surface. By this method conical internal surfaces may also be ground.

Internal grinding with transverse feed may also be performed by the planetary method, i.e. when the primary motion and both auxiliary motions are carried out by the grinding wheel.

Grinding time for internal grinding with transverse feed is calculated according to eqn. (15.4).

Centreless internal grinding with longitudinal feed is depicted in Fig. 15.10. The workpiece 3 rests on two discs 1, which play the role of guides with adjustable distance between axes. The workpiece is pressed to guides 1 and driven by the driving wheel 4, while the grinding wheel 2 is in contact with the hole surface and performs the cutting work.

Grinding time for this case is calculated according to eqn. (15.2).

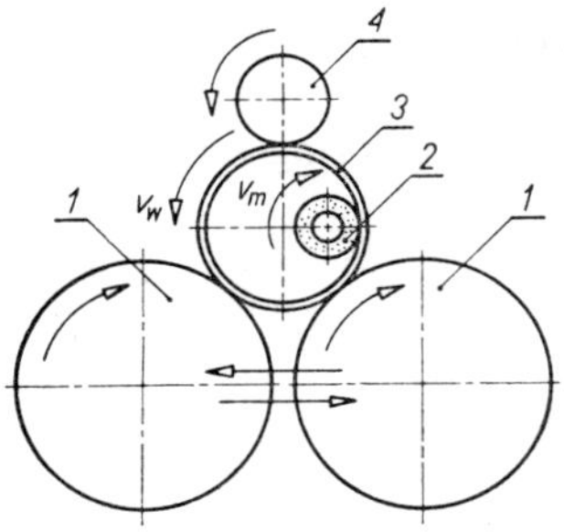

Fig. 15.10 Centreless internal grinding with longitudinal feed: 1—adjusting discs, 2—grinding wheel, 3—workpiece (bush), 4—driving disc (AGR/ILXT)

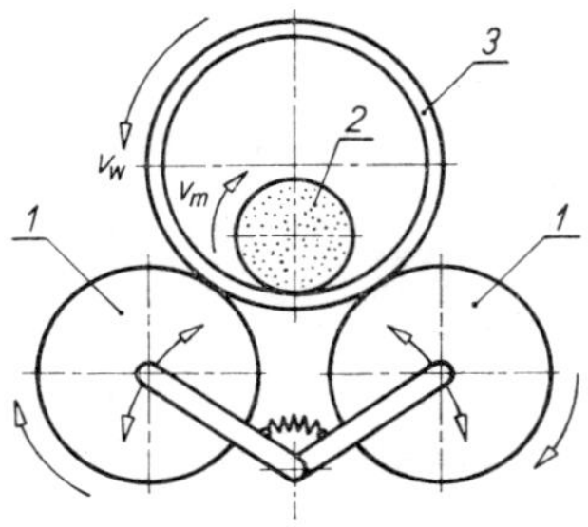

Fig. 15.11 Centreless internal grinding with transverse feed: 1—adjusting discs, 2—grinding wheel, 3—workpiece (AGR/ILY)

Figure 15.11 illustrates diagrammatically centreless internal grinding with transverse feed. Workpiece 3 rests on discs 1 with the distance between axes suitably adjustable to workpiece diameter. Rotary motion of speed v_w is imparted to the workpiece by one of the adjustable wheels playing the role of the driving wheel. Grinding wheel 2 is inside the workpiece, in contact with its internal surface, and carries out the cutting work at a speed v_m.

Grinding time for this type of grinding is calculated from eqn. (15.4).

Cutting conditions for internal grinding should be reduced by 20–40% relative to those for external grinding.

15.1.2.6 *Surface grinding*

Plain (*peripheral*) and *face grinding* of surfaces other than surfaces of revolution, including plain surfaces, may be performed according to the kinematic configuration of milling and vertical turning.

The configuration of surface grinding on machine tools with rectilinear table feed (analogy to milling machines) is shown in Fig. 15.12.

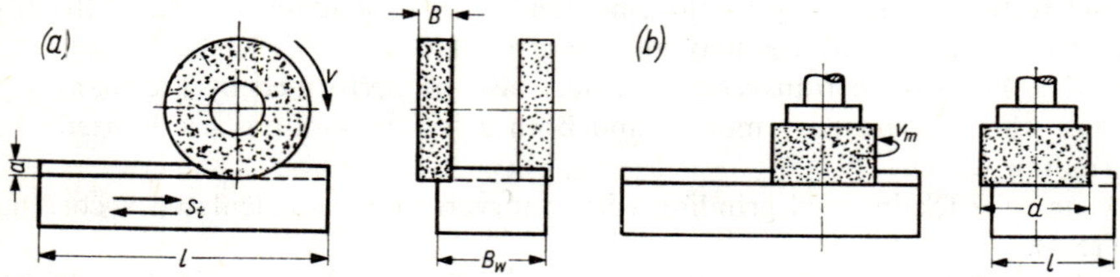

Fig. 15.12 Surface grinding with rectilinear reciprocating feed: (a) plain (periphery) grinding (AGO/PD), (b) face grinding (AGO/PF)

In this case the workpiece carries out rectilinear reciprocating motions constituting rectilinear feed at the rate s_t. The grinding wheel performs the primary motion and rectilinear transverse feed s_b. The latter may be step-wise (after each stroke) or continuous, in the direction of workpiece width B_w.

Such motions are carried out both in the case of plain grinding (Fig. 15.12a) and of face grinding (Fig. 15.12b).

Figure 15.13 shows surface grinding on grinding machines working on the same

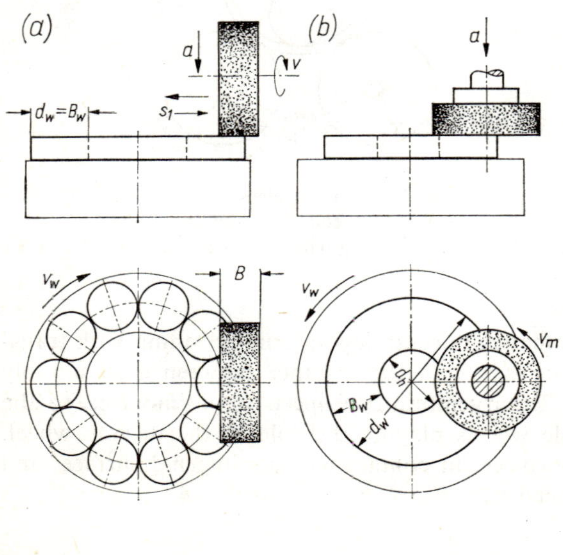

Fig. 15.13 Surface grinding on a rotating table: (a) plain grinding, (b) face grinding

principle as vertical lathes, i.e. instead of transverse feed there is rectilinear centripetal or centrifugal feed. This feed is unnecessary in the following cases, when:
— in plain grinding the width of wheel periphery is larger than workpiece width,
— in face grinding the grinding wheel diameter is larger than workpiece width.
Also, curvilinear grinding of surfaces other than surfaces of revolution or form grinding by means of a special profiled grinding wheel proceeds according to kinematics just described. An example of such grinding is shown diagrammatically in Fig. 15.14.

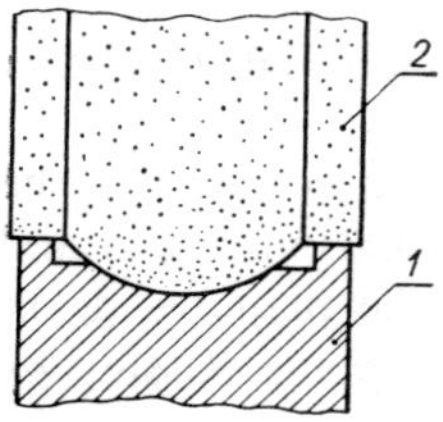

Fig. 15.14 Curvilinear (form) grinding: 1—workpiece, 2—grinding wheel

When grinding surfaces other than surfaces of revolution, including plane surfaces, the following formulae are used for the calculation of grinding time:
— in plain grinding on grinding machines with rectilinear reciprocal auxiliary motion

$$t_m = i\frac{l_a+l+l_o}{s_t}\frac{B_w}{B} \quad \text{min} \tag{15.13}$$

where: B_w—workpiece width in mm, B—grinding wheel width in mm, i—number of passes, and other denotations as before;
— in face grinding on grinding machines with rectilinear reciprocal motion

$$t_m = i\frac{C\,d+l}{s_t}\frac{B_w}{d} \quad \text{min} \tag{15.14}$$

while the value of coefficient C in this and the following formulae is adopted within limits $C = 1{\cdot}1\text{–}1{\cdot}2$; the ratio of workpiece width to grinding wheel diameter B_w/d is generally not a whole number and therefore, in calculations, is rounded off upwards to the nearest whole number;
— in plain grinding on grinding machines with rotating table

$$t_m = i\frac{C\,B+B_w}{s_t} \quad \text{min} \tag{15.15}$$

— in face grinding on grinding machines with rotating table

$$t_b = i\frac{C\,d+B_w}{s_t} \quad \text{min} \tag{15.16}$$

15.1.2.7 *Special applications of grinding*

Special applications of grinding comprise such cases as tool sharpening, gear machining etc.

15.2 Undeformed chip and its elements

15.2.1 Characteristic of cutting by a single grain

15.2.1.1 *Elements of undeformed chip*

As already mentioned in Chapter 14, corners of abrasive grains may be considered as multisided pyramids with a spherical top of radius r.

Thus the corner, being a tool point, will be characterized in a determined symmetrical cross-section:

— by top rounding radius r, shortly: top radius,

— by included angle 2ε.

In further considerations it will be assumed, for the sake of simplicity, that the corner has the form of a cone with a spherical top.

In an arbitrary case the axis of symmetry of an abrasive grain corner may be inclined in relation to the reference system (consisting of grain rotation radius and two straight lines perpendicular to it, one of them situated in the plane of grain rotation) by angles: η—in the plane of grain rotation, λ—in the plane passing by the grain rotation radius and perpendicular to rotation plane.

If it is assumed that the depth of grinding by the given grain will be a_z and the corresponding feed s_z, then the undeformed chip cut by this grain will have a variable thickness h_z.

In Fig. 15.15 is shown the longitudinal cross-section of an undeformed chip cut by a single grain and the corresponding parameters in the plane of grain rotation.

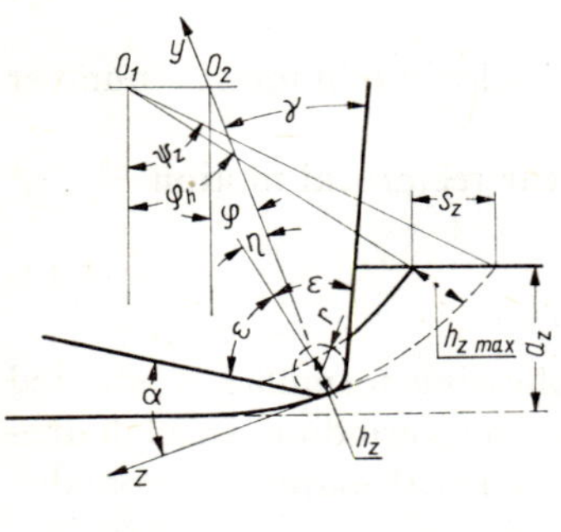

Fig. 15.15 Underformed chip cut by a single grain

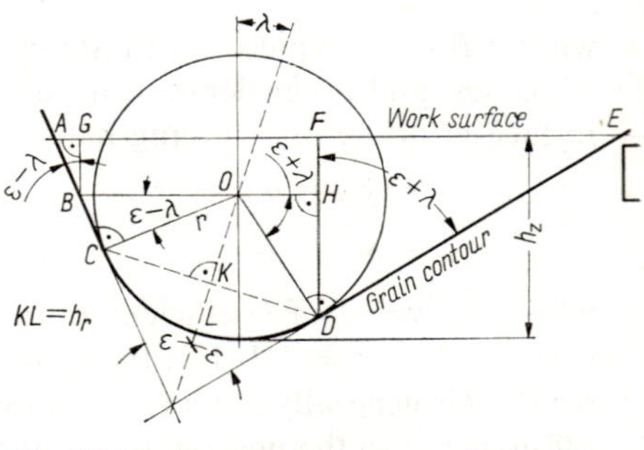

Fig. 15.16 Transverse cross-section of an undeformed chip cut by a single grain

Similarly as in the case of milling, it may be assumed with sufficient accuracy* that the undeformed chip thickness is determined by the relationship

$$h_z = s_z \sin\varphi \tag{15.17}$$

where φ—angle of grain rotation.

The maximum thickness of the undeformed chip corresponds to angle φ_h, approximated as $\varphi_h = \psi$, while ψ is the angle of contact between tool point and workpiece material, also called the *tool point working angle*. The mean value of the thickness

* The approximation consists here, among others, in neglecting the influence of the inclination of corner symmetry axis at angles η and λ on the value of undeformed chip thickness.

of the undeformed chip cut by a single grain h_{zm} is defined as half the maximum thickness ($h_{zm} = 0{\cdot}5h_{z\,\max}$).

In order to determine the transverse cross-section of the undeformed chip, the latter is cut along y–y, i.e. both the grain and material is cut by the plane y–x (where x is perpendicular to y and z).

Such a cross-section is presented in Fig. 15.16. It follows from this figure that if the axis of symmetry is inclined in this plane at the angle λ, then the area of transverse cross-section of the undeformed chip cut by a single grain will be equal to that of plane figure *AEDCA*. This area will be the sum of areas:

— of triangles *DFE* and *BGA*,
— of triangles *BCO* and *DHO*,
— of rectangle *GFHB*,
— and of circular sector *ODC*.

With the use of relationships shown in Fig. 15.16 and on the assumption that the influence of η is negligible, the formula for the area of transverse cross-section of undeformed chip cut by a single grain is found as

$$A_z = \frac{1}{2}\tan(\varepsilon+\lambda)\{h_z - r[1-\sin(\varepsilon+\lambda)]\}^2 + \\ + r(h_z - r)\left[\frac{1}{\cos(\varepsilon+\lambda)} + \cos(\varepsilon+\lambda)\right] + (h_z - r)^2\frac{\tan(\varepsilon-\lambda)}{2} + \\ + r^2\left[\frac{\sin 2(\varepsilon+\lambda)}{4} + \pi\left(\frac{1}{2} - \frac{\varepsilon}{180^\circ}\right) + \frac{\tan(\varepsilon-\lambda)}{2}\right] \tag{15.18}$$

In the case when the grain is positioned symmetrically, i.e. when $\lambda = 0$, the following formula is obtained

$$A_z = \left[h_z - r\left(1 - \frac{1}{\sin\varepsilon}\right)\right]^2 \tan\varepsilon - r^2\left(\cot\varepsilon - \pi\frac{90^\circ - \varepsilon}{180^\circ}\right) \tag{15.19}$$

It follows from Fig. 15.16 and eqns. (15.18) and (15.19) that:

— when the thickness of the undeformed chip cut by a single grain increases, so does the transverse cross-section A_z,
— when the included angle ε of the grain increases, so also does the undeformed chip cross-section A_z,
— when top rounding radius r increases, so also does the transverse cross-section A_z.

Values r and ε generally vary within narrow limits, while the thickness of cutting by abrasive grain may vary within wide limits. It may thus be considered that the value of A_z decidedly influences h_z.

For this reason all factors affecting the value of h_z, i.e. s_z and $\varphi(\psi)$, indirectly influence the transverse cross-sectional area of the undeformed chip.

The angle λ which may vary within limits from 0° to $\pm(90° - \varepsilon)$ insignificantly influences the value of transverse cross-sectional area A_z. Therefore, for all practical purposes, the value of angle λ may be neglected and approximate calculations carried out according to eqn. (15.19).

With the use of Fig. 15.16, formula* for the width of the undeformed chip cut by a single grain can be derived:
— in the case when $\lambda \neq 0$

$$b_z = 2\left[\frac{h_z \sin\varepsilon \cos\varepsilon + r(\cos\varepsilon \cos\lambda - \sin\varepsilon \cos\varepsilon)}{(\cos\varepsilon \cos\lambda)^2 - (\sin\varepsilon \sin\lambda)^2}\right] \tag{15.20}$$

$$b_z = \frac{(h_z - r)\sin 2\varepsilon + 2r \cos\varepsilon \cos\lambda}{\cos^2\varepsilon \cos^2\lambda - \sin^2\lambda \sin^2\varepsilon} \tag{15.20a}$$

— in the case when $\lambda = 0$

$$b_z = 2\left[h_z \tan\varepsilon + r\left(\frac{1}{\cos\varepsilon} - \tan\varepsilon\right)\right] \tag{15.21}$$

It is evident from eqns. (15.20) and (15.21) that:
— when h_z and r increase, the value of b_z rises proportionately,
— when ε and λ increase, so does the value of b_z.
In the case when $\lambda = 0$ and $\varepsilon = 45°$, i.e. when the included angle $2\varepsilon = 90°$, the value of b_z is expressed by

$$b_z = 2(h_z + 0{\cdot}41r) \tag{15.22}$$

which means that for $h_z = r$, $b_z = 2{\cdot}82r$.

In similar conditions, i.e. when $\lambda = 0$ and $\varepsilon = 45°$, the value A_z [according to eqn. (15.19)] will be

$$A_z = (h_z + 0{\cdot}41r)^2 - 0{\cdot}215r^2 \tag{15.23}$$

or, when $h_z = r$, then $A_z = 1{\cdot}785r^2$.

15.2.1.2 *Point working angle and length of point working travel in different types of grinding*

In the case of surface grinding, equations involving the parameters of motion of the top of the given point (Fig. 15.17) may be written in the following form:

$$x = 0{\cdot}5d \sin\varphi \pm s_\varphi \tag{15.24}$$

$$y = 0{\cdot}5d(1 - \cos\varphi)$$

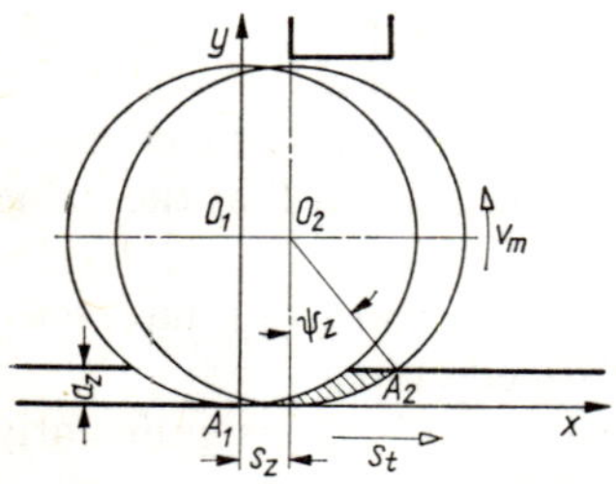

Fig. 15.17 Point working angle in the case of surface grinding

* It is evident from Fig. 15.16 that $b_z = AE = AG + BO + OH + FE$. Sections adding up to this total are found from right-angled triangles seen in the drawing.

where: d—diameter of point rotation circle, φ—rotation angle, s_φ—workpiece travel during point rotation through angle φ.

In order to determine the length of point travel L_ψ along the workpiece, $\varphi = \psi$ and $s_\varphi = s_{z\psi}$ should be assumed, then

$$\frac{s_\psi}{s} = \frac{\psi}{2\pi}; \qquad s_\psi = \frac{s}{2\pi}\psi \quad \text{mm} \tag{15.25}$$

where s—workpiece travel during one revolution of grinding wheel, in mm.

If the rate of workpiece feed is s_t mm/min, then

$$s = \frac{s_t}{n} = \pi d \frac{s_t}{60 v_m} \quad \text{mm} \tag{15.26}$$

where: n—rotational speed of grinding wheel in rev/s, v_m—peripheral speed of the grain in m/s, d—grinding wheel diameter in m.

The substitution of eqns. (15.25) and (15.26) and of the equation $\varphi = \psi$ into eqn. (15.24) yields:

$$x = 0{\cdot}5d\left(\sin\psi \pm \frac{s_t}{60 v_m}\psi\right)$$
$$y = 0{\cdot}5d(1-\cos\psi) \quad \text{mm} \tag{15.27}$$

As ψ is very small (usually $< 1°$), it may be assumed that

$$\psi \approx \sin\psi_z = \sqrt{1-\cos^2\psi_z} \tag{15.28}$$

Angle ψ_z may be determined from the formula for the non-right-angled triangle O_2CO_1 (Fig. 15.18)

$$\sin\frac{\psi_z}{2} = \sqrt{\frac{(p-b)(p-c)}{b\,c}}$$

where p—half the periphery, determined by

$$p = \frac{d+d_w}{2} - \frac{a}{2}$$

and the sides b and c are

$$b = \frac{d}{2} \qquad e = \frac{d+d_w}{2} - a$$

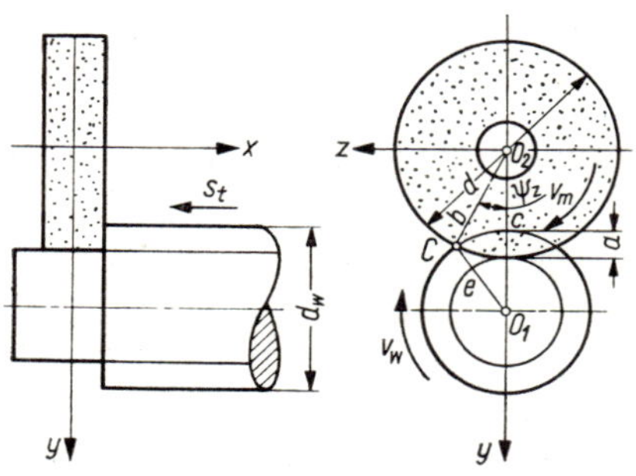

Fig. 15.18 Point working angle in the case of cylindrical grinding

For small angles it may be assumed that $\sin\varphi_z = 2\sin(\psi_z/2)$ and thus

$$\sin\varphi_z = 2\sqrt{\frac{a(d_w-a)}{d(d+d_w-2a)}}$$

Omitting the depth of cut a as negligible in comparison with the workpiece and grinding wheel diameters yields

$$\sin\psi_z = 2\sqrt{\frac{a\,d_w}{d(d+d_w)}} \tag{15.28a}$$

In surface grinding $d_w = \infty$, or

$$\sin\psi_z \approx 2\sqrt{a/d} \tag{15.28b}$$

The length of point working travel (or length of point contact) L_ψ may be calculated by integrating the formula

$$\mathrm{d}L_\psi = \mathrm{d}x^2+\mathrm{d}y^2 \tag{15.29}$$

On the assumption that ψ is very small ($\psi < 1°$) the integration yields

$$L_\psi \approx \left(1\pm\frac{s_t}{60v_m}\right)\sqrt{d\,a_z} \quad \text{mm} \tag{15.30}$$

In the case of grinding surfaces of revolution the equations of point motion are as follows (Fig. 15.18)

$$z = 0{\cdot}5d\sin\psi \pm s_\psi$$

$$y = 0{\cdot}5d(1-\cos\psi) \tag{15.31}$$

$$x = \frac{s}{2n}\psi_s$$

where ψ_w—workpiece working angle (angle of workpiece rotation during the time of grinding wheel rotation through angle ψ).

Using similar methods as in the case of surface grinding and taking into account that

$$\psi_w = \frac{v_w}{60v_m}\frac{d}{d_w}\psi \quad \text{rad} \tag{15.32}$$

where [see eqn. (15.28)]

$$\psi = 2\sqrt{\frac{d_w\,a_z}{d(d+d_w)}} \quad \text{rad} \tag{15.33}$$

yields, after suitable calculations, the following formula for the length of point working travel along the workpiece

$$L_\psi = \sqrt{\left(1\pm\frac{v_w}{60v_m}\right)^2+\left(\frac{s_t}{60v_m}\right)^2}\sqrt{\frac{d\,d_w\,a_z}{d+d_w}} \quad \text{mm} \tag{15.34}$$

The '+' sign in the term $1\pm(v_w/60v_m)$ corresponds to conventional grinding, and the '−' sign to climb-cut grinding. The term under the first radical sign illustrates the influence of rotary and feed motion of the workpiece on the length of working travel.

If the point working travel L_ψ is to be obtained in mm, v_w and s_t should be expressed [in eqn. (15.34)] in m/min, v_m in m/s, and d, d_w and a in mm.

As an example, Fig. 15.19 gives the length of point working travel L_ψ in external grinding of surfaces of revolution with longitudinal feed. In this drawing the influence of rotary and feed motions of the workpiece upon working length is neglected. The error thus committed does not exceed 1% of L_ψ.

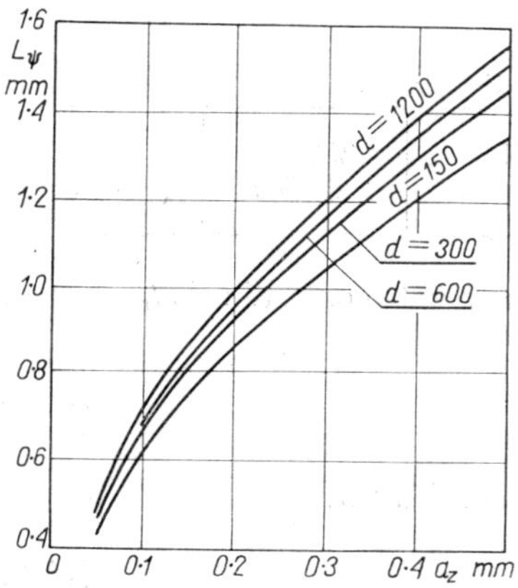

Fig. 15.19 Dependence of working travel of the grain grinding depth in the grinding of cylinders using grinding wheels of various diameters (for d_w = 50 mm)

In the case of internal grinding of surfaces of revolution with longitudinal feed, the following formulae may be obtained

$$\psi_z = 2\sqrt{\frac{D\,a}{d(D-d)}} \quad \text{rad} \tag{15.35}$$

where: D—hole diameter in mm, a—grinding depth in mm, d—grinding wheel diameter in mm

$$L_\psi = \sqrt{\left(1 \pm \frac{v_w}{60v_m}\right)^2 + \left(\frac{s_t}{60v_m}\right)^2}\sqrt{\frac{D\,d\,a_z}{D-d}} \quad \text{mm} \tag{15.36}$$

Workpiece working angle ψ_w in this case of grinding amounts to

$$\psi_w = \frac{d}{D}\frac{v_w}{50v}\psi_z \tag{15.37}$$

In practice, eqn. (15.36) is applied in a simplified form by assuming that the term under the first radical is equal to unity.

For surface grinding using the grinding wheel face, N. F. Gulev [15/2] derived the following formula for the length of point working travel

$$L_\psi = 0{\cdot}5d\psi_z\sqrt{1 + \left(\frac{s_t}{60v_m}\right)^2} \quad \text{mm} \tag{15.38}$$

In practice, the term under the radical sign is usually assumed to be equal to unity, since the feed rate is very small relative to the cutting speed.

15.2.2 CHARACTERISTICS OF CUTTING WITH A GRINDING WHEEL

15.2.2.1 *Number of grains or grinding wheel surface and grain spacing*

Abrasive grains are irregularly spaced in the grinding wheel. Their number per unit grinding wheel volume is so great that according to the laws of statistics a certain average grain spacing may be expected.

Thus, for example, according to research conducted by the Central Scientific Research Institute for Abrasives and Grinding (TsNILASh) the number of aluminium oxide grains of a total weight of 1 g varies within the range of 4750–41 800 (Table 15.2) for grain numbers from 40(36) to 15(80).

Table 15.2 *Number of corundum grains of a total weight of* 1 g

Grain number	40(36)	32(46)	25(60)	16(80)
Number of grains in 1 g z_0	4750	10 000	30 000	41 800

If it is assumed that the average theoretical distance between grains is l_t, then the number of grains on length L

$$z_L = \frac{L}{l_t} \tag{15.39}$$

and when both sides of the equation are raised to the third power

$$z_L^3 = \frac{L^3}{l_t^3} \tag{15.40}$$

In this formula the term L^3 determines the volume V of a cube cut out of the grinding wheel and z_L^3 the number of grains contained in this volume. It may thus be assumed that

$$z_V = z_L^3 = Q_V z_0 \tag{15.41}$$

where: Q_V—weight of the abrasive contained in volume V of the grinding wheel, in g, z_0—number of grains of the total weight of 1 g.

Using eqn. (14.2) one can write that

$$Q_V = \frac{V_z}{100} V \frac{\gamma}{1000} = \frac{62-2L_z}{100} V \frac{\gamma}{1000} \quad \text{g} \tag{15.42}$$

where: V_z—percentage content of abrasive, in per cent, V—grinding wheel volume in mm³, γ—specific weight of the abrasive in g/cm², L_z—grain number.

The use of relationships (15.40), (15.41) and (15.42) permits one to determine the average theoretical distance between grains in a grinding wheel*

$$l_t = \sqrt[3]{\frac{100\,000}{(62-2L_z)z_0}} \quad \text{mm} \tag{15.43}$$

* N. I. Volskii gives a different formula for the calculation of the theoretical number of grains in 1 mm³, i.e.

$$z_t = \frac{0{\cdot}0268\sqrt[3]{V_z^2}}{a_{av}^2}$$

where: V_z—abrasive content in grinding wheel volume in percent, a_{av}—average value of the characteristic dimension of grain in mm.

Table 15.3 *Average theoretical distances between abrasive grains in a grinding wheel*

Grinding wheel structure number	Grain number 40(36)	32(46)	25(60)	16(80)
	Average theoretical distance between grains l_t mm			
4	0·465	0·364	0·252	0·225
5	0·475	0·369	0·255	0·228
6	0·479	0·373	0·26	0·231
7	0·484	0·378	0·263	0·234
8	0·489	0·38	0·268	0·241

With the use of eqn. (15.43) average theoretical distances between grains were calculated and the results listed in Table 15.3.

These results show that the influence of structure on grain spacing is much less than the influence of grain size.

Actual distances l_z between grains may differ quite markedly from the theoretical average. This can be demonstrated empirically by making an imprint of the grinding wheel surface on a strip of transparent celluloid. By adjusting indentation depth it is possible to learn from the imprint how many points participate in the cutting process on a determined area.

In practice, the actual distance between grains undergoes considerable oscillations and the total number of grains (imprints) per unit area is generally smaller than that theoretically computed.

This results from the fact that grains are distributed in the grinding wheel in a non-uniform way and do not form regular layers on cylindrical surfaces whose common axis coincides with that of the grinding wheel rotation.

15.2.2.2 *Number of grains cutting simultaneously*

The number of grains cutting simultaneously z_c may be calculated from

$$z_c = \frac{A_a}{l_z^2} = \frac{L_\varphi b}{l_z^2} \tag{15.44}$$

where: A_a—the active surface of the grinding wheel, i.e. the surface in contact with the workpiece material, in mm², l_z—average actual distance between grains on the active surface in mm, b—grinding width in mm, L_φ—length of point working travel.

It is evident from eqn. (15.44) that the number of grains cutting simultaneously depends, among other things, on the actual average grain spacing on grinding wheel surface.

On the basis of studies conducted by the contact imprint method* it has been established that the actual distances between grains l_z are always greater than those calculated according to eqn. (15.43)

$$l_z = e\, l_z; \quad \text{where} \quad e > 1 \tag{15.45}$$

* J. Kaczmarek: 'Przyrząd do sprawdzania stanu ściernic' (Device for the checking of grinding wheel condition), Biuletyn IOSS, Przegląd Mechaniczny 12, 1956.

In the case of silicon carbide grinding wheels, within the range of medium and open structures and for grain sizes between 40(36) and 16(80) the value of e in eqn. (15.45) changed from 2·5 and 1·1, the average (most frequently appearing) value being 1·75.

Figure 15.20 represents the relationship between grain size and the theoretical and actual number of grains cutting simultaneously per 1 mm² surface of silicon carbide grinding wheel of structure 5. Measurements were taken for an imprint depth of about 0·05 mm.

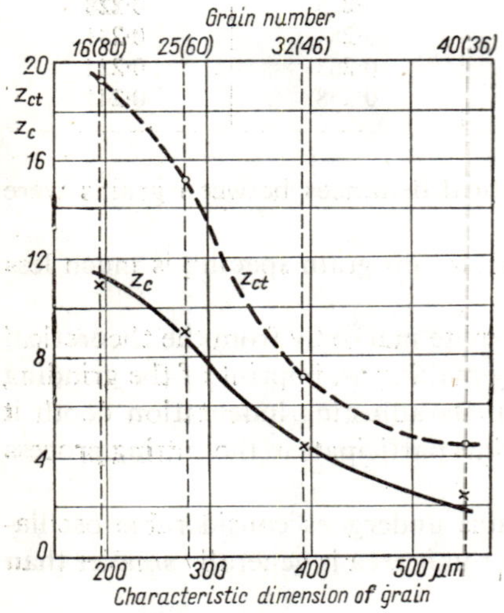

Fig. 15.20 Dependence of the theoretical (t_{ct}) and actual (z_c) number of grains cutting simultaneously on grain size (per 1 mm²)

It is obvious from Fig. 15.20 that with diminishing grain dimensions the number of grains cutting simultaneously increases, but not to the extent which would result from calculations based on theoretical grain spacing. Values of z_c given in the diagram should be further reduced if the grinding depth is smaller than the adopted imprint depth, i.e. 0·05 mm.

15.2.2.3 *Mean thickness of the undeformed chip cut by a grinding wheel*

By mean thickness of undeformed chip cut by a grinding wheel h_{om} is understood the instantaneous undeformed chip thickness being the sum of instantaneous thicknesses of undeformed chips cut by all grains cutting simultaneously and projected upon the grinding wheel rotation plane

$$h_{om} = \sum_{i=1}^{i=z_c} (h_{zm})_i \tag{15.46}$$

The value of the mean thickness of the undeformed chip cut by a grinding wheel may be calculated from

$$h_{om} = \frac{Q_V}{A_A}$$

where: Q_V—volumetric rate of metal removal by grinding in mm³/min, A_c—grinding wheel surface taking part in cutting work during 1 minute, shortly called the minute surface of grinding wheel in mm²/min.

Rate of metal removal in grinding is determined by

$$Q_V = 1000a\, B_a v_w \sqrt{1+\left(\frac{s_t}{v_w}\right)^2} \tag{15.48}$$

where: B_a—active width of grinding wheel, of a value equal to feed per revolution in mm, s_t—rate of longitudinal feed in m/mm, other denotations are as before.

In eqn. (15.48) the resultant speed of auxiliary motion is taken into account, i.e. both of workpiece rotary motion v_w and of rectilinear feed motion s_t.

The active minute surface of grinding wheel may be calculated from

$$A_c = B_f\, v \quad \text{mm}^2/\text{min} \tag{15.49}$$

where: B_f—actual active width of grinding wheel.

Approximating,

$$B_f = B - 2r \quad \text{mm} \tag{15.50}$$

where r—rounding radius of grinding wheel faces.

Since $r \ll B$, it may be assumed that $B_f = B$ and therefore, in subsequent formulae no difference will be made between B and B_f.

Adoption of the entire grinding wheel width as the active width is motivated by the fact that the grinding wheel has no continuous cutting edge. As a result of this e.g. in external cylindrical grinding with longitudinal feed $s = B_a = \frac{1}{3}B$ (mm/rev), after the grinding wheel has moved in relation to work surface by the length B_a, a part of material to be ground may still remain on this section (Fig. 15.21). The

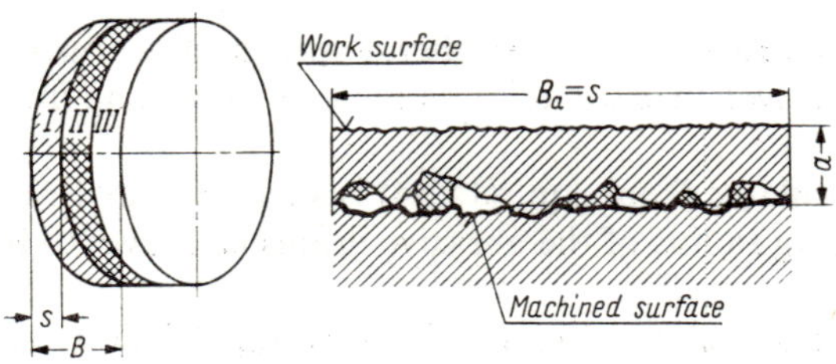

Fig. 15.21 Consecutive layers cut by sections I, II and III of the grinding wheel

passage of the second (II) and third (III) part of the grinding wheel during two subsequent revolutions of the workpiece causes, as a result of a different grain distribution, the removal of the remaining allowance. Thus the entire grinding wheel width takes part in the cutting process, although mean depths of cut on the particular grinding wheel sections are not identical.

The difference will be greater, the higher is the ratio

$$\frac{B}{B_x} = c_B \tag{15.51}$$

The value of ratio c_B indicates how many times a point of work surface will contact the grinding wheel surface during one pass.

The second factor in eqn. (15.49) is v, the resultant grinding speed, defined by

$$v = \sqrt{(60\,000v_m+1000v_w)^2+1000s_t^2}$$

$$= 60\,000v_m\sqrt{\left(1+\frac{v_w}{60v_m}\right)^2+\left(\frac{s_t}{60v_m}\right)^2} \tag{15.52}$$

Consequently

$$A_c = 60\,000B\,v_m\sqrt{\left(1\pm\frac{v_w}{60v_m}\right)^2+\left(\frac{s_t}{60v_m}\right)^2} \tag{15.53}$$

Substitution of (15.53) and (15.48) into (15.47) yields

$$h_{om} = \frac{a}{60}\,\frac{B_a}{B}\,\frac{v_w}{v_m}\sqrt{\frac{1+\left(\frac{s_t}{v_w}\right)^2}{\left(1\pm\frac{v_w}{60v_m}\right)^2+\left(\frac{s_t}{60v_m}\right)^2}}$$

If the square root term is denoted by $1/U$, then this formula will have the form

$$h_{om} = \frac{a}{60}\,\frac{B_a}{B}\,\frac{v_w}{v_m}\,\frac{1}{U}\quad\text{mm} \tag{15.54}$$

Equation (15.54) indicates that the value of the mean thickness of the undeformed chip in grinding is directly proportional to the quotient of peripheral speeds of the workpiece and grinding wheel, on B_a/B ratio and the depth of cut. It is inversely proportional to the factor U taking into account the rotary and feed motion of the workpiece.

In practice $U = 1$ usually since the ratio of the peripheral speeds of the workpiece and grinding wheel is very small. Even in extreme cases the error resulting from this assumption does not exceed 2%.

In most cases the value of h_{om} calculated from eqn. (15.54) is smaller than 1 μm. On the other hand, the average rounding radius of the abrasive grains is known to be 6–10 μm.

Thus, if all points would cut with undeformed chip thickness h_{om}, friction and plastic compression would prevail over the cutting process, while actually the latter prevails. This leads to the conclusion that not all theoretically active cutting points do cut in reality. Some of them just slide over the work surface or scratch it. As a result of this, a correspondingly greater actual thickness of undeformed chip falls to the remaining cutting points.

15.2.3 Mean thickness of the undeformed chip cut by a single point

The thickness of the undeformed chip cut by single grains may vary because of the disorderly, random distribution of points, i.e. grains of the grinding wheel. The large number of grains simultaneously participating in the cutting process and the endeavour to determine at least approximately the thickness of the undeformed chip cut by a single grain justifies the introduction of the notion of mean thickness of the undeformed chip cut by a single grain, i.e.

$$h_{zm} = \frac{h_{om}}{z_\psi} \tag{15.55}$$

where z_ψ—average number of points (grains) distributed on grinding wheel surface on a length equal to that of working travel.

If the average actual distances between grains are l_z, then

$$z_\psi = \frac{L_\psi}{l_z} \tag{15.56}$$

where L_ψ—length of contact between the point and work material, in mm.

When it is considered that for various kinematic types of grinding the length L_ψ is expressed by the different eqns. (15.30), (15.34), (15.36) or (15.38), the corresponding formulae may be derived for the mean thickness of the undeformed chip cut by a single point.

In the case of external cylindrical grinding with longitudinal feed, when the relationships (15.34), (15.54), (15.55) and (15.56) are taken into account, such a formula will be as follows

$$h_{zm} = l_z \frac{a}{60} \frac{B_a}{B} \frac{v_w}{v_m} \frac{1}{U} \sqrt{\frac{d+d_w}{d\, d_w\, a_z}} \frac{1}{\sqrt{\left(1+\frac{v_w}{60v_m}\right)^2 + \left(\frac{s_t}{60v_m}\right)^2}} \tag{15.55a}$$

After denotation

$$M = \sqrt{\left(1 \pm \frac{v_w}{60v_m}\right)^2 + \left(\frac{s_t}{60v_m}\right)^2}$$

one may find that

$$U = \frac{M}{\sqrt{1+(s_t/v_w)^2}} = \frac{M}{W} \tag{15.57}$$

In addition, with the assumption that the depth of grinding by a single grain and by the grinding wheel are equal, $a_z = a$, eqn. (15.55) may be written in the form

$$h_{zm} = \frac{l_z}{60} \frac{B_a}{B} \frac{v_w}{v_m} \sqrt{a \frac{d+d_w}{d\, d_w}} \frac{W}{M^2} \quad \text{mm} \tag{15.58}$$

In calculations for practical purposes it may be assumed that $W = 1$, that $\frac{s_t}{60v_m} = 0$ and that $\left(\frac{v_w}{60v_m}\right)^2 = 0$ and thus the value

$$\frac{60v_m}{v_w} M^2 \approx \frac{60v_m \pm 2v_w}{v_w} \tag{15.59}$$

With these simplifications and considering that $B_a = s$, the formula for the mean thickness of the undeformed chip cut by a single grain in the case of cylinder grinding will be

$$h_{zm} = \frac{v_w}{60v_m \pm 2v_w} \frac{s}{B} l_z \sqrt{a \frac{d+d_w}{d\,d_w}} \quad \text{mm} \tag{15.60}$$

A similar procedure yields formulae for h_{zm} in:
— internal cylindrical grinding with longitudinal feed

$$h_{zm} = \frac{v_w}{60v_m \pm 2v_w} \frac{s}{B} l_z \sqrt{a \left(\frac{D-d}{D\,d}\right)} \quad \text{mm} \tag{15.61}$$

— surface grinding using the cylindrical surface of grinding wheel

$$h_{zm} = \frac{s_t}{60v_m \pm 2s_t} \frac{B_a}{B} l_z \sqrt{a \frac{1}{d}} \quad \text{mm} \tag{15.62}$$

— surface grinding using the face of grinding wheel

$$h_{zm} = \frac{s_t}{60v_m + s_t} \frac{B_c}{B} l_z \frac{a}{d\psi} \quad \text{mm} \tag{15.63}$$

where ψ is in radians.
It follows from the formulae given above that in all types of grinding considered:
— when the values of v_w, s (or B_a), l_z and a increase, then the value of mean thickness h_{zm} increases,
— when the values of v_m and B increase, then h_{zm} decreases.
The influence of grinding wheel diameter depends on which speed—peripheral or rotational—is constant.

In the case of cylindrical grinding at a constant cutting speed (v_w = const), increasing the grinding wheel diameter causes very slight reduction of h_{zm}. On the other hand, when n = const, then, with increasing wheel diameter, thickness h_{zm} decreases almost hyperbolically.

In other types of grinding an increase of d also causes a reduction of h_{zm}.

Increasing workpiece diameter causes:
— a decrease of h_{zm} when v_w = const,
— an increase of h_{zm} when n_w = const.
Since the grinding process is influenced considerably by the value of h_{zm}, a knowledge of the influence of the various factors upon this value is of practical importance.

Consideration of eqns. (15.60), (15.61) and (15.62) allows one to conclude that the relation between the values of the mean thickness of the undeformed chip cut by a single grain in external cylindrical grinding ($\dot{h}_{zm}$), internal cylindrical grinding ($\mathring{h}_{zm}$) and surface grinding ($\bar{h}_{zm}$) is as follows:

$$\dot{h}_{zm} : \mathring{h}_{zm} : \bar{h}_{zm} = \sqrt{\frac{1+d/d_w}{d}} : \sqrt{\frac{1-d/D}{d}} : \sqrt{\frac{1}{d}} \tag{15.64}$$

It can be concluded from eqn. (15.64) that under identical grinding conditions the thickness of the undeformed chip cut by a single grain is greatest in external cylindrical grinding and smallest in internal cylindrical grinding.

15.3 Physical and technological relationships in grinding

15.3.1 GRINDING FORCES AND POWER

The resolution of cutting forces in grinding may be carried out in the same way as in milling, i.e. in a machine tool system and in a system determined by vectors of motion (this system is bound up with grinding wheel points).

However, this differentiation is essential here only in the case of surface grinding using a grinding wheel face, where the length of working travel may be considerable ($L_{\psi\,max} = 0{\cdot}5\pi\, d$). In other cases of grinding, in view of the very small length of contact, it is assumed that the QRS system coincides with the XYZ system.

In the latter system the resultant cutting force acting on the workpiece may be resolved into tangential (peripheral) component F_v, thrust component F_p and feed component F_f.

In this way both the cutting force acting on a single grain and the resultant cutting force of all points cutting simultaneously may be resolved.

It has been established empirically that the ratio of the particular components differs from that appearing in other cutting methods.

The F_p/F_v ratio varies within limits of 1–3 or even more if grinding is carried out using worn abrasive grains.

As an example, the dependence of F_p/F_v ratio (in cylindrical grinding with longitudinal feed, using aluminium oxide grinding wheels of 63(24)–32(46) grain size and $L_z = 14$ and $L_z = 5$ structure) on grinding time proportional to the number of passes i is shown in Fig. 15.22. The increase of this ratio results from the progressive wear of grinding wheel points.

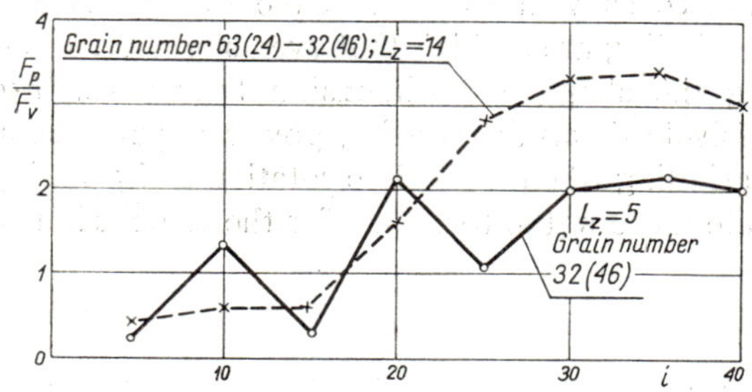

Fig. 15.22 Variations of F_p/F_v ratio depending on the number of passes in external cylindrical grinding. Conditions of measurement: workpiece material—alloy steel of HRC = 35–40 kg/mm², v_w = 15 m/min, s = 8 mm/rev, a = 0·02 mm, aluminium oxide grinding wheel with ceramic bond

The fact that in grinding $F_p \geqq F_v$ can be explained mainly by the circumstance that the particular abrasive grains cut with high negative rake angles, and often with the rounded part of point top. Experiments show that in other cutting methods, e.g. in turning, a reduction of rake angle from $\gamma = +10°$ to $\gamma = -10°$ causes an increase of force F_p by 150%. In addition, this results from burnishing instead of cutting of the material by a considerable number of points.It may then be assumed approximately that only the force of friction $T = \mu s_v$ acts in the tangential direction. It has been shown [15/2] that the value of friction coefficient μ of corundum grain on steel is approximately 0·13, on cast iron μ = 0·16–0·17, and on copper μ = 0·2–0·23. These data correspond to normal loads of F_p = 10–100 g per point.

It is thus understandable that a reduction of undeformed chip thickness in the range $h_z < h_{z\,min}$ causes a rapid increase of the F_p/F_v ratio. This is illustrated in Fig. 15.23 using the results of Maslov [15/2].

Force F_f amounts to 1/6–1/4 of force F_v. Although absolute values of component cutting forces are low as compared with other cutting methods, their knowledge is necessary not only for the evaluation of force and power characteristic of the grinding process, but also for practical purposes.

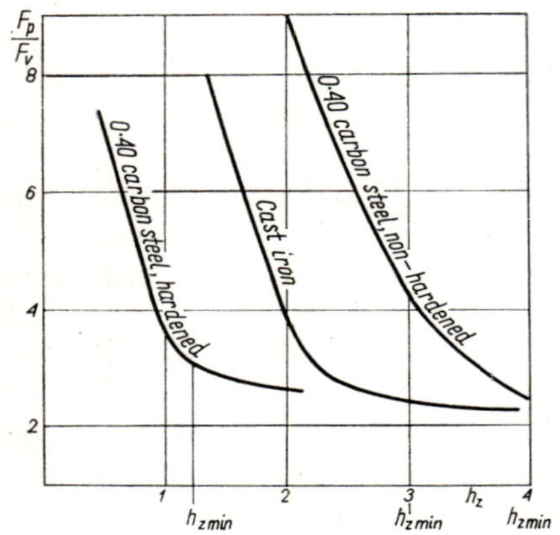

Fig. 15.23 Dependence of F_p/F_v ratio on the thickness of the undeformed chip cut by a single grain h_z

The greatest component F_p plays an important role in machining accuracy, causes elastic deflection of the workpiece as well as of machine tool parts and assemblies. Accuracy requirements in grinding are so high that even slight deformations may result in errors exceeding the shape and dimension tolerances. The component F_v and cutting speed decisively determine cutting power requirements.

In the case of external cylindrical grinding when workpiece rotation and grinding wheel rotation are imparted by separate motors, the formulae for the calculation of cutting power are as follows:
for grinding wheel motor

$$N_{meg} = \frac{F_v v_m}{1{\cdot}36 \times 75} = \frac{F_v v_m}{102} \quad \text{kW} \tag{15.65}$$

for workpiece motor

$$N_{mew} = \frac{F_v v_w}{6120} \quad \text{kW} \tag{15.66}$$

The value of F_f is required in order to calculate a suitable rigidity for the feed drive.

Formulae for peripheral cutting speed

The mean value of peripheral cutting force F_v is expressed by

$$F_v = F_{vz} z_c \quad \text{kg} \tag{15.67}$$

where: F_{vz}—mean peripheral force acting on a single point in kg, z_c—number of active points cutting simultaneously.

The value of F_{vz} in turn may be expressed as

$$F_{vz} = k_s A_z = 0{\cdot}5 C_z h_{zm}^{1-k} b_z \quad \text{kg} \tag{15.68}$$

where: $k_s = C_z/h_{zm}^k$—mean specific cutting force in kg/mm², $A_z \approx 0{\cdot}5 h_{zm} b_z$—transverse cross-section of undeformed chip per point in mm², C_z—a constant taking into account the influence of all factors not appearing separately in the formula, h_{zm}—thickness of undeformed chip cut by a single point in mm, b_z—width of undeformed chip cut by a single point in mm, k—exponent.

Substitution of expressions for h_{zm} (15.60) and b_z (15.21) into eqn. (15.68) yields, in cylinder grinding, after suitable computations and simplifications,

$$F_{vz} = C_z \left(\frac{v_w}{60 v_R \pm 2 v_w} \right)^{1-k} a^{0{\cdot}5(1-k)} l_z^{1-k} \left(\frac{d + d_w}{d\, d_w} \right)^{0{\cdot}5(1-k)}$$

$$\times \left(\frac{s}{B} \right)^{1-k} \left[\sqrt{2r h_r - h_r^2} + (h_z - h_r) \tan \varepsilon \right] \quad \text{kg} \tag{15.69}$$

where h_r—denoted as KL in Fig. 15.16.

Substitution of the expressions (15.69) and (15.44) into eqn. (15.67) and of the expression for L_φ (15.34), simplified by neglecting the influence of rotary and feed motion of the workpiece, yields (on the assumption that $B_a = B$)

$$F_v = C_z \left(\frac{v_w}{60 v_m \pm 2 v_w} \right)^{1-k} l_z^{-k-1} a^{1-0{\cdot}5k} \left(\frac{d_w\, d}{d + d_w} \right)^{0{\cdot}5k} s^{1-k} B^k$$

$$\times \left[\sqrt{2r h_r - h_r^2} + (h_z - h_r) \tan \varepsilon \right] \quad \text{kg} \tag{15.70}$$

Using a similar method, one can also derive formulae for peripheral cutting speed in internal and surface grinding.

The practical value of eqn. (15.70) for peripheral cutting speed consists in showing what factors the value of resistance—and thus also of cutting force—depends on. Taking into account that l_z depends on grinding wheel structure and other features and C_z, first of all, on the properties of the workpiece material, the conclusion may be drawn that all parameters characterizing the cutting process and tool directly or indirectly influence the cutting resistance.

Research conducted on cutting forces allows one to conclude that the extent of influence of these factors is not equal. In addition, many factors contained in eqn. (15.70) vary within narrow limits. As a consequence, in practical calculations simplified approximate formulae of a statistical and empirical character are applied.

For external grinding of surfaces of revolution of steel and cast iron workpieces using aluminium oxide grinding wheels of grain number 32(46), medium soft, of structure $L_z = 5$, the formula

$$F_v = C_z v_w^{0{\cdot}7} s^{0{\cdot}7} a^{0{\cdot}6} \quad \text{kg} \tag{15.71}$$

has been established [15/2], where the values of C_z are as follows:

— for hardened steel $C_z = 2{\cdot}2$,
— for non-hardened steel $C_z = 2{\cdot}1$,
— for cast iron $C_z = 2{\cdot}0$.

Cutting with large-pore grinding wheels meets with less cutting resistance than cutting with dense, or medium-structure wheels.

In addition to the empirical formulae of the type given above, formulae of the type (15.68) are sometimes applied, where the value of k_s is taken to be of the order of 10 000–20 000 kg/mm^2.

This calculation method may only be considered as approximate in view of the very great inaccuracy in the determination of k_s.

15.3.2 Grinding wheel wear and life

The description of abrasive tool wear and life given in Chapter 14 fully applies to grinding wheels.

The starting point for the determination of machining conditions and calculation of life speed is eqn. (14.13).

This formula may be applied in the case when $v_m \leqq 40$ m/s in grinding with a coolant and $d = 150$–350 mm, in several types of grinding. Values of C_v and exponents for those types are listed in Table 15.4 and the recommended values of T in Table 15.5.

Table 15.4 *Values of constants in the formula for peripheral life speed of the workpiece or feed rate in grinding*

Surfaces being ground	Workpiece material	Characteristic of grinding wheel	C_v	b_{vw}	m	e_v	u_v
External surfaces of revolution; cylindrical grinding with longitudinal feed	Non-hardened steel hardened steel cast iron	Corundum 40(36)CM2 Corundum 32(46)CI Corundum 40(36)CM2	0·055 0·850 0·071	0·3	0·5	1·2	1·0
External surfaces of revolution; cylindrical grinding with transverse feed	non-hardened steel hardened steel cast iron	Corundum 40(36)CI Corundum 40(36)CI Corundum 40(36)CI	0·950 0·850 1·235	0·3	0·35	0·5	0·0
Internal surfaces of revolution; longitudinal feed	non-hardened steel hardened steel cast iron	Corundum 40(36)CMI Corundum 40(36)CNI Corundum 32(46)CMI Corundum 25(60)CMI Corundum 32(46)CI Corundum 40(36)CMI	0·054 0·050 0·050 0·054 0·058 0·070	1·0	0·5	0·9	0·9
External surfaces of revolution; centreless grinding with longitudinal feed	non-hardened steel hardened steel	Corundum 25(60)CM2 Corundum 32(46)CI Corundum 25(60)CTI Corundum 32(46)CNI Corundum 25(60)CM2 Corundum 32(46)CI Corundum 25(60)CI	14 15·5 12·75 16·5 11·5 13·5 10·5	0·3	0·5	1·0	1·0
Plane surfaces, rectilinear feed of table	non-hardened steel	Corundum 40(36)CM2 Corundum 40(36)CI Corundum 40(36)CTI	2·5 2·34 2·0	0·0	0·7	0·75	1·0

Table 15.5 *Recommended values of grinding wheel life* [15/3]

Item	Grinding variety		T, min
1	External grinding of surfaces of revolution	through-feed grinding	40
2		grinding with longitudinal feed to a stop	30
3		grinding with transverse feed	30
4		grinding of stepped shafts	20
5	Centreless grinding	through-feed grinding with longitudinal feed	60
6		grinding with transverse feed	30
7	Internal grinding		10
8	Surfacing using grinding wheel periphery with longitudinal feed		25
9	Surfacing using grinding wheel periphery, with transverse feed		10
10	Form grinding		10

In surface grinding $v_w \approx s_t$.

In the case of cylindrical external and internal grinding and surface grinding the value of $s_B = B_a/B$ or s/B should be substituted for s.

The ratio $s_B = s/B$ is called the *relative feed.*

15.3.3 Surface quality and the properties of the surface layer

15.3.3.1 *Surface roughness*

Factors influencing the quality of work surface may be divided into: factors connected with the workpiece, factors connected with the grinding wheel and factors connected with the cutting conditions.

The chemical composition (type) of the material and its structure is of considerable importance from a workpiece point of view.

As far as surface quality is concerned, non-ferrous metals are generally more difficult to machine—under identical cutting conditions—than the ferrous ones. This is more evident, the more ductile is the non-ferrous metal. However, the conclusion should not be drawn that it is impossible to grind non-ferrous metals to the same surface quality as e.g. steel. Provided the optimum choice of all grinding conditions has been made, non-ferrous metal components may have a surface quality of the same order as steel components.

When grinding steel its structural components and the mechanical properties connected with them exert a marked influence on surface quality.

Studies have revealed that steels of a martensitic structure may be machined to a better surface finish than steels of a troosto-sorbitic or a pearlitic structure.

Thus, for example, after surface grinding using an aluminium oxide grinding wheel of medium hardness and grain number 25(60), in the machining of a 0·45 carbon steel of troosto-sorbitic structure with ferrite inclusions, roughness number 8 is obtained. This means that surface irregularities are, on average, twice as large as when grinding hardened high-speed steel of a fine martensitic structure under identical cutting conditions (v_m = 28 m/s, a = 0·005 mm, s_t = 10 m/min).

When external grinding of surfaces of revolution the surface finish is influenced by workpiece diameter. With increasing diameter (and constant peripheral speed) surface quality improves. To explain the reasons for such influence it should be remembered that the mean thickness of the undeformed chip cut by a single grain decreases with increasing d_w. As a result of the random distribution of abrasive grains on grinding wheel surface, irregularities may remain on certain areas of the machined surface which have not been removed by consecutive grains. These irregularities will be larger, the greater is the undeformed chip thickness.

In Fig. 15.24 a diagram is shown representing the dependence of mean total height of surface irregularities on workpiece diameter, according to studies conducted by Maslov and Ignatov [15/4]. These studies concerned external grinding with an aluminium oxide grinding wheel, medium soft, of grain size 32(46). The workpiece was made of 0·50 carbon steel. Grinding conditions: $v_m = 25$ m/s, $v_w = 30$ m/min, $s = 0{\cdot}3B$.

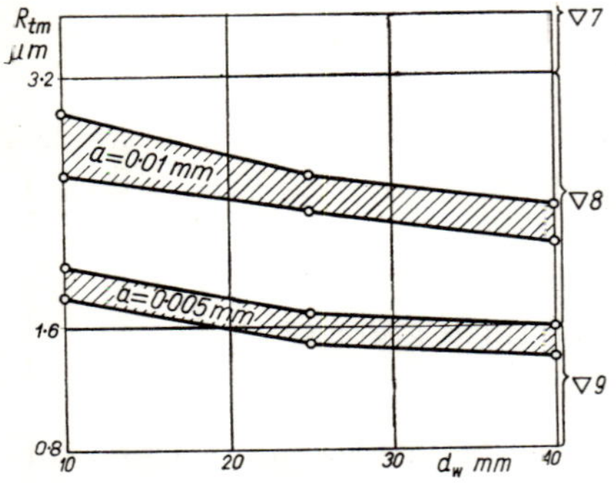

Fig. 15.24 Dependence of mean total height of surface irregularities R_{tm} on workpiece diameter d_w

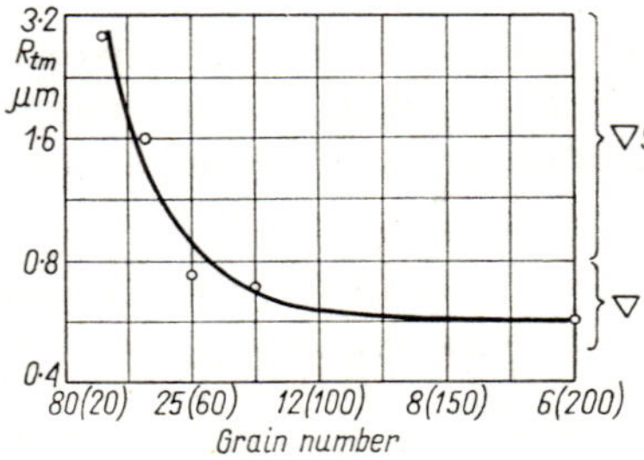

Fig. 15.25 Dependence of mean total height of surface irregularities R_{tm} on grain size in the grinding process

Factors connected with the grinding wheel comprise: grinding wheel material, abrasive grain size, bond, hardness, method of wheel dressing, degree of wheel wear and wheel diameter.

The type of abrasive material does not exert a marked influence on surface finish. Under constant grinding conditions smaller surface irregularities can generally be obtained by using silicon carbide (carborundum) on ceramic bond. Thus e.g. applying silicon carbide in grinding chromium coatings allows one to halve the height of surface irregularities as compared with those resulting from the use of aluminium oxide under the same conditions. This result is explained by the fact that silicon carbide grains have sharper cutting edges and are more easily chipped off than aluminium oxide grains, without reaching the worn stage.

A much more important influence is exerted by abrasive grain size. The smaller it is, the higher becomes the surface quality. For instance, when grinding chromium steels particularly great changes of surface finish are caused by changes of grain size within the range from 40(36) to 16(80). Further reduction in grain size only slightly improves the surface quality (Fig. 15.25).

Of the bonds used in practice, grinding wheels with rubber bonds produce the best surface quality. The bond of such wheels participates in the machining operation, simultaneously polishing the machined surface.

Grinding wheel hardness influences surface roughness in such a way that a hardness both higher and lower than the optimum value for the given workpiece material and machining conditions causes a deterioration of surface quality.

The method of grinding wheel dressing also influences the surface finish. This results from the fact that all irregularities appearing on the grinding wheel surface in consequence of its dressing are later reproduced on the machined surface of the workpiece. Experiments indicate that by appropriate, careful dressing of the grinding wheel surface roughness numbers as high as 10–12 may be obtained even with grinding wheels of grain size 32(46)–25(60).

There exist optimum values of feed rate, depth of cut and the number of passes of the dressing tool, yielding the highest surface quality. Thus e.g. in the case of dressing an aluminium oxide grinding wheel of grain number 32(46), medium soft, used for grinding hardened steel, optimum values are: dressing tool feed rate $s_{td} = 0{\cdot}2$ m/min, depth of cut $a_d = 0{\cdot}05$ mm, number of passes $i_d = 4$.

As time passes, and thus as the wear of the grinding wheel proceeds, the surface quality at first improves. After a certain number of passes the degree of surface roughness becomes stabilized. Even completely worn grinding wheel grains do not cause any changes of roughness measured in the direction of longitudinal feed. On the other hand, streaks and scratches appear, resulting in a wavy finish of the machined surface.

The influence of grinding wheel diameter on surface finish is practically negligible (within a small range of diameter values), since, at constant cutting speed, diameter variations cause only slight changes in the thickness of the undeformed chip cut by a single grain. Therefore, if the grinding wheel diameter is in the range $d = 250$–350 mm, it may be assumed not to have any influence on surface quality.

With respect to grinding conditions, peripheral speed of the grinding wheel and of the workpiece, feed rate, number of spark-out passes, depth of cut, and the coolant influence surface finish.

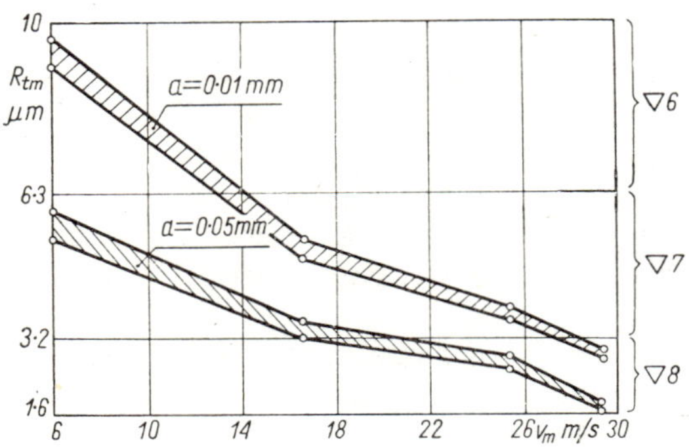

Fig. 15.26 Relationship between the mean total height of ground surface irregularities R_{tm} and cutting speed. Machined material: manganese steel. Grinding conditions: $v_m = 25$ m/s; $s = 0{\cdot}3B$, grinding wheel EB46K [15/4]

The influence of grinding wheel peripheral speed on surface finish is quite considerable. Figure 15.26 shows the dependence of mean total height R_{tm} of irregularities of a ground surface of hardened manganese steel (C = 0·5%, Mn ≈ 1%) workpiece on the peripheral speed of an aluminium oxide grinding wheel EB of grain size 32(46), hardness K, and with a ceramic bond. The peripheral speed of the workpiece was $v_w = 30$ m/min and the longitudinal feed rate $s = 0{\cdot}3B = 12$ mm/rev [15/4].

It is evident from this diagram that a reduction of grinding wheel peripheral speed is unfavourable, as it causes a deterioration of surface quality. This disadvantageous action may be explained in the following way: when cutting speed is reduced, the thickness of the undeformed chip cut by a single grain increases and the amount of heat in the undeformed chip diminishes, and so does the cutting temperature. These conditions are conducive to an increase of surface roughness.

The action of workpiece peripheral speed is the opposite of this. The greater is this speed the lower the surface quality, because of the increase of undeformed chip thickness. The quantitative influence of workpiece peripheral speed on mean total height R_{tm} of surface irregularities is illustrated in Fig. 15.27. This diagram concerns grinding under conditions described in the caption to Fig. 15.26.

In the case of surface grinding the feed rate s_t exerts a similar influence on surface quality as workpiece peripheral speed when grinding of surfaces of revolution.

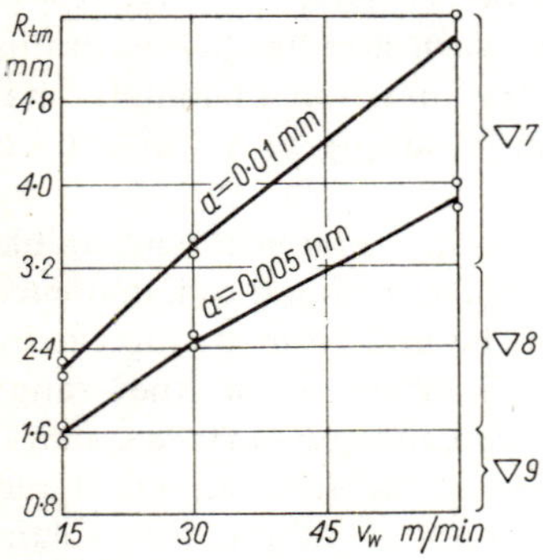

Fig. 15.27 Dependence of mean total height of surface irregularities R_{tm} on workpiece peripheral speed v_w [15/4]

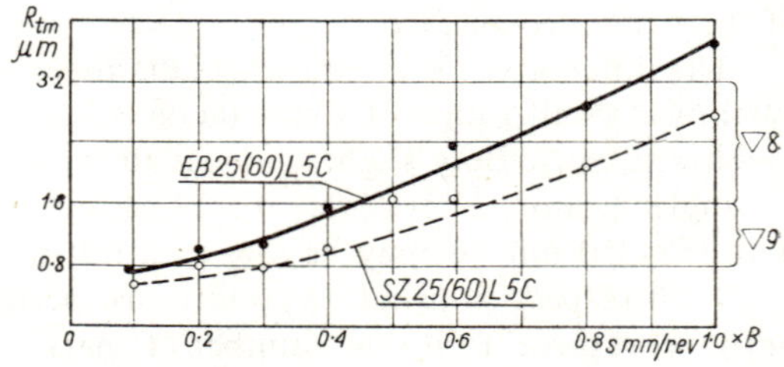

Fig. 15.28 Influence of the longitudinal feed s on mean total height R_{tm} of irregularities of a ground surface of revolution

Increasing the feed, expressed in fractions of grinding wheel width, causes a deterioration of surface finish (Fig. 15.28). This results mainly from the fact that as a consequence of an increase in the B_a/B ratio the number of contacts between the grinding wheel grains and the particular points of the workpiece is reduced.

This in turn decreases the chance of removal of irregularities left over by the preceding cutting points. The influence of B_a/B ratio is particularly evident in the range of $B_a/B > 0{\cdot}3$.

In the range of $B_a/B < 0{\cdot}3$ roughness variations diminish. Therefore in the case of finish grinding the adoption of a B_a/B ratio not exceeding 0·3 is desirable (in the case of grinding of surfaces of revolution $B_a/B = s/B$).

The number of spark-out passes (with nominal depth of cut $a = 0$) influences the surface finish in a similar way. It is known from production engineering practice that a suitable number of spark-out passes improves surface quality considerably. This may be explained just as in the case when the influence of B_a/B ratio was discussed. The greater is the number of spark-out passes, the better chance for a contact between abrasive grains and the irregularities left over from the pass with a nominal depth of cut $a > 0$. Consequently, the greater is the number of passes, the higher surface quality is obtained. Both the grinding theory and production practice indicate that after a certain number of spark-out passes the following ones are of no avail. This is obvious from Fig. 15.29. A stabilization of surface roughness is usually reached after 10–18 sparking-out passes.

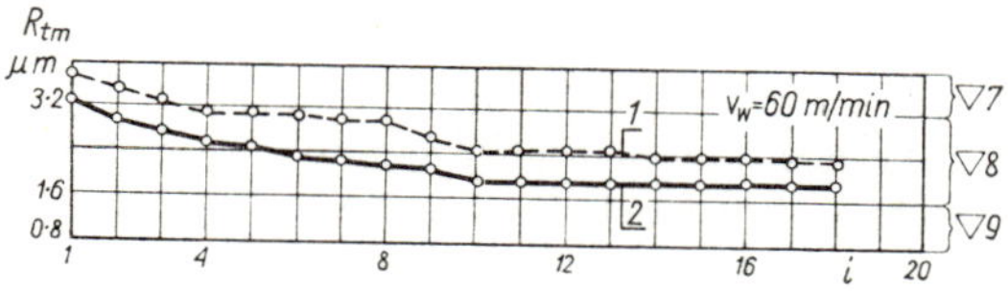

Fig. 15.29 Dependence of mean total height of surface irregularities R_{tm} on the number of spark-out passes: 1—after 10 min of grinding, 2—after 2 min of grinding. Workpiece material: hardened manganese steel. Grinding conditions: grinding wheel EB32/46/K5C, $v_m = 25$ m/s, $s = 0{\cdot}3B$ mm/rev, $v_w = 60$ m/min [15/14]

The influence of the depth of cut on the quality of ground surfaces is remarkable, unlike in other methods of machining by cutting. Increasing the nominal depth of cut involves a greater depth of cut of particular grains or the depth of their penetration into the workpiece material. It is therefore understandable that surface irregularities grow with cutting depth, whose influence is illustrated by some of the diagrams given above (Figs. 15.24, 15.25).

The cutting fluid favourably influences the surface finish since the cooling, lubricating and washing action of the fluid:

(a) reduces friction of grains which are not cutting but burnishing the material,

(b) washes chipped grains and bond particles off the material surface, thus reducing the possibility of scratches appearing.

In the case of grinding water is generally used as coolant, with 2% of soda added for grinding wheels with ceramic bond and 0·5–1% for those with phenolic resin bond. Also, emulsion with a low content of oil (up to 1·5%) may be applied as a coolant and lubricant.

Grinding with liquid cooling very favourably influences the dimensional and shape accuracy. Although the cutting temperature in the undeformed chip zone practically does not change (as testified by sparks appearing also in wet grinding), the workpiece temperature does not rise since almost all cutting heat passing to the workpiece material is taken away by the coolant.

The experimental evidence available does not yet permit one to establish exact quantitative relationship between ground surface finish and grinding conditions.

However, it has been established that in the majority of cases those quantitative relationships may be expressed by formulae of an exponential type, i.e. similar to those used in other machining methods.

Thus e.g. Maslov in his studies on grinding chromium steel surfaces of revolution with a EB32(46) K5C grinding wheel, and using a coolant, proposed the formula

$$R_{tm} = 3{\cdot}6 v_w^{0{\cdot}6}\, s^{0{\cdot}45}\, a^{0{\cdot}4} \quad \mu m \qquad (15.72)$$

In grinding manganese steel surfaces of revolution the following formulae were obtained:

for hardened steel (HRC = 52–54 kg/mm²)

$$R_{tm} = \frac{1960 v_w^{0\cdot68}\, s^{0\cdot75}\, a^{0\cdot56}}{v_m^{1\cdot61}} K_z K_i \quad \mu\text{m} \tag{15.73}$$

for non-hardened steel (HB = 210 kg/mm²)

$$R_{tm} = \frac{80 v_w^{0\cdot71}\, s^{1\cdot17}\, a^{0\cdot55}}{v_m^{0\cdot51}} \quad \mu\text{m} \tag{15.74}$$

Factors appearing in formula (15.74) take account of the influences: K_z—of grain size, K_i—of the number of spark-out passes and of workpiece peripheral speed. Thus e.g. for EB32(46)H grinding wheel $K_z = 1{\cdot}0$, while for EB25(60)L wheel $K_z = 0{\cdot}5$ at $v_m = 8$ m/s and $K_z = 0{\cdot}9$ at $v_m = 30$ m/s.

Formulae (15.73) and (15.74) are valid in the case when $s \leqq 0{\cdot}3B$, $a \leqq 0{\cdot}01$ mm and $v \geqq 25$ m/s.

K_i values are: 0·41 for $v_w = 15$ m/min and 0·57 for $v_w = 60$ m/min.

The formulae given above, although they are not of a universal character, approximately determine the order of magnitude of the influence of particular grinding parameters upon the quality of machined surfaces.

15.3.3.2 *Properties of surface layer machined by grinding*

In grinding, as in other methods of machining by cutting, conditions arise under which a thin surface layer changes its properties. The reasons for this are:
(a) plastic deformations resulting from point work,
(b) cutting heat generation, changes of temperature and its non-uniform distribution in the surface layer.

Variations of surface layer properties are mainly revealed in increased hardness and in the appearance of residual (final) stresses.

The depth of the hardened layer varies—depending on material properties and grinding conditions—from a few micrometres to slightly more than one tenth of a millimetre. Such deformations are considered large compared with the mean depths of cut. In the case of grinding hardened carbon tool steel it has been established that the depth of the deformed layer can reach about 0·16 mm.

Experiments indicate that in the case of smaller depths of cut the hardness of the layer situated nearest to the surface may exceed the original hardness of the material. On the other hand, for greater depths of cut the hardness of layers nearest to the surface falls below the original value.

It follows from Matalin's studies [15/5] that there is a rather distinct relationship between the hardness and residual stresses. As the former grows, the latter increase in the compressive sense, and vice-versa, while with decreasing hardness, the stresses become increasingly tensile. This is illustrated in Fig. 15.30a and b, when grinding a hardened carbon tool steel.

The type and value of final stresses and the character of hardness changes in the surface layer depend on almost all conditions of the grinding process. These relationships have not been sufficiently studied so far. Further studies of these problems are of both theoretical and practical importance, since their results would permit one to favourably influence the properties of surface layer.

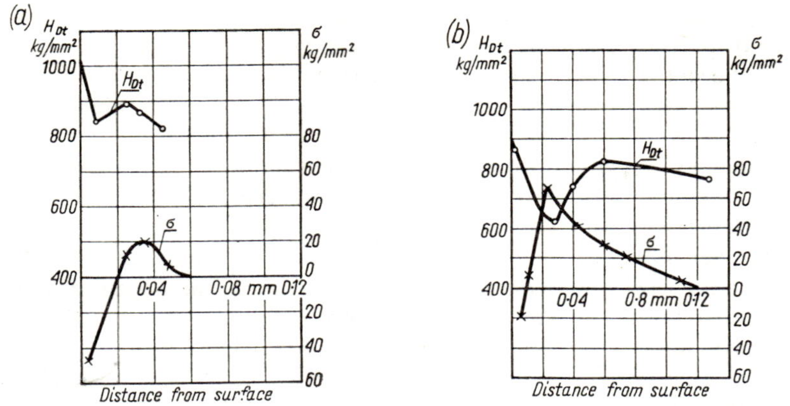

Fig. 15.30 Dependence of microhardness H_{Dt} on permanent stresses in the surface layer after grinding of steel: (a) $v_w = 29$ m/min, $a = 0{\cdot}01$ mm, $s_t = 1{\cdot}88$ m/min; (b) $v_w = 29$ m/min, $a = 0{\cdot}005$ mm, $s_t = 1$ m/min

Just as properly chosen grinding conditions may lead to an improvement of useful qualities of the surface layer, so improperly chosen conditions may cause surface damage and, consequently, rejection of the given part.

Damage to the ground surface most often encountered comprises so-called grinding burns, microgaps, and microracks, and spalling of the surface layer under load combined with kinetic friction.

The appearance of burns (locally or over the entire surface) results in tempering or re-hardening. Those processes may be caused by an unsuitably chosen time of contact between workpiece material and grinding wheel, rate of temperature increase and of cooling.

On average, it may be assumed to a first approximation that the increase of temperature in the surface layer proceeds at a rate of 50 000–100 000°C per second and that the cooling rate is 800–1000°C/s. These rates depend on many factors, such as wear of abrasive grains, cooling method, feed rate etc.

When the time of grain work or the value of cutting temperature or both these factors grow excessively, a softening of the surface layer occurs. Then, depending on cooling intensity and the temperature the process of re-hardening or of tempering may follow.

As a result of these processes additional stresses arise in the surface layer. In the case when these stresses exceed the strength of the material, microcracks (microgaps) may appear or spalling of the surface layer may take place. Grinding microgaps are a consequence of exceeding the tensile strength, while spalling is the consequence of exceeding the compressive strength.

15.4 Intensification and selection of grinding conditions

15.4.1 Some methods of high-output grinding

In order to intensify the grinding process, grinding methods featuring special characteristics are being developed and applied. Some of these will be mentioned here as examples.

Grinding at considerably raised cutting speeds and, simultaneously, suitably increased peripheral speeds of the workpiece is called *high-speed grinding*.

Until recently the cutting speed in grinding of surfaces of revolution did not exceed 30 m/s and workpiece peripheral speed 20–25 m/min, in the case of high-speed grinding the former amounts to 40–90 m/s and the latter to 90–120 m/min.

Such high speeds may be maintained only when using sufficiently strong grinding wheels and special machine tools. Therefore, in high-speed grinding rubber-bonded grinding wheels with a high tensile strength are applied.

A considerable increase of cutting speed influences the grinding wheel work. The phenomenon of the *dynamic hardness* occurs, which means the appearance of symptoms characteristic for grinding at a conventional cutting speed (v_m = 25–30 m/s), but with a harder grinding wheel than that actually used in high-speed grinding. The causes of dynamic hardness increase have already been explained before, and thus only a diagram (according to Yashcheritsyn [15/6]) illustrating the relationship between cutting speed and dynamic hardness will be given here (Fig. 15.31). It follows clearly from this diagram that the higher the cutting speed, the softer the chosen grinding wheel should be. The influence of cutting speed is greater in the case of fine-grained grinding wheels.

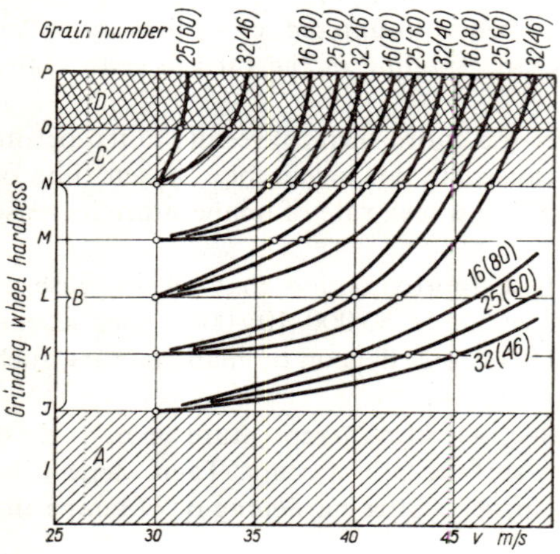

Fig. 15.31 Variations of dynamic hardness of grinding wheels depending on cutting speed v_m. Grinding ranges: A—with ample chipping of grains, B—normal, C—with pores becoming gummed up, D—with frictional vibrations [15/6]

In Fig. 15.31 various grinding ranges are given: for ample (A) and normal (B) chipping of abrasive grains, for the case of gumming up of grinding wheel pores (C) and for the case of vibrations caused by friction (D). Ceramic bond grinding wheels of structure number 12 were used. However, only qualitative conclusions may be drawn from this diagram, as the use of grinding wheels of other makes might yield quantitatively differing results.

Generally, the characteristic features and advantages of high-speed grinding are as follows:

(a) $v_w(s_t)$ or B_a may be increased without increasing the thickness of the undeformed chip cut by a single grain,

(b) for the same or only slightly increased v_w, a and B_a values the life of grinding wheel points is extended and machined surface quality improved,
(c) the rate of metal removal is increased.

Grinding with large-pore grinding wheels is principally a further evolution of the high-speed method.

The characteristic features of high-speed grinding with large-pore grinding wheels are as follows:

(a) grinding wheel cooling is much more effective,
(b) the grinding wheel contains considerably less bond, thus the friction between bond and work surface occurs to a lesser degree and this in turn favourably influences the surface finish,
(c) grinding wheel hardness and strength do not diminish but even increase, since the amount of abrasive in the wheel is also reduced,
(d) due to a lesser amount of abrasive involved, its consumption decreases,
(e) as a consequence of the lesser amount of abrasive, the average actual distances between grains l_z are larger, which causes an increase of undeformed chip thickness $h_{z\,av}$; to compensate for this influence, it is necessary to increase the cutting speed correspondingly,
(f) surface quality does not deteriorate since a suitably finer abrasive may be used.

Grinding with large-pore grinding wheels is not yet commonly practiced, but owing to its advantages it stands a good chance of further development.

Another grinding method worked out at the Polish Institute of Machining by Cutting (IOS) is called *grinding under elastic pressure* or, more exactly, under adjustable pressure. It is applied in the sharpening of lathe tools.

The characteristic feature of this method is self-dressing of the grinding wheel during work. This was obtained by optimization of the grinding conditions and their maintenance during the operations.

Beside commonly determined grinding conditions, such as cutting speed, longitudinal feed rate, depth of cut, cooling etc., particular attention was paid to the optimum specific surface pressure.

Constant optimum surface pressure was ensured by pressing the workpiece to the grinding wheel using a spring (with adjustable force, depending on the size of the surface being ground).

Under optimum conditions of grinding sintered carbides: silicon carbide grinding wheel (grain number 32(46), hardness J, structure number 5, ceramic bond), $v_m = 12–16$ m/s, $s_t = 4–8$ m/min, $a = 0{\cdot}04–0{\cdot}06$ mm, cooling by means of 10% emulsion (cooling rate 30 l/min) and unit surface pressure $p = 10–20$ kg/cm^2, an increased rate of metal removal and reduced machine time, compared with conventional methods was achieved.

The method of grinding under optimum elastic pressure indicates that grinding under conditions of self-dressing may yield a faster rate of metal removal than grinding operations involving the process of grinding wheel wear.

A type of grinding under adjustable pressure is the method developed and applied by Heald Machine Co. of USA [15/8] and called Controlled Force Grinding, shortly: CF.

The latter method differs from the formerly described IOS method in that unit pressures in grinding are not of the order of magnitude which would cause the self-dressing process to occur. In the CF method the grinding wheel wears as grain corners become rounded off and abraded.

The spindle itself constitutes the element which measures and controls the pressure.

An optimum pressure causing suitable elastic deformations of the spindle is experimentally determined for the particular workpiece material, taking into account accuracy and quality requirements.

The CF method is used in internal grinding, e.g. in the grinding of antifriction bearing rings, replacing conventional methods with adjustable feed. This method simultaneously reduces machining time (by reducing the number of spark-out passes), and ensures a high accuracy and surface quality.

15.4.2 Selection of cutting conditions in grinding*

In the case of grinding with grains wearing down to a conventional degree of wear, the basis for the selection of cutting conditions should be point life computed according to principles discussed in Chapter 14. Point life values recommended for practical purpose may be sufficiently accurate in other machining methods, but not in grinding. This follows from an insufficient amount of study devoted so far to the dependence of point life on various grinding parameters.

In the case of grinding with self-dressing of the grinding wheel some other index than T_e should be chosen for the determination of economical cutting conditions. The *relative volumetric rate of metal removal* K_V is recommended as such an index.

The value of K_V characterizes the grinding process. The higher K_V is, the better the choice of cutting conditions and grinding wheel.

As an example, in Fig. 15.32 values of K_V are presented, as obtained in experimental grinding of some metals with aluminium oxide grinding wheels EB40(36) of hardness K and M, and with silicon carbide wheels SC40(36) of the same hardness. Graphical representation of the results of these experiments give an indication of the order of magnitude and scatter of the relative rate of metal removal.

Thus e.g. in the case of grinding sintered carbides the value of $K_V = 0{\cdot}2\text{–}1$.

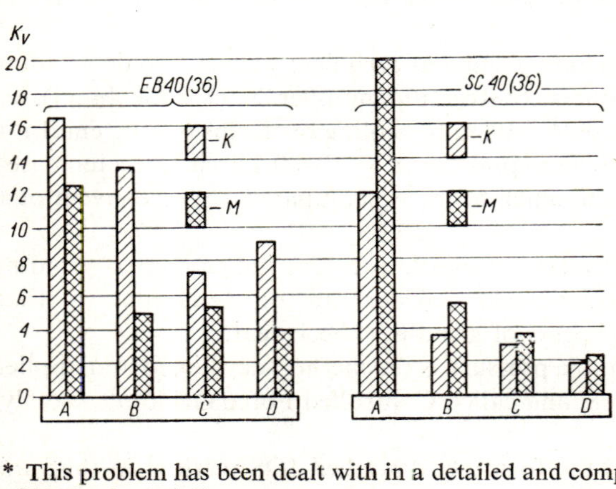

Fig. 15.32 Relative rate of metal removal when grinding: A—of cast iron, B—of 0·45 carbon steel, C—of chromium-nickel-molybdenum steel, D—of fungsten tool steel, with EB40(36) and SC40(36) grinding wheels of hardness K and M

* This problem has been dealt with in a detailed and comprehensive way by L. Burnat in Chapter 5 of his book [15/13].

The value of factor K_V, however, cannot be the only and final criterion for the selection of cutting conditions.

The decisive index of an economically correct choice of cutting conditions should be the unit cost. There exists a value of K_V for which the cost of the operation reaches its minimum (Fig. 15.33). This value is called the *economical coefficient of the metal removal rate* (by analogy to T_e) and denoted by K_e. The K_e value may differ from $K_{V\,max}$ obtained.

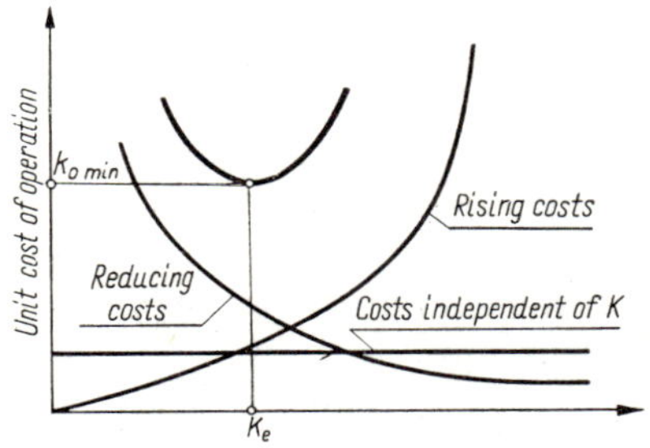

Fig. 15.33 Relationship between the unit cost of operation and the coefficient of relative rate of metal removal

The principles of grinding wheel choice follow from the description of the grinding process.

The abrasive material is chosen depending on its mechanical strength properties. Aluminium oxide is used in grinding hardened and non-hardened steels, malleable cast iron and soft bronze. Silicon carbide is used in grinding sintered carbides, grey cast iron, and bronze and aluminium castings. When grinding sintered carbides the use of green silicon carbide is recommended.

The size of the abrasive grain is chosen depending on the surface quality required, on the properties of the work material and on the area of contact between grinding wheel and workpiece. The higher the surface quality requirements the finer is the required abrasive grain (higher grit size number).

In grinding ductile and soft materials grinding wheels with coarser grain should be used, while in grinding hard and brittle materials finer grain is applicable.

For larger work surfaces coarse grained grinding wheels are recommended.

Hardness should be chosen carefully, since its optimum degree depends on the type of workpiece material and on grinding conditions. In principle, a hardness should be aimed at, which results in the highest value of coefficient K_V in work involving self-dressing of the wheel. The coarser grained the grinding wheels, the harder they should be. On the other hand, the longer is the path of contact between the points and workpiece material, the lesser should be the grinding wheel hardness (for the sake of reducing the probability of burns and seizures). Therefore, in the case of surfacing, especially when using the face of the grinding wheel, the latter should be softer than in the case of grinding surfaces of revolution.

In most cases a ceramic bond is used. Phenolic resin and rubber bonds are recommended in the case of high wheel speeds or when using very fine-grained grinding wheels or when a high surface quality is required.

The selection of structure should mainly depend on grinding conditions. In finishing and profiling the use of grinding wheels of dense structure is indicated, while an open structure is appropriate in the case of machining soft and ductile metals, particularly in surfacing using the grinding wheel face. In all other cases grinding wheels of medium structure are used.

In choosing a grinding wheel it should be remembered that the final criterion of this choice is an operational test, as only then do all features of the wheel interact.

In Table 15.6 recommendations are listed for the choice of grinding wheels for various grinding operations.

The selection of cutting conditions in grinding depends, to a higher degree than in chip removing methods, on descriptive and qualitative recommendations.

Cutting speed selection is limited, first of all, by the tensile strength of the grinding wheel. Therefore, attention should be paid to the permissible rotational speed as given on each grinding wheel (or on the accompanying attestation).

Other limitations in the choice of cutting speed are imposed—less frequently—by the type of work material, surface quality and point life requirements etc. It should be remembered in choosing the cutting speed that it exerts a distinct influence on the so-called dynamic hardness of the grinding wheel.

The selection of workpiece peripheral speed (v_w) or feed rate (s_t) is performed according to previously given formulae.

Grinding allowances are determined within the limits of 0·05–0·3 mm on each side, divided into 60–80% for rough grinding and 20–40% for finish grinding.

Depth of cut is adopted within a relatively narrow range. Excessive depth of cut results in heavy point loads and increases the cutting forces, thus unfavourably influencing the machining accuracy. On the other hand, too small a depth of cut makes grinding difficult and increases friction which may cause burns. Usually, for initial passes $a = 0{\cdot}01$–$0{\cdot}3$ mm is adopted and for finishing passes $a = 0{\cdot}005$–$0{\cdot}015$ mm.

Grinding width B_a (in grinding surfaces of revolution $B_a = s$) is expressed in fractions of grinding wheel width B. The value of this fraction depends mainly on the required surface finish and machining accuracy and on the rigidity of the m–w–t system. The higher is the desired surface quality the smaller should be the B_a/B ratio. The same measure is recommended in the case of less rigid m–w–t systems.

In the case of rough (initial) grinding $B_a = (0{\cdot}3$–$0{\cdot}5)B$ is taken if $d_w \geqq 20$ mm and $B_a = (0.7$–$0{\cdot}85)B$ if $d_w < 20$ mm. In the case of finish grinding the values $B_a = (0{\cdot}2$–$0{\cdot}3)B$ for steel and $B_a = (0{\cdot}25$–$0{\cdot}4)B$ for cast iron are applied.

The rigidity of the m–w–t system and its vibration resistance have a particular importance for machining accuracy and surface finish in grinding and therefore the choice of cutting conditions recommended above should be checked by a practical test in operation.

The succession of the choice of cutting conditions in grinding should be as follows:

(a) division of allowance among the roughing and finishing operations, determination of the number of cutting and spark-out passes and of the depth of cut,
(b) selection of longitudinal feed (or grinding width B_a) expressed in fractions of grinding wheel width,
(c) selection of cutting speed,
(d) calculation of peripheral speed of the workpiece or of feed rate (s_t—in surface grinding).

Table 15.6 *Recommended characteristics of grinding wheels for various grinding operations and certain work materials*

Grinding type	Grinding wheel features	Work material: Non-hardened steel	Hardened steel	High-speed steel	Cast iron
Cylindrical external grinding of surfaces of revolution	Abrasive (1)	EB	EB; EA	EB; EA	SC
	Grain number (2)	40(36)–32(46)	40(36)/32(46)– –40(36)/25(60)	40(36)/46– –32(46)/25(60)	40(36)–32(46)
	Hardness (3)	K–N	H–K	K–L	J–L
	Bond (4)	C	C	C	C
	Structure (5)	5	5	5	5
Centreless external grinding	(1)	EB	EB	EB	SC
	(2)	32(46)–25(60)	25(60)–15(80)	25(60)–15(80)	32(46)–25(60)
	(3)	L–M	J–L	J–L	K–L
	(4)	C	C	C	C
	(5)	5	5	5	5
Internal grinding of surfaces of revolution	(1)	BE	EB; EA	EB; EA	SC
	(2)	40(36)–32(46)	32(46)–25(60)	40(36)/25(60)– –32(46)/25(60)	40(36)–32(46)
	(3)	J–K	H–K	J–K	J–K
	(4)	C	C	C	C
	(5)	5	5	5	5
Surfacing using grinding wheel periphery	(1)	EB	EB; EA	EB	SC
	(2)	40(36)–32(46)	40(36)–32(60)	32(46)–25(60)	40(36)–32(46)
	(3)	J–L	H–J	H–J	J–K
	(4)	C	C	C	C
	(5)	8	8	8	8
Surfacing using grinding wheel face	(1)	EB	EB	EB	SC
	(2)	100(16)–40(36)	63(24)–40(36)	63(24)–32(46)	100(16)–63(24)
	(3)	J–M; H–L	H–K	O–P	J–L; J–M
	(4)	B	B; C	B	C; B
	(5)	8	8	8	7

As a final step, the calculation of cutting force and grinding time is made according to previously given formulae.

REFERENCES

15/1 SLONIMSKII, V. I.: 'Teoriya i praktika beztsentrovogo shlifovaniya' (Theory and Practice of Centreless Grinding), Mashgiz, Moskva 1949

15/2 MASLOV, E. V.: 'Osnovy teorii shlifovaniya metallov' (Elements of the Theory of Metal Grinding), Mashgiz, Moskva 1951

15/3 Min. Mashinostroi. S.S.S.R.—N. Byuro Tekhn. Norm. — Normy stoikosti shlifovalnykh krugov i rozkhoda almaznogo instrumenta (Standards of grinding wheel life and diamond tool wear), Mashgiz, Moskva 1953

15/4 MASLOV, E. N., and IGNATOV, B. A.: 'Novye issledovaniya v oblasti chistoty shlifovannoi poverkhnosti. Novye issledovaniya v oblasti obrabotki rezaniem metallov i plastmass' (New Research in the Field of Ground Surface Quality. New Research in the Field of Machining Metals and Plastics by Cutting), Mashgiz, Moskva 1958

15/5 MATALIN, A. A.: 'Prichiny vozniknoveniya ostatochnykh napryazhenii'. Lonitomash. Kniga 34. 'Kachestwo obrabotannykh powerkhnostei' (Causes of the Appearance of Final Stresses. Lonitomash, Volume 34. Quality of Machined Surfaces), Mashgiz, Moskva 1954

15/6 YASHCHERITSYN, P. J.: 'Szlifowanie szybkościowe' (High-speed Grinding), Translation from Russian, PWT, Warszawa 1955

15/7 KACZMAREK, J., and MARKOWSKI, S.: 'Podstawy sprężystego ostrzenia ściernego noży z płytkami z węglików spiekanych' (Elements of elastic sharpening of tools with sintered carbide bits by abrasion), Mechanik 10 and 11, 1960.

15/8 HALM, R. S.: 'Controlled-force grinding—a new technique for precision internal grinding', ASME, paper No 63–WA–21, Nov. 1963

15/9 Collective work edited by MASLOV, E. N.: 'Osnovnye voprosy wysokoproizvoditelnogo shlifovaniya' (Basic Problems of High-Output Grinding), Mashgiz, Moskva 1960

15/10 LUR'E, G. B.: 'O rezhushchei sposobnosti shlifoval'nykh krugov' (On the cutting ability of grinding wheels), Abrazivy 17, 1956

15/11 EILMES, J., and KACZMAREK, J.: 'Wpływ niektórych warunków szlifowania węglików spiekanych na powstawanie pęknięć i mikropęknięć' (Influence of some conditions of cutting sintered carbides on the appearance of cracks and microcracks), Zeszyty Naukowe PIOS 20, 1965

15/12 SHAW, Mc C., and FARMER, D. A.: 'Analysis of the abrasive machining of plane surfaces', ASME, paper No 64–WA/Part 4, Dec. 1964

15/13 BURNAT, L.: 'Szlifowanie i dogładzanie ścierne metali' (Metal Grinding and Abrasive Superfinishing), WNT, Warszawa 1962

CHAPTER 16

Honing

16.1 General characteristics of smoothing operations

Smoothing is the term used for abrasive processes carried out by means of abrasive tools called hones (abrasive sticks); there may be various working motions, the grinding speed being much higher than the cutting speed.

Depending on the kinematics of the working motions and their aims, the following smoothing processes are distinguished:

(a) honing or smoothing with long-stroke motions,
(b) superfinishing or smoothing with short stroke oscillatory motions,
(c) lapping by means of laps.

A common feature of all these smoothing methods is that they are intended for finishing operations whose principal aim it is to improve surface quality and only partially to correct the accuracy of shape and dimensions. Besides, all types of smoothing contribute to endowing the surface layer with qualities desirable from the practical point of view.

Therefore smoothing, as well as other abrasive processes, should find a wide application. All the more so, since the necessary technically justified machining allowances are very small (of the order of the mean total height of surface irregularities left over by the previous machining operation).

16.2 General characteristics of honing

Honing is a smoothing process in which the tool carries out a rotary and a reciprocating motion (at one or two frequencies) while the workpiece does not perform any working motions.

The honing tool is called a *hone* or *abrasive stick*. As a rule, honing is carried out by means of heads with a number of hones mounted on them depending on the diameter of the work surface.

The length of stroke is shorter than that of the work surface, as during the operation hones should not leave the latter. However, stroke length must be sufficient for honing the entire length of the hole (Fig. 16.1). The relationship between the lengths of stroke, work surface and hone is determined from

$$k = l - l_h + l'_{o1} + l_{o2} \tag{16.1}$$

Honing may be carried out for a very large range of diameters and lengths of work surfaces: beginning with very small diameters, of the order of a few millimeters, to such values as 1 m and lengths of between 1 cm and 20 m.

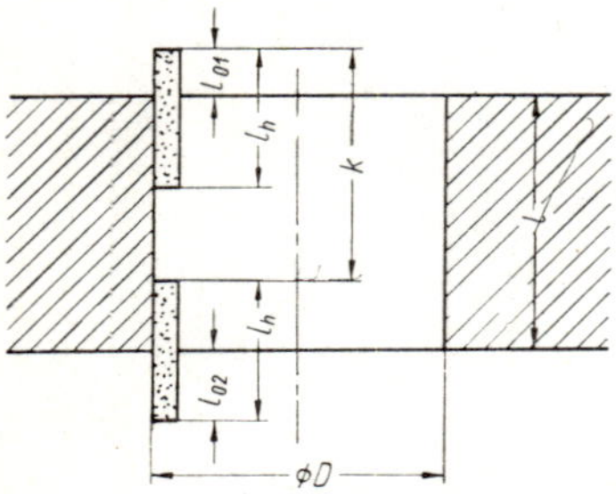

Fig. 16.1 Stroke length dependence on the lengths of work surface, hone and its overruns

Normally, applications are those involving surfaces of the order of a few tens of centimetres. This is the reason for calling this operation *long-stroke smoothing*.

In cases when the honing head stroke is not very long, vertical honing machines are used.

In the case of very long strokes the height of production shop limits the range of vertical honing. Horizontal honing machines are then used, and possible deflections which may be caused by the force of gravity are prevented by appropriate guiding of the honing head.

Honing applications comprise:

— honing of cylindrical and tapered holes,
— honing of external surfaces of revolution in the form of cylinders and cones,
— honing of certain formed surfaces (e.g. of gear teeth) and plane surfaces; these applications are not as frequent as the preceding ones.

As compared with grinding, honing is characterized by:

— participation in the cutting process of a number of cutting grains hundreds of times greater, the machined surface thus being many times larger,
— cutting speeds reduced to 1/50–1/150 of speeds used in grinding,
— possibility of obtaining a better quality of machined surface (roughness numbers 9 to 13).

Honing is the most widely used of all the smoothing types of abrasive processes, as it enables one to obtain a high accuracy of form and dimension and a good surface quality. It also ensures a good quality of the surface layer resulting from—among other factors—low cutting temperatures.

An important advantage of honing is the possibility of obtaining a very high dimensional accuracy almost independently of the kinematic and dynamic properties of the machine tool.

Relatively high rates of metal removal should also be mentioned among the advantages of honing. In this respect honing is considerably better than lapping although the latter has been known for a longer time.

An increase of production capacity output and reduction of machining costs may be obtained by the application of honing, since several through holes may be machined simultaneously provided that the workpiece design and the length of the honing machine stroke allow it. This particularly concerns cases when the length of the work surface is shorter than its diameter.

Blind holes may also be honed but this requires that the hole be extended and enlarged at the end, thus enabling the hones to overrun the work surface in both directions. Blind hole honing indicates the versatility of the process.

In addition to the most common type, characterized above, other types of honing are being developed.

One of them consists in imparting an oscillatory motion of a frequency of about 20 Hz to the honing head in addition to its usual reciprocating motion*. This increases the productivity of the honing process although it complicates the honing head and machine design.

Another type aims at improving the surface quality and accuracy. This type is applied as finish-honing in a double-cut honing process. It consists in honing without rotary motion, only with reciprocating motion, by means of hones arranged in such a way as to cover the entire circumference.

16.3 Honing kinematics

Working motions in honing are:
- a uniform rotary motion at a rotational speed n and peripheral speed $v_p = \pi Dn$;
- reciprocating motion at a stroke frequency f and an average linear speed $s_t = 2kf$.

Composition of both these speeds results in the cutting speed

$$v = \sqrt{v_p^2 + s_t^2}$$

In the case of honing the peripheral speed may not be considered equal to the cutting speed, because the value of linear speed s_t can be of the same order as that of the peripheral speed or even higher.

Cutting speed direction is defined by angle φ measured between the tangent to the path of grain motion at the given point and the plane perpendicular to the axis of rotary motion. The following relationships are therefore valid:

$$\begin{aligned} v_p &= v \cos\varphi \\ s_t &= v \sin\varphi \end{aligned} \tag{16.2}$$

The composition of rotary and reciprocating motion should occur so that the abrasive grains of the hones should not move along their own trails at all, or at least may do so only after a suitably large number of strokes. This can be achieved if stroke length is not a multiple of the helical pitch of the motion path, i.e. when (acc. to Fig. 16.2):

$$k = (m+e)H = (m+e)\pi D \tan\varphi \tag{16.3}$$

* This variety has kinematics similar to that of oscillatory superfinishing. The reason for classifying it as honing are only the quantitative differences, i.e. parameters more typical of honing than of oscillatory superfinishing.

where: m—a whole number; $|e| < 1$—a number smaller than unity (usually $|e| < 0{\cdot}2$) indicating through what part of the circumference the trails of the given point are displaced.

The length of that part of the circumference through which the trails of the given point are displaced is determined by the relationship

$$s = e\pi D \tag{16.4}$$

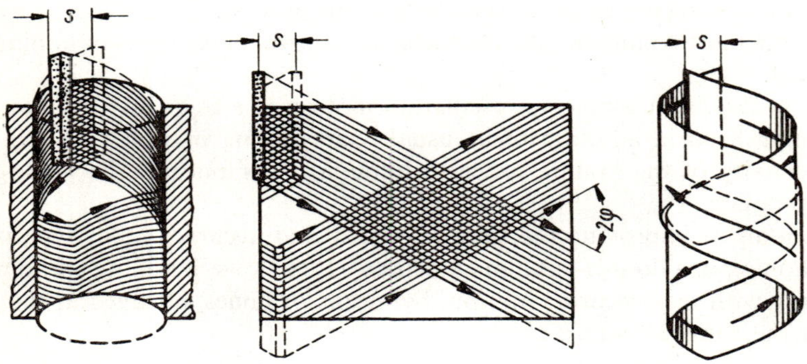

Fig. 16.2 System of paths of abrasive grain motion in the honing process

The value of s should not be greater than the circular pitch between hones, minus hone width b

$$s \leqq \frac{\pi D}{z} - b \tag{16.5}$$

where: z—number of hones, b—hone width.

It follows from eqn. (16.3) that

$$\tan\varphi = \frac{k}{(m+e)\pi D} \tag{16.6}$$

At the same time the following relationship occurs

$$\tan\varphi = \frac{s_t}{v_p} = \frac{2kf}{\pi Dn} \tag{16.7}$$

From eqns. (16.6) and (16.7) it follows that

$$\frac{n}{f} = 2(m+e) \tag{16.8}$$

It is desirable that $1/e$ be a prime number, as then the condition is satisfied that grains do not follow their own trails.

If k and D are given and the value of φ is assumed, the calculation of

$$m+e = \frac{k}{\pi D \tan\varphi} \tag{16.9}$$

may then be performed.

Honing time is shorter, the larger is the number of strokes f per time unit, the higher is the peripheral feed rate s and the greater is the average depth of cut a. The latter value grows with—among other factors—the peripheral speed v_p. This can be expressed by

$$t_m = \frac{1}{f} \frac{\pi D}{s} \frac{q}{a} \tag{16.10}$$

Equation (16.10) may be read as follows: honing time is equal to the product of the time of a single stroke $1/f$, the number of strokes corresponding to the circumference of work surface $\pi D/s$ and the number of passes q/a.

16.4 Honing tools (honing heads)

16.4.1 Honing head construction

Honing heads have three functional parts:

— the body, also constituting a shank connecting the head with the machine tool spindle and a mounting for the hones,
— the set of hones, constituting the working part,
— device for the adjustment of reach and pressure exerted on the hones; it is placed inside the honing head body.

From the point of view of the cutting process an important role is played by the method of aligning the spindle axis and honing head axis with the axis of the work surface, by the method of exerting pressure on the hones and by the cutting properties of the hones themselves.

As far as the alignment of spindle and head axes and work surface axis are concerned, two methods are distinguished:

— the workpiece is clamped in a stationary position and the honing head is self-aligning (Fig. 16.3a) due to a special design solution,
— the head is clamped rigidly to the machine tool spindle and the workpiece may align itself according to the head position (Fig. 16.3b).

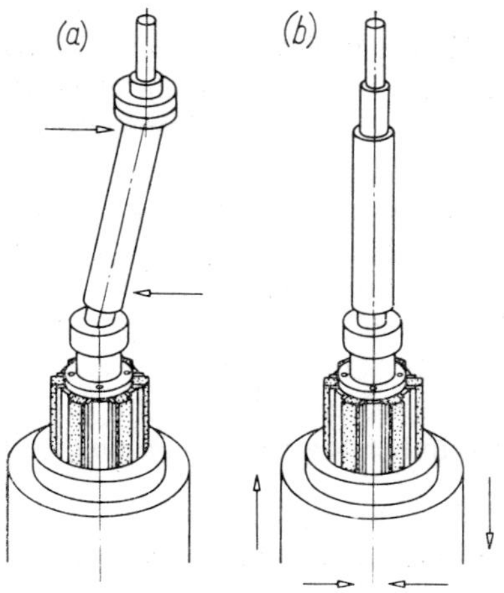

Fig. 16.3 Methods of aligning the honing head and spindle axes with the axis of the hole being machined: (a) self-aligning honing head, (b) rigid honing head and self-aligning workpiece [16/1]

The method involving rigid head and self-aligning workpiece is less frequently applied, as in general it is only suitable for light and small components and requires the use of a special fixture for the workpiece. However, it enables one to obtain a high bore accuracy.

The method involving a self-aligning head prevails in design solutions, as it reduces the time of workpiece setting by not demanding precise alignment, even in the case of strict accuracy requirements.

As far as the setting of hones and adjustment of their pressure is concerned, there exist many and varied design solutions. From the point of view of basic features of these mechanisms two extreme cases are distinguished. One of them consists in setting the hones to the required dimension by adjusting the grinding depth without ensuring a constant pressure exerted on the hones. In practice intermediate solutions are most numerous.

With a view to the correctness and good effects of honing, the second design should be considered more favourable. All the more so, as it permits one to adjust the pressure at the beginning and at the end of the honing process. The unit pressure should be higher at first, with regard to the increase of the rate of metal removal, and lower at the end, with regard to the necessary surface quality.

The greatest influence on the honing process is exerted by the selection of the hones which are discussed below.

16.4.2 Hones

Abrasive materials used in the production of hones are the same as those used in the production of grinding wheels. However, synthetic diamond* hones are used more frequently than diamond wheels.

Recommendations for the particular type of abrasives are similar to those applied in grinding:

— when honing steel—special kind of aluminium oxide and green silicon carbide (carborundum),

— when honing cast iron and non-ferrous metals—green silicon carbide (carborundum).

The size of abrasive grains used in the production of hones varies within wide limits, from 32(46) (the characteristic grain dimension $a \approx 400$ μm) to micro-dust of 14 μm ($a \approx 14$ μm). This is the consequence of the application of honing to improve both the accuracy of form and dimensions and the surface quality.

Grain numbers up to about 3(320) are applied in operations and cuts aimed at improving the accuracy, when larger allowances are to be removed. Grains of the smallest size (micro-dust) are used to improve surface quality only.

It may be stated tentatively that in the machining of steel and cast iron:

— hones of grain number 10(120) yields roughness number 7 when sharp and 8–9 when partly worn,

— hones of grain number 3(320) yield roughness number 8 when sharp, and 9–10 when somewhat worn,

* During the last few years in some industrial countries borazon (boron nitride) has been introduced, particularly for honing austenitic steels.

— hones of grain number 28 μm yield roughness number 9 when sharp and 10–11 when somewhat worn,
— hones of grain number 20 μm yield roughness number 11 when sharp and 12–13 when somewhat worn,
— hones of grain number 14 μm and finer yield roughness number 12 when sharp and 13 when worn.

Hone bond does not play an important role in the honing process itself, but bond strength is essential as the permissible rotational speed and the number of strokes per unit time depend on it.

In addition, it is desirable that the bond should have polishing properties.

For these reasons hones with an organic bond (such as phenolic resins) are becoming increasingly popular. They are characterized by a high strength $\sigma \approx 700$ kg/cm^2.

Hone hardness is normally within the limits from L to P. The following recommendations are valid in the choice of hardness:
— a higher hardness should be chosen in the case of large machining allowances, rigid workpieces and high unit pressures,
— a lower hardness should be chosen when workpieces are not sufficiently rigid, work surfaces are discontinuous (intermittent) and accuracy requirements high.

It should also be remembered that the *relative dynamic hardness* changes depending on cutting speed.

Hone dimensions are standardized. With regard to the cutting process the hone length (l_h) and width (b) are of importance.

The larger is the area of contact between hone and work surface, the more material is being cut. Hone length normally amounts to $\frac{1}{3}$–$\frac{3}{4}$ of work surface length so that with overruns within the limits of $\frac{1}{3}$–$\frac{1}{2}$ of hone length the unit pressure should not change to such an extent as to cause shape errors.

Hone life depends on many factors and varies within wide limits. It can be considerably extended if grinding occurs under conditions of partial self-dressing. The latter takes place most often at the beginning of the honing process when the roughness resulting from the preceding operation is still substantial. Thus a small area of actual contact combined with constant pressure causes the actual values of unit pressure to increase markedly.

Hone life is most often determined in a statistical way. The number of operations which the hone is capable of performing is used to characterize hone life.

It is assumed approximately that standard hones:
— if made of aluminium oxide or silicon carbide are capable of performing 100–500 operations, depending on machining allowance and other factors,
— if made of diamond dust are capable of performing 5–50 thousand operations within the allowance range of up to 30 μm.

16.5 Production engineering possibilities and the selection of honing conditions

The values of machining allowances for honing depend, on the one hand, on the method of workpiece preparation, i.e. of the preceding operations, and, on the other hand, on the objective of the operation.

Semi-finished products which are to be honed are usually drilled and reamed or bored and ground. In such cases roughness number 5–8 may be expected.

Under these conditions the thickness of the layer to be removed by honing is generally not more than double the mean total height of surface irregularities. Thus the machining allowance for honing is

$$q_h \approx B\, R_{tm} \tag{16.11}$$

where: R_{tm}—mean total height of irregularities of the surface to be honed; B—a number determined empirically: for roughness numbers 5–6 its value is $B = 2$, and for roughness numbers 7–8 it is $B = 1{\cdot}5$. Certain deviations of B values depend on the kind of work material.

In the case when honing is intended not only to improve surface quality, but also the accuracy of form and dimensions, machining allowances must be larger. They are usually contained within limits of 20–100 μm.

In these cases the principle of dividing the honing process into *preliminary* and *finish honing* is applied, while suitable tools and machining conditions are selected. If the total allowance does not exceed 50–60 μm, *single-cut honing* is permissible.

Tentative conditions of single-cut honing are given in Table 16.1. It concerns holes of dimensions $D \geqq 40$ mm and $l \geqq 100$ mm. Machining conditions are given for typical representatives of 4 groups of materials i.e. for non-hardened carbon steel, hardened steel, grey cast iron and bronze, all of medium hardness characteristic of such materials.

In addition to already known symbols, in Table 16.1 the machining allowance for single-cut honing aimed at surface finish and surface layer quality improvement is denoted by q_{h1}.

It has also been assumed that starting with roughness number 5 only roughness number 9 may be attained economically by single-cut honing, and starting with roughness number 6—only 9 or 10.

In order to apply Table 16.1 as approximate data and to adapt them to different conditions it is necessary to use the general relationships between the most important machining conditions and honing results.

It should be added to the general indications concerning the choice of hone grain size, that the influence of grain size depends to a large extent on working time or on grain wear. The dependence of the linear rate of metal removal and of the attainable roughness number on the time of honing using hones of various grain numbers is shown in Fig. 16.4. The relationship illustrated there is as follows:

- — the rate of metal removal by honing diminishes with time in an approximately hyperbolic way, independently of grain number,
- — the mean total height R_{tm} decreases with honing time for any grain size until, as a result of wear, some burning and scratching of the surface by single grains begins to occur; from then on surface irregularities increase,
- — the use of finer grained hones yields lower rates of metal removal and improved surface quality; the difference between the influences of various grain sizes decreases with honing time.

Particularly important for the honing process as for all other abrasive processes is the unit pressure. Its influence upon the linear rate of metal removal and on surface roughness is illustrated by Fig. 16.5.

From the relationships presented there, the following conclusions may be drawn:

— with an increase of unit pressure the metal removal rate grows almost proportionately over a wide range,

— reduction of unit pressure below a certain value has almost no further influence on the decrease of surface roughness; this is the range where cutting stops and only friction and surface scratching occur;

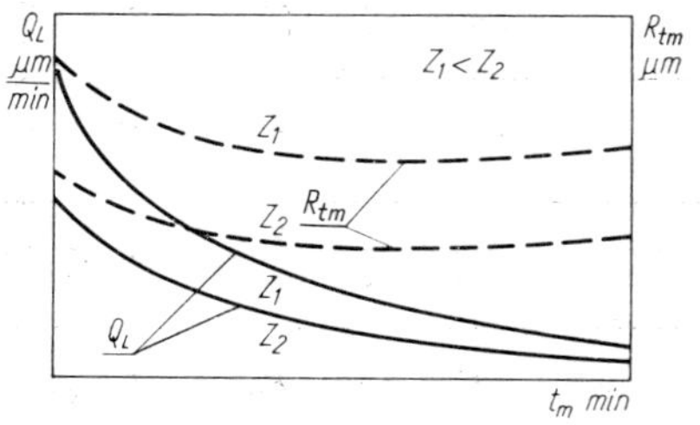

Fig. 16.4 Dependence of the linear rate of metal removal Q_L and of the mean total height of surface irregularities R_{tm} on honing time t_m for various grain sizes

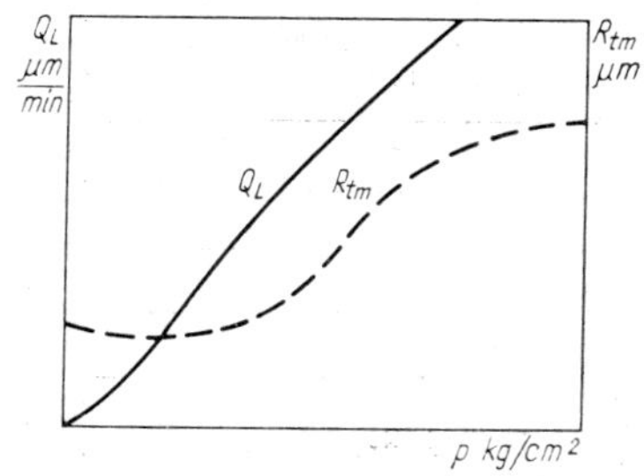

Fig. 16.5 Dependence of the linear metal removal rate Q_L and the mean total height R_{tm} on unit pressure p

— above a certain value of unit pressure its influence on surface roughness diminishes, as in this range self-dressing begins to play its role, stabilizing the conditions of shaping of the machined surface;

— within the honing range in which the grinding process occurs normally and the hones work under conditions of wear, an increase of unit pressure causes surface quality to deteriorate.

The unit pressure value is chosen:

— higher, when machining allowances are relatively large and surface quality requirements low, hones are hard, coarse-grained and large, and the speed of reciprocating motion rather high,

— lower, when conditions and requirements are just the opposite of those listed above.

When choosing unit pressure values one should remember that the actual unit pressure is always higher than its nominal value when the honing process begins and lower when it ends. This is the result of the actual contact area increasing with honing time.

The cutting speed components exert a twofold influence on the rate of metal removal and on the surface roughness: by the variations of their absolute values and by the changes of the resultant cutting speed direction.

Increasing the peripheral speed of hones causes an almost linear growth of metal removal rate. On the other hand, surface roughness becomes slightly reduced. This general relationship is illustrated in Fig. 16.6.

Increasing the speed of reciprocating motion influences the metal removal rate in honing in a way similar to that of peripheral speed, but this influence is more intensive, while surface roughness grows with speed s_t. Therefore, changes of metal removal rate not accompanied by those of surface quality are better effected by

Table 16.1 *Tentative machining conditions for single-cut honing* (for $d \geqq 40$ mm and $l \geqq 100$ mm)

Required roughness number	Machining conditions		Semi-finished products: roughness numbers (▽) and machining allowances (q_h)															
			▽5 (R_{tm} < 20 μm)				▽6 (R_{tm} < 10 μm)				▽7 (R_{tm} < 6·3 μm)				▽8 (R_{tm} < 3·2 μm)			
			q_{h1} = 0·04				q_{h1} = 0·02				q_{h1} = 0·01				q_{h1} = 0·006			
			Non-hardened steel	Hardened steel	Cast iron	Copper alloys	Non-hardened steel	Hardened steel	Cast iron	Copper alloys	Non-hardened steel	Hardened steel	Cast iron	Copper alloys	Non-hardened steel	Hardened steel	Cast iron	Copper alloys
▽9	Hone	Abrasive material	EA	EA	SZ	SZ	EA	EA	SZ	SZ	EA	EA	SZ	SZ	EA	EA	SZ	SZ
		Grain number	10(120)	8(150)	10(120)	5(240)	8(150)	180*	8(150)	4(280)	180*	5(240)	180*	3(120)	5(240)	3(320)	5(240)	3(320)
		Hardness	P	O	P	O	P	O	P	O	P	O	P	O	P	N	O	O
	Unit pressure kg/cm²		10	10	8	4	9	10	7	4	8	9	6	3	7	9	5	3
	Kinematics	v, m/min	45	35	60	75	40	30	60	70	35	26	62	65	30	22	64	60
		$\varphi°$	30	11	25	6	28	10	23	6	26	9	20	5	24	8	17	5
		v_p, m/min	39	34	54	74	35	29	55	69	32	25	58	65	27	21	61	60
		p_t, m/min	23	7	25	8	19	5	23	7	15	4	21	6	12	3	19	5
	Cutting fluid		N† 100	N 100	N 100	W‡	N 90	N 90	N 90	W	N 90	N 90	N 100	W	N 80	N 80	N 90	W
▽10	Hone	Abrasive material					EA	EA	SZ	SZ	EA	EA	SZ	SZ	EA	EA	SZ	SZ
		Grain number					5(240)	3(320)	5(240)	3(320)	5(240)	3(320)	5(240)	3(320)	3(320)	28 μm	3(320)	28 μm
		Hardness					N	O	P	O	O	N	O	N	O	N	N	N
	Unit pressure kg/cm²						8	9	6	3	7	8	5	3	6	8	4	3
	Kinematics	v, m/min					40	33	62	70	35	28	64	66	35	26	66	62
		$\varphi°$					26	10	20	6	24	9	17	5	21	8	14	4

		v_p,m/min p_t,m/min					36 18	32 6	58 21	69 7	32 14	27 4	61 19	66 6	33 13	25 4	64 16	62 4
	Cutting fluid						N 80	N 80	N 90	E§	N 80	N 80	N 90	E	N 70	N 70	N 90	E
▽11	Hone	Abrasive material Grain number Hardness									EA 28 μm N	EA 28 μm N	SZ 28 μm N	SZ 20 μm N	EA 28 μm N	EA 20 μm N	SZ 28 μm N	SZ 20 μm N
	Unit pressure kg/cm²										6	7	4	3	5	7	3	2
	Kinematics	v, m/min $\varphi°$ v_p,m/min p_t,m/min									40 21 37 14	30 9 29 5	66 14 64 16	68 4 68 5	40 18 38 12	28 9 27 4	68 11 67 13	64 4 64 5
	Cutting fluid										N 70	N 70	N 80	E	N 60	N 60	N 80	E
▽12 to ▽13	Hone	Abrasive material Grain number Hardness									EA 20 μm M	EA 14 μm M	SZ 20 μm M	SZ 14 μm N	EA 20 μm M	EA 14 μm M	SZ 20 μm L	SZ 14 μm N
	Unit pressure kg/cm²										5	6	3	2	3	6	2	1
	Kinematics	v, m/min $\varphi°$ v_r,m/min p_t,m/min									45 18 43 14	33 10 32 6	68 11 67 13	70 4 70 5	45 15 44 12	30 9 29 6	70 8 69 10	66 4 66 5
	Cutting fluid										N 60	N 60	N 75	E	N 55	N 55	N 75	E

* Old denotation acc. to Norton classification.
† N denotes paraffin and the following number—its percentage in the cutting fluid (the rest consisting of oils paraffin hydrocarbons).
‡ W denotes water.
§ E denotes 4% emulsion of oil.

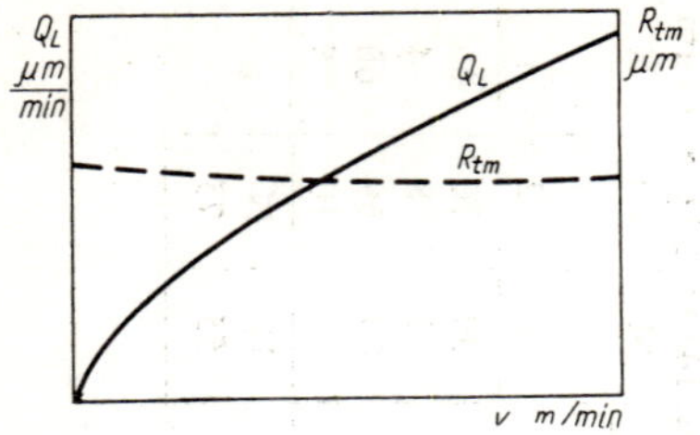

Fig. 16.6 Dependence of the linear metal removal rate Q_L and the mean total height R_{tm} on the peripheral speed of hones v_p

varying the peripheral speed. On the other hand, changes of surface quality may be caused by varying the speed of reciprocating motion.

Changes of v_w and s_t values should be introduced in such a way as to obtain not only the desired results of honing, but also the recommended value of cutting speed and its appropriate direction.

Ranges of optimum cutting speed values have been established empirically for certain work materials:

— for constructional carbon steels $v = 36\text{–}48$ m/min
— for hard steel alloys and hardened steel $v = 22\text{–}36$ m/min
— for grey cast iron $v = 60\text{–}70$ m/min
— for non-ferrous and light alloys $v = 60\text{–}75$ m/min.

The angle determining the cutting speed direction is also half the angle of intersection of lays in the honing process. The value of this angle depends on the interrelationship of the cutting speed components, as shown by eqn. (16.7).

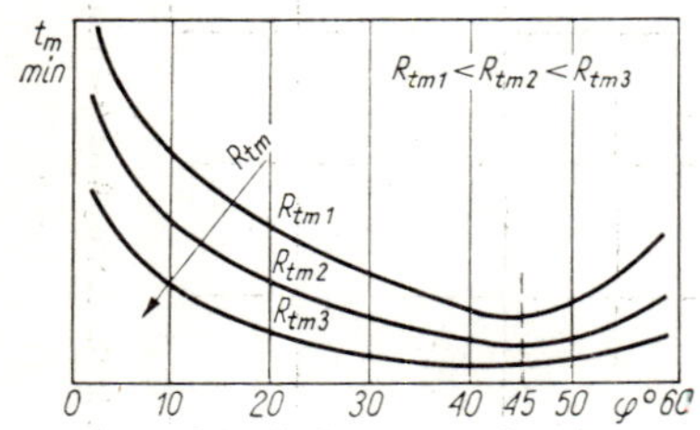

Fig. 16.7 Dependence of the honing time t_m, up to the moment when various mean total heights of surface irregularities R_{tm} are attained, on the angle of path of resultant grain motion (cutting speed direction)

Experiments indicate that angle φ exerts its influence on the rate of metal removal and on surface finish independently of the influence of the cutting speed components.

The general character of this influence is illustrated by Fig. 16.7. The honing time needed to obtain the desired surface quality is the measure of metal removal rate. It is evident from Fig. 16.7 that as angle φ grows, the time t_m necessary to attain the desired surface finish decreases. This time is longer, the smaller is to be the mean total height of surface irregularities R_{tm}. When angle φ passes the value of 45°, machining time begins to increase again.

Thus the optimum value of cutting speed direction angle is 45°. This is accurate in the case when the tool marks on the work surface before honing are parallel to the axis of rotation or to a plane perpendicular to it. If the preliminary machining tool marks form a different lay then the optimum value of angle φ differs from 45° correspondingly.

The practical application of the optimum value of angle φ is usually impossible for technical reasons (among others, because of excessive mass forces of the reci-

Table 16.2 *Ranges of practically applied cutting speed directions in honing* [16/1]

Workpiece material	Cut	v_p/s_t ratio	Cutting speed direction angle $\varphi°$
Grey cast	preliminary finishing	2–5 4–8	26–11 14–7
Non-hardened steel	preliminary finishing	1·5–2·5 2·5–40	34–22 22–14
Hardened steel	single	5–7	11–8
Bronzes	single	10–16	6–35

procating motion). Therefore practically permissible and satisfactory values of φ have been determined experimentally. These values are given according to Lur'e [16/1] in Table 16.2.

In the selection of a cutting fluid the principles discussed in the general part of the book are applied. Because of the lack of more extensive studies on the influence of various liquids, mainly mixtures of paraffin and oils are used in practice. Increasing the oil content leads to an improvement in surface quality while the metal removal rate is reduced.

It should be expected that the assortment of honing coolants will be widened, as paraffin has some disadvantages, such as being harmful to health and easily inflammable.

An improvement in surface finish may be obtained in the honing process by providing ample amounts of cutting fluid to the machining zone and by filtering the fluid.

REFERENCES

16/1 Lur'e, G. B.: 'Metody otdelki detalei abrazivnymi instrumentami' (Methods of Producing Parts by Means of Abrasive Tools), Trudrezervizdat, 1958

16/2 Woodridge: 'An experimental investigation into the honing process', Production Engineer, 11, 1963

16/3 Żebrowski, H.: 'Niektóre zagadnienia gładzenia otworów osełkami karborundowymi i elektrokorundowymi' (Some problems of honing holes by means of silicon carbide and aluminium oxide hones), Mechanik, 11, 1964

16/4 Żebrowski H.: 'Gładzenie otworów osełkami diamentowymi' (Honing of holes by means of diamond hones), Mechanik, 12, 1964

16/5 Fragin, I. J., and Safranow, W. G.: 'Diamentowe gładzenie (honowanie) otworów w przedmiotach ze stali hartowanych' (Diamond honing of holes in workpieces made of hardened steel), Mechanik, 8, 1964

16/6 Fragin, I. E.: 'Osobennosti khoningovaniya sinteticheskimi i estestvennymi almazami. Khoningovanie'. Iz knigi: Obrabotka mashinostroitel'nykh materialov almaznym instrumentom (Particular Features of Honing by Means of Synthetic and Natural Diamonds. Honing. From the book: Machining of Mechanical Engineering Materials by Means of Diamond Tools), Izd. Nauka, Moskva 1966

16/7 Yashchenko, P. F.: 'Opyt khoningovaniya vysokoprochnykh stalei bruskami iz sinteticheskikh almazov' (Experiments in Honing-strength Steels by Means of Synthetic Diamond Hones), Izd. Nauka, Moskva 1966

16/8 Collective work: 'Rabotosposobnost' bruskov iz sinteticheskikh almazov pri khoningovanii' (Working Capability of Synthetic Diamond Hones in Honing), Izd. Nauka, Moskva 1966

CHAPTER 17

Oscillatory superfinishing

17.1 General characteristics of oscillatory superfinishing

Oscillatory superfinishing—shortly called *superfinishing*—is a smoothing method in which the working motions are as follows:

— oscillatory motion of the tool, i.e. reciprocating motion of short stroke and high frequency,
— rotary motion of the workpiece,
— feed motion of tool or workpiece.

Direction of the oscillatory motion is parallel to the axis of workpiece rotation.

Feed motion may be parallel to the workpiece axis (*straight superfinishing* of external and internal surfaces of revolution) or perpendicular to it (*radial superfinishing*). If there is no feed motion, the operation is called *plunge superfinishing*.

In Fig. 17.1a–c external straight superfinishing, plunge superfinishing and radial superfinishing are presented.

By assumption, oscillatory superfinishing is a machining method whose exclusive aim it is to improve surface finish and the quality of the surface layer. A correction of shape and dimension accuracy is not aimed at.

Consequently, superfinishing allowances are still smaller than those for honing.

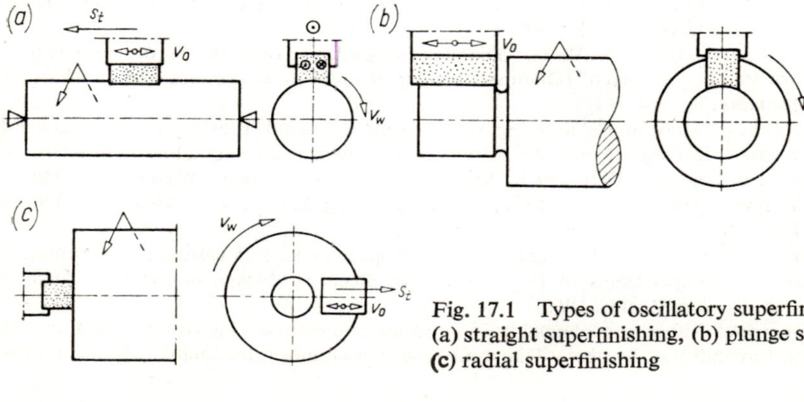

Fig. 17.1 Types of oscillatory superfinishing: (a) straight superfinishing, (b) plunge superfinishing, (c) radial superfinishing

In principle, they are equal to the mean total height of surface irregularities resulting from preliminary machining plus a part of the surface layer damaged in the latter operation. Thus superfinishing allowances are often contained within the limits of dimensional tolerance.

Oscillatory superfinishing offers the greatest possibilities of obtaining high-quality surface finish up to and including roughness number 14.

Of the useful characteristic of an object with superfinished surfaces, particularly advantageous is its wear resistance. A superfinished surface does not wear easily because its irregularities are very low and, consequently, its load carrying capacity is relatively high. The approximate range of roughness numbers and of load carrying capacity of the profile of superfinished surfaces—as compared with ground and turned surfaces—is shown in Fig. 17.2.

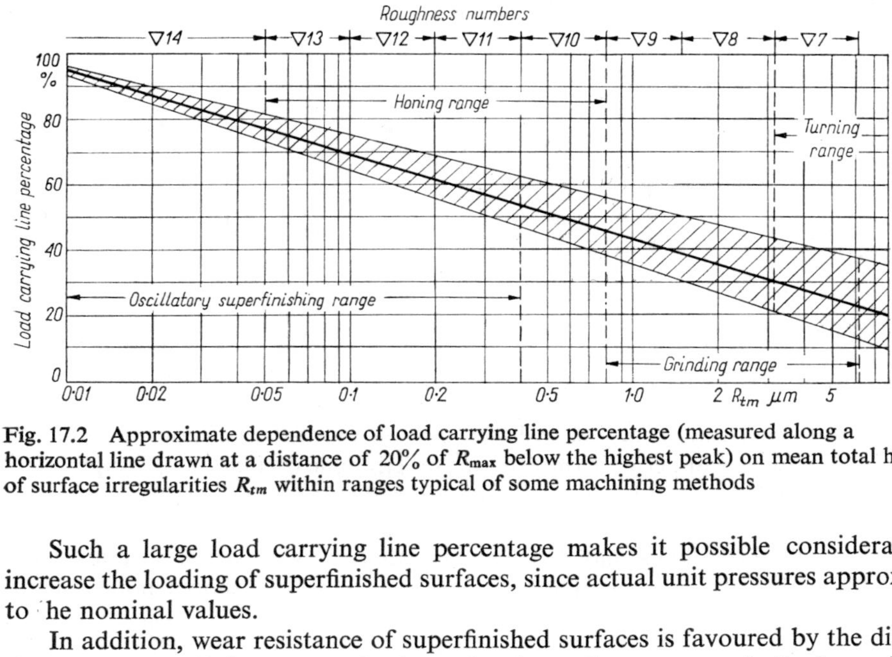

Fig. 17.2 Approximate dependence of load carrying line percentage (measured along a horizontal line drawn at a distance of 20% of R_{max} below the highest peak) on mean total height of surface irregularities R_{tm} within ranges typical of some machining methods

Such a large load carrying line percentage makes it possible considerably to increase the loading of superfinished surfaces, since actual unit pressures approximate to he nominal values.

In addition, wear resistance of superfinished surfaces is favoured by the distribution of post-machining stresses in the surface layer. In most cases in the layer situated closest to the surface and exposed to wear, compressive stresses appear which increase its wear resistance.

When compared with grinding, oscillatory superfinishing shows the same differences and the same characteristic features as honing (see Section 16.1).

17.2 Superfinishing kinematics and superfinishing time

In Fig. 17.3 the path of grain motion in straight oscillatory superfinishing is projected on a developed cylindrical work surface.

The superfinishing hone should start its work in such a position that a large portion of hone length will be in contact with the workpiece surface. However, some part of the hone must protrude beyond this surface. The length of this part l_a should not be smaller than the feed per revolution, i.e. workpiece travel in relation to the oscillating part of the superfinishing head during one revolution of the workpiece. This condition is recorded as

$$l_a \geqq s \tag{17.1}$$

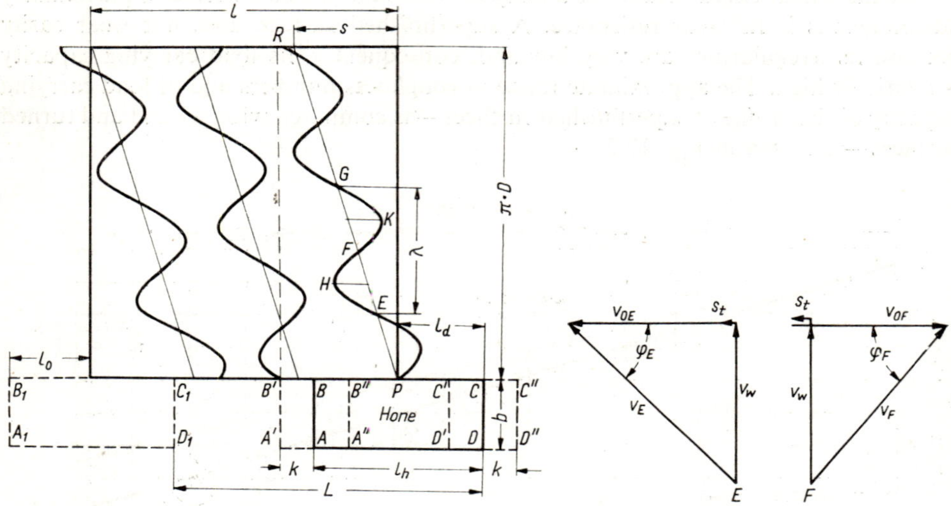

Fig. 17.3 Path of grain motion on work surface and cutting speed components in straight oscillatory superfinishing

The composition of rotary motion and feed motion yields the path of grain motion *PR*. As a consequence of periodic (mostly sinusoidal) hone oscillations of amplitude k and wave length λ the resulting grain path motion P will pass the points *E–H–F–K–G–R*.

The amplitude (stroke) k may be controlled while the wave length λ results from the condition that the number of waves along the periphery should not be a full multiple

$$(m+e)\lambda = \pi D \tag{17.2}$$

where: m—number of full lengths of oscillation wave on the periphery, e—a number contained within the limits $0 < e < 1$, usually taken as $e \approx 0{\cdot}5$.

Since

$$f\lambda = n(m+e)\lambda = \pi D n = v_w \tag{17.3}$$

it follows that the number of strokes per minute or the frequency of oscillations f will be

$$f = n(m+e) \tag{17.4}$$

The relationship shown in eqn. (17.2) expresses the requirement that—particularly in plunge superfinishing—grains should not move along their previous paths.

The cutting speed of straight superfinishing constitutes the vectorial sum of the peripheral speed of the workpiece v_w, oscillatory motion speed v_o and longitudinal feed rate $s_t = s\,n$ and is determined by

$$v = \sqrt{v_w^2 + (v_o + s_t)^2} \tag{17.5}$$

In the case when the oscillations correspond to a sinusoidal motion, then

$$v_o = 2\pi fk \cos(2\pi ft) \tag{17.6}$$

The maximum value of oscillation speed occurs for hone displacements equalling zero, e.g. for points E and F. In Fig. 17.3 the composition of vectors resulting in the cutting speed is presented, while the sense of v_{oE} and v_{oF} is taken into account.

Maximum cutting speed occurs at point E (and generally speaking, at points where the sense of $v_{o\,max}$ agrees with that of the longitudinal feed rate vector) and thus

$$v_{max} = \sqrt{v_w^2 + (2\pi fk + s_t)^2} \tag{17.7}$$

Minimum cutting speed will occur when $v_o = 0$ or

$$v_{min} = \sqrt{v_w^2 + s_t^2} \tag{17.8}$$

Equations (17.5), (17.7) and (17.8) are applied in the calculation of cutting speed, usually for design purposes. In production engineering the oscillatory motion is characterized by the average speed determined by

$$v_o = 2kf \tag{17.9}$$

It is evident from the comparison of eqns. (17.9) and (17.6) that the maximum speed of oscillation

$$v_{o\,max} = \pi v_o \tag{17.10}$$

Taking into account eqn. (17.9) one may write the formula for average cutting speed

$$v_{av} = \sqrt{v_w^2 + (2fk + s_t)^2} \tag{17.11}$$

or, taking into account the equations for v_w and s

$$v_{av} = \sqrt{(\pi Dn)^2 + (2fk + sn)^2} \tag{17.12}$$

By convention, if it is not specially indicated which cutting speed is to be taken into consideration, then from a production engineering point of view the average speed is to be understood. Its value is calculated according to eqn. (17.12) which means that $v = v_{av}$ is not accepted.

The angles of intersection of superfinishing marks are varied and changing since the sinusoidal path of one grain intersects the paths of other grains forming different intersection angles.

Because of the variety of these angles, to characterize the system of superfinishing marks on a machined surface, intersection angles corresponding to two hone positions are conventionally considered:

— angle of intersection of grain path (mark) with the direction of generation on work surface when the superfinishing hone is in its central position (displacement equal to zero). This angle is denoted by φ and its value is determined by the relationship (Fig. 17.3)

$$\tan\varphi = \frac{v_w}{s_t \pm v_o} \approx \frac{v_w}{v_{o\,max}} \tag{17.13}$$

— angle of intersection of paths (marks) of two grains, when these paths are displaced in the direction of generation by half the stroke (amplitude) of oscillation and have a phase shift $e = 0{\cdot}5$ (i.e. by half the wave length). Of two path intersection angles which may be determined (Fig. 17.4), angle 2ψ is chosen. The value of path intersection angle is determined by

$$\tan\psi = \frac{k}{4}\,\frac{12}{\lambda} = 3\,\frac{k}{\lambda} \tag{17.14}$$

since the co-ordinates of the point of intersection are: $k/2$ and $\lambda/12$.

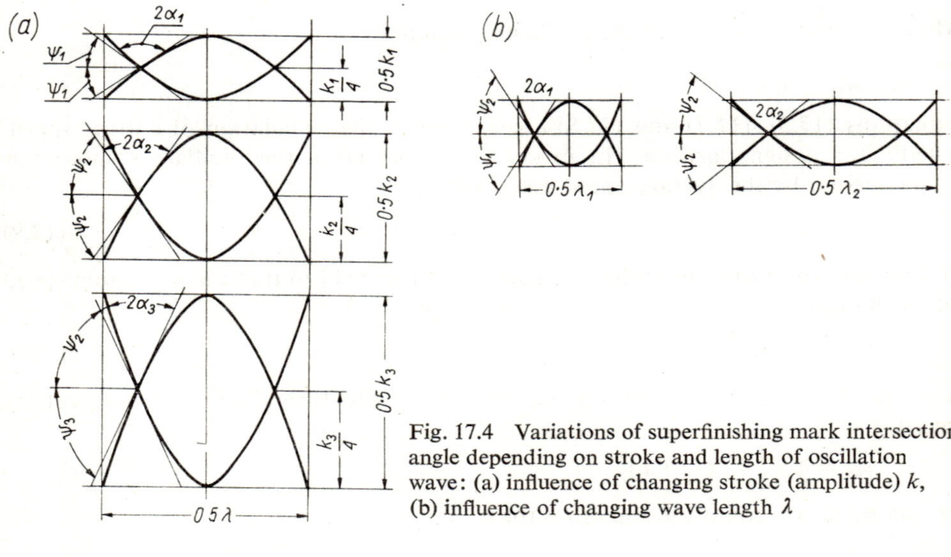

Fig. 17.4 Variations of superfinishing mark intersection angle depending on stroke and length of oscillation wave: (a) influence of changing stroke (amplitude) k, (b) influence of changing wave length λ

It follows from eqn. (17.14) that the value of path intersection angle depends in a directly proportional way on oscillation stroke and in an inversely proportional way on wave length, i.e. on n/f ratio and work surface diameter. The influence of variations in stroke k and wave length λ are illustrated in Fig. 17.4a and b, respectively.

Taking into consideration eqns. (17.13), (17.9), (17.10), the equation for peripheral speed of the workpiece and eqn. (17.14), one may write an equation showing the relationship between angle φ and ψ

$$\tan\varphi = \frac{3}{2}\,\frac{D}{\lambda}\,\frac{n}{f}\cot\psi \tag{17.15}$$

In view of the complicated character of the system of superfinishing marks, optimum values of angles may be determined empirically.

Superfinishing time of a surface of length L is calculated according to

$$t_m = \frac{q}{a}\frac{L}{sn} = \frac{q}{a}\frac{\pi DL}{1000v_w s} \quad \text{min} \tag{17.16}$$

where: q—total machining allowance in min, a—depth of cut in mm, L—length of work surface in mm, D—diameter of work surface in mm, s—longitudinal feed in mm/rev, n—rotational speed in rev/min, v_w—peripheral speed of the workpiece in mm/min.

The influence of oscillatory motion on superfinishing time is expressed indirectly by the value of superfinishing depth (in a single pass). The higher is the oscillation frequency, and the greater the stroke length and active surface area of the hone, the greater is the depth of cut.

The superfinishing time [eqn. (17.16)] may be expressed verbally as follows: superfinishing time is equal to the product of the number of passes (q/a) and work surface area (πDL) divided by surface area being superfinished during one minute ($1000v_w s$).

17.3 Selection of oscillatory superfinishing conditions

The most important criteria for the optimization of oscillatory superfinishing are:
— the required surface quality and, consequently, load carrying capacity of the surface,
— surface rate of metal removal by superfinishing,
— relative wear of hones or relative surface rate of metal removal.

The selection of semi-finished products, hones, kinematic conditions, unit pressure and cutting fluid should be made with regard to these criteria.

The choice of semi-finished products should take into account, first of all, the appropriate accuracy of dimensions and shape, as well as surface quality, since the machining allowance results from these factors.

It should be considered as a general rule that the quality of surfaces to be superfinished must correspond to roughness number 8 or 7.* Form and dimension accuracy should be kept within such tolerances that after superfinishing the class of accuracy would remain the same as before, although the roughness number will have reached the values 12–14.

In cases when the machining allowance must exceed 8 μm it is recommended that the superfinishing operation be performed in two passes: the preliminary and the finishing one.

Improving surface quality over roughness number 8 is already uneconomical, as it yields no marked effects in superfinishing.

Of the shape errors which should be avoided in the semi-finished products, the most undesirable is waviness, particularly of considerable wave length or amplitude.

Hone selection should be carried out according to the same general principles as for honing.

* This rule concerns the present state of the superfinishing technique. Future progress may allow superfinishing of much coarser surfaces, e.g. those of roughness numbers 5–6, to become economical.

Only the choice of hone dimensions and number should be dealt with in a different way.

For small work surface lengths, hones of the same length are recommended. The machining of such surfaces is then undertaken by plunge superfinishing.

For straight superfinishing longer hones should be selected, the larger is the longitudinal feed (s) then the smaller is the width of the hones and their number in the superfinishing head.

The phenomenon of hone vibrations, considerably deteriorating surface quality, limits the possibility of increasing hone length. In addition, the factor limiting hone length is the trend towards versatile heads. Short hones may be used for straight superfinishing of longer surfaces, but long hones are rarely suitable for superfinishing much shorter work surfaces.

Hone width may not be smaller than half the oscillation wave length. This means that the larger is the n/f ratio, the wider should be the hone. It is recommended to aim at a hone width

$$b \approx (0{\cdot}3\text{–}0{\cdot}4)\lambda \tag{17.17}$$

Excessive hone width is not recommended since it worsens the conditions of surface wetting and lubrication and, in addition, the number of cutting grains does not grow in proportion to width increase. In practice, therefore, widths of 20–25 mm are not exceeded. If this does not satisfy the condition given in eqn. (17.17), then the oscillation wave length should be changed correspondingly, or else two or more narrower hones should be used instead of a single larger one.

The number of hones is determined mainly from the designer's point of view. For example, the central angle ε (Fig. 17.5), comprising all hones on the work periphery, should not be too wide, as this would increase the total pressure and also the

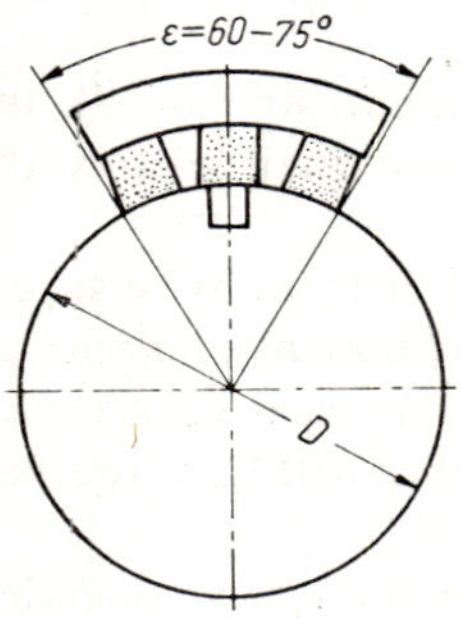

Fig. 17.5 Arrangement of three hones permitting one to superfinish a shaft with a keyway

bulk of the superfinishing head. In practice, an angle ε of less than 75° is usually adopted.

The number of abrasive grains used in the production of superfinishing hones are contained within the range from 3(320) to 7 μm which corresponds to characteristic grain dimensions from 30 to 5 μm.

Ceramic or phenolic resin bonds are used. The addition of graphite to the bond contributes to the achievement of better surface quality.

The hardness of superfinishing hones is chosen according to the same principles as in honing. The softest hones, from H to J, are used in superfinishing hardened steels. Non-hardened constructional steel, cast iron and non-ferrous alloys are superfinished using hones of hardness K to O, while hardness from I to L is considered most suitable in the case of light alloys.

In the choice of hardness the range of cutting speed should be taken into account, and particularly that of the rotational speed of the workpiece. The higher is the speed v_w the softer may be the abrasive sticks. So-called relative or dynamic hardness of the hones may also be influenced by suitable adjustment of the rotational speed of the workpiece.

The structure of superfinishing hones should be contained in the range 6–8 for light alloys and 8–10 for ferrous metals and copper alloys (bronzes).

Unit pressure is one of the most important factors controlling the oscillatory superfinishing process. By assumption, superfinishing should be performed under small unit pressures.

It has been established experimentally that optimum values of unit pressure are contained within the ranges:

1–5 kg/cm² when superfinishing steel and cast iron,
0·5–2·5 kg/cm² when superfinishing non-ferrous metals,
0·2–1·0 kg/cm² when superfinishing light alloys.

The selection of unit pressure within the optimum range is carried out while taking into consideration the known influence of unit pressure on superfinishing.

There is an optimum value of pressure with regard to surface quality. When this value is exceeded, surface quality deteriorates as a result of the increasing depth of cut of particular grains. Pressures not attaining the optimum value have a similar effect, since friction and surface scratching begin to take the place of cutting. Insufficient pressure may also cause frictional vibrations.

The rate of metal removal by superfinishing increases almost in proportion to the growth of actual unit pressure. At the start of superfinishing operations the actual unit pressure decreases because of an enlarged contact surface area. This reduction of unit pressure is usually sufficient to obtain the desired surface quality.

The relative rate of metal removal by hones, i.e. the ratio of the rate of metal removal by superfinishing to hone wear, attains a maximum value for a certain value of unit pressure.

In Fig. 17.6 is shown the dependence of the rate of metal removal Q_L, the relative rate of metal removal K_L and mean total height of surface irregularities R_{tm} on unit pressure. This diagram illustrates the principle that the optimum range of unit

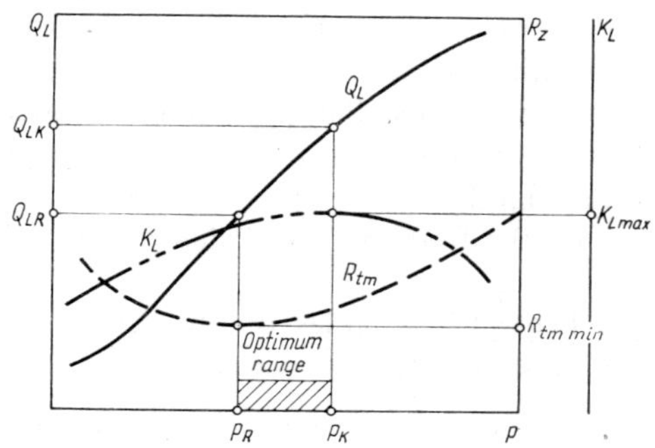

Fig. 17.6 Dependence of mean total height of surface irregularities R_{tm}, rate of metal removal Q_L and the relative rate of metal removal K_L on unit pressure p

pressure values is situated within the limits p_R and p_K. Depending on the remaining conditions of machining, p_K may be greater or smaller than p_R.

The choice of kinematic factors in superfinishing is made difficult by the complex character of their influence. In order to find the optimum values of these factors several studies have been conducted. It follows from them that the arrangement of machining marks indicates in a synthetic way the influence of kinematic factors on surface quality and metal removal rate.

In Table 17.1 the range limits are listed which may be recommended with regard to surface quality and the rate of metal removal. Data contained in this Table are useful in the selection of kinematic conditions to be applied in superfinishing.

Table 17.1 *Recommended range limits of some kinematic factors with regard to surface quality and metal removal rate*

Kinematic factor	Recommended range limits with regard to:	
	metal removal range	surface quality
e	0·5	0·2
$\psi°$	45	15
$\varphi°$	25	70
k mm	$\frac{1}{3}\lambda$	$\frac{1}{12}\lambda$

This selection should begin with the choice of oscillation frequency and stroke, determining oscillation speed. It may generally be recommended that oscillation speed should be as high as possible.

Limitations in the choice of upper values of f and k result from calculations and from dynamic phenomena, i.e. mass forces and vibrations. Their influence deteriorates surface quality and increases hone wear.

In successful designs up to 3000 oscillation strokes per minute are obtained without harmful consequences. However, most often superfinishing devices working at f = 800–2000 strokes/min are encountered.

Stroke length k is most frequently adopted within the limits of 2–4 mm. In successful designs good results have been obtained applying stroke lengths up to 6 mm in the superfinishing of steel and up to 8 mm in that of non-ferrous metals.

It may be established statistically that the average speed of oscillation is most often contained within the range of v_o = 6–9 m/min.

Experiments indicate that increasing the speed of oscillatory motion causes an almost proportional growth of the rate of metal removal by superfinishing while hone wear increases only slightly. On the other hand, surface quality improves if the increment of speed v_o results from increasing the frequency and deteriorates if this increment is caused by increasing stroke length.

Using eqn. (17.13) and the values given in Table 17.1 enables one to determine the range of recommended values of peripheral speed of the workpiece

$$v_w = (1{\cdot}5\text{–}8{\cdot}5)v_o$$

Lower v_w values are adopted at the start of the superfinishing operation, in the case of coarser surfaces to be superfinished or when surface quality requirements are not very high (up to roughness number 12). In order to improve the surface finish, upper values of oscillation speed are applied. In the case of soft workpiece materials, upper values of v_w are raised to $(12–15)v_o$.

It should be taken into consideration that increasing the peripheral speed v_w leads to accelerated hone wear. These symptoms are more distinct, the harder is the workpiece material.

The few studies conducted so far on the influence of feed rate on surface quality and the rate of metal removal by superfinishing, indicate the existence of an optimum feed rate. However, different values are given as this optimum, ranging from 0·2 mm/rev to 8 mm/rev. It may be assumed as a general indication for the choice of feed rate that higher feed values should be selected in the case of large hones (particularly length), of high oscillation frequency and of low peripheral speed of the workpiece.

An important factor in superfinishing is the machining time. As in other abrasive processes, the metal removal rate and roughness reduction reduces with time. The form of this kind of relationship is shown in Fig. 17.7.

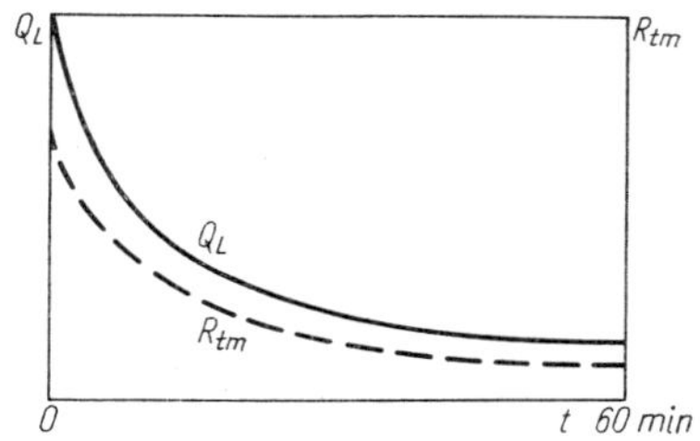

Fig. 17.7 Dependence of the rate of metal removal by superfinishing Q_L and mean total height of surface irregularities R_{tm} on machining time t

These relationships indicate that a good surface finish may be obtained even if machining conditions have not been selected quite correctly, provided that the superfinishing process is suitably prolonged. However, such a method is incorrect and leads to a deterioration of surface layer qualities, as after a certain time hones become worn and rub against the work material. This may be accompanied by a rise of work surface temperature which favours the appearance of tensile stresses.

The use of a cutting fluid in superfinishing aims—as in honing—at lubricating and washing the work and hone surfaces.

Paraffin mixed with spindle oil of a viscosity of 2·5–3·5°E is the most commonly used cutting fluid. Mixing proportions depend on work material:

— for non-hardened steel 70% of paraffin and 30% of oil,
— for hardened steel 85% of paraffin and 15% of oil,
— for cast iron 90% of paraffin and 10% of oil,
— for non-ferrous metals 80% of paraffin and 20% of oil.

In view of the disadvantages of the use of paraffin, the replacement of the latter by new harmless cutting fluids should be aimed at.

A rational application of cutting fluids in superfinishing requires careful filtration, possibly in two stages, i.e. by a magnetic and percolating method in addition to decantation. Experiments indicate that the attainment of roughness number 13 under normal manufacturing conditions is very difficult if appropriate filtration of cutting fluid is not applied.

REFERENCES

17/1 Lur'e, G. B.: 'Metody otdelki detalei abrazivnymi instrumentami', Razd.: Otdelochnaya obrabotka abrazivnymi bruskami—superfinish (Methods of Producing Parts by Means of Abrasive Tools. Chapter: Production Machining by Means of Abrasive Hones—Superfinishing), Trudrezervizdat, 1958

17/2 Górski, E.: 'Obróbka gładkościowa', Rozdz. VI: Dogładzanie (superfinish) (Smoothing. Chapter VI: Superfinishing), WNT, Warszawa 1963

17/3 Butorin, G. I.: 'Superfinishirovanie pri kolebaniyakh bruskov s povyshennoi chastotoi' (Superfinishing with hones oscillating at a high speed), Stanki i Instrument, 12, 1963

17/4 Markowski, S.: 'Badania doboru optymalnych warunków dogładzania oscylacyjnego zewnętrznych powierzchni obrotowych' (Studies on selecting optimum conditions of oscillatory superfinishing of external surfaces of revolution), Zeszyty Naukowe AGH — Przegląd Naukowo-Techniczny, 16, 1965

17/5 Keller, C., Filek, T., and Sawka, Z.: 'Dogładzanie oscylacyjne' (Oscillatory superfinishing), Prace IOS, Seria: Materiały Instruktażowe, 65, 1966

17/6 Uzarowicz A.: 'Zbadanie celowości zastosowania dogładzania oscylacyjnego do ostrzy narzędzi skrawających z węglików spiekanych' (Study of the usefulness of applying oscillatory superfinishing to sintered carbide cutting tool points). A doctor's thesis at the Department of Mechanics of Warsaw Institute of Technology, 1966

CHAPTER 18

Lapping

18.1 Definition and types of lapping

Lapping is an abrasive process carried out by means of loose abrasive in which the role of the cutting points is played by abrasive grains finding momentary support of the laps. The latter perform the function of tool shanks (Fig. 18.1).

As far as the kinematics are concerned, lapping is characterized by the type of relative motions of the lap and work surface. However, it is desirable that the particular grains should not follow their own previous paths on the machined surface.

Lapping may be classified from various points of view.

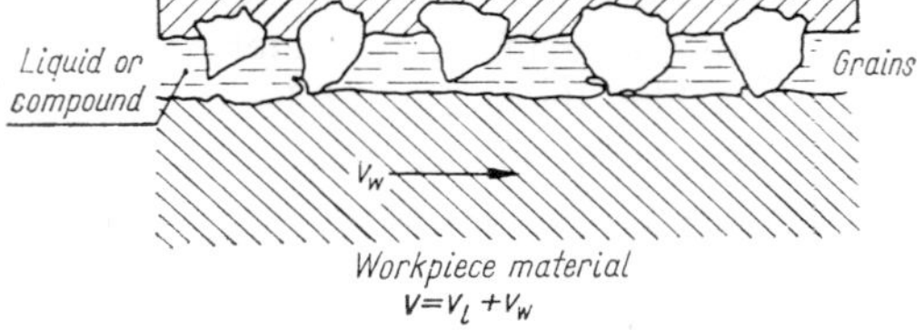

Fig. 18.1 Diagram of the lapping process: P—total pressure, p—unit pressure, A—nominal contact surface area, v_l—lap speed, v_w—workpiece speed, v—cutting speed

With regard to the degree of mechanization *hand* and *machine lapping* are distinguished. The latter is performed on special machine tools called *lapping machines* or by means of special tooling.

With regard to the type of machining process *mechanical* and *chemo-mechanical lapping* are distinguished, the latter being performed with the use of chemically active liquids.

Lapping may also be divided with regard to the way the abrasive and lap co-operate, viz. with laps charged with abrasive either freely or forcedly.

Lap charging means the pressing of abrasive grains into the lap material. Forced charging takes place if the grains are pressed into the lap material before the lapping process starts. This is done by means of special burnishing tools of a hardness higher than that of the laps. Forced charging is applied mainly to distribute the grains evenly over the entire surface of the lap and to ensure that grains of approximately equal dimensions do the work and protrude equally from the lap.

Free charging of the lap is performed during the machining process by delivering portions of abrasive to the machining zone and by pressing the lap against the work surface.

With regard to the form of work surface, rotary lapping (of surfaces of revolution), plane lapping and profile lapping of mating surfaces (i.e. when the mating part plays the role of the lap) are distinguished.

18.2 Laps and abrasive mixture

18.2.1 Laps

The selection of a lap must take into account the properties of the material and requirements concerning shape.

The kind of lap material is particularly important for the lapping process.

For the cutting process in lapping with loose abrasive to take place correctly, the particular grains must be partly driven into the lap surface for a certain time. Lap material then plays a role similar to that of the bond in a grinding wheel. Only the grains embedded in the lap can perform the cutting.

Therefore, lap hardness and its surface structure is of particular importance. Too hard a lap causes excessive rapid chipping of abrasive grains and the periods during which the grains remain embedded are very short. This results in faster wear of the abrasive mixture and a reduction in metal removal rate. In addition, some of the grains are then driven into the work material, thus deteriorating the machined surface quality.

Too soft a lap causes the abrasive grains to be driven too deeply and permanently into the lap material; as grain corners become worn, the rate of metal removal diminishes and the quality of surface finish and of the surface layer deteriorates.

A correctly chosen lap material hardness enables the grains to be temporarily supported, but also to change their positions repeatedly. Thus each grain may cut with several corners instead of one. This favours longer abrasive life, increase in metal removal rate and the improvement of surface finish and surface layer quality.

In view of the requirements listed above, laps used for machine lapping are mostly made of ferritic cast iron of hardness 120–160 HB or pearlitic cast iron of hardness 160–200 HB. The former are better suited for free charging, the latter for forced charging.

In addition to cast iron steel, copper alloys, lead, certain plastics, mirror glass and wood are also used as lap materials. The rule is observed that lap hardness should be higher than that of the workpiece material, if a soft abrasive is used.

As far as shape is concerned, a lap should satisfy two requirements: its shape and dimensions should make possible the lapping of a work surface with the necessary accuracy and an even distribution of abrasive compound over the entire work surface.

To fulfil the second requirement, a suitable pattern of grooves is cut on the lap surface. Sufficient depth of these grooves permits several lap regrindings without repeated groove-cutting.

The permanence of lap form depends mainly on the distribution of pressure on the lap surface and on the wear resistance defined by the *wearability*. With regard to the latter the best material is cast iron. Cast steel laps show a wearability higher by 25% and the wearability of copper laps is even 2·5 times higher than that of cast iron ones.

With regard to the rate of metal removal by lapping better results are generally obtained with steel or copper laps, since particular grains remain embedded in the same place for a longer time.

Examples of laps used for lapping holes, external surfaces and threads are shown in Figs. 18.2 and 18.3.

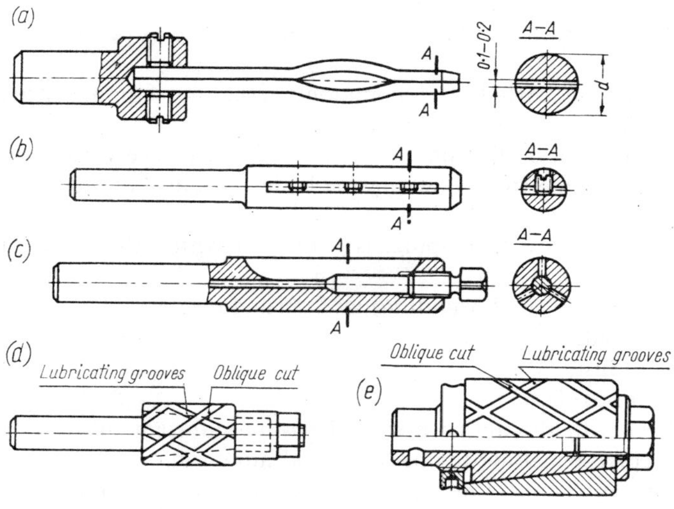

Fig. 18.2 Examples of laps used for lapping holes (acc. to [18/6]): (a) spring-action lap for very small holes (1–4 mm in diameter) made of spring wire (lap wire of a diameter equal to that of the preliminary hole is ground to less than half its cross-section when wound helically around a shaft, (b) and (c) expanding laps for small holes, (d) and (e) laps with expanding cast iron bush for large holes

18.2.2 Abrasive mixture

The abrasive mixture in lapping used most often is in the form of compounds. However, slurries containing abrasive suspended in a cutting fluid are also applied.

The abrasive mixture may be characterized by:

— properties of the abrasive,
— properties of lubricant or cutting fluid,
— degree of concentration and regeneration (dosing).

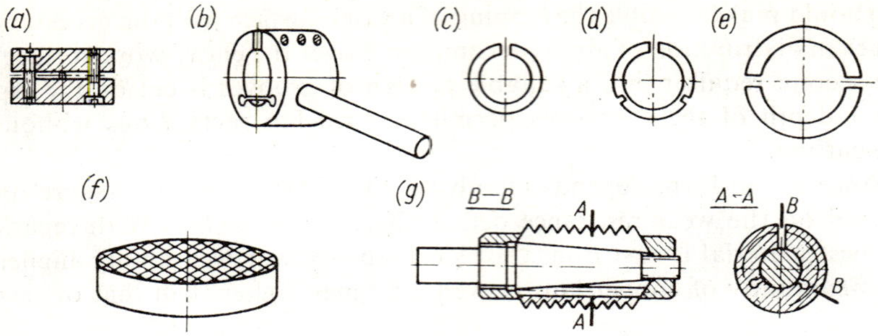

Fig. 18.3 Examples of laps used for lapping shafts, plane surfaces and threads (acc. to [18/6]: (a) plate-type lap for very small shafts, e.g. thread measuring cylinders, (b) lap for small shafts; screws visible at the top of the lap are intended: the side screws for clamping, the central one for expanding, (c) and (d) bushes for shaft lapping holders, (e) split bush for lapping necks and large shafts, (f) lapping block with grooves for abrasive grain, used for hand lapping of plane surfaces, (g) lap with cast-iron expanding bush, used for lapping of large threaded holes

Abrasives used in lapping are characterized—as in other abrasive processes—by the type and size of the grains.

Aluminium oxide and silicon carbide are commonly used for lapping steel, cast iron, non-ferrous metals and sintered materials. In addition, boron carbides (mostly for lapping sintered carbides), diamond dust (for hardened steels and sintered carbides) and chromium oxide (for obtaining the highest quality of steel and copper alloy surfaces) are employed.

The size of abrasive dust, as used in lapping, is contained within the limits of characteristic grain dimensions:

100–5 μm for silicium carbide (carborundum) and corundum,
60–5 μm for boron carbide,
5–0·5 μm for diamond dust,
2–1 μm for chromium oxide.

The compound to be mixed with the abrasive should ensure sufficient lubricating properties of the mixture, correct melting temperature and adhesion.

Of the lubricating compounds the following are used in the preparation of abrasive mixtures: grease, tallow, stearin and wax. In some cases colloidal graphite is added to improve the lubricating properties.

Lubricating liquids in which the abrasive dust is suspended should have—just like the compounds—a suitable density, lubricating properties and adhesion. Machine oil, rape oil or other oils and paraffin are mainly used as lapping fluids.

Certain oils and special additives to abrasive slurries are surface-active. Their influence is evident in finish lapping when the depth of cut of particular grains is very small.

A commonly known compound is the composition of the surface-active compound produced by the State Institute of Optics of the USSR called GOI compound (after the first letters of the Institute's Russian name). This composition is as follows: 74% of chromium oxide, 1·8% of silica gel, 10% of stearin, 10% of hydrolyzed fat, 2% of olein (fatty acid), 0·2% of sodium bicarbonate and 2% of paraffin.

Abrasive mixtures are characterized, first of all, by the degree of concentration, volumetric or gravimetric, on which depends the number of grains per unit work surface area (1 cm^2).

In the light of studies presently conducted, for each set of machining conditions there is an optimum concentration of mixture (with regard to the rate of metal removal by lapping), i.e. an optimum ratio of abrasive volume to the compound or cutting fluid volume. The amount of fluid or compound reaches the optimum value when it suffices for maintaining an extremely thin but continuous film between the lap and work surface, wetting all grains involved.

The optimum degree of concentration depends on unit pressure. It has been established empirically that increasing the concentration necessitates a corresponding increase of unit pressure, if the maximum rate of metal removal under given conditions is to be attained. However, after a certain value of abrasive concentration has been exceeded, a further increase in unit pressure does not result in a further increase in metal removal rate.

This relationship is illustrated in Fig. 18.4. It follows from this diagram that there exists an optimum unit pressure and an optimum degree of concentration at which the maximum rate of metal removal is obtainable.

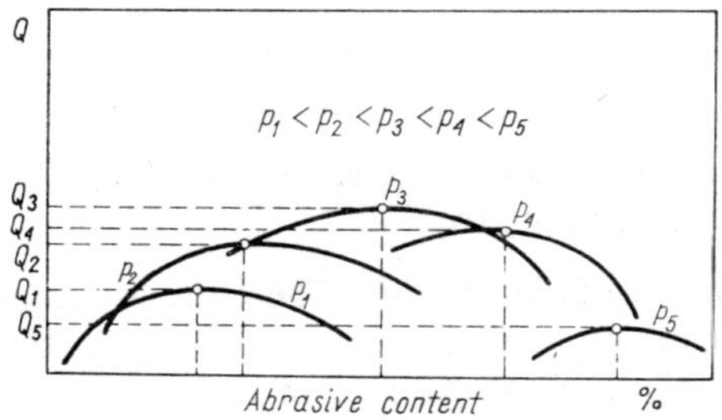

Fig. 18.4 Dependence of the rate of metal removal on abrasive concentration and unit pressure (for constant grain size)

So far no sufficiently comprehensive research has been conducted and, consequently, no numerical values for particular workpiece materials and abrasive mixtures can be recommended. Therefore, certain experience is necessary for the determination of optimum lapping conditions in any particular operation.

A similar situation concerns the optimum composition of abrasive mixtures and lap materials. Many different compositions are employed in practice. Certain tentative indications pertaining to the selection of laps and main components of abrasive compounds or slurries are contained in Table 18.1.

Table 18.1 *Selection of laps and abrasive mixtures for various types of lapping*

Kind of lapping	Abrasive material	Grease or oil	Lap
Preliminary lapping of machine parts	Aluminium oxide or silicon carbide (carborundum) [grain number 16(80)–10(120)]	Machine oil with paraffin to obtain semi-liquid consistency of the compound	Materials softer than workpiece material: ferritic cast iron, copper, brass, lead

Table 18.1 *continued*

Kind of lapping	Abrasive material	Grease or oil	Lap
Finish lapping of machine parts	Aluminium oxide, silicon carbide (carborundum), boron carbide, diamond dust [grain size smaller than 8(150)]. Besides: chromium oxide, ferric oxide	For steel and cast iron: machine oil, paraffin. For copper alloys: tallow, vaseline, with graphite additives. For light alloys: rape oil, paraffin	Materials harder than workpiece material: pearlitic cast iron, chromium steel, hardened steel, glass
Lapping of cutting points	For high-speed steel: aluminium oxide. For sintered carbides: silicon carbide (carborundum), boron carbide [grain size smaller than 8(150)] and diamond dust (grains 1–5 μm)	Paraffin oil, stearin, with oil and graphite additives	Cast iron of lamellar pearlitic structure (HB ≈ 180–200 kg/mm²)

18.3 Production engineering possibilities and the choice of lapping conditions

18.3.1 Machining allowances and lapping accuracy

The main purpose of lapping is to obtain a surface of high quality. Therefore, machining allowances could be of the order of the mean total height of surface irregularities left over from the previous machining operation. Since the rate of metal removal by lapping is relatively low, surfaces intended for lapping are machined to roughness numbers 8–6 which means that lapping allowances may be 10 μm or less. Actually, allowances of 10–20 μm are adopted, as the maximum heights of irregularities should be taken into consideration.

In addition to improving surface quality, lapping may also be used to correct the shape and improve dimensional accuracy of the workpiece. In such cases, depending on the errors involved, lapping allowances are increased. In view of the necessary machining time and the cost of the operation, an allowance value of 0·1 mm is usually not exceeded.

If lapping is to correct the accuracy of workpiece form and dimensions, it should be carried out in several passes, from preliminary to finish lapping. This means that initial passes should involve the use of coarser grains at higher unit pressures and the final passes the application of fine grains and lower pressure.

High machining accuracy is obtainable by lapping. It depends, however, on many factors, such as the method and time of lapping, initial accuracy of the workpiece, kinematic and geometric accuracy of the lap, etc. It is therefore assumed, on a statistical basis, that the attainable dimensional accuracy is generally within the range of ±0·5 μm.

Similarly, a high quality of surface finish may be obtained by lapping. Depending on initial surface roughness and the type and conditions of lapping, the latter may yield surfaces of roughness numbers 11 to 13. A lapped surface is usually matted. For lustre, additional polishing is necessary or lapping under conditions similar to those of polishing.

18.3.2 Unit pressure

The most important factors influencing the lapping results are: unit pressure, abrasive grain size, abrasive concentration and lapping speed.

Practically applied unit pressures are contained within the limits of 2–5 kg/cm² for aluminium oxide and 0·5–2·5 kg/cm² for silicon carbide (carborundum) in preliminary lapping and within the limits of 0·3–1·2 kg/cm² in finish lapping.

Requirements concerning surface finish are decisive for the choice of unit pressure value. A typical relationship between mean total height of the surface irregularities R_{tm} and unit pressure p is shown in Fig. 18.5. In the same diagram a typical dependence of the linear rate of metal removal on unit pressure is presented.

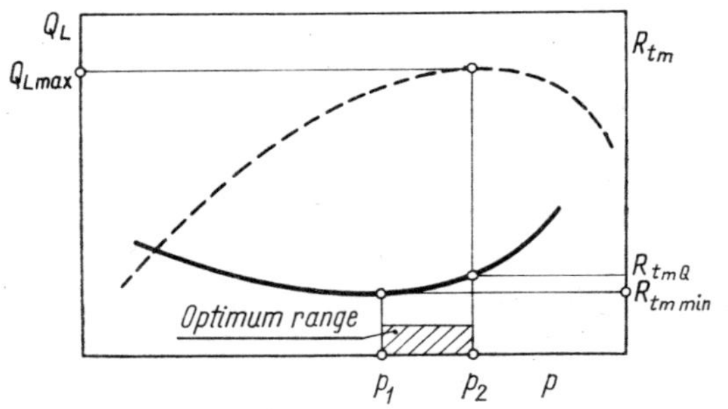

Fig. 18.5 Typical dependence of mean total height of surface irregularities R_{tm} and the rate of metal removal by lapping Q_L on unit pressure p

In general, the optimum unit pressure with regard to surface quality (p_1) is lower than the optimum pressure with regard to the rate of metal removal (p_2). The optimum range of unit pressure values is contained within the limits p_1–p_2, and particular values should be chosen depending on the required surface quality:

— if the permissible mean total height of irregularities R_{tm} is contained within the limits $R_{tm\,min} \leqq R_{tm} < R_{tmQ}$ then unit pressures should be chosen accordingly, in the range from p_1 to p_2;

— if the permissible mean total height of irregularities R_{tm} is greater than R_{tmQ}, then $p = p_2$ is chosen.

18.3.3 Abrasive grain size

Abrasive grain size strongly influences the surface quality and metal removal rate. The selection of appropriate grain size in lapping, under constant unit pressure and other machining conditions, is made taking into account its influence on the rate of metal removal and on surface finish. This may be explained by means of Fig. 18.6. Typical relationships are presented there, between the linear rate of metal removal and mean total height of surface irregularities R_{tm} on the one hand, and the size of abrasive grains on the other hand. Surface roughness increases monotonically with grain size.

This relationship may be approximated with sufficient accuracy by

$$R_{tm} = K\,a^{x} \tag{18.1}$$

where: K—coefficient taking into account the influence of all factors not appearing separately in the formula; a—characteristic dimension of the grain; x—exponent determined empirically.

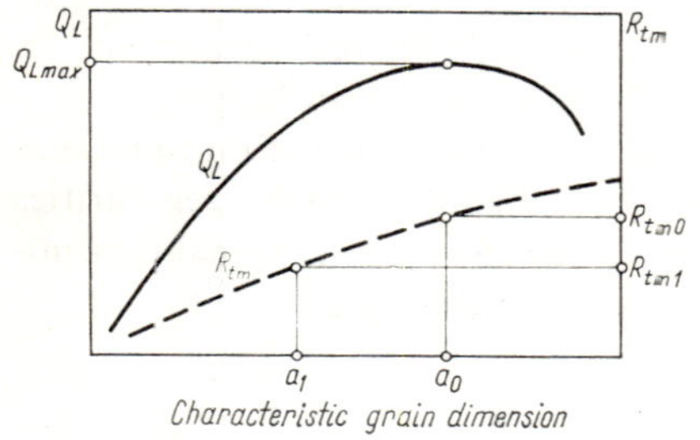

Fig. 18.6 Typical dependence of linear rate of metal removal by lapping Q_L and mean total height of lapped surface irregularities R_{tm} on the characteristic grain dimension a

On the other hand, the rate of metal removal (under constant pressure) reaches its maximum for a certain grain size (a_0). It follows that surface quality requirements allowing, i.e. if the permissible height of irregularities $R_{tm} \geqq R_{tm0}$, grains of dimension $a = a_0$ should be selected.

If $R_{tm} < R_{tm0}$ then a suitable grain size should be chosen to obtain the desired surface quality. For example, if the mean total height of irregularities R_{tm1} is to be attained, the grain dimension fulfilling this condition is $a_1 < a_0$.

Values of a_1 approaching a_0 may be obtained in such cases by suitable choice of unit pressure and the grain concentration in the abrasive mixture (see Fig. 18.4).

18.3.4 Interdependence of unit pressure, abrasive concentration and grain size

Interdependence of optimum values of unit pressure, degree of abrasive concentration in the mixture and abrasive grain size—with regard to the rate of metal removal—may be shortly explained as follows, while making use of certain simplifications.

The volume of material removed by cutting during a determined time and from a determined surface area is the sum of volumes removed by all cutting grains. This means that it depends on the number of such grains and on the volume removed by a single grain.

Increasing the abrasive concentration in the mixture, at a constant grain size and unit pressure, results in an increase of the number of grains per unit area, but also reduces the pressure exerted on a grain, thus decreasing the volume cut by the latter. This indicates that an optimum concentration should exist, for which the sum of volumes cut by the working grains will reach its maximum. At a concentration sufficiently high in relation to the optimum value a reduction of the metal removal rate should occur, because the volumes cut by the particular grains diminish quicker than the number of working grains increases. At a concentration sufficiently low in relation to its optimum value a reduction of the metal removal rate should occur, as the pressure exerted on the grains exceeds the compressive strength of a certain number of them and these are crushed. This, and the fact that the volume of material removed by the particular grains grows slower than the number of grains decreases, should result in a decrease of metal removal rate.

Experience—as mentioned previously—confirms this explanation.

Increasing—at a constant pressure and abrasive concentration—the characteristic dimension of the abrasive grains obviously increases the strength of the latter. Simultaneously, the surface area, necessary for the distribution of the same number of grains, grows larger. As a result, from a certain grain size upwards, the number of cutting grains per unit area does not increase any more, but begins to decrease, and at the same time increases the load acting on the particular grains. With further increase of grain dimensions, the number of grains diminishes in such a way that their loads exceed their compressive strength and they are crushing.

It follows from this description that optimum grain size should be chosen with regard to maximum rate of metal removal. Experiments also confirm this statement.

The increasing of the *nominal unit pressure* from very low values, at a constant concentration and size of abrasive grains, causes a greater depth of cut and thus a larger volume of material removed by the particular grains. This results in an increase of the total rate of metal removal. Beginning with a certain value of unit pressure, the load acting on the grains exceeds their compressive strength which results in crushing and disintegration. Thus the volume cut by a single grain decreases more and more, followed by a diminishing rate of metal removal. This phenomenon does not have a step-like character since the dimensions and strength of abrasive grains are not uniform (see Fig. 18.5).

18.3.5 Lapping time

The influence of abrasive working time on the lapping process is explained principally by two phenomena: wear of grain corners and grain crushing. The latter will always occur, since grain dimensions and compressive strength are not uniform. Only a part of the grains takes the total load, while for a certain number of grains their share of the load exceeds the breaking load value.

In Fig. 18.7 typical relationships are presented between the rate of metal removal Q_L, mean total height of machined surface irregularities R_{tm} and breaking up of

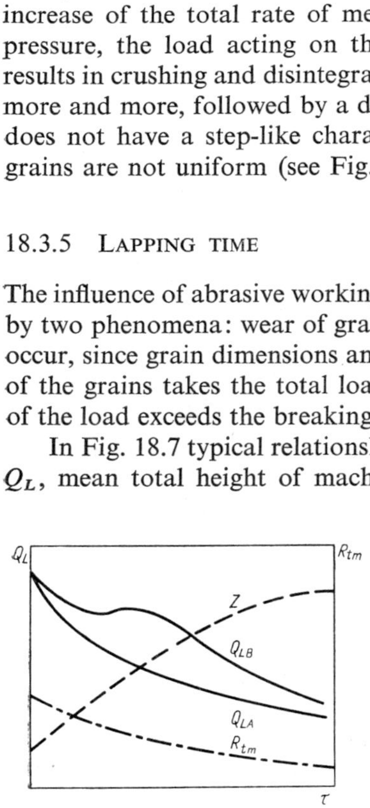

Fig. 18.7 Dependence of the rate of metal removal by lapping Q_L, mean total height of surface irregularities R_{tm} and grain size z variations on lapping time

the abrasive, as measured by actual grain size z, on the one hand and lapping time τ on the other hand.

The dependence of the rate of metal removal is presented in two versions Q_{LA} and Q_{LB}. The first version occurs when unit pressure is lower than the optimum value. The phenomena of grain corner wear and grain disintegration then have a reducing influence on the rate of metal removal. The second version corresponds to the case when unit pressure is at first higher than the optimum value. Thus only after partial disintegration of grains do the unit pressures attain the latter value. This is represented in the diagram by a temporary increase of the metal removal rate, with a following decrease of a monotonic character.

On the other hand, surface quality improves with lapping time in a monotonic way, although—after a certain time—these changes are very slight.

Similarly, the increase of the actual size of abrasive grains with time is also monotonic.

The determination of the necessary lapping time depends on fundamental requirements, i.e. permissible values of parameters concerning surface quality (R_{tm}) and linear loss of workpiece dimension (L_p). In Fig. 18.8 two versions of the combination of the requirements concerning R_{tmp} and L_p are presented with respect to the dependence of R_{tm} and L on lapping time.

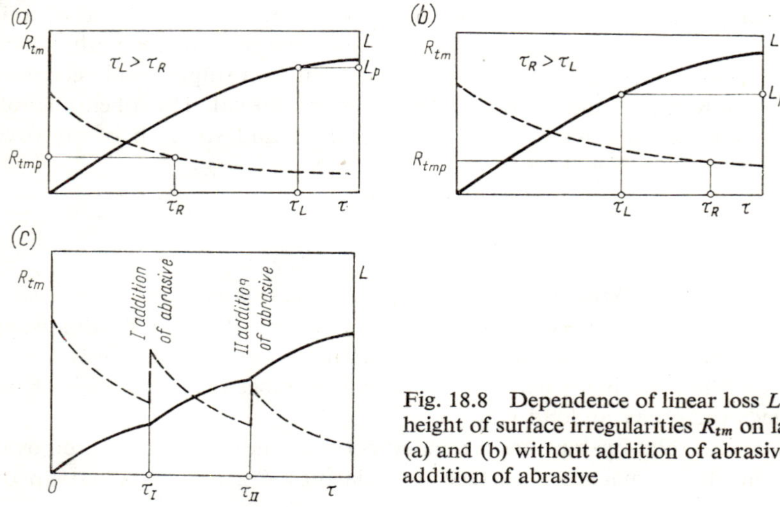

Fig. 18.8 Dependence of linear loss L and mean total height of surface irregularities R_{tm} on lapping time: (a) and (b) without addition of abrasive, (c) with addition of abrasive

The version of Fig. 18.8a indicates that the desired surface quality is obtained earlier than the permissible linear loss. In the version of Fig. 18.8b the opposite is true.

From the practical point of view the version of Fig. 18.8a is desirable and that of Fig. 18.8b is not, since, in general, further improvement of surface quality beyond the given requirements is permissible, while a linear loss exceeding the required value cannot be permitted. The conclusion follows that the choice of those machining conditions should make the version of Fig. 18.8a possible, and the smaller the difference between τ_R and τ_L the better.

Both versions concern lapping without addition of abrasive. In practice addition (dosing) of the abrasive mixture is applied. In such a case the dependence of linear loss and surface quality on lapping time changes its character, as shown in Fig. 18.8c.

18.3.6 Lapping speed

Lapping speed, i.e., approximately, the speed of the lap in relation to work surface, is one of the most effective factors permitting the control of the relationship between linear loss and surface quality.

The dependence of linear loss L and mean total height of lapped surface irregularities R_{tm} on lapping speed v is given in Fig. 18.9. It is evident from this diagram that if the lapping time necessary with regard to the required surface quality is longer than that necessary with regard to the desired linear loss (Fig. 18.8b), then a reduction

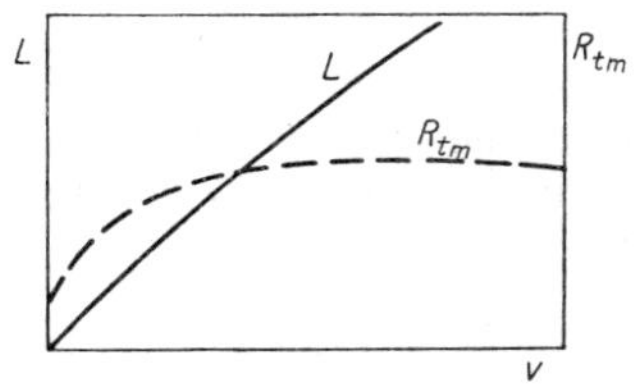

Fig. 18.9 Dependence of linear loss L and mean total height of surface irregularities R_{tm} on lapping speed

of lapping speed could be recommended. Linear loss decreases with speed reduction at a higher rate than does surface roughness and thus the equality $\tau_R = \tau_L$ or even inequality $\tau_L > \tau_R$ may be obtained.

In practice, the following speed ranges are applied in machine lapping of plane surfaces:

- for medium accurate lapping, i.e. for roughness numbers 8–10 and machining allowances up to 0·5 mm, cutting speed v = 200–400 m/min,
- for accurate lapping, i.e. for roughness numbers 10–12, and machining allowances up to 0·25 mm, cutting speed v = 100–250 m/min,
- for very accurate lapping, i.e. for roughness numbers 12–14 and machining allowances up to 0·04 mm, cutting speed v = 10–100 m/min.

In the case of form lapping the speed is reduced to $\frac{1}{2}$–$\frac{1}{3}$ of these values in view of lap wear.

The influence of lapping speed on lap wear is larger, the worse are the conditions of heat abstraction. Therefore, the higher is the cutting speed, the better cooling conditions should be provided.

18.3.7 Vibratory lapping

Lapping is one of the oldest known and applied abrasive processes. However, it is not widely used. Although a simple method, it is inferior to superfinishing as far as metal removal rate is concerned. Thus lapping is not the most economical process.

In order to increase the rate of metal removal by lapping, additional vibrations may be applied during the process. This method is called *vibratory lapping*. Metal removal rate may thus increase by some 30–40%, but simultaneously the height of surface irregularities grows by 50–100%. It follows that vibratory lapping may be considered only for preliminary lapping or in the case of low surface quality requirements.

In the case of vibratory lapping, abrasive mixtures containing boron carbide or diamond dust should be used with regard to abrasive life and metal removal rate. In Fig. 18.10 an example is shown of the dependence of the volumetric rate of metal removal by vibratory lapping Q_V on machining time τ when using diamond dust, boron carbide and silicon carbide.

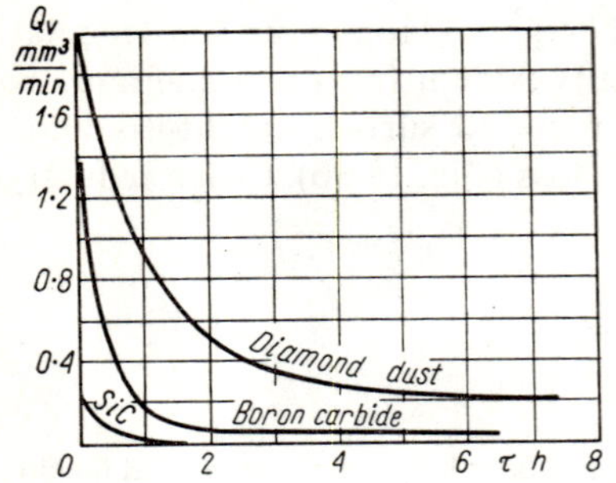

Fig. 18.10 Dependence of volumetric rate of metal removal by vibratory lapping Q_V on time τ (according to Serebrennik and Mushak [18/5])

It is evident from Fig. 18.10 that using diamond dust or boron carbide instead of silicon carbide (carborundum) is advantageous. The very high price of diamond dust and its limited supply are the reasons why its advantages have not yet been fully exploited.

REFERENCES

18/1 LUR'E, G. B.: 'Metody otdelki detalei abrazivnymi instrumentami' (Methods of Producing Parts by Means of Abrasive Tools), Trudrezervizdat, 1958

18/2 GRÖSSEL, H.: 'Docieranie' (Lapping), PWT, Warszawa 1956

18/3 GÓRSKI, E.: 'Obróbka gładkościowa' (Smoothing), WNT, Warszawa 1963

18/4 MASAHISE MATSUAMGA: 'Fundamental studies on laping', Bulletin of Japan Society of Grinding Engineers, Jap. Soc. of Grind. Eng. (JSGE) 5, 1966

18/5 SEREBRENNIK, Yu. B., and MUSHAK, A. S.: 'Rabotosposobnost' abrazivnykh i almaznykh poroshkov pri vibratsionnoi pritirke twerdosplavnykh ploskikh detalei'. Iz knigi pod red. E. N. MASLOVA: 'Obrabotka mashinostroitel'nykh materialov almaznym instrumentom' (Working Capability of Abrasive and Diamond Dust in Vibratory Lapping of Plane Parts Made of Hard Metals. From the book ed. by MASLOV, E. N.: Machining of Mechanical Engineering Materials by Means of Diamond Tools), Izd. Nauka, Moskva 1966

18/6 KUBIK, K.: 'Badanie docieraka płasko-równoległego w układzie planetowym' (Study on a plane-parallel lap in a planetary system), A doctor's thesis at the Department of Mechanics of Wrocław Institute of Technology, 1965

CHAPTER 19

Abrasive-blast treatment

19.1 Types of abrasive-blast treatment

Abrasive-blast treatment (ALA) is one of the abrasive processes using loose abrasives by a kinetic method.

The principle of abrasive blast treatment consists in the projection of abrasive grains by a high-speed jet of fluid (gas, liquid or both). Driven by this jet, grains attain such a high speed that, notwithstanding their small mass, their kinetic energy is sufficient to perform the cutting work.

The oldest type of abrasive-blast treatment is *sand blasting*, used in the cleaning of castings, forgings, heat-treated parts etc.

The working principle and construction of one type of such a plant may be explained by Fig. 19.1. In the working chamber 1 is the workpiece 3 on which a jet of compressed air and abrasive grains is directed from nozzle head 2. Abrasive grains, having performed their work, fall down a hopper and then, through automatic valve 4, to pressure vessel 5. From there the abrasive is again projected by the compressed air stream to the nozzle head.

Sand blasting owes its name to the fact that quartz sand is used as the abrasive.

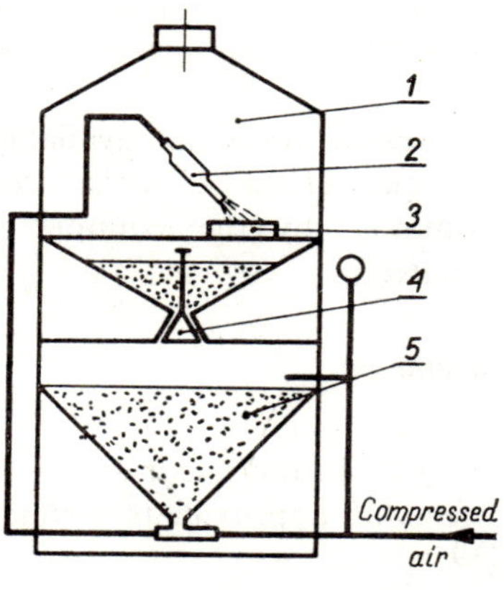

Fig. 19.1 Diagram of sand-blasting plant: 1—working chamber, 2—nozzle head, 3—workpiece, 4—valve, 5—pressure vessel

In addition to the type just described, sand blasting was formerly carried out in open space, the nozzle head being held by a worker in suitable protective clothing. This method had many disadvantages from the point of view of work safety and worker's health (dust generation, shrill hissing of the air, danger of becoming injured by rebounding grains) and therefore should not be applied, unless a vacuum reclaimer for the abrasive dust and debris is built into the gun head.

The abrasive-blast treatment type most widely known and used at present is the so-called *hydro-abrasive treatment*. A diagram of this process is shown in Fig. 19.2.

It differs from sand blasting in that the abrasive is contained in a tank with liquid 4, where a special agitator maintains the latter in suspension. Due to this action the liquid with abrasive, or the abrasive slurry, is uniformly pumped up or sucked by the injector into the nozzle. Compressed air is mixed with the abrasive slurry and this combined jet is driven on to the work surface.

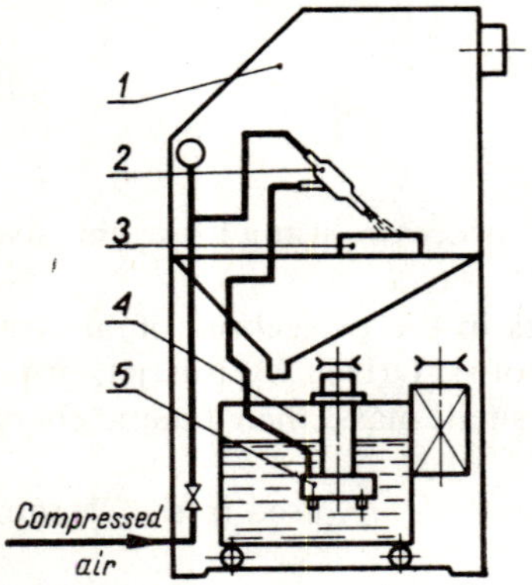

Fig. 19.2 Diagram of a hydro-abrasive treatment plant: 1—working chamber, 2—nozzle head, 3—workpiece, 4—tank with abrasive slurry, 5—pump

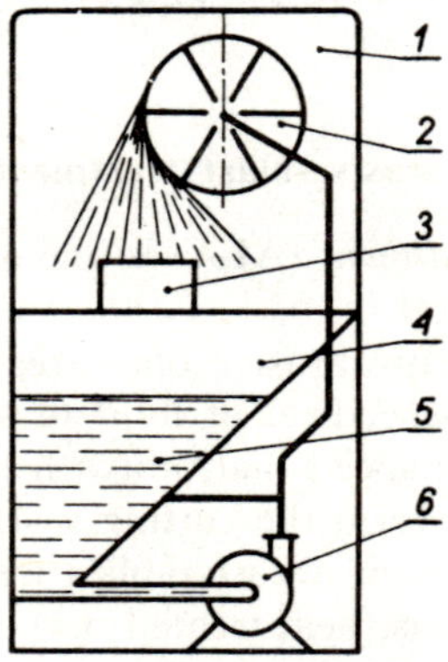

Fig. 19.3 Diagram of a centrifugal blasting plant: 1—working chamber, 2—impeller, 3—workpiece, 4—tank with abrasive slurry, 6—pump

More distinct features characterize the type of abrasive-blast treatment called *centrifugal blasting*. A diagram of a centrifugal blasting plant is shown in Fig. 19.3.

In this type of ALA process the abrasive slurry is fed to the blades of an impeller which accelerates the mixture due to the centrifugal force. Control of the jet, i.e. of its distribution and direction, is performed by suitable setting of the mixture entry to the impeller eye.

Centrifugal blasting may be employed where compressed air is not available.

Notwithstanding the differences in plants and techniques of the ALA varieties, their principles and essentials may be discussed in common, by the example of hydrodynamic treatment, which is the most studied to date.

19.2 Characteristic motions and the work of grain and jet

19.2.1 Grain motion in the jet

Grain projected by the jet does not acquire the latter's speed instantaneously.

Application of the principles of fluid mechanics yields an approximate formula for grain speed v_z in a jet moving at constant speed [19/1]

$$v_z = u \frac{ut}{ut + \frac{8}{3}\frac{r}{C_o}\frac{\varrho_z}{\varrho_f}} = u \frac{ut}{ut+K} \tag{19.1}$$

while

$$K = \frac{8}{3}\frac{r}{C_o}\frac{\varrho_z}{\varrho_f} \tag{19.2}$$

where: u—jet speed, r—radius of a sphere circumscribed on the grain, C_o—coefficient of form drag of the grain moving in the jet, ϱ_z—grain density, ϱ_f—fluid density.

One of the conclusions that follow from eqns. (19.1) and (19.2) is that grain acceleration in the jet will be greater, the higher is the density of the jet fluid. This is illustrated by Fig. 19.4 which represents the relationship between grain travel l_z necessary for the grain to reach a certain grain-speed-to-jet-speed ratio and the type of jet fluid.

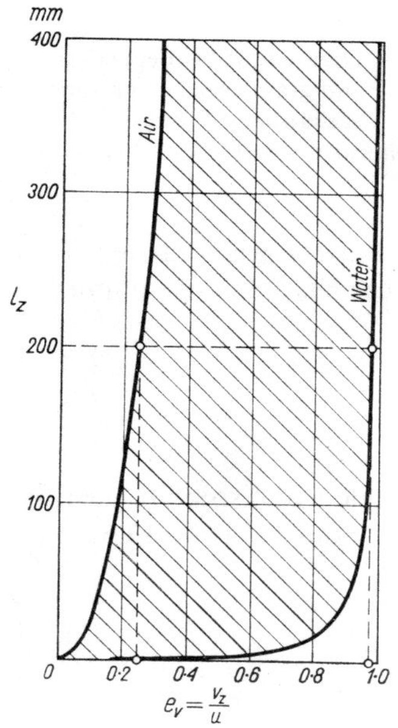

Fig. 19.4 Travel l_z of a spherical grain necessary for the latter to reach a given grain speed to jet speed ratio $e_v = v_z/u$ for grain motion in an air jet and a water jet

In the case of an air jet it is necessary for the grain to travel 200 mm before it acquires a speed $v_z = 0{\cdot}23u$. In the case of a water jet the same travel will impart to the grain a speed $v_z \approx 0{\cdot}96u$.

In mixed jets, consisting of air and water, grain acceleration will yield intermediate values (within the shaded area in Fig. 19.4), but higher than those obtained by an air jet.

This justifies the use of liquid-and-gas (e.g. water-and-air) jets.

Equation (19.1) concerns stationary flow, while in reality the jet squirts into the air medium in which expansion and slowing-down as a result of medium resistance take place. Consequently, the grains present in the jet are accelerated only up to a certain speed, and then their speed decreases along with jet deceleration.

The absolute speed of the grain in a jet decelerated by medium resistance may be defined with sufficient accuracy by

$$v_z \approx u_0 \frac{(u_0 - a\,t)t}{u_0 t + K} = u_0 \frac{u\,t}{u_0 t + K} \tag{19.3}$$

where: u_0—initial speed of the jet (at the moment when it squirts into the medium), a—deceleration (negative acceleration) caused by medium resistance, u—actual jet speed.

A graphical interpretation of eqn. (19.3) is given in Fig. 19.5.

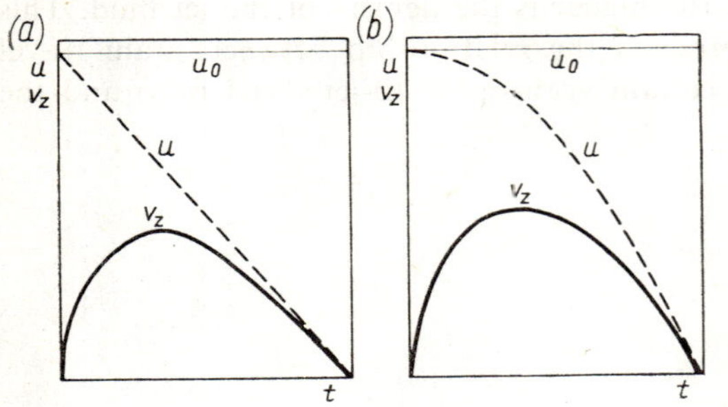

Fig. 19.5 Diagram of the dependence of jet speed u and grain speed v_z on time for: (a) linear deceleration of the jet, (b) non-linear deceleration of the jet

These explanations indicate that the speed of an abrasive grain carried by the jet should be distinguished from jet speed. Only the former has a decisive importance for the effectiveness of blast treatment.

19.2.2 Cutting work performed by a single grain

A cutting grain of mass m falls on the work surface at angle $\varkappa$ at a speed v_z measured at the moment when the grain top contacts the surface at point A_0 (Fig. 19.6).

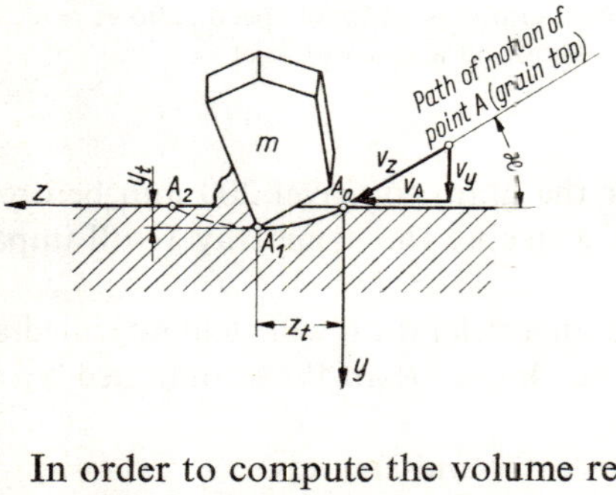

Fig. 19.6 Diagram of cutting work being performed by a single abrasive grain

In order to compute the volume removed by a single grain, the following simplifying assumptions are adopted:

— the grain does not rotate during its motion,
— the surface of transverse cross-section of the undeformed chip cut by a single grain has the form of a triangle whose surface area is

$$A = \frac{b\,y}{2} = \frac{e}{2}\,y^2 \tag{19.4}$$

where: b—base of the triangle, y—its height, $e = b/y$,
— specific cutting force k_s is constant, regardless of undeformed chip thickness,
— the ratio of component forces $F_y/F_z = w_y$ is constant,
— cutting begins when a point of contact between grain and workpiece material occurs.

Depending on the relation between the energy of the falling grain and the cutting work to be performed, the grain: (1) either will get stuck in the material (in Fig. 19.6 after reaching point A_1 will have lost all its energy and stopped), (2) or will cut a determined layer (in Fig. 19.6 reaching point A_2) and leave the material while still preserving some of its energy or having lost all of it.

Assuming that the path of grain motion in the y–z system has the form of a section of parabola, one may calculate the volume removed by a single grain in both cases [19/1]:
— in the case of a grain getting stuck in the material

$$V_z = \frac{m\,v_z^2}{w_y\,k_s\,s}\,\frac{w_y}{6}\cos^2\varkappa = \frac{m\,v_z^2}{6w_y\,e}\cos^2\varkappa \tag{19.5}$$

— in the case of a grain completing its cutting work

$$V_z = \frac{m\,v_z^2}{w_y k_s e}\left(\sin 2\varkappa - \frac{6}{w_y}\sin^2\varkappa\right) \tag{19.6}$$

Equations (19.5) and (19.6) become identical when $w_y = 6\tan\varkappa$ which means that:
— for angle of fall $\varkappa \leqq \arctan(w_y/6)$ cutting occurs according to eqn. (19.6),
— for angle of fall $\varkappa > \arctan(w_y/6)$ sticking of the grain occurs according to eqn. (19.5). Thus the limiting value of the angle of fall is e.g. 18·5° for $w_y = 2$ and 40° for $w_y = 5$.

Equating $\mathrm{d}V_z/\mathrm{d}\varkappa = 0$ indicates that the maximum volume of removed material will occur for

$$\varkappa_{\mathrm{opt}} = \frac{1}{2}\arctan\frac{w_y}{3} \tag{19.7}$$

It is evident from eqn. (19.7) that the optimum value of the angle of fall depends on w_y. Studies indicate that a reduction of rake angle of the falling grains from 25° to −30° causes a 4·5–5 fold increase of w_y, i.e. from $w_y = 2$ to $w_y = 9$–10.

In the case of abrasive-blast treatment rake angles below −30° may be reckoned with, so that w_y may take on values exceeding 10.

Values of $\varkappa_{\mathrm{opt}}$ for some w_y values are given in Table 19.1.

Table 19.1 *Values of optimum angle of fall of abrasive grains*

w_y	2	3	4	5	10
$\overset{\circ}{\varkappa}_{\mathrm{opt}}$	17	23	27	30	37

19.2.3 Geometric characteristics of the jet

Depending on the form of nozzle outlet, or so-called *outlet guide*, the jet of abrasive slurry may acquire various shapes. In order to characterize the geometric features of the jet, its main cross-section will be considered.

The main cross-section of the jet (Fig. 19.7) is obtained drawing a plane perpendicular (or normal) to the workpiece surface and passing through the outlet guide axis and thus also by the jet axis.

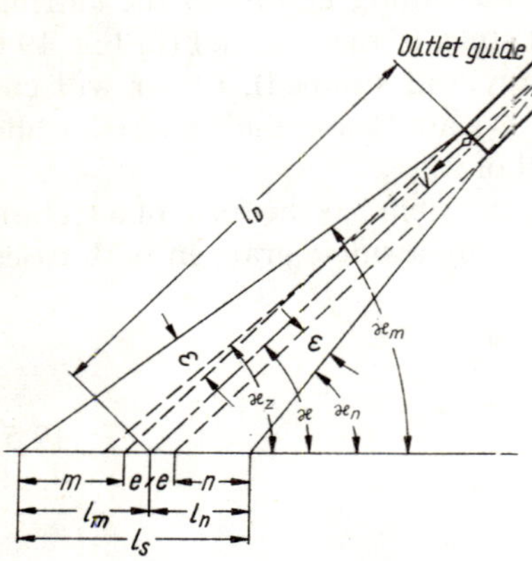

Fig. 19.7 Main cross-section of the jet

In the main cross-section the following geometric features are distinguished:

— jet divergence angle 2ε, characterizing jet expansion (assumption of linear expansion is sufficiently accurate for practical calculations),

— spray angle $\varkappa$, i.e. the angle of fall of grains moving along the jet axis; this is the angle between the jet axis and the tangent to the work surface,

— extreme angles of fall:

$$\varkappa_m = \varkappa - \varepsilon \qquad \text{and} \qquad \varkappa_n = \varkappa + \varepsilon \tag{19.8}$$

limiting the range of angles of fall of grains contained in the jet,

— jet length or nozzle distance L_D,

— spray length l_s whose value may be calculated from Fig. 19.7 as: $l_s = l_m + l_n = m + 2e + n$ or

$$l_s = D\left[\frac{1}{\sin\varkappa} + \sin\varepsilon\left(\frac{w + 0{\cdot}5\cot\varkappa}{\sin(\varkappa-\varepsilon)} + \frac{w - 0{\cdot}5\cot\varkappa}{\sin(\varkappa+\varepsilon)}\right)\right] \tag{19.9}$$

where: D—outlet guide diameter, $w = l_D/D$.

It is evident from eqn. (19.9) that the spray length depends on outlet guide diameter, on jet length expressed as a multiple of this diameter (w), on spray angle and jet divergence angle.

It is known from experience that within the range of practically applied pressures and outlet guide designs, jet divergence angles in general do not exceed 12°. Within these limits the relationship between spray length and jet divergence angle is almost linear

$$w = 1{\cdot}5 + |\varepsilon^\circ| \tag{19.10}$$

or for jet divergence angle $2\varepsilon = 12°$ the spray length is about $7{\cdot}5D$.

19.2.4 Kinematic parameters of abrasive-blast treatment

By analogy with other machining methods in abrasive-blast treatment the primary and auxiliary working motions are distinguished.

The primary working motion is the motion of abrasive grains in the jet. Since grains move in the jet at various speeds depending on their distance from the jet axis, it is assumed that the speed of primary motion, or cutting speed, is the speed of grains moving along the jet axis (their angle of fall is the spray angle).

The auxiliary working motion is the feed motion of the workpiece material in relation to the jet. It is characterized by feed rate s_t and feed direction.

Up-blasting feed rate and up-blasting spray, when the direction of the horizontal component of cutting speed v_H is directed in a sense opposite to that of s_t (Fig. 19.8), are distinguished from conventional blasting feed rate and convential blasting spray, when the sense of vector v_H agrees with that of vector s_t.

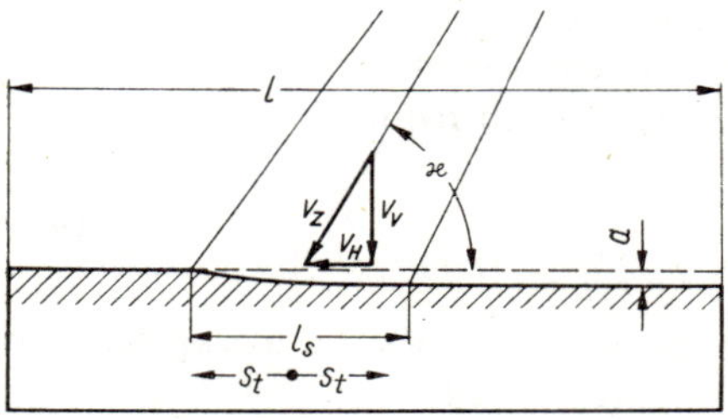

Fig. 19.8 Denotations of abrasive-blast treatment parameters

Spraying time t_s is the time during which grains of the moving jet of spray length l_s fall on a given point of the work surface

$$t_s = \frac{l_s}{s_t} \quad \text{min} \tag{19.11}$$

Abrasive blast time of a single pass t_{m1} in abrasive-blast treatment is calculated according to

$$t_{m1} = \frac{l}{s_t} + 2t_s = \frac{l+2l_s}{s_t} \tag{19.12}$$

where l—length of work surface.

Depth of cut a is the thickness of a layer of material removed in a single pass. Depth of cut is not uniform over the spray surface.

The character of depth changes may be studied by means of so-called *meniscus test*, i.e. directing the jet on the work material for some time. A cavity will be formed whose geometric parameters are given in Fig. 19.9:

b_m—spray width, calculated from

$$b_m = D+2l_D \tan\varepsilon \tag{19.13}$$

b_0—spray width at the spot of greatest depth,
a_0—maximum spray depth,
a_m—depth at half the spray length.

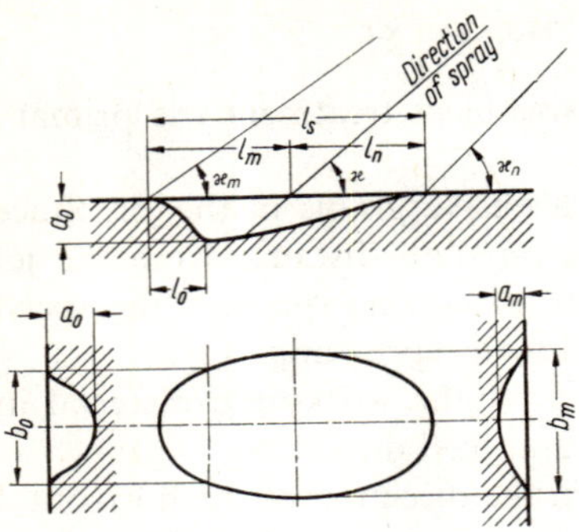

Fig. 19.9 Shape and cross-sections of cavity resulting from local spraying by means of a nozzle with outlet guide of circular cross-section

When the feed motion is engaged, the difference between a_0 and a_m becomes smaller, the faster is the feed rate. Therefore, the depth of cut in abrasive-blast treatment is denoted by a and measured in the main cross-section plane as shown in Fig. 19.8.

Depth of cut is greater, the slower is the feed rate.

Abrasive blast time t_m determines the time of removal of the allowance q from the given surface of dimensions B and L and is calculated from

$$t_m = \frac{q}{a}\left(\frac{B+2bm}{c\,b\,m}\right) t_{m1} = \frac{q}{a}\left(\frac{B+2bm}{c\,b\,m}\right)\frac{l+2l_s}{s_t} \tag{19.14}$$

In eqn. (19.14) c denotes by what part of the spray width does the workpiece move in each successive pass. Thus c characterizes transverse feed.

19.3 Technical and economic indices of abrasive-blast treatment

19.3.1 Classification of technical and economic indices

As in other methods of machining and abrasive processes, the following technical and economic indices are used to characterize and evaluate abrasive-blast treatment: the rate of metal removal, the quality of machined surface and of the surface layer, tool wear (abrasive mixture) and efficiency.

Metal removal rate in the case of abrasive-blast treatment is determined by the following formulae:

— volumetric rate of metal removal

$$Q_V = \frac{l\,B\,a}{t_m} = \frac{B\,a}{1+l_s/l}\,s_t \tag{19.15}$$

— surface rate of metal removal

$$Q_S = \frac{B\,l}{t_m} = \frac{B}{1+l_s/l}\,s_t \tag{19.16}$$

— linear rate of metal removal

$$A_L = \frac{a}{t_m} = \frac{1}{l}\,\frac{a}{1+l_s/l}\,s_t \tag{19.17}$$

The linear rate of metal removal should be distinguished from the penetration speed v_a

$$v_a = \frac{a}{t_s} = \frac{a}{l_s} s_t \tag{19.18}$$

Surface quality and the surface layer quality after abrasive-blast treatment are defined by the same indices as after other methods of machining and abrasive processes (mean total height of surface irregularities R_{tm}, waviness, structure, hardening, stresses).

Abrasive wear is characterized by the degree of disintegration and the life period:
— index of disintegration is the grain number applying to at least 50% of the abrasive at the given stage of wear;
— disintegration intensity which is evaluated according to the ratio of the difference between the initial and the actual characteristic dimension of the main fraction of abrasive to the time of abrasive work;
— economical and productive life.

The first two indices are seldom used in practice, as their determination requires time-consuming measurements.

The efficiency of abrasive blast treatment is usually defined by the specific rate of metal removal, i.e.

$$q_V = \frac{Q_V}{N} \quad \text{mm}^3/(\text{kW min}) \tag{19.19}$$

where N means the power supplied by the mains.

The technical and economic indices mentioned above depend on many factors as follows
— factors connected with the composition of the abrasive slurry,
— factors connected with nozzle shape,
— factors connected with jet control,
— factors connected with the workpiece material.

19.3.2 Influence of factors connected with the composition of the abrasive slurry

The results of the process depend to a high degree on the type and properties of the abrasive.

With regard to the rate of metal removal an abrasive material whose grains have corners with small angles and roundness radii is to be preferred. Abrasive material density influences the metal removal rate in two ways. On the one hand, density increase causes an increase of grain mass, and on the other hand it reduces the speed of the grains in the jet [eqns. (19.1) and (19.3)]. As a result, changes of kinetic energy of the grains caused by the changes of grain material density, may have different characters depending on their combination with other conditions of the treatment.

Of the practically used abrasive materials, silicon carbide (carborundum) should be considered best as regards metal removal rate which exceeds that obtainable with aluminium oxide (other conditions being identical) by 20–40%. On the other hand, quartz sand yields a metal removal rate equal to 0·4–0·5 of that of aluminium oxide.

Abrasive wear must also be considered. A reduction of metal removal rate to one half of the initial value occurs in the case of silicon carbide in a time 50% longer than in the case of aluminium oxide, since abrasive disintegration intensity is connected with grain strength (see Chapter 14).

The influence of abrasive material on surface quality may be observed only within the range of very smooth surfaces and small grain sizes (microdust). Thus, for example, abrasive-blast treatment using 20 μm aluminium oxide grains yielded a mean total height only half as large (R_{tm} = 1 μm) as that obtained with similar silicon carbide grains (R_{tm} = 2 μm).

As far as efficiency is concerned, materials yielding a higher rate of metal removal are more advantageous, as power consumption is practically independent of the type of abrasive material used.

Notwithstanding the numerous advantages of silicon carbide it is not the material most commonly applied in abrasive-blast treatment. The reason for this is the cost of silicon carbide (twice that of aluminium oxide). Therefore the cost of unit surface area treated with silicon carbide is higher.

The grain number characterizes grain size, mass, angles, and the roundness radii of the corners, and thus must influence the treatment process.

Studies indicate that for constant working pressure of the air there is always only one optimum grain size for maximum rate of metal removal. In Fig. 19.10 the result of research on the influence of grain size on metal removal rate for two different air pressures is presented. The occurence of maximum metal removal rate for the

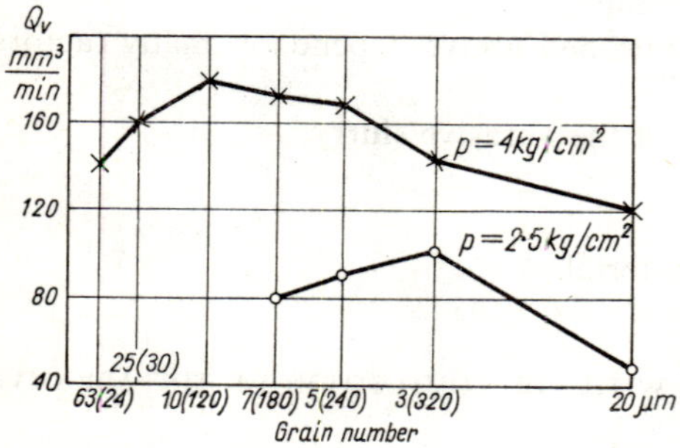

Fig. 19.10 Diagram of grain size influence on volumetric rate of metal removal Q_V for various values of air working pressure p

optimum grain size is explained in the following way. Reduction of the size of abrasive grain causes its kinetic energy to decrease as a result of its lower mass, but at the same time it causes a certain increase of that energy as a result of the accelerated motion of the grains in the jet. Grain size reduction is accompanied by a reduction of corner top radii and angles and this in turn results, on the one hand, in a smaller transverse cross-section of the undeformed chip and, on the other hand, in an increase in the specific cutting force. Simultaneously, the value of thrust force decreases in relation to the tangential force.

As a result of these opposing influences, for any working pressure there is only one determined size of grains whose properties give the optimum rate of metal removal. The higher is the working pressure, the larger are the optimum characteristic dimensions of the abrasive grains.

This hypothesis has been confirmed by studies on metal removal rate under constant pressure and for various grain sizes, depending on the working time of the abrasive (Fig. 19.11). For optimum and smaller grain sizes monotonic curves of the relationship between metal removal rate and working time are obtained [curves for grain numbers 10(120) and 5(240)], with the optimum size curve occupying the highest position in the diagram. On the other hand, curves for larger-than-optimum grain sizes are non-monotonic [curve for grain size 210(36)]. This is explained as

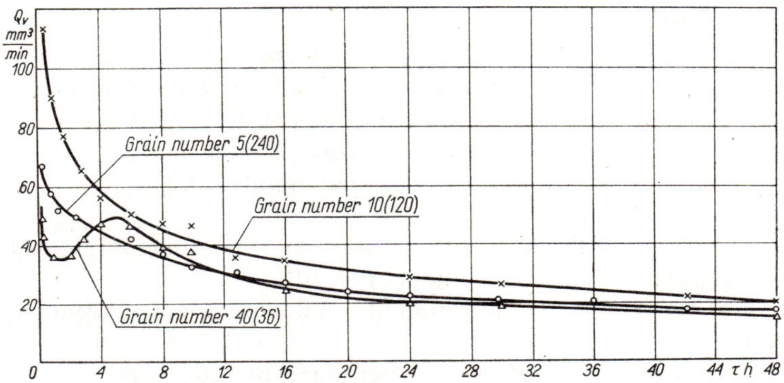

Fig. 19.11 Diagram of the dependence of metal removal rate in abrasive-blast treatment on abrasive working time for various grain sizes. Conditions under which the studies were conducted: plant SDU-400, working pressure 3 kg/cm², aluminium oxide EA, mixture ratio 1:4, spray angle 45°, jet length 60 mm, feed rate 100 mm/min, workpiece material 0·55 carbon steel, air temperature 20°C, abrasive quantity 12 kg

follows: in the phase when, as a result of disintegration, optimum size grains form a majority in the abrasive composition, metal removal rate increases, while further disintegration causes a monotonic decrease of this rate.

Abrasive dust wear grows more intense with grit size. These differences decrease when abrasive working time is sufficiently prolonged (Fig. 19.11).

Surface quality depends to a very high degree on grain size. In Fig. 19.12 the dependence of mean total height of irregularities of a steel workpiece surface on abrasive grain size is presented. As is evident from the diagram, the reduction of grain size causes a monotonic decrease of surface roughness. The reasons for this are that depth of cut and undeformed chip cross-section are smaller in the case of smaller

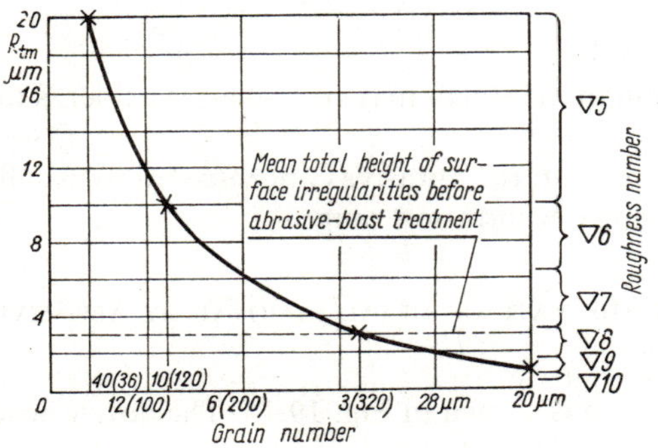

Fig. 19.12 Diagram of the dependence of mean total height of surface irregularities R_{tm} on the size of abrasive grains used in abrasive-blast treatment

grains, and in addition, the larger is the number of grains per unit slurry volume, the more of them fall on a unit surface area.

The carrier liquid exerts a twofold influence on the effects of abrasive-blast treatment.

On the one hand, the liquid, consisting of water with anticorrosive additives, has a much greater density than air. This contributes to a higher acceleration of the abrasive grains in the jet and, consequently, the grains attain a higher speed on the same jet length, thus increasing the rate of metal removal.

On the other hand, the carrier liquid, when spreading over the surface, fills its cavities and forms a film which impedes the striking action of the grains. Due to this liquid film, bulges and tops of surface irregularities are the first to be affected by the treatment and therefore surface quality improves. Experiments indicate that a water–air jet permits one to obtain, as an average, a roughness number higher by 1, as compared with the effects of an air jet.

The surface layer of the carrier liquid slightly reduces the rate of metal removal, but the positive effect of grain acceleration in the jet prevails.

The proportions of abrasive slurry composition also influence the metal removal rate and surface quality.

Abrasive slurry composition (volumetric) is determined by concentration coefficient $K_V = A/C$ (where A denotes abrasive content and C the carrier liquid).

Experience indicates that metal removal rate grows with abrasive concentration. However, this increase gradually becomes smaller and smaller and further increases in concentration may cause either a reduction of metal removal rate or disturbances in the functioning of the nozzle head.

A reduction in metal removal rate occurs when the energy of the compressed air jet is not sufficient to suitably accelerate the excessive quantity of grains in the abrasive, but the amount of liquid is large enough to ensure the continuity of mixture feed.

Disturbances in the functioning of the nozzle head are caused by such a degree of abrasive concentration that the amount of liquid is too small to maintain the continuity of the inflow of the abrasive mixture.

The optimum coefficient of weight concentration of the abrasive K_w with regard to metal removal rate and the efficient functioning of the nozzle head depends on grain size and under average conditions of treatment:

— for grain number 10(120) $K_w \approx 0{\cdot}4$–$0{\cdot}5$
— for grain number 6(200) $K_w \approx 0{\cdot}6$–$0{\cdot}8$
— for grain number 28 μm $K_w \approx 0{\cdot}8$–$1{\cdot}0$.

This means that the abrasive concentration in the slurry may be higher, the finer is the abrasive dust.

With regard to surface quality improvement, the abrasive concentration is usually reduced to the lower limit of the optimum concentration range.

19.3.3 Influence of nozzle head on the rate of metal removal by abrasive-blast treatment

A simplified diagram of a nozzle head design is shown in Fig. 19.13. The nozzle head type represented there is the so-called *central nozzle head*, as the air is fed by an internal conduit. In the opposite case the nozzle head is said to be of the concentric type.

The nozzle head comprises air nozzle 1, chamber 2 in which air mixes with abrasive slurry, and outlet guide 3 forming and directing the jet.

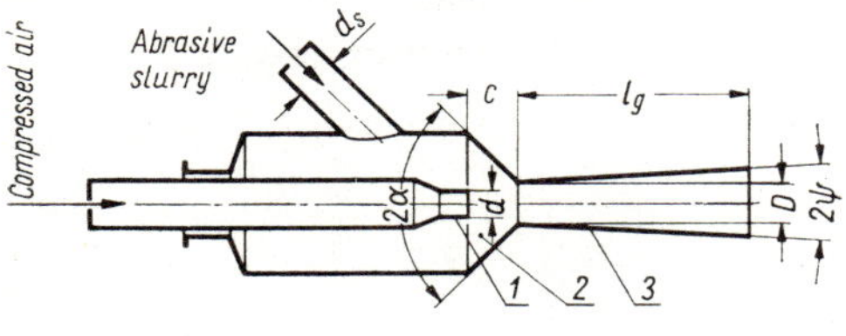

Fig. 19.13 Simplified diagram of a nozzle head: 1—nozzle, 2—mixing chamber, 3—outlet guide

The air nozzle and mixing chamber are characterized by the following parameters: d—outlet diameter of air nozzle, D—inlet diameter of outlet guide, α—inclination angle of mixing chamber walls, c—mixing chamber length.

The characteristic dimensions of the outlet guide are—besides the inlet diameter—guide length l_g and divergence angle 2ψ.

The air pressure and inflow have an important influence on the rate of metal removal. A pressure rise causes an increase of jet speed, resulting in an increase in kinetic energy and metal removal rate. Therefore the use of high working pressures is recommended. However, technical difficulties and plant costs grow progressively with the pressure, therefore abrasive-blast treatment plants usually work at mains pressure, i.e. at the pressure usually maintained in industrial mains (4–6 kg/cm^2).

The amount of inflowing air should be in keeping with that of the inflowing abrasive slurry. The correct choice of their proportion depends on inflow pressure and feed conduit diameters. In Fig. 19.14 the changes of metal removal rate Q_V are plotted against the quantity of abrasive slurry q fed at constant air inflow. In this

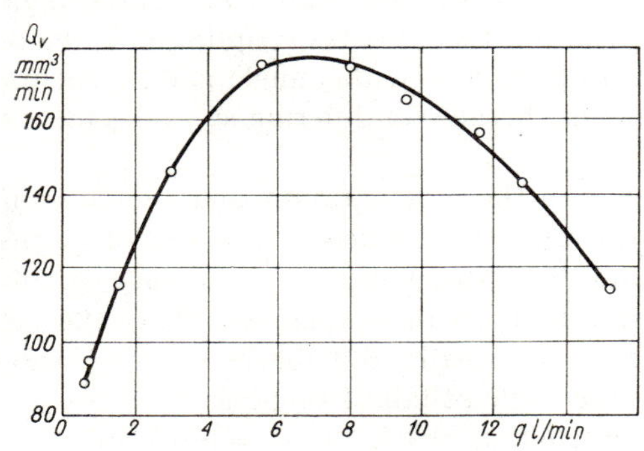

Fig. 19.14 Diagram of the dependence of metal removal rate Q_V on the inflow q of abrasive slurry fed into the mixing chamber. Conditions of measurements: aluminium oxide EA, grain number 6(200), $K_V = 0{\cdot}2$, $d = 6$ mm, $l_D = 50$ mm, $\varkappa = 45°$, $D = 10$ mm, $l_g = 80$ mm, $p = 4$ kg/cm^2, $c = 10$ mm, $\psi = 0°$

case the optimum inflow of the mixture was 6–8 l/min. The geometric parameters of the outlet guide should be carefully chosen, as they have a distinct influence on metal removal rate.

In particular, the inlet diameter of the guide should be selected with regard to the air nozzle outlet diameter. The importance of the choice of the D/d ratio is illustrated in Fig. 19.15 in which the appearance of an optimum D/d for various methods of feeding the abrasive slurry is clearly visible.

Also, the outlet guide length has its optimum value. Under normal conditions of treatment the optimum guide length, with regard to the rate of metal removal, is $l_g = (6\text{–}8)\ D$.

Maximum metal removal rate is obtained with cylindrical guides or those with angles of convergence up to 2° or with angles of divergence not exceeding 2°.

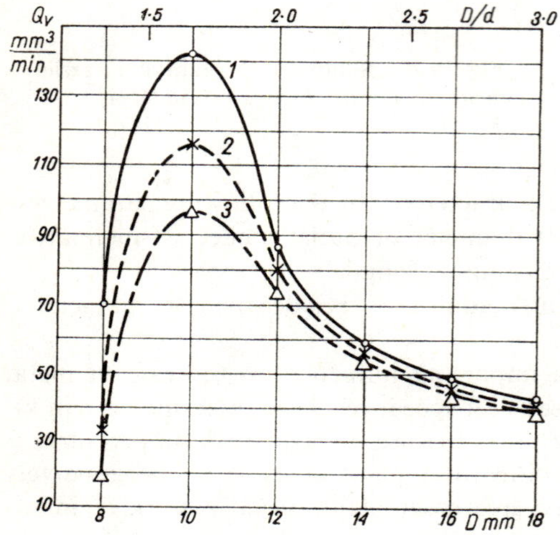

Fig. 19.15 Diagram of the dependence of metal removal rate Q_V on outlet guide diameter D for three different methods of feeding abrasive slurry to nozzle head: 1—by pumping, 2—by gravity, 3—by injecting. Conditions of measurements: as given in the caption of Fig. 19.14 except for c = 2 mm (suction)

19.3.4 Influence of jet control

By jet control is understood the variations of spray angle, jet length, direction of spray shift, spray time or feed rate.

The influence of spray angle may be explained in a similar way to that of the angle of fall of a single grain. Experiments confirm that there exists an optimum value of spray angle with regard to metal removal rate. However, this angle is different for various workpiece materials (Fig. 19.16), mainly because of differing w_y values for the particular materials.

Optimum spray angle values are higher than those of optimum angles of fall for single grains. The reason is that during abrasive-blast treatment a number of grains rebound without doing cutting work, and the percentage share of such grains decreases with increasing spray angle. The validity of this explanation is confirmed by the fact that in spraying with flat outlet guides smaller differences occur between the optimum values of spray angle and of the angle of fall of single grains.

The existence of an optimum value of jet length is obvious in view of the distribution of grain speeds in the jet (Fig. 19.4).

Experiments fully confirm the theoretical considerations and optimum jet length l_D increases when spray angle $\varkappa$ is reduced (Fig. 19.17). Studies already conducted lead to the conclusion that optimum jet length is $l_D = (6\text{–}8)\ D$.

Spray direction, i.e. up-blasting or conventional-blasting spray, does not exert a visible influence on the technical and economic indices of ALA treatment.

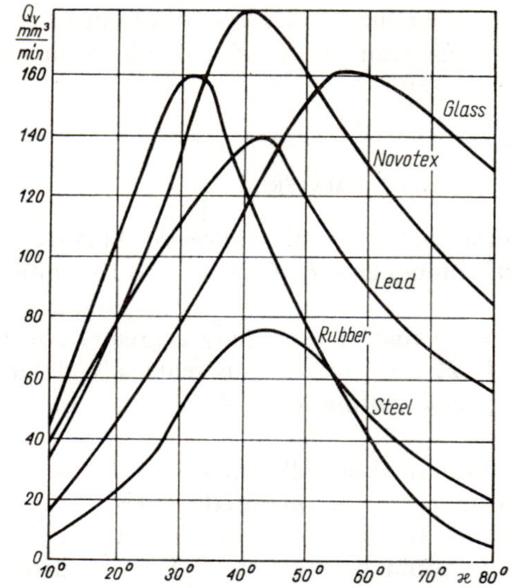

Fig. 19.16 Diagram of the dependence of metal removal rate Q_V on spray angle $\varkappa$ for various materials (acc. to Dickore [19/30])

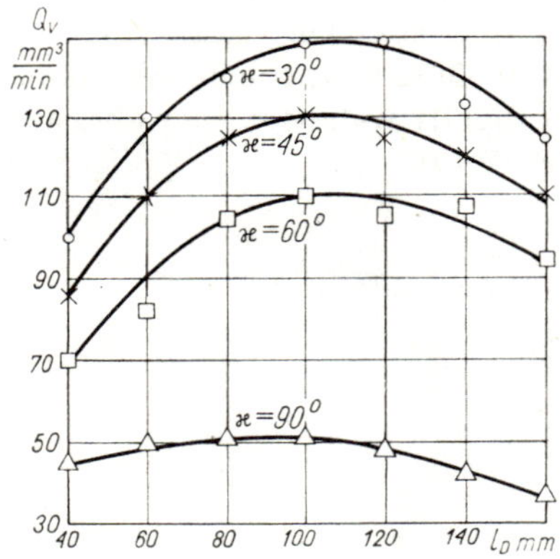

Fig. 19.17 Diagram of the dependence of metal removal rate Q_V on jet length l_D for various spray angles. Conditions of measurements: EA, grain number 10(120), $d = 14$ mm, $D = 16$ mm, $l_g = 75$ mm, $\psi = 0°$, $c = 10$ mm, $p = 3$ kg/cm²

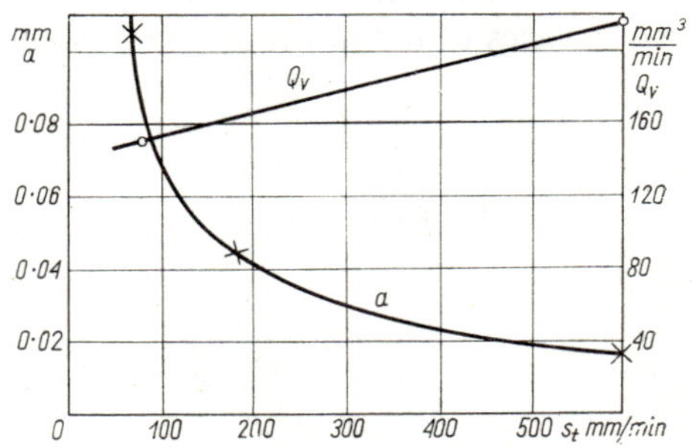

Fig. 19.18 Diagram of the dependence of the depth of cut a and volumetric rate of metal removal on feed rate s_t

Spray time or feed rate influences the metal removal rate, depth of cut and surface quality.

The most distinct effect of feed rate is its influence on the depth of cut a (Fig. 19.18). However, with an increase of feed rate s_t the metal removal rate Q_V also rises. On the other hand, with regard to surface quality there exists a certain optimum spray time. For example, when spraying constructional steel workpieces with an initial height of surface irregularities $R_{tm} = 10$–20 µm, the optimum spraying time with abrasive of grain number 10(120) was 1·5 s/cm.

It follows, that in the choice of feed rate either the required depth of cut or the required surface quality, for which there is an optimum spray time, should be taken into consideration.

19.3.5 Influence of type and state of workpiece material

The workpiece material influences the technical and economic indices of ALA treatment mainly through: cutting force, ratio of components $F_y/F_z = w_y$ and the initial quality of the surface.

Brittle materials, characterized by a higher value of w_y require a greater spray angle (Fig. 19.16). In the case of identical F_y/F_z ratio for various materials, differences in metal removal rate depend on the specific cutting force. An example of this may be steel and lead (Fig. 19.16).

All materials yielding high rates of metal removal, usually show a lower surface quality. Therefore in the case of soft, easily machined materials, finer abrasive should be used. Thus e.g. to obtain a mean total height of surface irregularities $R_{tm} \approx 4$ μm on brass, abrasive dust 4(280) should be applied, and on hardened steel—abrasive dust 10(120).

Experiments show that improving the surface quality by ALA treatment is economical only within a certain range of roughness numbers. This means that e.g. by a treatment with abrasives of grain numbers 12(100)–3(320) the roughness number may be economically increased by 2. This range is slightly wider for finer abrasives used in cases of initial roughness numbers 8–9 (Fig. 19.19).

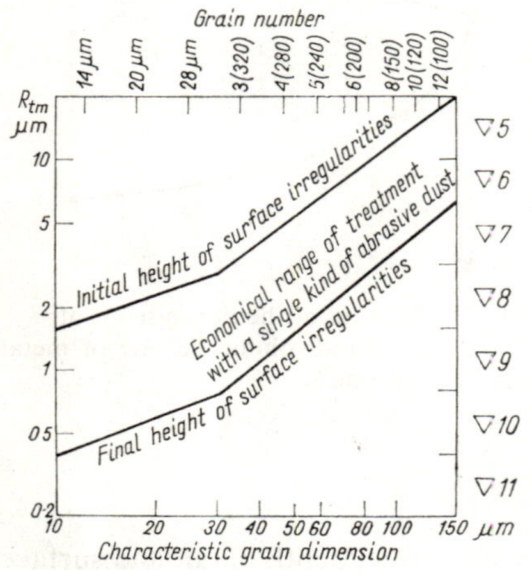

Fig. 19.19 Approximate range of abrasive-blast treatment by a single kind of abrasive for suitable initial and final height of surface irregularities (aluminium oxide treatment of steel)

The attainment of roughness numbers differing by more than 2 from the initial value is difficult, as then the phenomenon of the so-called *reproduction of irregularities* occurs, considerably extending the time of abrasive-blast treatment.

In the case when the necessity arises of improving the roughness number by more than 2, it is advisable to apply the treatment in two operations: first—by coarser-grained abrasive and second—by an abrasive of suitably finer grains.

REFERENCES

19/1 KĄCZMAREK, J., SIKORA, J., and CZYŻEWICZ, Cz.: 'Obróbka strumieniowo-ścierna' (Abrasive-blast Treatment), WNT, Warszawa 1963

19/2 BILIK, Sh.: 'Abrazivno-zhidkostnaya obrabotka metallov' (Abrasive-slurry Treatment of Metals), Mashgiz, Moskva 1960

19/3 DICKORE, G.: 'Strahl-Läppen', Schriftenreihe Feinbearbeitung, Das Industrieblatt, 1956

CHAPTER 20

Vibratory and barrel finishing

Vibratory finishing and barrel finishing have many features in common and they are used for similar purposes, which justifies their joint discussion.

20.1 Vibratory finishing

Vibratory finishing consists in placing the workpieces in an abrasive mixture and subjecting it all to vibrations of controllable frequency, amplitude and direction.

This is carried out by means of a plant, a typical diagram of which is shown in Fig. 20.1. In tank 1 lined with a layer of rubber 2, is the abrasive slurry 3 containing

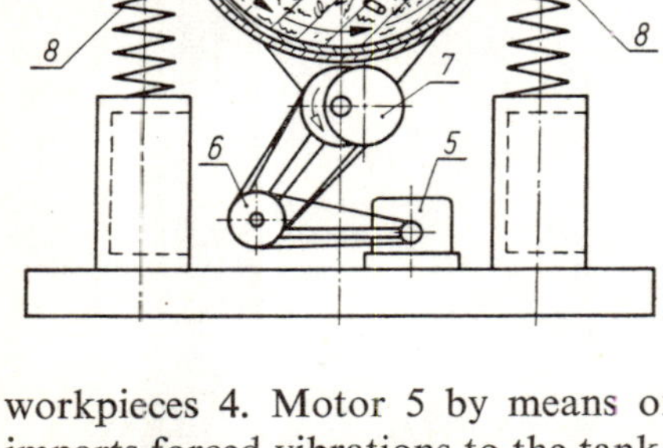

Fig. 20.1 Diagram of a vibratory finishing plant

workpieces 4. Motor 5 by means of belt transmission 6 drives eccentric 7 which imparts forced vibrations to the tank resting on elastic supports 8. By suitable choice of the eccentric and of the rigidity and damping of the particular supports it is possible to influence the value of amplitude and the horizontal and vertical speed components denoted as v_H and v_V, respectively.

On the basis of experience it has been established that the thickness of rubber lining should be 6–12 mm in order to ensure an economical tank life.

It is assumed, also on the basis of experience, that the most advantageous path of tank point motion is a curve similar to an ellipse, with its horizontal axis 10–30% longer than the vertical one. In the finishing of small workpieces, each weighing up to a few kilogrammes, a vibration frequency of 12–40 Hz is considered to be the optimum value, for a vertical amplitude of 8–4 mm.

Vibration excitation may be performed in various ways, the most common—because of its simplicity and efficiency—being the eccentric-type mass vibrator. Vibrators exciting tanks weighing (when loaded), up to 800 kg, need a driving power of 1·5–2·0 kW.

The abrasive slurry consists of an abrasive of a grain size depending on the necessary surface quality (from grain number 16(80) for cleaning to microdust 20 μm for finish-polishing), water and chemical additives. It has been determined empirically that the optimum range of water volume to abrasive volume ratio—with regard to surface quality, metal removal rate and abrasive life—is contained within the limits from 1:5 to 1:10.

Chemical additives are used to protect workpiece surfaces from corrosion (passivators) and as surface-active agents. In addition, the inclusion of foaming agents has a favourable effect on the vibratory finishing process, as small products of abrasion become suspended in the foam. Surface-active agents play a particularly important role in the case of vibratory polishing.

The rate of metal removal by vibratory finishing is determined by

$$Q_V = \sum_{i=1}^{i=z} A_{zi} v_{zi} \tag{20.1}$$

where: z_c—number of grains cutting simultaneously, A_z—transverse cross-section of undeformed chip cut by a single point, v_z—cutting speed of a single grain.

It follows from eqn. (20.1) that metal removal rate is higher:

— the larger is the number of grains cutting simultaneously, which is a consequence of applying fine-grained abrasive and a high abrasive concentration on large work surfaces;

— the larger is the surface area of the undeformed chip cut by a single point; such conditions occur when the unit pressure acting on the grain is high and the grains have wide apex angles and small top rounding radii; this means that large grains and best quality abrasives should be used;

— the higher is the cutting speed of the grains, this speed being the difference between workpiece speed (v_M) and the speed of the particular grains (v_m) coming into contact with the given workpiece

$$v_{zi} = v_M - v_{mi} \tag{20.2}$$

These considerations indicate that in order to determine optimum machining conditions it is necessary to know the relative importance of all factors influencing one of the three quantities appearing in eqn. (20.1) or directly the metal removal rate.

Fig. 20.2 shows the topographical distribution of the average charge mass motion speed v_c, dynamic force F_d and loss of workpiece mass Q_L.

The distribution of these quantities is not uniform at the particular points of circles *K*1, *K*2, *K*3 and *K*4 (Fig. 20.2). The greater is the distance from the centre of the circles, the higher are v_c, F_d and Q_L values.

Maximum values of the charge mass motion speed occur in all circles at a position of about 45°. This indicates a certain role of the motion resulting from the sliding of a part of the charge mass by gravity.

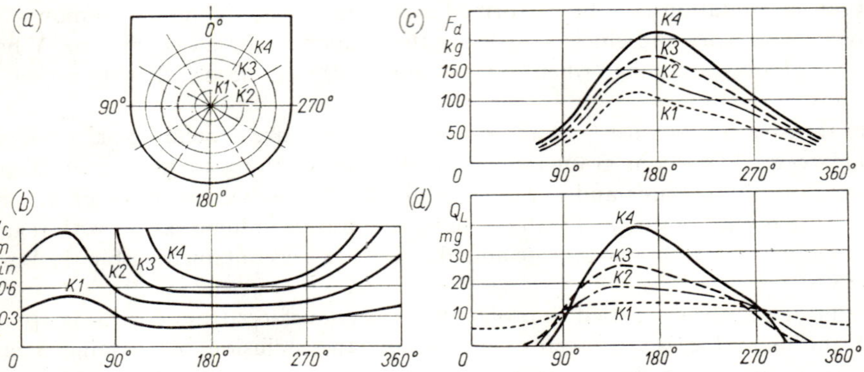

Fig. 20.2 Dependence of mixture motion speed, dynamic forces and workpiece mass loss on the position in the tank (acc. to measurements by Matsunaga and Hagiuda)

On the other hand, maximum values of dynamic forces occur between 135° and 180°. This indicates a considerable role of the charge mass which exerts the highest pressure in the 180° position when stationary and in a position between 135° and 180° when in motion.

The comparison of curves representing the workpiece mass loss (Q_L) with those representing the charge mass motion speed (v_c) and dynamic forces (F_d) shows that:

— dynamic forces considerably influence the workpiece mass loss (and thus also the metal removal rate),

— the charge mass motion speed has no visible influence on the workpiece mass loss.

From the fact that the position corresponding to maximum metal removal rate does not coincide with that corresponding to maximum dynamic forces, the following conclusion may be drawn: the cause of this shift is that in a position between 90° and 180° the highest cutting speeds occur, or the greatest differences between the speed of the workpiece and that of the abrasive slurry.

The influence of vibration frequency on metal removal rate results from the fact that an increase of the former accelerates the mass motion and makes dynamic forces grow. The symptom of this is the change of charge mass form and its surface (Fig. 20.3). As a result, in the range under study a monotonic, linear increase of the workpiece mass loss is observed when vibration frequency rises (Fig. 20.4) [21/4], [21/5].

Charge value, defined in fractions of tank volume V_r, exerts an influence on dynamic forces (increases the static pressure and mass).

The influence of charge volume on metal removal rate is represented in

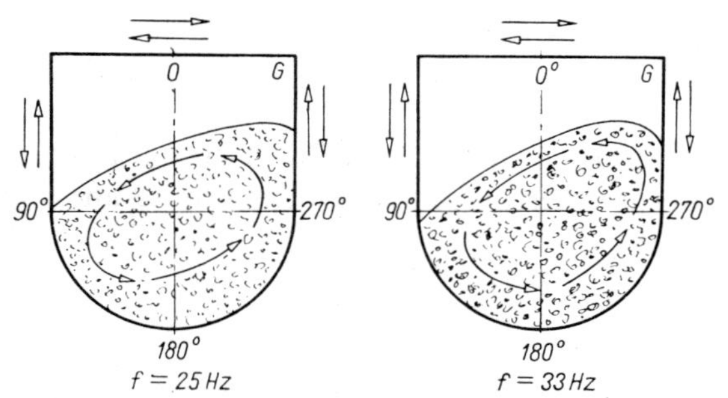

Fig. 20.3 Change of shape of charge surface depending on tank vibration frequency

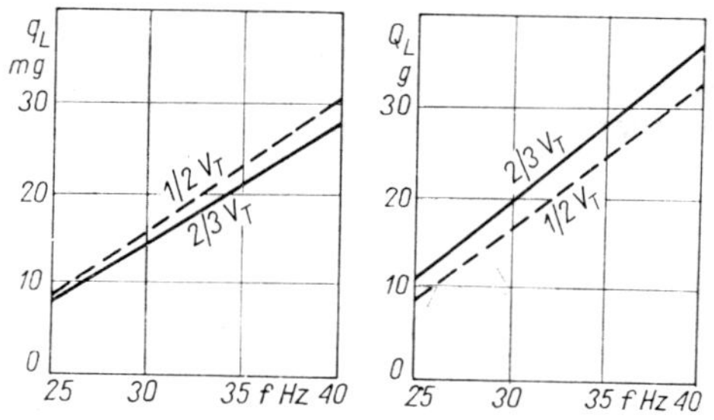

Fig. 20.4 Influence of vibration frequency on unit loss (q_L) and total loss (Q_L) of workpiece mass for two degrees of tank filling [20/4]

Fig. 20.4. Increasing the charge from $1/2V_T$ to $2/3V_T$ results in an increase of the total workpiece mass loss by some 15–20%. On the other hand, the mass loss of a single workpiece is reduced by some 8–10% when charge volume grows. The conclusion follows that for a charge increase of 33% the processing time should be extended by 10%. Thus charge increase may be considered favourable with regard to the rate of metal removal. In practice, charge volume must be limited to about 70% since for larger volumes the charge, when in motion, presses against the tank lid, is partially slowed-down and creates a stagnation zone in the corner *G* (Fig. 20.3).

The influence of working time on metal removal rate is an indirect one, and is connected with the abrasive disintegration process. Phenomena described in Chapter 19 (when discussing the abrasive wear process and its interdependence with metal removal rate) may then occur. However, the disintegration process in vibratory finishing is much slower than in abrasive-blast treatment. Therefore, during longer work periods an almost linear increment of the workpiece mass loss is observed.

Abrasive concentration influences the loss of workpiece mass in a similar way as in other processes conducted using loose abrasives. The higher is the abrasive concentration, i.e. the less workpieces are contained in the charge mass, the greater is the unit workpiece mass loss. On the other hand, the increment of the total workpiece mass loss is less than proportional to concentration increase.

As a consequence there exists an optimum abrasive concentration with regard to the time necessary to machine a single workpiece. Depending on the type of workpiece material and abrasive, this optimum degree of concentration is found empirically. In most cases the optimum values of concentration are comprised within the range 1:1·5–1:3·5.

The size and weight of the workpieces also exert an influence on metal removal rate. Larger workpieces cause an increase in dynamic forces and in the difference between the speed of the workpiece motion and the speed of the abrasive. As a result, the loss of workpiece mass increases, e.g. 2·5 times for a 5-fold increase of the weight of the workpiece.

Abrasive grain size (grain number) exerts a similar influence on metal removal rate and surface quality as in the case of other processes involving loose abrasive (e.g. abrasive-blast treatment, Chapter 19).

20.2 Barrel finishing

Barrel finishing consists in placing the workpieces in so-called barrels, i.e. in closed rotating containers, partially filled with abrasive mixture, and then imparting a rotary motion to the barrels.

The barrel cross-section usually has the shape of a regular polygon (Fig. 20.5) which facilitates the lifting of the abrasive together with the workpiece to the so-called *angle of repose* α.

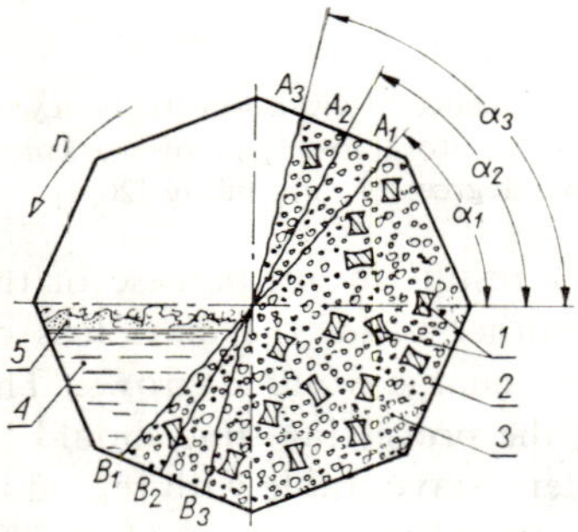

Fig. 20.5 Transverse cross-section of barrel with charge and the change of angle of repose at various rotational speeds: 1—workpiece, 2—barrel body, 3—abrasive, 4—cutting liquid, 5—foam

Under static conditions the angle of repose depends only on the state of the loose mixture (e.g. for dry sand, with grains uniform in size and shape, the angle of repose is approximately 45°). Under dynamic conditions the angle of repose depends on rotary speed and sliding resistance. The higher is the rotary speed, the higher the speed when sliding begins and the thicker is the sliding layer. In Fig. 20.5 points A, B, $A_2 B_2 A_3 B_3$ correspond to barrel rotational speeds n_1, n_2 and n_3, while $n_1 < n_2 < n_3$.

An excessive increase of the rotational speed of the barrel causes too violent falling and tumbling of the workpieces which may make them collide, and accelerates abrasive disintegration. Therefore it has been established empirically that the angle of repose α, measured at the point A, where sliding begins, should not be wider than 75°. Barrel rotation should be controlled accordingly.

Barrel finishing, commonly known as *barrelling*, is easy to perform, but its rate of metal removal is much lower than that of vibratory finishing. The cause of this is the relatively small space in which the effective cutting and abrasion work is carried out. If the rotary motions of the workpiece resulting from the changes of position of their centres of gravity are neglected, only the sliding layer may be considered as working space. The thickness of this layer depends on the difference between the dynamic angle of repose and the so-called *critical* or *static angle of repose* α_0 (Fig. 20.6). Thus the number of grains working simultaneously is considerably smaller than the number of grains cutting simultaneously under conditions of vibratory finishing where the entire volume of abrasive mixture may be considered as the working space. Also, attainable grain speeds in relation to the workpieces are lower under conditions of barreling than under those of vibratory finishing. On average, the metal removal rate in the former process is 1/4 to 1/8 of the rate obtained in the latter.

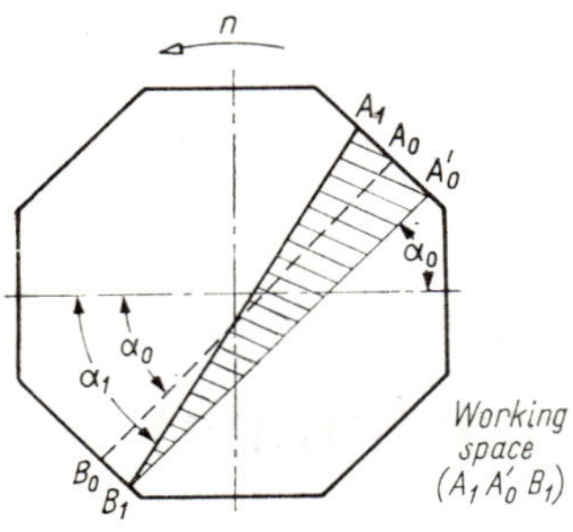

Fig. 20.6 Diagrammatic representation of working space variations

20.3 Application of vibratory and barrel finishing

Applications of vibratory and barrel finishing comprise:
— cleaning of castings, forgings and certain heat-treated workpieces,
— removing and levelling burrs and flashes (at the partings of casting moulds or segment dies) and the rounding of edges,
— improving the surface quality of the entire outer and inner surface (particularly in vibratory finishing),
— removing superfluous paints and electrodeposits etc.

Vibratory and barrel finishing should be applied, first of all, to workpieces of complicated shape, of which a suitable surface quality is required, but whose form and dimensional tolerances are not too narrow, as the characteristic feature of both finishing varieties under discussion is non-uniform removal of the allowance. Spots of higher curvature are machined with greater intensity, since higher unit pressures appear there.

Notwithstanding certain features limiting the scope of application, vibratory and barrel finishing should be considered as modern methods of mechanization mass-cleaning and smoothing of small formed parts. Consequently, these methods find increasingly wide application, particularly in foundries, forges, hardening shops and paint shops.

REFERENCES

20/1 Kartyshev, B. B.: 'Vibroabrazivnoe shlifovanie' (Vibratory Finishing), Mashinostroitel', 4, 1963

20/2 Rodziewicz, M.: 'Wygładzanie w pojemnikach luźnym ścierniwem' (Smoothing in tanks using a loose abrasive), Mechanik, 4, 1964

20/3 Szmidt, K., and Żak, T.: 'Wygładzanie powierzchni metali metodą Roto-Finish' (Smoothing of metal surfaces by the Roto-Finish method), Przegląd Mechaniczny, 1, 1960

20/4 Matsunaga, M., and Hagiuda, Y.: 'Vibratory finishing—Fundamental research', Metal Finishing, 9, 10, 1965

20/5 Babichev, A. P., and Dyachenko, V. I.: 'Opredelenie zavisimostei protsessa polirovaniya v vibriruyushchikh barabanakh' (Determining the relationships of the polishing process in vibrating barrels), Vestnik Mashinostroeniya, 1, 1965

CHAPTER 21

Impact grinding (Ultrasonic machining)

21.1 Principles of impact grinding and impact grinding equipment

21.1.1 PRINCIPLES OF IMPACT GRINDING

In the *impact grinding process* (ALI) carried out using a loose abrasive, the work of cutting, crushing and abrading is performed by abrasive grains periodically pressed or struck by a tool whose form is to be reproduced in the workpiece material. A diagrammatic representation of the principle of impact grinding is shown in Fig. 21.1.

It is easy to perceive that the rate of metal removal by this method will depend, first of all, on the percussion frequency of the tool striking a layer of abrasive grains situated between this tool and the workpiece material.

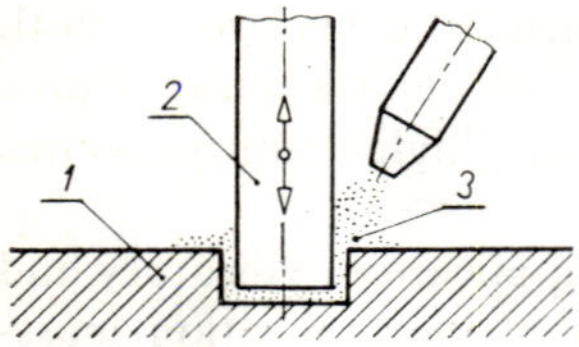

Fig. 21.1 Diagrammatic representation of the principle of impact grinding: 1—workpiece material, 2—tool, 3—abrasive mixture

Of all possible ways of increasing the tool motion frequency, the best results are obtained by taking advantage of the phenomenon of longitudinal magnetostriction. This phenomenon manifests itself in that the variations of a magnetic field in which a ferromagnetic object is placed create changes in the length of this object. If the magnetic field variations occur at ultrasonic frequency, so will the length changes.

In practice, energizing by 30 kHz generators causes the appearance of ultrasonic longitudinal waves (i.e. periodical elongations) in a ferromagnetic object. This is why the process is also called *ultrasonic machining*.

21.1.2 MAGNETOSTRICTOR

The phenomenon of magnetostriction occurs in special devices called *magnetostrictors*. A magnetostrictor (or magnetostriction transducer, Fig. 21.2) of the type used in impact grinding has a high-frequency winding 1 wound on a magnetostrictor core 5 and on special, so-called polarizing winding 2 wound on armature 3.

The method of operation of a magnetostrictor may be explained using Fig. 21.3.

Magnetostrictive capacity is characterized by the coefficient of magnetostrictive elongation

$$e_m = \frac{\Delta l}{l} \tag{21.1}$$

where: Δl—length increment, l—length of magnetostrictor core (Δl increases in an asymptotic way to a certain saturation value Δl_s, as field intensity increases).

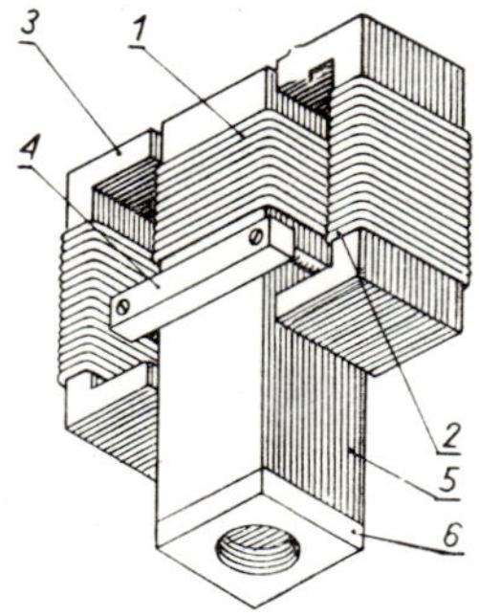

Fig. 21.2 Magnetostriction transducer applied in impact-grinding machines (of low power rating or up to 250 W): 1—high-frequency winding, 2—polarizing winding, 3—armature, 4—magnetostrictor holder, 5—magnetostrictor core, 6—attachment for amplitude transformer [21/1]

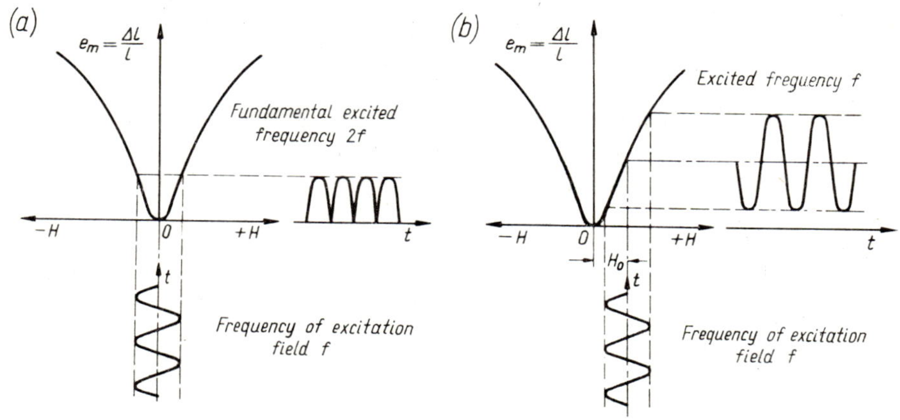

Fig. 21.3 Curves of magnetostriction excited by a variable magnetic field: (a) without magnetizing, (b) with magnetizing (after polarization) [21/2]

Figure 21.3a shows the relationship between the coefficient of magnetostrictive elongation and magnetic field intensity H. It is evident from this characteristic that elongation does not depend on the sign of the magnetic field. Therefore, a change of magnetic field intensity using a high-frequency winding (Fig. 21.3a) produces elongation changes of double that frequency. These elongations will not have a sinusoidal character. This can be overcome by magnetizing the transducer with direct current (in polarizing winding) so that the changes of magnetic field are displaced to the rectilinear part of the characteristic (Fig. 21.3b). This operation is called *polarization*. As a result of it, the elongation changes become sinusoidal and the amplitude is correspondingly increased.

If at the centre of the magnetostrictor length the *node of elongation wave* is situated, then the maximum value of elongation amplitude will be $\frac{1}{4}$ wavelength distant from this centre (Fig. 21.4). It follows that maximum elongation will correspond to magnetostrictor length $l = 0{\cdot}5\lambda$. Wavelength λ is calculated from

$$\lambda = \frac{c}{f} = \frac{1}{f}\sqrt{\frac{E}{\varrho}} \tag{21.2}$$

where: $c = \sqrt{E/\varrho}$—speed of sound in magnetostrictor material in cm/s, f—frequency of magnetic field changes in 1/s, E—Young's modulus in kg/cm², ϱ—density of magnetostrictor material in g/cm³.

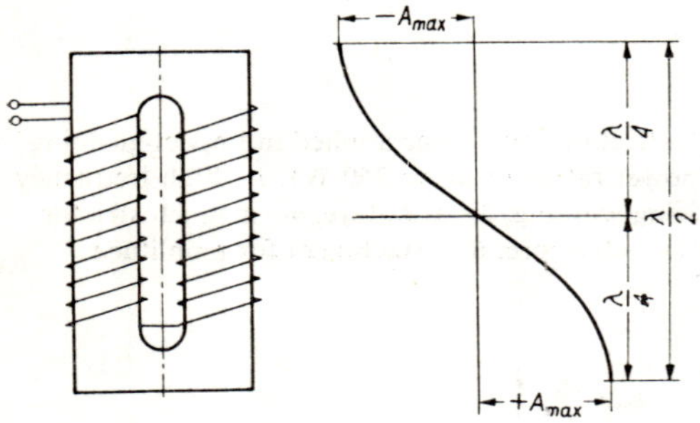

Fig. 21.4 Variation of the wave of elongation of a magnetostrictor along the latter's length

When the magnetostrictor material and length l are given, the condition (21.2) may be fulfilled only by the selection of a suitable frequency, called the *resonance frequency* f_r and calculated from eqn. (21.3)

$$f_r = \frac{1}{2l}\sqrt{\frac{E}{\varrho}} \tag{21.3}$$

21.1.3 CONCENTRATOR

Elongations of a magnetostrictor of length $l = 0{\cdot}5\lambda$ working in resonance are too small for practical applications, since they range from 0·001 to 0·1 μm (depending on the type of material).

Their amplitude may be increased by utilizing the so-called concentrators. A *concentrator* is a suitably shaped metal rod in which, due to its varying cross-section, a concentration of vibration energy occurs, causing a corresponding increase in amplitude.

A tool is mounted at the end of the concentrator. In order to transmit to the tool the greatest possible amplitude, a concentrator must have a length equal to a multiple of half the wavelength of sound in the concentrator material. When this condition is satisfied, the entire ultrasonic device works in resonance, i.e. with the maximum amplitude of tool vibrations.

The degree of amplitude increase, or the so-called *coefficient of amplification* K_A, is defined as

$$K_A = \frac{A_T}{A_M} \tag{21.4}$$

where: A_T—elongation amplitude transmitted by the concentrator to the tool, A_M—elongation amplitude transmitted by the magnetostrictor to the concentrator.

The value of amplification coefficient depends on the shape and size of the concentrator faces.

The amplification coefficient for concentrators of an exponential form is

$$K_{Ae} = \frac{d_M}{d_T} \tag{21.5}$$

where: d_M—diameter of concentrator face on the side of the magnetostrictor, d_M—diameter of concentrator face on the side of the tool.

A similar coefficient for a cylindrical two-stage concentrator is

$$K_{Ac} = \left(\frac{d_M}{d_N}\right)^2 \tag{21.6}$$

The amplification coefficient for a conical shape is lower than K_{Ae} and for a catenoidal shape K_{Ak} it is of an intermediate value $K_{Ae} < K_{Ak} < K_{Aw}$.

It follows that the most advantageous shape for a concentrator is a cylinder.

Still better results may be obtained by applying multi-stage concentrators, each stage having a length $l = 0{\cdot}5\lambda$.

In Fig. 21.5, as an example, a two-stage (exponential and cylindrical) concentrator 2 is presented (together with magnetostrictor 1), as used in the ultrasonic impact grinding machine. The tool 3 is mounted in the lower part of concentrator 2.

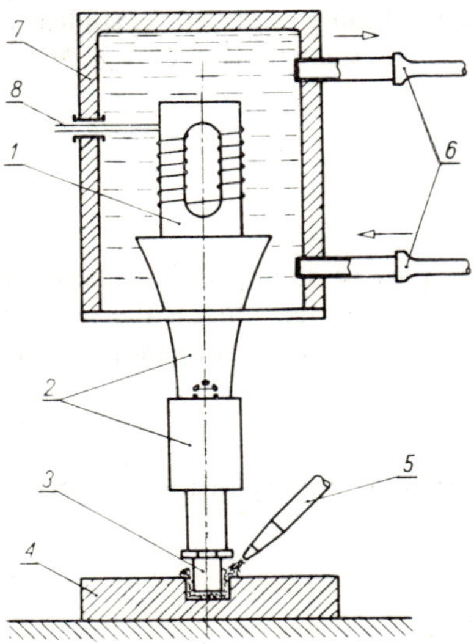

Fig. 21.5 Diagram of ultrasonic device: 1—magnetostriction transducer, 2—concentrator, 3—tool, 4—workpiece, 5—supply of abrasive slurry, 6—supply and draining of cooling water, 7—cooling cylinder, 8—leads energizing the transducer winding [21/3]

21.1.4 Materials for ultrasonic device construction

The selection of materials for the particular component parts is especially important for correct and economical functioning of the ultrasonic device.

Good magnetostrictive materials should be characterized by a high value of:
(a) the coefficient of magnetostrictive elongation in the state of saturation

$$e_{ms} = \frac{\Delta l_s}{l} \tag{21.7}$$

(b) the coefficient of magneto-mechanical coupling k_r which is equal to the square root of the ratio of mechanical energy obtained to magnetic energy supplied

$$k_r = \sqrt{\frac{E_w}{E}} \tag{21.8}$$

Relatively good values of these coefficients are found in:

— alfer (13% Al, 87% Fe) $e_{ms} = 40\times10^6$ $k_r = 0{\cdot}28$
— hypernik (50% Ni, 50% Fe) $e_{ms} = 25\times10^6$ $k_r = 0{\cdot}20$
— permalloy (40% Ni, 60% Fe) $e_{ms} = 25\times10^6$ $k_r = 0{\cdot}17$
— permendur (49% Co, 2% V, 49% Fe) $e_{ms} = 9\times10^6$ $k_r = 0{\cdot}20$

Concentrator materials should have a high fatigue strength. With regard to this feature a very good material is the alloy of titanium (94%) with aluminium (5%) whose fatigue strength $Z_{rc} = 7{\cdot}2\times10^9$ dyn/cm². However, it is not used because of its relatively high price. The commonly applied material is aluminium bronze (containing 10% of aluminium and called hidurax) whose fatigue strength $Z_{rc} = 1{\cdot}85\times10^9$ dyn/cm² and inexpensive marine bronze whose $Z_{rc} = 1{\cdot}5\times10^9$ dyn/cm².

Tool tips must be made of materials having, first of all, a high resistance to wear (low wearability) and also a high fatigue strength. Wearability depends on the mating material. Therefore the selection of tool materials also depends on workpiece material. Thus, for example, the best tool materials for the ultrasonic machining of glass are tungsten carbides, copper or silver steel. When machining sintered carbides the lowest relative wear occurs for tools made of silver steel (0·87% C, 8% Ni) or chromium-nickel steel (18% Cr, 8% Ni).

21.2 Quantities characterizing impact grinding

21.2.1 Kinematic quantities

There are several types of impact grinding. Examples of some of these types are given in Fig. 21.6. The primary motion in all types of impact grinding is the vibrating (reciprocating) tool motion, of a sinusoidal character. Thus the tool motion speed will be

$$v_T = v_{max} \sin\omega t \tag{21.9}$$

where: $\omega = 2\pi f$—angular frequency of vibrations; f—frequency of tool vibrations, while

$$v_{max} = \omega A = 2\pi f A \tag{21.10}$$

where A—amplitude of vibrations.

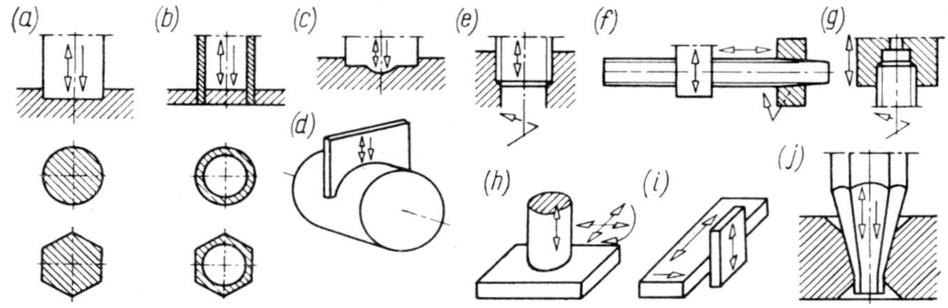

Fig. 21.6 Kinematic diagrams of several types of ultrasonic machining: (a), (b) impact grinding of round and formed holes, blind or through; (c) engraving, (d) parting, (e), (f) internal threading, (g) external threading, (h), (i), (j) lapping and polishing of faces, side and internal surfaces [21/3]

Since the grains are not of equal size and the tool moves the grains and exerts a pressure on them at various points of its own motion, a more characteristic quantity is the average tool speed

$$v_{T\,\mathrm{av}} = \int_0^{\frac{T}{2}} \frac{v_{\max} \sin\omega t}{\pi}\, \mathrm{d}t \tag{21.11}$$

which, when equation (21.10) is taken into account, yields

$$v_{T\,\mathrm{av}} = 4fA \quad \mathrm{mm/s} = \frac{4fA}{1000} \quad \mathrm{m/s} \tag{21.12}$$

As an example, for the average values applied: $f = 20$ kHz and $A = 0{\cdot}04$ mm, the resulting speeds are: $v_{T\,\mathrm{av}} = 3{\cdot}2$ m/s $= 192$ m/min and $v_{\max} \approx 5$ m/s $= 300$ m/min.

It may be assumed that the speed of the grains when entering the material will approach the value $v_{T\,\mathrm{av}}$, but will be lower than $v_{\max}$. As in practice $v_{\max}$ does not exceed 10–11 m/s, and $v_{T\,\mathrm{av}} \leqq 7$ m/s, it is obvious that the cutting speeds of the grains are considerably lower than in the case of grinding.

Feed motions in different kinematic types (Fig. 21.6) may be rectilinear (longitudinal and transverse) and rotary.

In the case of impact grinding of holes, which is the most commonly applied type of ultrasonic machining, the feed motion is rectilinear and longitudinal (parallel to the axis of the ultrasonic device). In this case the feed rate will be equal to the linear rate of metal removal or to the speed of the tool entering the material

$$s_t = Q_L = v_t = \frac{\mathrm{d}l}{\mathrm{d}t} \tag{21.13}$$

since, unlike in drilling, the entry speed decreases as the cavity becomes deeper.

For the determined time interval

$$v_t = \frac{\Delta l}{\Delta t} \tag{21.14}$$

where: Δl—increment of impact grinding depth, Δt—time.

The impact grinding time will be

$$t_m = \frac{L}{v_t} \tag{21.15}$$

On the other hand, in the case of ultrasonic lapping, the impact grinding time is calculated from

$$t_m = \frac{L}{s_t}\frac{q}{a} \tag{21.16}$$

where: L—machined length including approach and, possibly, overrun, q—machining allowance, a—depth of cut (in a single pass).

21.2.2 Technical and economic indices

The typical technical and economic indices of impact grinding are:
- — the rate of metal removal, particularly the linear rate of metal removal,
- — abrasive and tool wear,
- — machining errors,
- — surface quality.

The metal removal rate and abrasive wear have been discussed in Chapter 19.

Tool wear is defined by relative wear, either volumetric (m_V) or linear (m_L).

In the case of machining to toleranced dimensions the shape-and-dimensional wear (m_s) of the tool is determined by

$$m_s = \frac{\varepsilon}{\Delta} \tag{21.17}$$

where: ε—error above or below the nominal dimension, Δ—permissible deviation of the toleranced dimension in the direction of actual error.

Machining errors are defined as follows:

— *shape error* (in transverse cross-section)

$$b = \frac{W-w}{2} \tag{21.18}$$

where: W—dimension of cavity (for a round hole $W = D$), w—the corresponding tool dimension (for a round hole $w = d$);

— *distortion error*

$$i = W_{\max} - W_{\min} \tag{21.19}$$

(In the case of a round hole the distortion error is that of ovalization).

— *convergence error* (in longitudinal cross-section)

$$c = \frac{W_1 - W_2}{2l} \tag{21.20}$$

where: W_1—(larger) dimension of cavity at one end, W_2—(smaller) dimension of cavity at the other end.

Surface quality is determined as in other machining methods.

21.2.3 Factors influencing the process of impact grinding

Factors influencing the technical and economic indices of impact grinding may be divided into the following groups:
- factors connected with the workpiece material, in particular tensile and compressive strength, hardness and ductility,
- factors connected with the machine tool, such as the accuracy of feed and other motions, and the static and dynamic rigidity;
- factors connected with the tool, and in particular the properties of the tool material (hardness, wearability, fatigue strength), the tool manufacturing accuracy and mounting method;
- factors connected with the abrasive mixture, viz.: the type of abrasive and its properties, cutting fluid, its components and properties, abrasive concentration in the slurry (see Chapter 19) and the quantity and method of supplying the slurry to the machining zone;
- machining conditions and requirements, such as:
  vibrations frequency f in Hz,
  vibration amplitude A in mm or in μm,
  unit pressure p in kg/cm^2 or in kg/mm^2,
  surface area F in mm^2 and its position,
  hole depth l in mm and its shape.

21.3 Dependence of technical and economic indices on various factors

21.3.1 Metal removal rate

The rate of metal removal by impact grinding depends, first of all, on the frequency of tool vibration, unit pressure and the size of the work surface, and also on the abrasive and workpiece material.

The influence of tool vibration frequency on the linear rate of metal removal may be demonstrated by studying the dependence of Q_L (or v_t) on the average tool speed $v_{T\,\mathrm{av}}$ which should vary in proportion to frequency changes (eqn. 21.19). Dependence of Q_L on $v_{T\,\mathrm{av}}$ is, within the range of average tool speeds $v_{T\,\mathrm{av}} = 0\text{–}6$ m/s, almost rectilinear. This means that the linear rate of metal removal should be proportional to vibration frequency.

Actually it has been established that in the low frequency range ($f < 400$ Hz) the metal removal rate rises more than proportionately and in the range of frequencies $f > 10$ kHz—less than proportionately to the increase of vibration frequency [21/3].

The influence of tool vibration amplitude A on the linear metal removal rate is very distinct and may be approximated by an exponential function with its exponent approaching 2. This relationship is illustrated by an example from research [21/1] given in Fig. 21.7.

However, the relationship represented in Fig. 21.7 is valid only in the range of amplitudes which are smaller or only slightly larger than the abrasive grain dimensions. Further increases in amplitude lead to the occurence of splashing which causes a reduction of the number of active abrasive grains and results in a decrease in metal removal rate.

In practice, with regard to the range of grain sizes of abrasive materials used, amplitudes vary within the limits of 0·04–0·08 mm.

Studies on the influence of constant unit pressure on the linear rate of metal removal revealed the occurrence of a distinct maximum of this rate for a certain optimum pressure value. This is illustrated in Fig. 21.8 [21/3]. Such a relationship

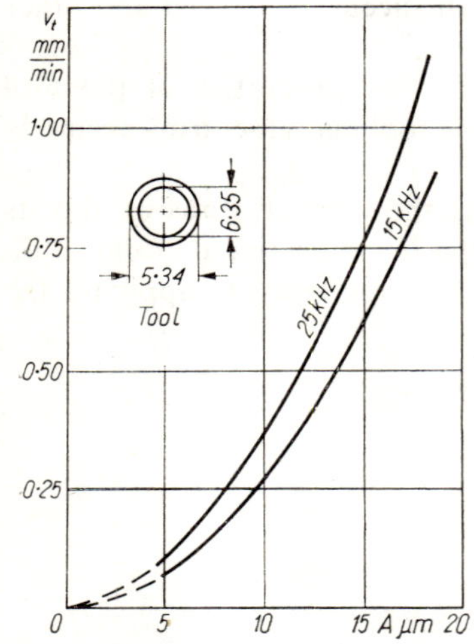

Fig. 21.7 Dependence of the speed of tool penetration into the material v_t on the amplitude of tool vibrations A [21/1]

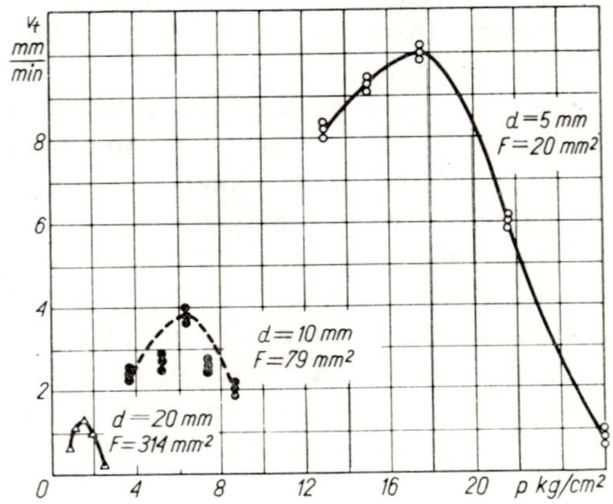

Fig. 21.8 Dependence of the speed of tool penetration into the material v_t on unit pressure p for various surface areas subject to impact grinding [21/3]

should be explained as follows: within the range of pressures lower than the optimum, the force pressing the grains into the material is too small and the volume of material chipped by particular grains diminishes. On the other hand, within the range of unit pressures exceeding the optimum value, damping is too strong and gradually the tool ceases to break away from the grains, thus preventing their changes of position. As a result, the volume of chipped material decreases.

The machined surface area also influences the linear rate of metal removal. As is evident from Fig.21.8, the larger is this area, the lower is the position of the curves $v_t = f(p)$ and the lower are the values of optimum pressure. The relationship between optimum unit pressure and the machined surface area has a hyperbolic form (Fig. 21.9).

When optimum unit pressures are applied, it is found that the maximum rate of metal removal corresponds to a certain optimum value of surface area. In Fig. 21.10, as an example, the relationship $Q_V = f(F)$ is presented, as determined on the 1000 W impact grinding machine. It is evident from this figure that in this case the optimum value of F is $\sim$ 300 mm² which corresponds to the optimum unit pressure $p = 2$ kg/cm² (acc. to Fig. 21.9).

The optimum value of machined surface area related to the power supplied to the transducer (from the mains) may be considered an index for the evaluation of impact grinding machines. The larger is the optimum surface area for the given

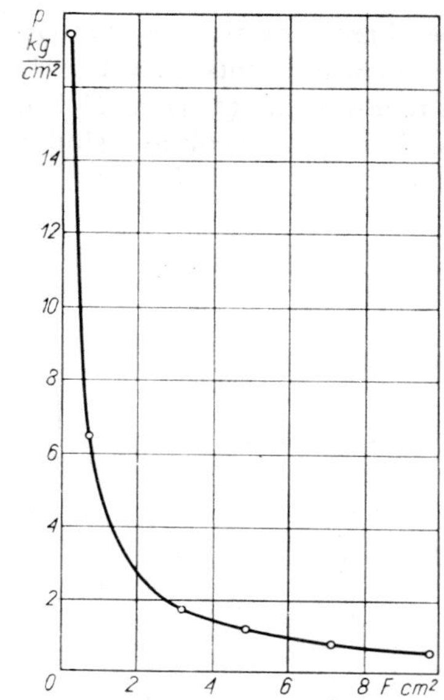

Fig. 21.9 Dependence of optimum unit pressure on the surface area subject to impact grinding [21/3]

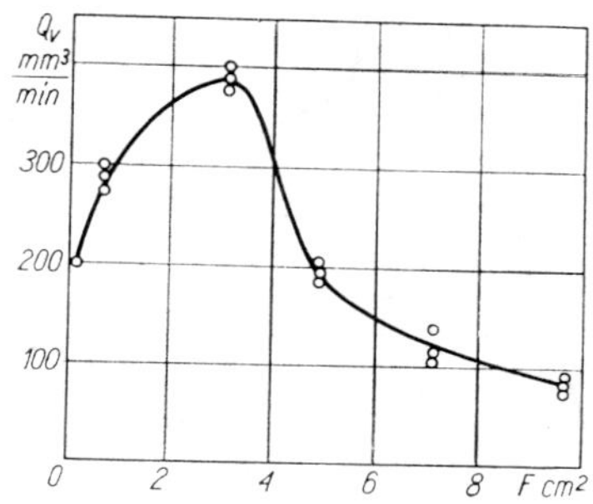

Fig. 21.10 Dependence of the volumetric rate of metal removal Q_V by means of an impact grinding machine on the surface F subject to impact grinding [21/3]

power, the better is the design and construction of the machine tool. On the basis of the characteristics of existing designs it may be stated that in the best impact grinding machines the value of optimum machined surface area in mm² is about 40% of the power in watts (i.e. $F_{opt} \approx 0{\cdot}4N$).

Of the known abrasive materials diamonds may not be used in impact grinding (because of excessively high cost) nor may aluminium oxide, emery and other materials of inferior hardness (because of insufficient metal removal rate).

The two remaining abrasives, i.e. boron carbide and silicon carbide (carborundum) show the best usability for impact grinding purposes. Boron carbide yields higher rates of metal removal in the initial period, but undergoes disintegration within a shorter time. The metal removal rate obtained with silicon carbide is about 15% lower in the machining of glass, about 33% in the case of tool steels and about 35% in the case of sintered carbides.

The influence of abrasive grain size on metal removal rate shows a close analogy to a similar relationship in abrasive-blast treatment. All other conditions being constant, Q_V acquires its maximum value for a certain optimum grain size. In the example given in Fig. 21.11 the maximum value of the volumetric rate of metal removal was obtained for grain number 10(120).

Connected with the relationship discussed above is the influence of working time of the abrasive on metal removal rate. Because of the progressive disintegration of abrasive dust, similar changes of metal removal rate will occur as in abrasive-blast treatment (Chapter 19) but at a faster rate as a result of the higher frequency of grain impacts.

Of the remaining factors influencing the rate of metal removal, particular attention should be paid to the depth of the machined hole. Measurements indicate that increasing the hole depth l causes a decrease of metal removal rate Q_V (Fig. 21.12). The reason for this is that the deeper the tool reaches, the more difficult and slower is the exchange of abrasive from underneath this tool.

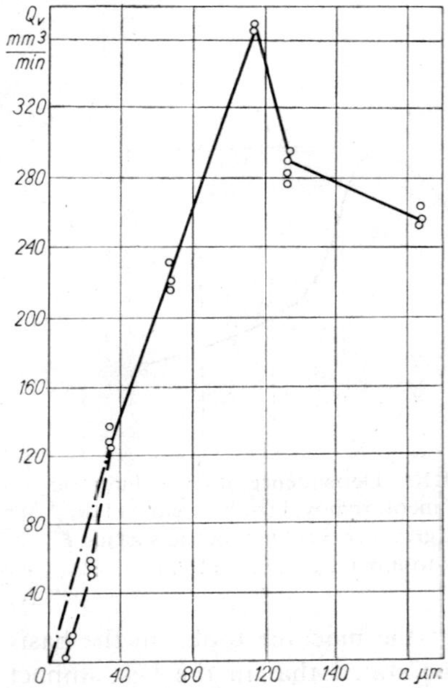

Fig. 21.11 Dependence of the rate of metal removal by impact grinding Q_V on the characteristic dimension of the abrasive grain a [21/3]

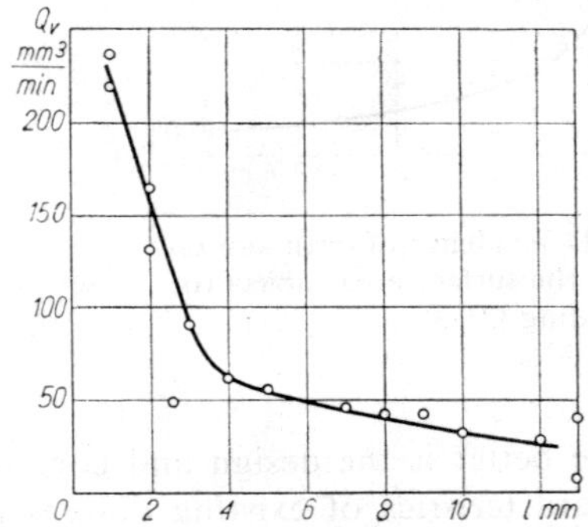

Fig. 21.12 Dependence of the rate of metal removal by impact grinding Q_V on the depth of the machined hole l [21/3]

The abrasive concentration in the slurry has a distinct influence on metal removal rate. Within the range in which it is possible to maintain the homogeneity and continuity of slurry supply, i.e. up to a volumetric concentration of about 40%, metal removal rate is observed to rise almost linearily with abrasive concentration.

In practice, a volumetric concentration of about 30–35% of silicon carbide in cutting fluid (water with anticorrosive additives) is maintained.

A reduction of volumetric concentration from 35% to 10% causes a decrease of metal removal rate by as much as about 50%. During the machining operation a change of concentration often occurs. It is mostly a decrease due to the abrasive dust settling on the machine tool table. In view of the considerable influence of abrasive concentration on metal removal rate, the actual concentration should be checked at certain time intervals.

21.3.2 Tool and abrasive wear

Tool wear in the process of impact grinding is generally slight if the tool material is harder than the workpiece material and quite negligible if the former is only a little softer or harder than the abrasive. Thus, for example, when impact grinding glass with a steel tool the relative wear is of the order of 1%, while the use of sintered tungsten carbide tool and boron carbide dust reduces the relative wear to about 0·1%.

On the other hand, tool wear increases considerably when both the abrasive grains and workpiece material are harder than the tool material. For example, when machining sintered tungsten carbide using boron carbide [grain number 12(100)] and a brass tool, the relative wear was 140%.

This indicates that the process of abrasion and chipping by abrasive grains occurs in parallel on the tool and work surfaces. The intensity of this process depends on the strength properties of the material.

Tool wear shows mainly in parts having the strongest curvature (smallest rounding radii), e.g. on the corners of a tool having the shape of a rectangular prism.

The possibilities of accurate machining are limited mainly by the relative form-and-dimensional wear. Therefore, the higher are the accuracy requirements, the more wear-resistant must be the tool.

Abrasive wear shows, first of all, in grain disintegration and in the resulting decrease of metal removal rate.

For the determination of the economical and productive life of the abrasive, equations from Chapter 19 (on abrasive-blast treatment) may be used, with constants found in an empirical way.

The disintegration process in impact grinding is much more intense than in abrasive-blast treatment. In addition, the amount of abrasive dust is generally smaller in the former process than it is in the latter. This results in the economical and productive life of an abrasive in impact grinding being only a small fraction of such life in abrasive-blast treatment.

21.3.3 Dimensional accuracy and surface quality

Of all factors influencing the impact grinding process, the accuracy of the machine tool, its rigidity and vibration-resistance, as well as abrasive grain size and tool wear have the greatest influence on machining accuracy and machined surface quality.

Experiments reveal that the assembly guiding the magnetostrictor–concentrator system has a considerable influence on dimensional accuracy. For this reason the acceptance inspection of an impact grinding machine has no lesser importance than that of conventional machine tools of high cutting accuracy.

However, the greatest influence is exerted on the indices of machining accuracy by the size of the abrasive grains.

In Fig. 21.13 is given, by way of example, a diagram representing the relationship between hole shape and grain size. The difference between the actual measured dimension and the nominal dimension of a round hole is

$$d_{act} = d_{nom}+2(a+\Delta) \tag{21.21}$$

where: a—characteristic dimension of a grain of the basic fraction in mm, and $\Delta = b-a$.

The value of Δ may be approximated by the empirical formula

$$\Delta \approx M+H\,a \tag{21.22}$$

In the example: $M = 0$; $H = -0{\cdot}9986$.

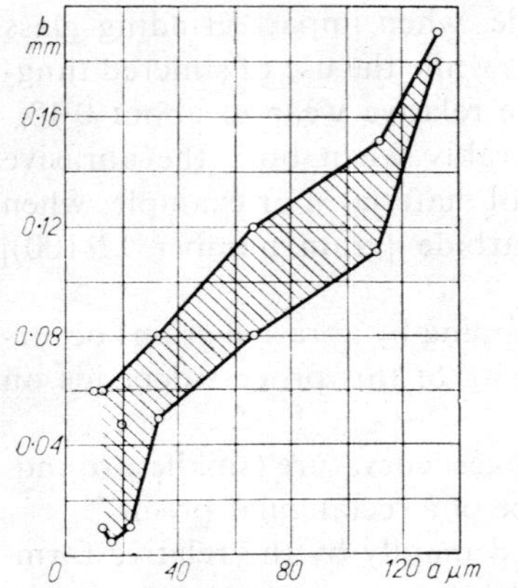

Fig. 21.13 Dependence of shape change b on the characteristic dimension a of silicon carbide (carborundum) grains in impact grinding [21/3]

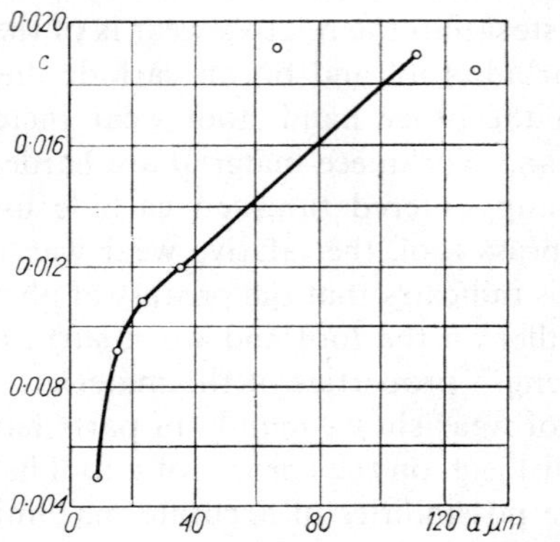

Fig. 21.14 Dependence of hole convergence c on the characteristic dimension a of SiC grains in impact grinding [21/3]

The dependence of ovality error on grain size, in the case of hole machining, may be expressed by the empirical formula

$$u = K a^n \tag{21.23}$$

In the machining of holes e.g. in glass, within the diameter range of 5–20 mm, $K = 0{\cdot}0035$ and $n = 0{\cdot}66$.

It follows from eqn. (21.23) that with increasing grain size the ovality error grows less than proportionately.

The error of convergence varies in a similar way. In Fig. 21.14, by way of example, the dependence of convergence on grain size is presented, for constant hole length l.

The conclusion follows from these relationships that in the design and construction of a tool for a hole of given nominal dimension, suitable dimensional correction with regard to the given grain size should be taken into account. Thus, for example, when the shape error is taken into consideration, the tool diameter should be reduced by the quantity b:

$$d_m = d - b \tag{21.24}$$

where: d_m—manufacturing diameter of the tool, d—nominal diameter of the hole, b—shape change assumed on the basis of previous measurements.

In impact grinding the quality of the machined surface depends mainly on the abrasive grain size and the workpiece material.

In Fig. 21.15, by way of example, a typical diagram of the relationship between mean total height of surface irregularities R_{tm} and grain size is presented, for the case of impact grinding of glass and tungsten carbide surfaces. The continuous lines in the diagram refer to the side walls of the holes and the dashed lines to the hole bottom.

As is evident from the diagram, the irregularities of the side wall surfaces of the cavities are considerably larger than those of the bottom. This results from the side walls being scratched by grains entering and leaving the machining zone.

In addition, Fig. 21.15 also shows an important influence of workpiece material. The more brittle and soft is this material, the greater is its resulting roughness, since the deeper are the cavings left by the particular grains.

The influence of grain size on surface quality has a similar character and causes as in the cases of other abrasive processes, especially abrasive-blast treatment.

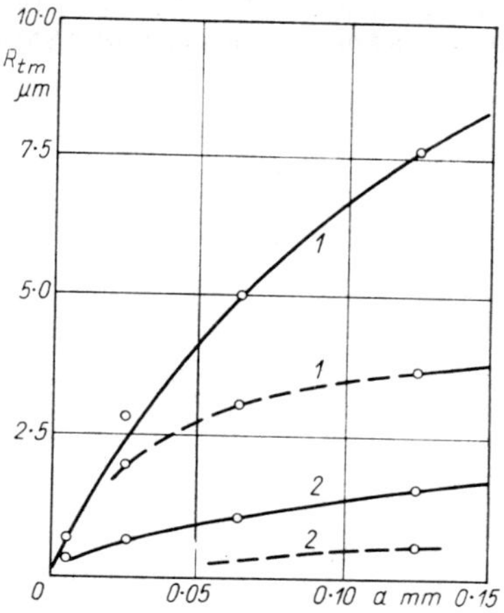

Fig. 21.15 Dependence of mean total height of surface irregularities R_{tm} after impact grinding on the characteristic dimension of abrasive grains a: 1—for holes in glass, 2—for holes in sintered carbides

21.4 Applications of impact grinding

The application of impact grinding is determined by its characteristic features.

Impact grinding may be applied economically, first of all, to:
- hard and brittle materials,
- shallow cavities and cuts.

As has been mentioned before, the work surface depends on machine tool power. However, it may be stated generally that impact grinding is applied when machining small surfaces (less than 1000 mm^2).

Impact grinding is most profitable if applied to:
- parting and cutting out of formed elements of semiconducting materials (germanium and quartz); the advantage of these applications is the high material removal rate as compared with conventional methods, high accuracy and a minimum of rejects due to cracking, since impact grinding is practically a cold process and thus does not cause the appearance of temperature cracks, occurring e.g. in conventional grinding,
- engraving on glass (e.g. radio set scales, graduations etc.) and on hardened steel and sintered carbides (e.g. dies),
- manufacturing of spinning nozzles (spinnerets) made of ceramics, metals and minerals,
- parting and machining of precious stones including diamonds (by means of diamond or borazon dust).

The trend toward accurate machining and the growing percentage of hard-to-machine materials being used favours further widening of the scope of impact grinding applications.

REFERENCES

21/1 NEPPIRAS, E. A.: 'Ultrasonic machining', Metalworking Production, **100**, 27–31, 33, 34, 1956

21/2 MATAUSCHEK, J.: 'Einführung in die Ultraschalltechnik', Verlag Technik, Berlin 1957

21/3 KOPS, L.: 'Obróbka i obrabiarki ultradźwiękowe' (Ultrasonic machining and machine tools), Prace IOS, 15, 1961

21/4 DMOCHOWSKI, J.: 'Obróbka ultradźwiękowa' (Ultrasonic Machining), WNT, Warszawa 1962

21/5 KACZMAREK, J., KOPS, L., and SHAW, M. C.: 'Ultrasonic grinding economics', J. of ASME, Paper 66, 1966

PART 4

Outline of the theory and techniques of erosion machining

CHAPTER 22

Spark machining

22.1 Definition and classification of erosion machining methods

Mechanical energy is not the sole form of energy which lends itself to shaping machine parts and other products. Shaping can also be effected by means of the energy of electrical discharges, the energy of chemical changes, and the energy borne by ions, electrons, and photons.

Shaping methods employing energy other than mechanical have many distinctive features. On the other hand, these methods have the following features in common:

— shaping by removal of surplus material (the machining allowance) in the form of minute particles which are indistinguishable to the unaided eye, in some cases these particles being molecules or even ions,

— a capability for shaping even the most troublesome of materials, by liquation (fusion), evaporation, or chemical processes.

In general, methods of shaping machine parts and other products by processes removing material in the form of particles which are difficult to distinguish by the unaided eye are called *erosion machining*.* In particular, erosion machining is the term applied to hitherto employed shaping methods utilizing non-mechanical forms of energy.

Erosion machining may be classified from various points of view:

— division into methods of erosion machining according to how the machining allowance is removed, i.e. bearing in mind what kind of energy is used and what form of erosion it causes,

— division into kinematic types according to the kinematics of the tool and the workpiece.

The names of the kinematic types are formed by analogy with the methods and kinematic types of machining by cutting, except that the term drilling is replaced by hole sinking (rotary or not),

* This term is a matter of convention since, according to the old definition, *erosion* is the process of material being removed by the action of mechanical forces. In the sense employed here, erosion is a process wherein material is removed in the form of particles which defy distinction by the unaided eye, regardless of the kind of forces and the form of energy used to remove those particles.

— division into methods according to the degree of accuracy and order of machining (as in the case of machining by cutting).

Other criteria employed in machining by cutting can also be used in an analogous manner for classification.

The division into methods of erosion machining is represented schematically in Table 22.1. According to this table, erosion machining (EM) subdivides into:

— electro-discharge machining (EDM), which subdivides further into: electro-machining by spark, or electrospark machining (EDS); and electro-discharge machining by impulse, or electro-pulse machining (EDI);

Table 22.1 *Modes of electro-erosion machining*

E	Erosion machining		
EDM	Electro-discharge machining	EDS	Electro-discharge machining by spark
		EDI	Electro-discharge machining by impulse
ECM	Electro-chemical machining	ECE	Electrochemical machining by electrolysis
		ECA	Electrochemical machining by abrasion
		ECI	Chemical and spark machining
EBM	Electro-beam machining	EBI	Electro-beam machining by ions
		EBE	Electro-beam machining by electrons
		EBP	Electro-beam machining by photons (by laser)

— electrochemical machining (ECM), in which we distinguish: electrochemical machining by electrolysis, or electrolytic machining (ECE); electrochemical machining by abrasion, or chemical-abrasive machining (ECA); and chemical and spark machining (ECI);

— electro-beam machining, or beam-erosion machining (EBM), in which we can distinguish: electro-beam machining by plasma jet, i.e. by ions, or ion-beam machining (EBI); electro-beam machining by electrons, or electron-beam machining (EBE); and electro-beam machining by photons, or laser machining (EBP). In addition to the aforementioned methods and types of erosion machining there are some other intermediate varieties which constitute a combination of varieties of erosion machining and machining by cutting and by abrasion.

22.2 Electric discharges and erosion

The process of spark discharge from cathode to anode, when both electrodes—at constant separation—are immersed in a dielectric, may be described in simplified terms as follows.

Cold emission of electrons occurs when a voltage is applied to the cathode, and this in turn produces a state of ionization in a particular space. At a given voltage, the ionization ends at a certain distance from the cathode since dielectrics display a considerable capacity for attenuating the ionization process, i.e. they bring about deionization.

An increase in the voltage extends the ionization zone and intensifies the ionization. At a certain moment, the state of ionization becomes sufficient for a discharge, i.e. a flow of charge from cathode to anode.

The voltage at which discharge is initiated is called the *limiting voltage*.

A similar discharge phenomenon can be attained if, with the voltage remaining constant, the electrodes are drawn closer together. The distance at which discharge begins is called the *limiting electrode separation*.

Discharge may occur at constant voltage and constant electrode separation if the properties of the dielectric are altered. This can be done, for instance, if a liquid dielectric is pumped out and replaced with another with a steadily increasing amount of suspended graphite; discharge then takes place at a certain moment. The conducting particles of the suspension change the state of the dielectric, forming so-called *conduction bridges*.

As a result of the discharge there is a brief concentration of electrical and mechanical energy of electrons on the anode. Owing to this, very high temperatures are produced in the immediate vicinity of the discharge. Partial melting, and even vaporization, of the metal of the anode ensues. The electrodynamic forces and the large gradient of internal stresses due to the temperature field cause the molten metal to be ejected into the dielectric.

Such, in simple terms, is the process of erosion by electric discharge which results in a characteristic *crater*, also referred to as *pit* or *meniscus* (Fig. 22.1), being formed at the site of the discharge. The crater produced by a single discharge can be treated approximately as a spherical segment. With this assumption, the crater is characterized by diameter and depth.

The erosion products, among which almost spherical pieces of fused metal predominate (Fig. 22.2), confirm that thermal processes and the liquation of anode metal which they bring about play the principal role in the process of erosion by electric discharge.

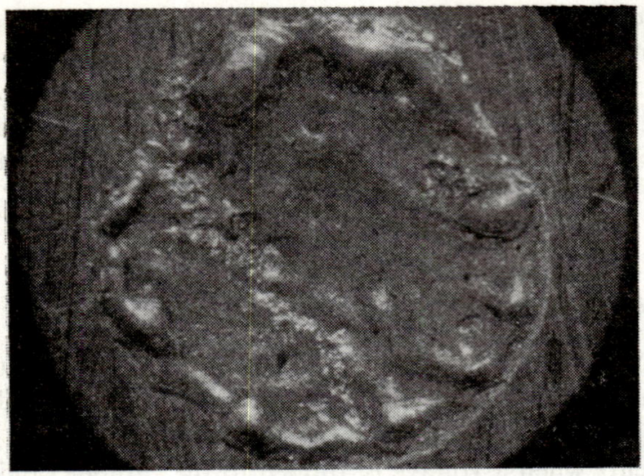

Fig. 22.1 Crater (pit) after single spark discharge (magnification 500 ×)

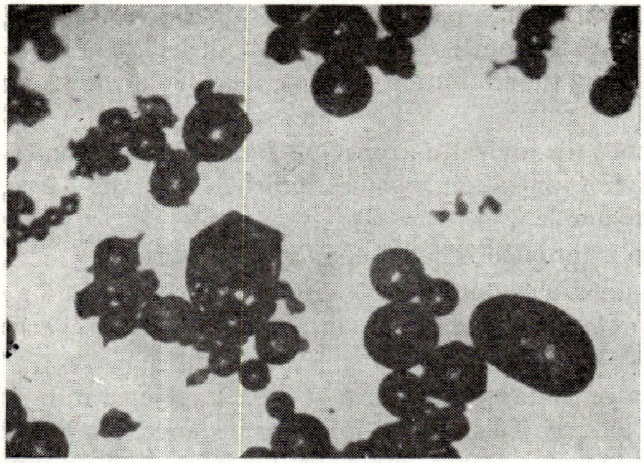

Fig. 22.2 Products of spark erosion (magnification 200 ×)

22.3 Electro-spark machining

22.3.1 THE CHARACTERISTICS OF THE SPARK-EROSION PROCESS

The principal criterion for the classification of electrodischarge machining (EDM) is the method used to generate current pulses leading to discharges and the mode and behaviour of the discharges.

Electro-spark machining (EDS) (Table 22.1) is effected in relaxation circuits. The method of generating current pulses in a system of the relaxation type is called *dependent generation*. The oldest and most common relaxation system, also called a *Lazarenkov* or *RC circuit*, is depicted in Fig. 22.3.

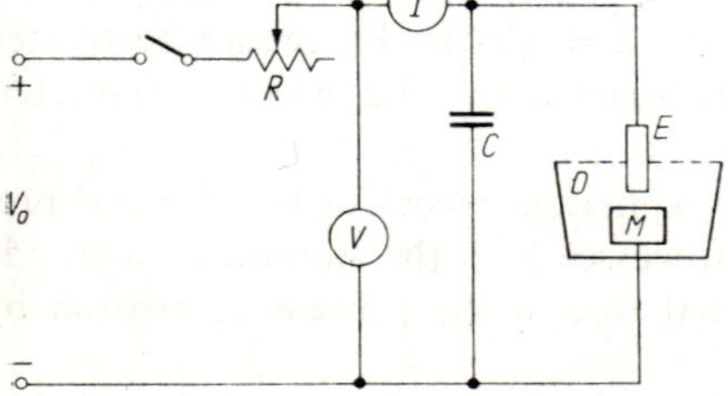

Fig. 22.3 Diagram of RC circuit for electric-spark machining (Lazarenkov)

The RC type of relaxation system has two circuits:
- charging circuit which consists of a source of direct current at a voltage of V_0, ohmic resistor R_i, and capacitor C;
- discharging circuit, which comprises: a capacitor C, a working or tool electrode (also called eroder) E (the cathode), the electrode being machined M (anode), and between the electrodes a dielectric D which sets up a particular resistance.

The increase in voltage across the capacitor C in the charging circuit is expressed by the formula

$$V = V_0(1-e^{-t/(RC)}) \tag{22.1}$$

where: V—the instantaneous voltage across the capacitor plates, V_0—the voltage of the current source, t—the charging time, R—the resistance in the charging circuit, C—capacitance.

Similarly, the formula for the instantaneous charging current is

$$i = \frac{V_0}{R} e^{-t/(RC)} \tag{22.2}$$

If in eqns. (22.1) and (22.2) we replace the resistance R with the resistance R_g between the electrodes, i.e. the resistance of the tool-work gap (or spark gap), we obtain suitable formulae for the instantaneous charging current and voltage.

The charging and discharging processes in an RC system are depicted in Fig. 22.4. The time that the capacitor takes to rise from 0 to the limiting voltage V_{lim} is called the *charging time* t_{ch}. The time the voltage takes to fall off from V_{lim} to 0 is

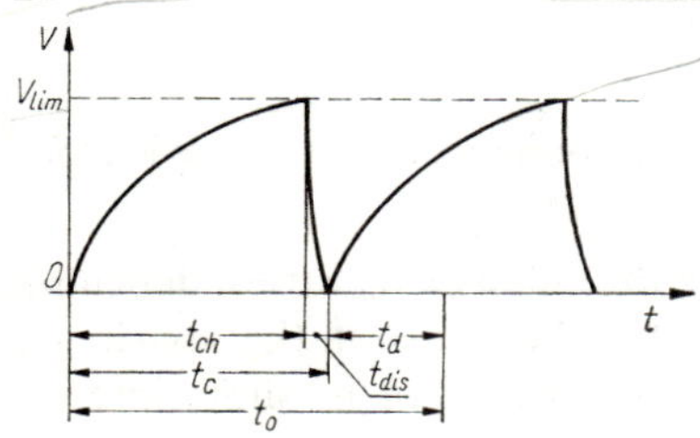

Fig. 22.4 Charge and discharge voltage during generation of RLC type discharges

called the *discharging time* t_{dis}. The time which elapses from the conclusion of the discharging process to the deionization of the tool-work gap is called the *deionization time* t_d.

It is seen from Fig. 22.4 that the cycle time is defined by

$$t_c = t_{\text{ch}}+t_{\text{dis}} \tag{22.3}$$

and the discharge period, by

$$t_0 = t_c+t_d = t_{\text{ch}}+t_{\text{dis}}+t_d \tag{22.4}$$

The inverse cycle time gives the frequency of discharge

$$f = \frac{1}{t_c} = \frac{1}{t_{\text{ch}}+t_{\text{dis}}} \tag{22.5}$$

Actual discharging circuits comprise not only the ohmic resistance R_i but also a capacitance R_g and an inductance L_g. Therefore, in accordance with electric circuit theory discharges may be aperiodic or oscillatory. This depends on the ratio of the values of R_g and L_g and C_g [22/3].

The discharge is:

— aperiodic when $R_g \geqq 2\sqrt{L_g/C_g}$ (22.6)

— oscillatory when $R_g < 2\sqrt{L_g/C_g}$ (22.7)

It follows from these conditions that in the case of aperiodic discharges the instantaneous discharge voltage and current will take on only positive values (Fig. 22.4).

In the case of oscillatory discharge the voltage and current change sign. An example of such a discharge is shown in Fig. 22.5. In this case useful work by erosion is obtained only during the period t_u when the voltage attains sufficiently high positive values.

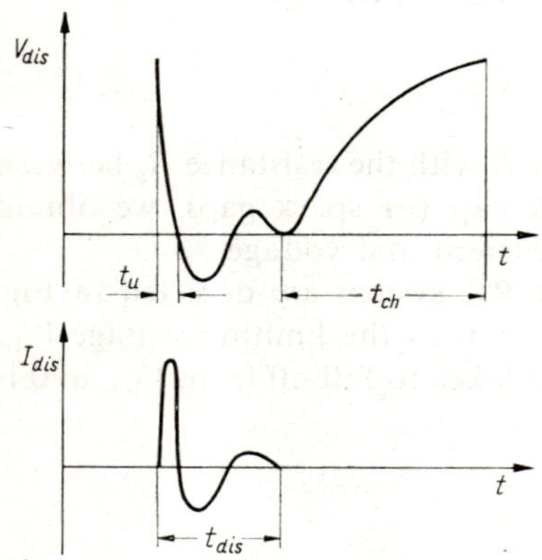

Fig. 22.5 Aperiodic discharges in RC type generator

Under the conditions of electro-spark machining the value of R_g varies, depending on the processes of ionization and deionization and the state of the spark gap. Accordingly, there may be a change of types of discharge even in short intervals of time.

The frequency of discharge in an RC system can be varied, with constant source voltage and constant R and C, only by changing the limiting voltage, that is to say, by changing the state of the spark gap.

The extent to which the cycle time can be reduced is determined by the requirement that the charging time be suitably longer than the deionization time, i.e.

$$t_{ch} \geqq z\, t_d \tag{22.8}$$

where $z \geqq 1$ and depends on the ionization state of the spark gap and on the intensity with which the dielectric is deionized.

In actual equipment used in practice the quantities discussed above take on the following values:

— voltage of current source $V = 60\text{–}250$ V;

— size of spark (or work) gap $s = 0{\cdot}01\text{–}0{\cdot}8$ mm;

— average value of discharge current I_{dis} up to 300 A;

— charging time $t_{ch} = 10^{-2}\text{–}10^{-5}$ s;

— discharging time $t_{dis} = 10^{-9}$–10^{-5} s;
— discharge frequency f, up to 100 kHz.

In most cases the discharging time is only an insignificant fraction of the charging time. Hence, the discharge frequency may be sufficiently approximated if only t_{ch} is taken into account. Accordingly, if V_{lim} is assumed to be constant, then by eqn. (22.1) [22/4]

$$e^{-\frac{t_e}{RC}} = 1 - \frac{V_{lim}}{V_0} = a \qquad (22.9)$$

where a—a constant dependent on the ratio V_{lim}/V_0.
Therefore

$$t_{ch} = -RC \ln a \qquad (22.10)$$

The approximate value of the discharge frequency may thus be calculated from

$$f \approx \frac{1}{t_{ch}} = -\frac{1}{RC \ln a} \qquad (22.11)$$

or, more accurately

$$f = \frac{1}{t_{dis} - RC \ln a} \qquad (22.11a)$$

The energy of a single discharge can be evaluated by

$$E_i = \frac{CV_0^2}{2}(1-a)^2 = \frac{CV^2}{2}(1-e^{-t_{ch}/(RC)})^2 \qquad (22.12)$$

22.3.2 Spark generators (relaxation generators)

In addition to the aforementioned conventional system of discharge generation, called an *uncontrolled* or *dependent relaxation generator*, several other types of relaxation generators are in use [22/5].

The *RLC generator* is a variant of the uncontrolled relaxation type. The diagram of the charging and discharge system is given in Fig. 22.6.

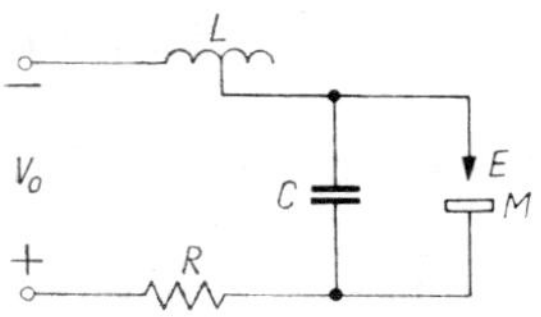

Fig. 22.6 Diagram of RLC type generator

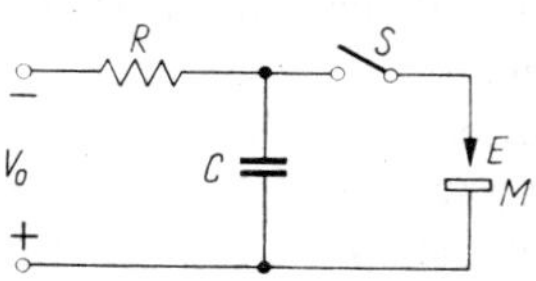

Fig. 22.7 Diagram of controlled RC type generator

An advantage of this generator over the RC type is a higher energy efficiency. In RC generators not more than half of the electric energy is dissipated as heat in the resistor in the charging circuit. In LC generators the charging is oscillatory and not exponential. Consequently, the maximum voltage across the capacitor is twice that of the current source, and the charging time becomes shorter, whereby the discharge frequency can be increased and hence so can the erosion rate. With a suitable choice of inductance, the charging process is retarded in the initial phase and considerably accelerated in the final phase. As a consequence, it is possible to charge the capacitor even during the deionization period.

Another, improved version of the relaxation (dependent) generator is the *controlled RC generator* (Fig. 22.7).

Any increase in the frequency of RC or RLC generators is limited by the deionization time. For when the discharge is completed, the spark gap still remains ionized for some time and this substantially reduces the limiting voltage. Too early charging then leads to premature discharges of lower energy and in some cases, even to arcing.

In order not to lose time for advancing the deionization, the discharge circuits may be disconnected for this interval of time. A switch S is used for this purpose (Fig. 22.7). This function may be performed by electronic valves, thyratrons, ignitrons, etc.

The type of controlled relaxation generator described above has not, however, found widespread application since the switch constitutes a limitation on the current and if one wants to maintain the current at a sufficiently high level, a very large number of valves (as many as 1000) must be connected in parallel, and this increases costs while reducing reliability of operation.

A characteristic feature of all relaxation systems is that the frequency of the charging and discharging cycles depends on the state of spark analysis determining the limiting voltage. Hence the name of dependent generator.

22.4 Electric-pulse machining

22.4.1 THE CHARACTERISTIC OF PULSE EROSION

Pulse generators, also known as independent generators, operate on a principle different from that of the relaxation type. A feature common to all the types of pulse generator is that they contain no capacitor. Hence, there is no division into charging and discharging circuits. Contrary to the relaxation generator, impulse generators do produce pulses at a certain rate, regardless of the state of the spark gap (thus the name of *independent generator*).

Electric discharges in systems with a pulse generator are brief flows of current from a source whose voltage is not less than the limiting voltage of the spark gap.

Electro-erosion machining with the aid of pulse generators is called *electric-impulse machining*.

The variation of the voltage and current during a pulse is plotted in Fig. 22.8.

The moment the pulse starts, the voltage drops from a value of V_0 to $V_{\lim_0}$ in the initial state of the spark gap. As the pulse persists, the limiting voltage decreases further; this can be approximated by the empirical formula

$$V_{\lim} = V_{\lim_0} - 2b\,t \tag{22.13}$$

where: $V_{\lim}$—the instantaneous value of the limiting voltage, $V_{\lim_0}$—the initial value of the limiting voltage, b—a coefficient determined experimentally, and t—the duration of the impulse.

The effective limiting voltage or effective impulse voltage is expressed by the formula

$$V_{\lim_s} = V_{\lim_0} - b\,t \tag{22.14}$$

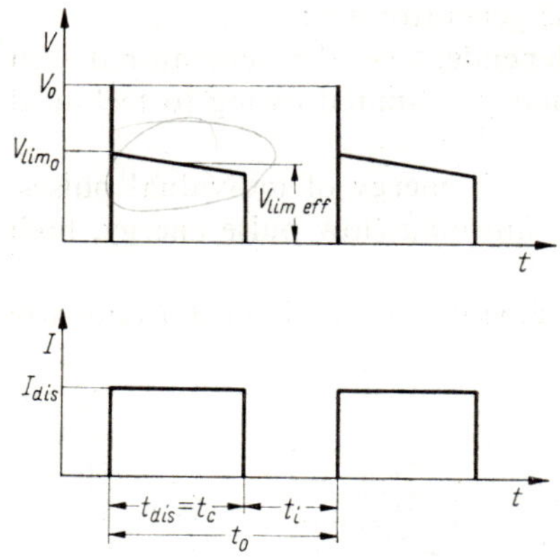

Fig. 22.8 Simplified plot of pulse voltage and current as a function of pulse duration

The current remains almost constant during the pulse. The pulse rate is defined by the relationship

$$f = \frac{1}{t_0} = \frac{1}{t_{\text{dis}}+t_p} \tag{22.15}$$

where t_p—the pulse interval.

It is seen from this formula that the rate depends on the pulse duration and the pulse interval. The highest pulse rate, for a given pulse duration, occurs when the pulse interval is reduced to the value of the deionization time, i.e.

$$f \leqq \frac{1}{t_{\text{dis}}+t_d} \tag{22.16}$$

The deionization time can be approximated as a function of the pulse current. In view of the relatively brief duration of the pulses, it may be assumed to a first approximation that the deionization time is directly proportional to the current I_{dis}:

$$t_d \approx m\, I_{\text{dis}} \tag{22.17}$$

where m—a coefficient of proportionality found experimentally.

Accordingly, eqn. (28.16) may be rewritten as

$$f \leqq \frac{1}{t_{\text{dis}}-m\, I_{\text{dis}}} \tag{22.18}$$

The energy of a single pulse in electric-pulse machining is calculated as

$$E_i = V_{\text{lim}_s}\, I_{\text{dis}}\, t_{\text{dis}} \tag{22.19}$$

and, taking account of eqn. (22.14) this can be rewritten

$$E_i \approx I_{\text{dis}}(V_{\text{lim}_0}\, t_{\text{dis}} - b\, t_{\text{dis}}^2) \tag{22.20}$$

22.4.2 Pulse generators [22/5], [22/6]

The first type of pulse generator is the *rotary generator* (Fig. 22.9). Current pulses are generated by a multipolar d.c. machine.

The characteristic features of the machine pulse generator are:

— generation of pulses at constant frequency, dependent on the generator design and speed; consequently, the increase in frequency is limited owing to technical considerations (*f* below 1000 Hz);

— relatively high efficiency in rough machining (high energy of individual pulses, low frequency) and relatively low efficiency in finishing (low pulse energy, high frequency).

In view of the foregoing features, the machine generator is used most frequently only in conjunction with some other type of generator.

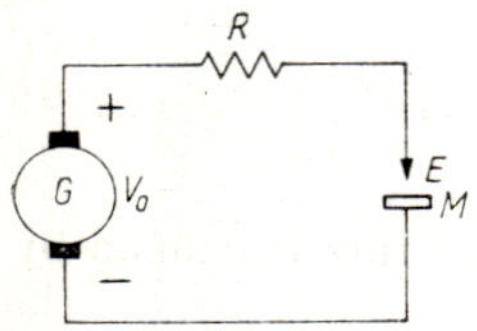

Fig. 22.9 Diagram of a rotary pulse generator

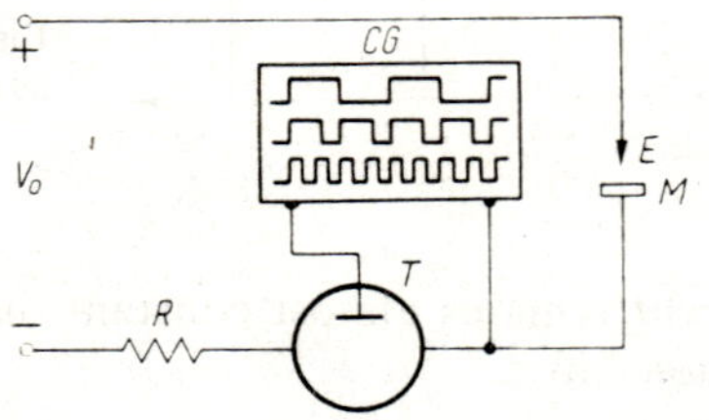

Fig. 21.10 Diagram of an electronically-controlled pulse generator

A further step in the development of generators for electro-erosion machining is the *controlled pulse generator* (Fig. 22.10).

The controlled generator differs from the machine generator with respect to the following features:

— it has a capability for varying the voltage, pulse duration, and pulse interval independently and over a wide range. The function of regulating these quantities is performed by a transistor *T*, controlled by a special control generator *CG* (Fig. 22.10);

— the element limiting the working voltage is a transistor which can usually operate without any special circuits from a current source up to 100 V and enables pulses with a voltage of the order of 15–30 V to be obtained.

Compared with relaxation generators, pulse generators have a much more advantageous ratio of operating time t_{dis} to pulse interval t_p. The ratio $t_{dis}/t_p = e_R$ in well-designed relaxation generators is $e_R = 0{\cdot}1–0{\cdot}01$. In the case of pulse generators, by contrast, the ratio is $e_P = t_{dis}/t_p = 1–10$. This comparison shows that pulse generators should exhibit a much higher efficiency.

However, in comparison to relaxation generators, controlled pulse generators are of a more complex construction and hence, more expensive. Accordingly, for economical considerations RC and RLC generators and machine generators are used in practice.

22.5 The technical and economic indices of electro-erosion machining

22.5.1 THE PARAMETERS AND INDICES OF ELECTRO-EROSION MACHINING

Among the factors which affect electro-erosion machining, i.e. the parameters of electro-erosion machining, are factors related to:

— generation of discharges, i.e. the electrical conditions; these include: V_0, V_{lim},

I_{dis}, R and R_{dis}, L and L_{dis}, C, t_{ch}, t_{dis}, etc. and the resultant shape of the discharges, and pulse energy and frequency;
— the dielectric in the spark gap, i.e. the chemical composition and physical properties of the dielectric, the degree of contamination with erosion products or other suspensions, the spark gap size s, and the rate of flow of the dielectric;
— the eroder (tool electrode), i.e. the chemical and physical properties of the eroder material, the shape, and dimensions of the eroder, and the kinematics of the eroder;
— the material being machined, such as the chemical and physical properties of the material, the shape and dimensions of the surface being machined, and the kinematics of the material being machined;
— the machine tool such as dynamic stiffness and stability, the follow-up capacity of the control system, accessibility to the machining area, the degree of mechanization and automation of the setting-up and auxiliary motions, etc.

The course, and especially the effects, of the electro-erosion machining process are characterized by means of technical and economic indices.

The principal technical and economic indices of electro-erosion machining include:
— the volume erosion rate (Q_V) and the unit or specific volume rate (q_V),
— the machining time (t_m),
— the relative eroder wear (m_V) and relative erosion rate (K_r),
— machining errors, which determine the class of machining,
— the geometric and physical properties of the surface layer after machining.

22.5.2 The electro-erosion machining rate

The volume erosion rate is defined by analogy with machining by cutting, i.e. it is the average volume of the material eroded from the workpiece per unit time. Usually, the volume rate is given in terms of volume of material removed per minute.

Use is also made of the concept of surface and linear erosion rates (Q_r and Q_L, respectively).

The erosion rate per unit power used for machining is called the *specific rate*. For instance, the *specific volume rate* is the ratio

$$q_V = \frac{Q_V}{E_i f} \tag{22.21}$$

The value of q_V is determined experimentally as a function of the factors affecting the process of electro-erosion.

The relation between the volume rate and the electrical conditions [22/4] can be expressed for electrospark machining by a formula in which use is made of relations (22.12), (22.11a), and (22.21):

$$Q_V = k_Q q_V \frac{C V_0^2}{2} \frac{(1-a)^2}{t_{dis} - R C \ln a} \tag{22.22}$$

where k_Q is a conversion factor, and the other notations are the same as before.

The relation $Q_V = f(a)$, as expressed by eqn. (22.22), is exemplified in Fig. 22.11. This plot is for $R = 32\,\Omega$, $C = 25\,\mu F$, $M = 0{\cdot}25 \times 10^{-4}$ F, $V_0 = 220$ V.

It follows from Fig. 22.11 that for given constant conditions, the highest erosion rate for the electrospark process can be attained only for a particular value of the ratio $V_{\text{lim}}/V_0 = a_{\text{opt}}$. In the given example, this is the value $a_{\text{opt}} = 0{\cdot}272$, which corresponds to a value of $V_{\text{lim}} = 160$ V.

In the case of electric-pulse machining the volume rate is expressed by a formula resulting from the combination of eqns. (22.21), (22.15), (22.17) and (22.20):

$$Q_V = k_Q q_V I_{\text{dis}} \frac{V_{\text{lim}_0} t - b\, t_{\text{dis}}^2}{t_{\text{dis}} + m\, I_{\text{dis}}} \tag{22.23}$$

The function expressed by eqn. (22.23) has a maximum value with respect to the pulse duration t_{dis}. The form of the curve of Q_V, f, and E_i is depicted in Fig. 22.12. This means that there exists an optimal value $t_{\text{dis}_{\text{opt}}}$ which also conditions the optimal values of E_i and f.

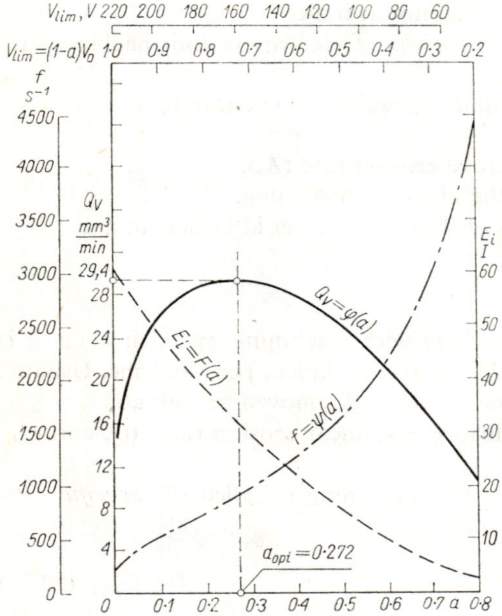

Fig. 22.11 Example of plot of energy of single discharges, frequency, and erosion rate versus the limiting discharge voltage

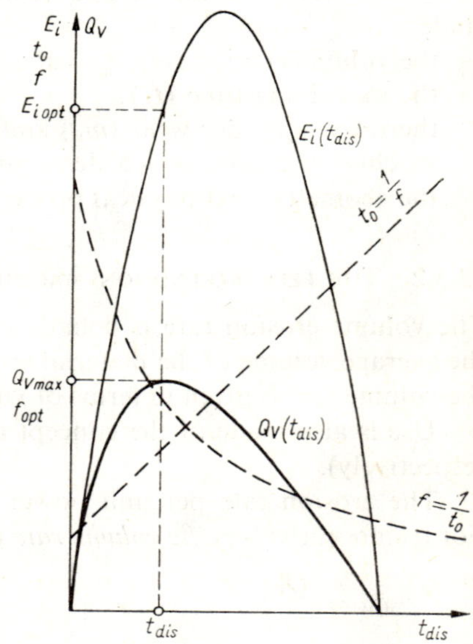

Fig. 22.12 Plot of pulse frequency f, pulse energy E_i, and volume erosion rate in electric-impulse machining

Electric-pulse machine tools of the best design achieve a rate of up to 10 000 mm³/min when machining steel.

The relationship expressed by eqns. (22.22) and (22.23) should serve as the principal basis for rational design of the electrical part of electro-erosion machine tools.

Among the most important factors associated with the dielectric are the chemical composition (type and grade) of dielectric and the degree of contamination. The spark gap size is of fundamental importance for the rate of electro-spark machining.

The dielectric used most frequently is paraffin (clear, in order to increase observability of the working zone). In some cases of rough machining, oils are also used.

Dielectric contamination has a deleterious effect, if optimal machining conditions have been selected. An example of the variation of the volume rate in electrospark machining as a function of the volume percent of dielectric contamination is shown in Fig. 22.13 [22/9].

With an increase in the amount of impurities in the dielectric, the erosion rate falls off, mainly because V_{lim} drops below the optimal value. For this reason filters for the dielectric are recommended.

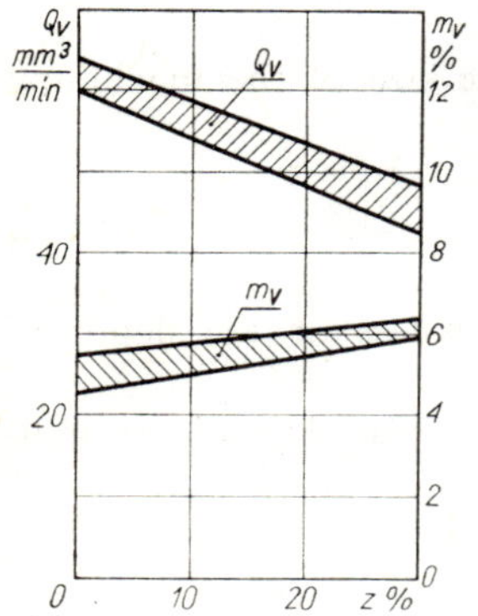

Fig. 22.13 Example of plot of erosion rate Q_V and volume wear coefficient m_V versus volume percent of paraffin contamination with erosion products.

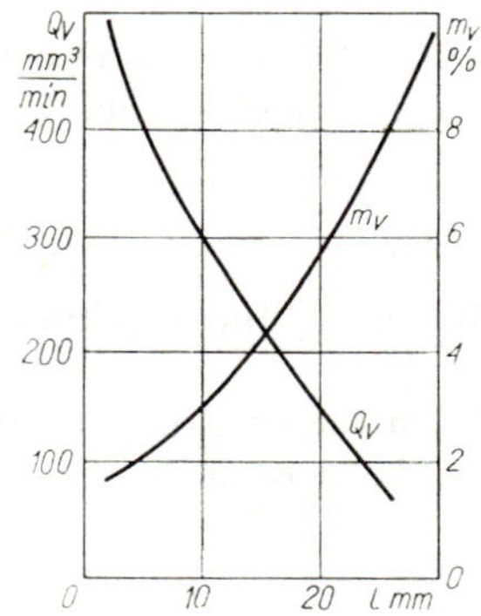

Fig. 22.14 Example of plot of erosion rate Q_V and volume coefficient of relative eroder wear m_V versus depth l of hole machined

In some cases, if the limiting voltage has been chosen improperly ($V_{lim} = V_{lim\,opt}$), a certain degree of contamination may result in the erosion rate increasing [22/10].

The choice of eroder is important for the erosion rate. It has been found by experience that the better the conductance of the eroder, the higher the rate. Relatively speaking, the best results with respect to erosion rate are obtained by using copper, copper-graphite, and graphite electrodes. Widespread use of the latter is hindered by the low mechanical strength of graphite.

The type of material being worked also affects the erosion rate. For instance, the erosion rate for sintered carbides is, under analogous conditions, approximately half that of hardened tool steel. The state of the material and the grades of the same kind of material have less influence on the rate.

The dimensions of the surface being machined also have little influence on yield, but for every set of electrical conditions there is a particular optimum area of machined surface.

This quantity should be given by the specifications of the machine tool.

In the case of electro-impulse machining, it turns out that large areas are optimum with respect to erosion rate. This, among other reasons, is why pulse generators are highly recommended for machining large surfaces.

The depth of hole sinking has a greater influence on the electro-erosion machining rate. As the depth l increases, the erosion rate Q_V decreases. A plot from investigations into such relations is given by way of example in Fig. 22.14.

The most important property of the machine tool as far as the erosion rate is concerned is the follow-up capacity of the automatic control system of the eroder. The follow-up characteristic depends on the sensitivity and speed of reaction to variations in the electrical conditions in the spark gap. This is particularly important for electro-spark machining, since it offers protection against short-circuits and stabilizes the limiting voltage.

22.5.3 Electro-erosion machining time

The electro-erosion machining time is calculated from the size of the machining allowance and the erosion rate, whence

$$t_m = \frac{V}{Q_V} = \frac{L}{Q_L} \tag{22.24}$$

where: V—the volume, L—the machining allowance thickness or the depth of penetration.

Since it depends on the erosion rate, the machining time also depends on individual factors, just as the erosion rate does.

22.5.4 Eroder wear

The *relative wear of the eroder* is extremely important in determining whether or not it is economical to employ electro-erosion machining. This wear is characterized by the ratio of the eroder volume consumed V_E to the eroded volume of material V_M, expressed in per cent,

$$m_V = \frac{V_E}{V_m} 100\% \tag{22.25}$$

The reciprocal of the relative eroder wear is called the *relative erosion rate* and is expressed by

$$K_V = \frac{100}{m_V} \tag{22.26}$$

Considerable strides have been made in the past dozen or so years in reducing eroder wear.

In the years 1950–55 the relative wear in electro-spark machining still came to 40–60%, whereas at present it may be less than 1% when materials of high erosion resistance are used for the eroders [22/5], [22/7].

In the case of electro-pulse machining, a relative wear of several tenths of one per cent is attainable when steel is machined with a graphite eroder. Accordingly, light metals and non-ferrous metals may be used for eroders in electric-pulse machining because, despite their relatively low erosion resistance, they are economical by virtue of the low cost of the eroder itself and the ease with which eroders are formed.

The aforementioned values of the relative wear apply to simple eroder shapes. In the case of more complex shapes, a much higher wear (2–10%, or more) should be expected. This results in accelerated wear at the corners and on high curvatures. In these cases it is advisable to employ eroder compensation [22/8]. This consists in those parts of the eroder subject to most rapid wear being suitably elongated (Fig. 22.15).

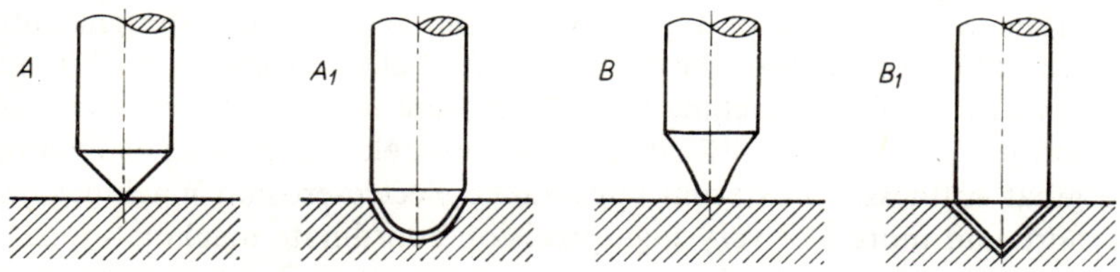

Fig. 22.15 Machining a conical cavity with an ordinary eroder (A, A_1) and compensated eroder (B, B_1)

Eroder wear is also caused by erosion due to discharges, especially of the oscillatory type. Consequently, eroder wear in electro-spark machining is higher than in electric-pulse machining. Abrasion of the soft eroder material by hardened erosion products recoiling off the workpiece and striking the eroder is considered to be another cause of eroder wear [22/1]. The electrode polarity* is also of some significance.

Erosion resistance [22/7] depends on a number of physical properties of the eroder material. In the case of electro-spark machining, the higher the melting point, thermal conductivity, Young's modulus, and Poisson's ratio and the lower the value of the coefficient of linear expansion, then the greater the erosion resistance.

All the factors affecting the process of electro-erosion machining have an influence on the eroder wear.

The relative eroder wear, being the ratio of the eroder wear to the material eroder, may take on different shapes. In many cases, the relative wear curve may have points of inflexion.

When there is a minimum relative wear, there is a basis for establishing an optimal value for a given factor with respect to relative wear.

An example of the dependence of m_V with a minimum is the dependence on the depth of the hole sunk (Fig. 22.14). On the other hand, the dependence of m_V on impurities in the dielectric is rather monotonic in the range investigated.

22.5.5 The accuracy and roughness of surfaces machined by electro-erosion

The accuracy of electro-erosion machining is evaluated by the same indices as in the case of machining by cutting or abrasion. These are dimensional errors, expressed in absolute values or in percentages.

The most important constituents of errors in electro-erosion machining are:
— shape errors caused by eroder wear and varying intensity of wear on the eroder,

* A distinction is made between straight polarity, when the eroder is the cathode, and reverse polarity, when the workpiece is the cathode. Straight polarity predominates in practical use.

— errors in setting up and guiding the eroder with respect to the material,
— the tool-work gap, increasing the final dimension by its thickness.

The influence of eroder wear on increasing the machining error may be eliminated by dimensional-shape compensation of the eroder as well as by the successive use of several eroders of the same shape. This second method is employed more frequently for very small dimensional tolerances.

The errors of eroder set-up and guidance have been reduced substantially in recent years by the incorporation of typical precision machine-tool designs into erosion machines. In the early years of the development of electro-erosion machining, the stiffness and precision of the machine tools built were too low. It was believed that inasmuch as it took place practically without any force, electro-erosion machining did not require great stiffness. In external appearance electro-erosion machines are now becoming more and more like very high-precision coordinate machine tools.

Allowance can easily be made in the eroder design for the influence of the tool-work gap since the sizes of tool-work gaps are known from numerous investigations and experiments. Depending on the machining conditions, the tool-work gap size varies from 0·01 to 0·5 mm.

Dimensional corrections to the eroder are made in the course of the production of the first workpiece.

Taking account of all the sources of error, one may evaluate the present accuracy capabilities of electro-erosion machining by means of aggregate errors of the order of $\pm 0{\cdot}05$ to $\pm 0{\cdot}01$ mm.

With respect to smoothness capabilities, in the range of high rates electro-erosion machining is inferior to machining by cutting. However, in the range of low machining rates the smoothness attainable by electro-spark machining (with relaxation generators) is characterized by a height of irregularities R_z of the order of 1 μm when machining sintered materials, and of the order of 2–4 μm when machining steel. Electric-pulse machining does not permit such a high class of smoothness to be achieved; accordingly, relaxation generators are more suited for use in smoothing operations.

The purity of the dielectric is of immense importance for macro- and micro-precision [22/9], [22/10]. For this reason, one or two stages of dielectric filtration are employed in advanced electro-erosion machines.

Figure 22.16 plots the heights of the irregularities for single discharges ($R_{i\,max}$) and multiple discharges, under normal machining conditions. It is seen from the plot that when the craters of successive discharges are made to overlap, the height of the irregularities (R_z) is reduced.

A reason why the roughness increases with the product $R\,C$, that is, with the time constant, is that as $R\,C$ increases so does the energy of the pulses. The greater the pulses, the larger the erosion pits (craters). A dependence of this type is illustrated by the example given in Fig. 22.17.

Just as cutting machine tools have distinct spindle speed stages, so electric-spark machine tools have distinct working stages, each characterized by a fixed value of the product $R\,C$, i.e. by a particular resistance and a particular capacitance. The higher the working stage, the smaller the capacitance and the greater the charging resistance.

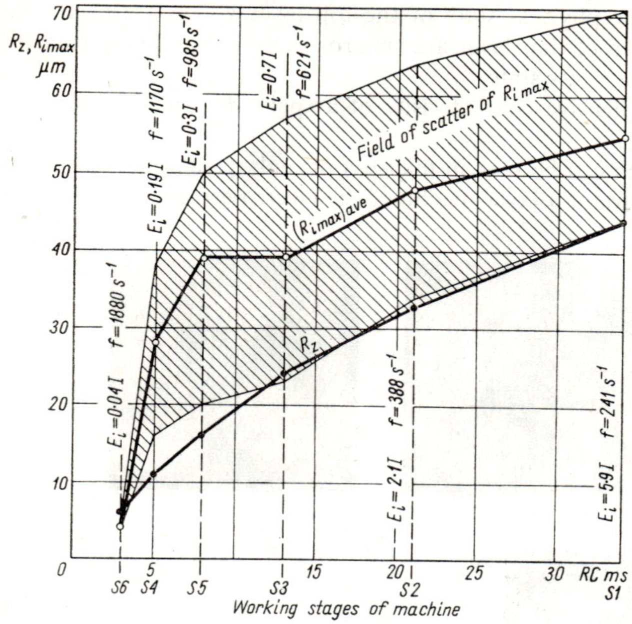

Fig. 22.16 Dependence of height of surface irregularities on electrical conditions of particular working stages and individual discharges ($R_{i\max}$) and surface machined at frequency appropriate to particular stage (R_z). Material machined—hardened tool steel; eroder material—copper

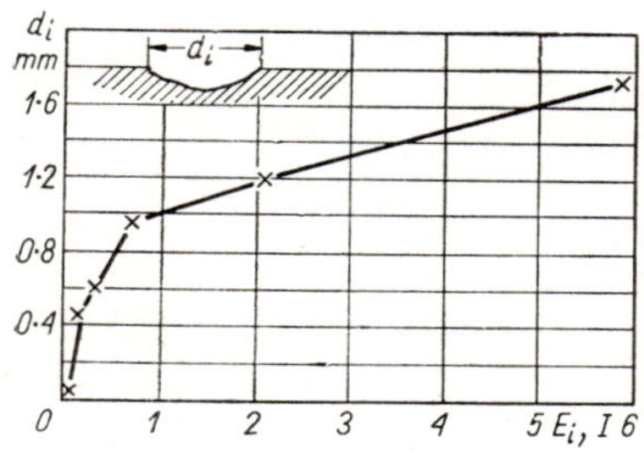

Fig. 22.17 Average diameters of pits (cavities)—from individual discharges. Material machined—hardened tool steel; eroder—copper

The roughness of surfaces machined by electro-erosion has an unoriented point structure. Roughness numbers up to and including eight can be evaluated by means of master roughness blocks machined by electro-erosion.

22.5.6 The properties of the top layer of surfaces machined by electro-erosion

The physical properties of the top layer after electro-erosion machining have not been sufficiently and comprehensively investigated thus far, and the results have not been fully generalized. The methodology of testing and evaluating the properties of the top layer is thus all the more important.

Among the features of the physical state of the top layer which are necessary for a comprehensive evaluation of that layer are: microstructure, hardening (plastic and strain hardening), and the stress state.

An example of a comprehensive characteristic of a spark-machined top layer* is given in Fig. 22.18 [22/15].

Three typical layers, which can be seen in microscopic enlargements, occur most frequently in the microstructure of a layer machined by electro-erosion.

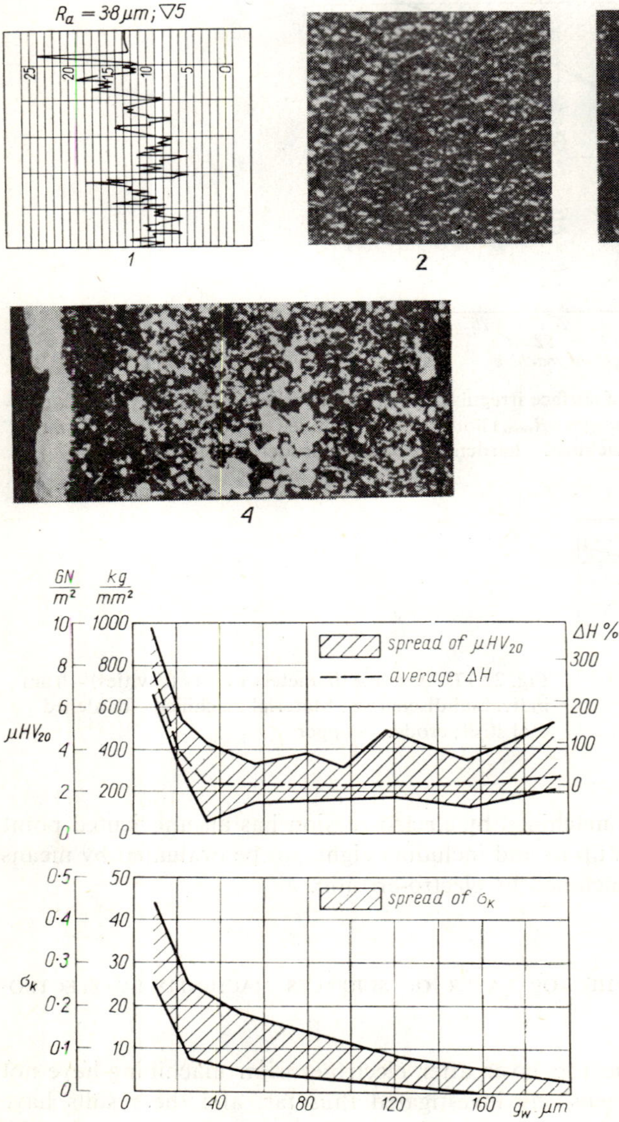

Fig. 22.18 Complex characteristic of surface layer of NC6 alloy tool steel, annealed after electro-spark machining (RC generator, $V_0 = 220$ V, $C = 2\mu F$, $R = 135\ \Omega$ after [22/15]. Magnifications: 1—vertical—1000×, horizontal—20×; 2—10×, 3—90×, 4—500×

* Cf. Chapter 6

The layer adjoining the surface, that is the subsurface layer, is called white owing to the white colour it displays in photographs after etching with 4% HNO_3. This layer is the melted part of the material being machined, with some inclusions of eroder material. The thickness of this layer, after electric-spark machining, varies from several micrometres to several hundredths of a millimetre. After impulse machining, this layer is more irregular, but in general is thicker. Its hardness is of the same order as, or somewhat greater than, that of the native material.

It should be noted that when a sintered carbide eroder, for example, is used, eroder material may be introduced into the subsurface layer. Consequently, the hardness and erosion resistance of that layer is enhanced. Therein lies the basis of electro-spark surface hardening.

The second layer is the zone of thermal effects. In the case of pearlitic steels, this layer displays a martensitic structure, i.e. it is in a hardened state. The hardness of this layer is much greater than that of the native material. Its thickness and the hardness distribution exhibit considerable variations depending on the machining conditions [22/12].

The next layer is also a result of thermal effects. In the case of pearlitic steels it constitutes a tempering zone, with reduced hardness as compared to the native material.

The state of stress after electric-spark machining may vary. The example shown in Fig. 22.18 should be considered rare. Most frequently, the electro-erosion process sets up in the subsurface layer tensile stresses [22/13] which decrease with the distance from the surface and at a certain depth, most often already in the native material, they change to compressive stresses.

After investigations by Aleksandrov [22/13], in Fig. 22.19 we present an example of what may be considered as a common distribution of tensile stress.

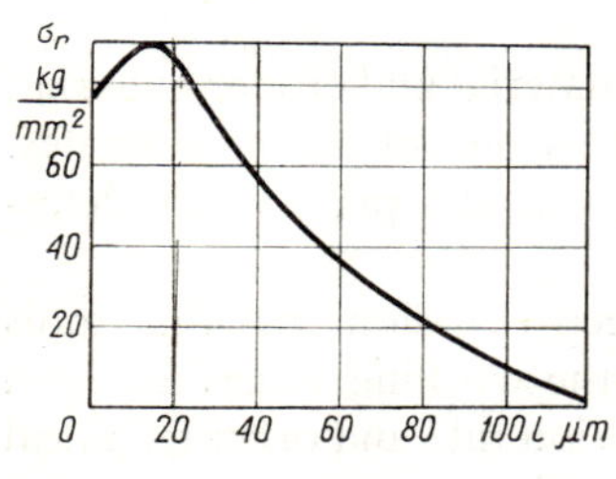

Fig. 22.19 Distribution of stress σ_r in the top layer of heat-resistant nickel steel machined by electro-erosion as a function of the distance l from the surface (discharge duration t = 1050 ms, pulse energy E_i = 0·1 J) after V. P. Aleksandrov [22/13]

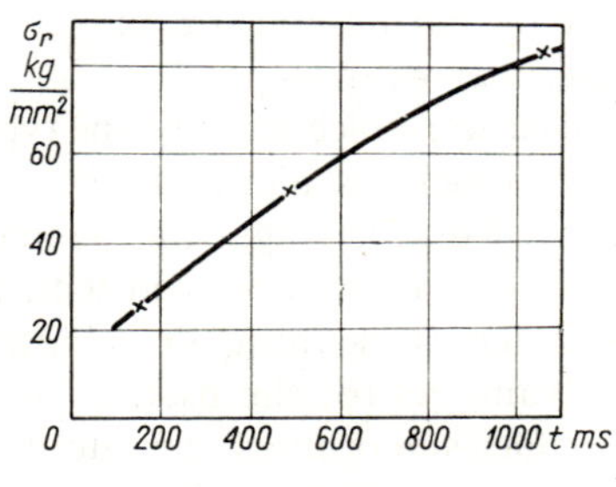

Fig. 22.20 Stress σ_r in top layer of heat-resistant nickel steel, machined by electro-erosion (pulse energy E_i = 0·2 J), at depth of 20 mm as a function of the pulse duration t (after V. P. Aleksandrov) [22/13]

With an increase in the energy of individual discharges, there is an increase in the tensile stresses (Fig. 22.20).

Accordingly, surface microcracks form, their depth being as much as 0·1 mm in the case of electric-pulse machining.

Fatigue testing [22/13] also shows that the longer the discharges and the higher the energy of the discharges, the greater the loss of strength as compared to a layer machined mechanically.

From this follow recommendations that preliminary, electric-pulse machining with a high erosion rate should be followed by electro-spark finishing. The allowance for finishing must be sufficiently large (not less than the depth of the microcracks), so that the strength properties be no worse than after mechanical machining.

22.6 The technique of electro-erosion machining

22.6.1 THE TECHNOLOGICAL CHARACTERISTIC OF THE MACHINE TOOL AND THE CHOICE OF MACHINING CONDITIONS

A successful outcome of the application of electro-erosion machining depends on how the machining capabilities and properties of this process are exploited.

The proper choice of machining conditions must, on the one hand, take into account the requirements of the designer with respect to precision and properties of the top layer, and on the other hand, must heed economical considerations, above all give the highest possible erosion rate for the process.

To facilitate proper choice of the machining conditions, electro-erosion machine tools should have, for each type of material and machine, and for each particular eroder, technological specifications which should give the following data for every working stage:

— the volume erosion rate Q_V (or possibly the linear erosion rate Q_L) in mm^3/min,
— the height of irregularities on the machined surface, R_z, in micrometres or in roughness numbers,
— the tool-work gap s, in mm,
— the taper (in boring holes) z, in %,
— relative eroder wear m_V, in %.

Such a specification concerns only one type of machined material and eroder material, and thus the roughness parameter R_z may be adopted as one of the coordinates. The individual working stages can be marked on the R_z-axis at a place which determines the surface quality obtained with this stage.

This manner of defining the characteristics of electro-erosion machine tools has great advantages for the user. It shows how dense the working stages are with respect to surface smoothness, the interdependence of erosion rate and other technical and economic indices attainable on the given machine tool.

An example of a technological characteristic of an electro-spark machine tool in graphical form is given in Fig. 22.21. The characteristic plotted in this figure is typical as to the shape of the various curves. Accordingly, some generalized remarks can be made on the basis of observation of the graphs. A characteristic feature is that the erosion rate falls off considerably with a decrease in roughness, i.e. with an increase in smoothness. At the same time, dimensional errors are diminished and the relative eroder wear changes.

In any choice of machining conditions on an electric-spark machine tool, the problem reduces to one of choosing the proper working stage since the machining indices are coupled with each other.

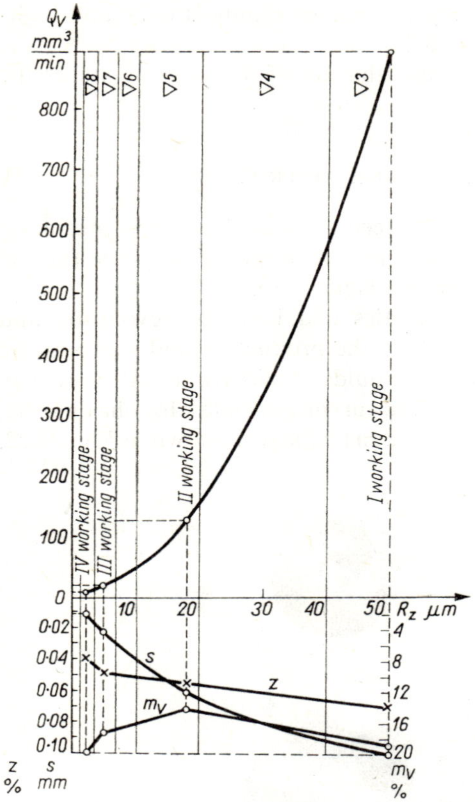

Fig. 22.21 Characteristic of electric-spark machine tool (electric-spark diesinking machine) for machining steel

The proper working stage is chosen according to the most stringent restrictions. For instance, when it is merely required that the surface quality correspond to roughness number 5, it also follows that this requirement can be satisfied only on the second working stage.

However, if it is also required that the hole taper must be no greater than 0·05, then the third working stage must be selected.

Another example of requirements is typical of preliminary (rough) machining when usually no other precision restrictions are made. In such a case, the working stage chosen is the one which permits the highest erosion rate (first stage).

However, if for economic considerations the requirement is for machining with the least relative eroder wear, the second stage should be chosen.

The spark-gap size at the particular working stages make appropriate eroder dimensioning possible, compensating for the increase in the cavity dimension by the spark gap.

The technological characteristics of an electro-erosion machine tool also permit evaluation of how rationally the working stages have been distributed with regard to the given material to be machined.

A rational distribution of the working stages should provide a capability for rough machining (roughness numbers 1–3), machining of medium precision (roughness numbers 4–5), precision machining (roughness numbers 6–7), and high-precision machining (roughness number 8 and higher).

The example given here indicates a good distribution of the working stages for machining steel.

22.6.2 Electro-erosion machine tools and their applications

Electro-erosion machine tools are commonly referred to as diesinking machines, although according to strict classification diesinking machines are only one type of electro-erosion machine tool (the one used most extensively).

Diesinking machines are used to make cavities and holes by revolution and otherwise. Typical applications in industry include the production and regeneration of dies, punching dies, drawing dies, and metal moulds. Apart from that, they may also be used for other purposes, such as machining curvilinear holes, for chasing, etc. Examples of workpieces machined by electro-spark machines are shown in Fig. 22.22.

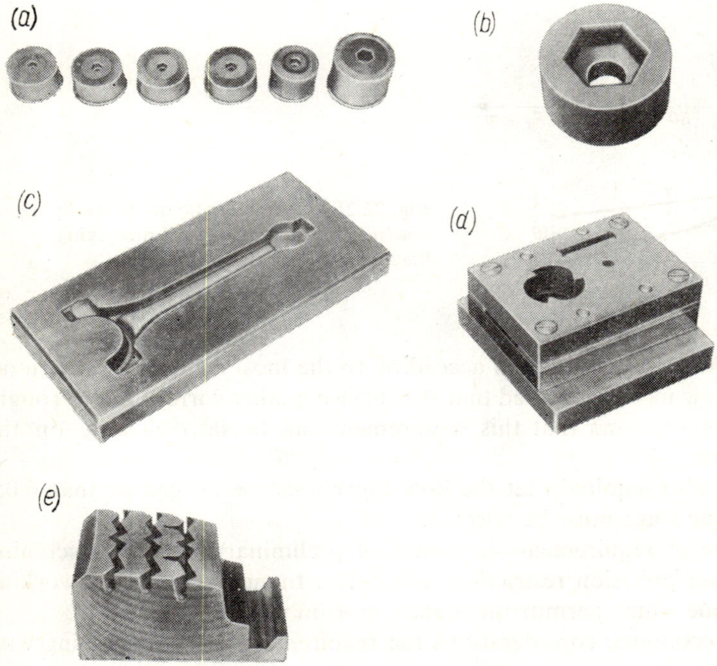

Fig. 22.22 Workpieces machined by electro-erosion

A specialized type of electro-erosion machine is the so-called disintegrators, i.e. electro-erosion machines for removing such broken tools as screw taps or drills from holes.

Other types of specialized electro-erosion machine tools, such as cutting-off machines and grinders, particularly tool grinders, are used less frequently.

Generally speaking, the harder and more difficult a given material is to machine

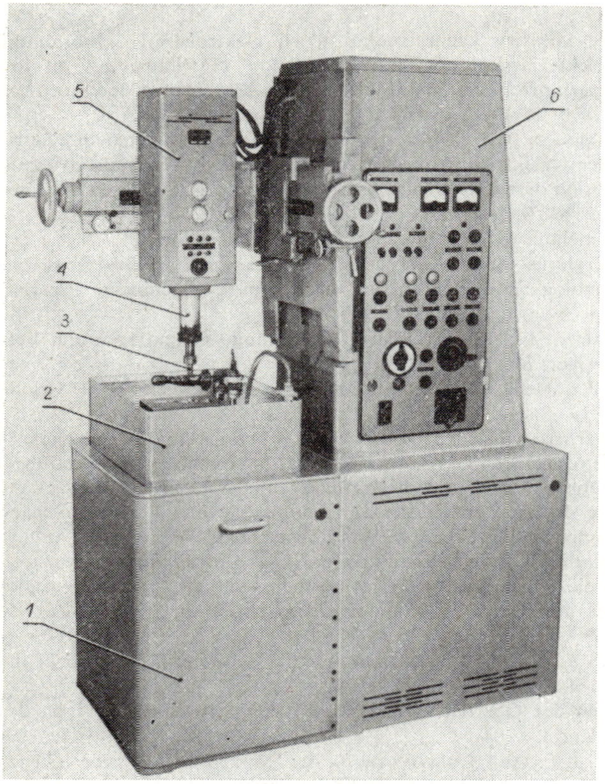

Fig. 22.23 Electro-erosion diesinking machine

by cutting, the more profitable it is to employ electro-erosion machining. For instance, electro-erosion machining of sintered carbides and hardened steel is economical.

The design of a typical electro-erosion diesinking machine is indicated in Fig. 22.23 which shows a photograph of a Polish-made 40-kW diesinking machine.

The machine base 1 contains the hydraulic equipment, dielectric tank, filters, and part of the electrical equipment. The table 2, which rests on the base, is surrounded by a tank into which the dielectric fluid flows during machining operations. In large electro-erosion machines the table is immobile, but in small ones, for machining small components, it may be movable as in milling machines.

The eroder 3 is held in a spindle which is driven by the control unit 5. The entire control unit is mounted so that it can be moved along ways, just as in precision jig boring machines. The housing 6 contains the electrical units, while the control panel is mounted on the front.

The current source is usually located outside the machine tool proper. In some designs, the machines have three distinct units: a mechanical unit, i.e. the actual machine tool, the electrical unit in the form of a console, and a hydraulic unit, i.e. the dielectric tank, pump, filters, etc.

Such a design makes operation more convenient but requires more floor space.

REFERENCES

22/1 ZOLOTYKH, B. N.: 'Osnovnye voprosy kachestvennoi teorii elektroiskrovoi obrabotki v zhidkoi srede. Problemy elektricheskoi obrabotki materialov' (Fundamentals of the qualitative theory of electric-spark machining in a liquid medium. Problems of the Electrical Machining of Materials), Izd. AN SSSR, Moscow 1962

22/2 LAZARENKO, B. R., and LAZARENKO, N. I.: 'Hidrodinamicheskaya teoriya iskrovoi elektricheskoi erozii metallov. Problemy elektricheskoi obrabotki materialov' (The hydrodynamic theory of the electric-spark erosion of metals. Problems of the Electrical Machining of Materials), Izd. AN SSSR, Moscow 1962

22/3 ALBIŃSKI, K.: 'L'usinage des métaux par électro-erosion', Dunod, Paris 1960*

22/4 KACZMAREK, J.: 'Niektóre kryteria technologicznej oceny obrabiarek elektroiskrowych' (Some criteria for the technological evaluation of electro-spark machines), Przegląd Mechaniczny, 13/61

22/5 RÜDINGER, O., and WINKELMAN, A.: 'Elektroerosive Bearbeitung—Grundlagen und Verfahrenstechnik, Industrie Anzeiger, 80, 1964

22/6 TRAUBE, G., and KUBE, M. J.: 'Neue Generatoren und neue Arbeitstechnik', Deutsche Maschinenwelt, 6, 1963

22/7 ALBIŃSKI, K.: 'Odporność elektroerozyjna elektrod roboczych w przypadku drążenia elektroimpulsowego' (Electro-erosion resistance of working electrodes in electric-pulse machining), Zeszyty Naukowe Instytutu Obróbki Skrawaniem (IOS), 16

22/8 POLAŃSKI, Z.: 'Elektroiskrowe drążenie metali erodami kompensacyjnymi' (Electro-spark Machining of Metals with Compensation Eroders), WNT, Warszawa 1965

22/9 SIWCZYK, M.: 'Badania przydatności różnych gatunków nafty jako płynów dielektrycznych oraz ich filtracje dla obróbki elektroiskrowej' (Studies on the usefulness of various grades of paraffin as dielectric fluids and their filtration for electro-spark machining), Zeszyty Naukowe IOS, 18

22/10 KURAFUJI, H., and SUDA, K.: 'Study on Electrical Discharge Machining I', Journal of the Faculty of Engineering, Univ. of Tokyo, 1, 1965

22/11 KACZMAREK, J.: 'Eigenschaften der Oberflächenschicht und deren Abhängigkeit von der Bearbeitungsart', CIRP-Annalen, 3, 1964

22/12 MOGILEVSKII, I. Z., LINETSKII, I. L., and CHEPOVAYA, S. A.: 'Makroskopicheskoe issledovanie izmenenii struktury poverkhnostnykh sloev stalei i splavov posle elektroiskrovogo rezaniya. Elektroiskrovaya obrabotka materialov' (Macrographic examination of changes in the structure of surface layers of steels and alloys after electro-spark machining. Electro-spark Machining of Materials), Izd. AN SSSR, Moscow 1963

22/13 ALEKSANDROV, V. P.: 'Issledovanie ostatochnykh napryazhenii dlitel'noi i ustalostnoi prochnosti zharoprochnykh materialov pri elektroiskrovoi obrabotke. Elektroiskrovaya obrabotka materialov' (Research on steady stress and fatigue strength of materials under electro-spark machining. Electro-spark Machining of Materials), Izd. AN SSSR, Moscow 1963

22/14 KACZMAREK, J.: 'L'usinage par étincelage', La Pratique des Industries Mécaniques, Septembre 1964

22/15 POLAŃSKI, Z.: 'Badania warstwy wierzchniej wybranych materiałów stosowanych na narzędzia do obróbki plastycznej na zimno kształtowane za pomocą drążenia elektroiskrowego' (Testing the surface layer of some materials used for cold plastic-working tools shaped by electro-spark machining), Zeszyty Naukowe Pol. Krakowskiej, Seria Mechanika, No. 8

* An enlarged and revised edition of Elektroerozyjna obróbka metali (Electro-erosion Machining of Metals), PWT, Warszawa 1954.

CHAPTER 23

Electrochemical machining

23.1 General characteristics

Electrochemical machining (ECM) is a process of shaping materials, which are conductors, by removing some of the material using appropriate chemical reactions under the conditions of electrolysis.

ECM is based on the general principles of chemical reactions and to some extent on the galvanostegy technique, i.e. on electroplating. Compared to electroplating, ECM has the opposite purpose, i.e. not to cover the anode with new layers of material but on the contrary to digest the top layers of the anode (hence the name *anodic digestion* or *etching*). Moreover, in the ECM process—in contrast to galvanostegy—there should be no cathode wear.

The electrochemical machining process is one of chemical reaction under the conditions of electrolysis, and thus must depend on the following factors:

— type of chemical reaction, conditional upon the choice of cathode and anode materials, and electrolyte,
— the degree of dissociation and concentration of the electrolyte, on which depends the resistance, or conductivity, of the electrolyte,
— the electrolyte temperature which determines the reaction rate,
— the voltage,

By a suitable choice of these factors, one can create machining conditions which enable an appropriately high machining rate to be achieved and, moreover, at the same time make it possible to attain a high degree of surface smoothness, and practically no wear of the cathode-tool, i.e. the eroder.

It should be added that the machining process takes place without plastic strain in the top layer and with only small forces. As a result, ECM is gaining increasing application in mechanical engineering; another factor contributing to this is the fact that the ECM can be used in various kinematical versions.

23.2 Definition and classification

From the point of view of kinematics electrochemical machining can be divided into many methods and types. Their names have in most cases been borrowed from machining by cutting, abrasion, and electro-erosion.

Classification of ECM methods according to the working motions, i.e. the main and auxiliary working motions, is like that for machining by cutting (Table 23.1).

Table 23.1 *Classification of methods of electrochemical machining by working motions*

Material \ Tool	●	↗	↻	screw motion
●	×	Hole sinking	×	Drilling (rotary hole sinking)
↗	Hole sinking	Shaping	Milling (surface grinding)	
↻	×	Turning super-finishing	×	Grinding
screw motion			Cylindrical grinding	Honing

● —immobile ↗ —rectilinear motion ↻ —rotation screw motion —screw motion

Four boxes in which no method can be entered (absence of a rectilinear component) are immediately excluded. The following entries can be put in the other boxes:

— *electrochemical hole sinking* when the tool moves rectilinearly and the workpiece remains immobile, or vice-versa; examples of applications of this method are given in Fig. 23.1.

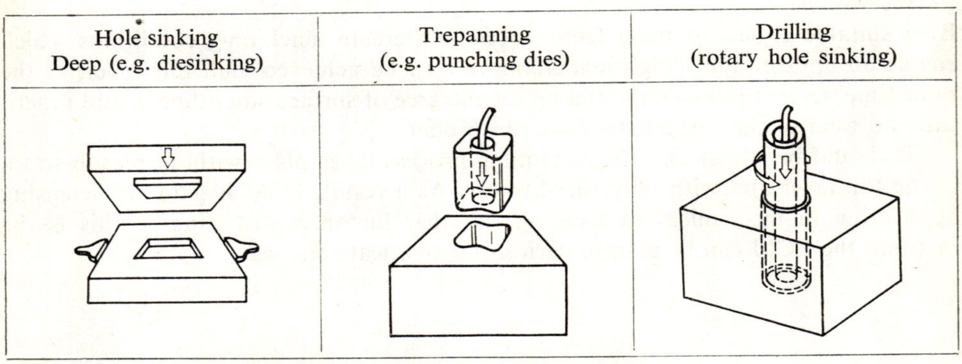

Fig. 23.1 Examples of electrochemical hole sinking and drilling

— *electrochemical drilling* or *rotary hole sinking* when the tool has a screw motion and the workpiece is motionless; this is illustrated in Fig. 23.1.
— *electrochemical shaping*, which is illustrated in Fig. 23.2, where the external and internal shaping versions are shown,

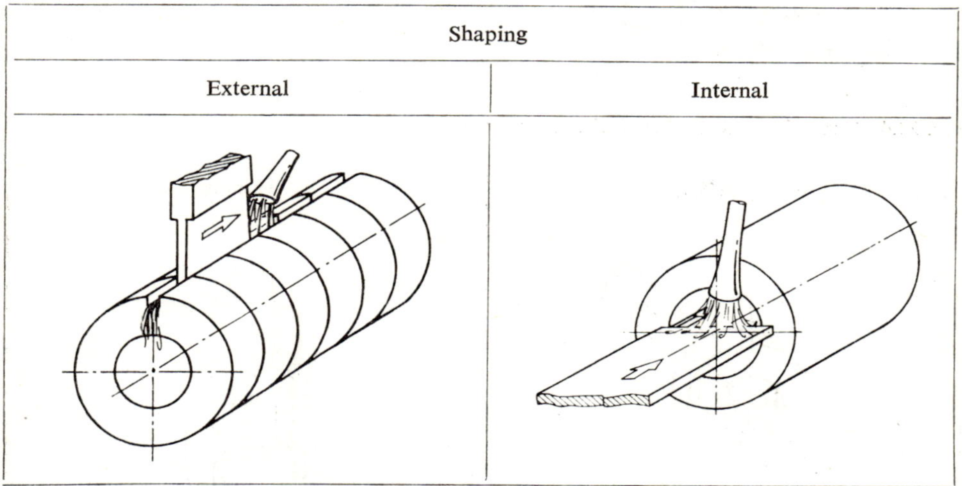

Fig. 23.2 Examples of electrochemical external and internal shaping

— *electrochemical milling*, which is illustrated in Fig. 23.3; this drawing shows the case of milling (or grinding) a cylindrical profile surface, and a case of parting (analogous to parting by milling cutters or abrasive disks),
— *electrochemical turning*, used most frequently as contour turning (Fig. 23.4),
— *electrochemical grinding*, which can be realized in various versions, e.g. as cylindrical and internal grinding or as face grinding. The latter is illustrated in Fig. 23.5.
— *electrochemical honing*, the scheme of which is given in Fig. 23.5.

The electrochemical machining methods discussed above are in different stages of development and have spread in varying degrees. Electrochemical hole sinking, which is typical of this kind of machining, has become most widespread.

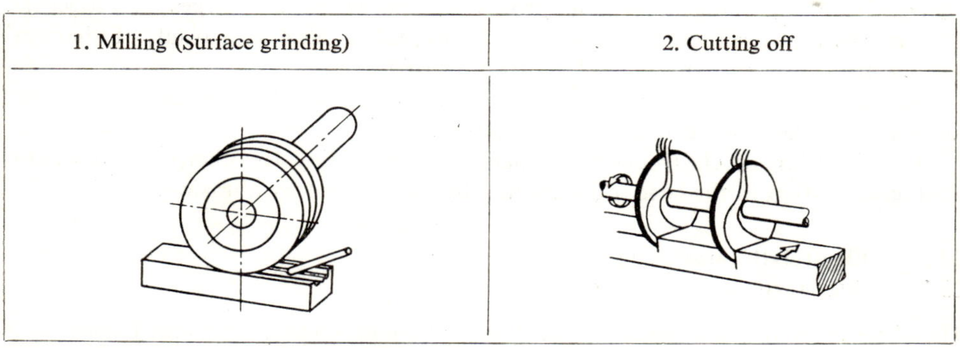

Fig. 23.3 Examples of cylindrical milling and cutting off

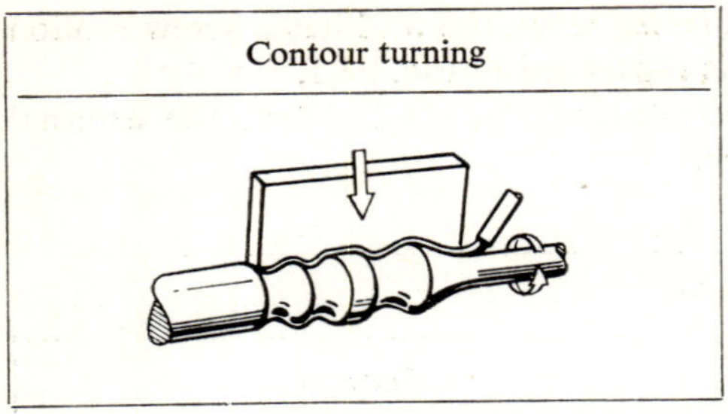

Fig. 23.4 Example of electrochemical contour turning

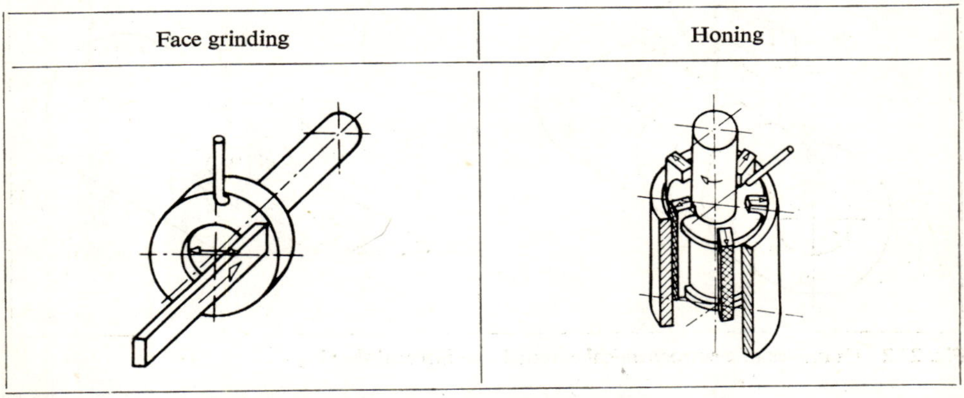

Fig. 23.5 Example of electrochemical face grinding and honing

With respect to the type of work done during electrochemical machining, a pure process of electrochemical machining as well as mixed processes may be distinguished.

Electrolytic machining (ECE) encompasses those methods of electrochemical machining in which use is made only of the energy of chemical reactions occurring under applied electric energy.

Chemical-abrasive machining (ECA) encompasses methods of machining in which chemical processes take place alongside mechanical-abrasive processes, that is to say, chemical and mechanical work is done.

Chemical and spark machining (ECI) comprises machining methods in which chemical and mechanical-abrasive processes, and spark discharges act together. This means that chemical and spark machining takes place as a result of chemical and electrical erosion, and partly, as a result of friction.

Electrochemical machining is classified according to the precision attainable, i.e. just as in the case of machining by cutting.

Classification of electrochemical machining according to other criteria is also analogous to that in the case of machining by cutting and by abrasion.

23.3 Electrolytic machining

23.3.1 Theoretical foundations

The theoretical foundations of *electrolytic machining* consist first and foremost of Ohm's law and Faraday's first and second laws (the first and second laws of electrolysis). In reference to the application of Ohm's law to electrolytic machining, it should be borne in mind that the total resistance consists of two components

$$R = R_e + R_p \tag{23.1}$$

where R_e—the ohmic resistance of the layer of electrolyte between the electrodes, R_p—the ohmic resistance of the oxide-salt layer polarized on the anode surface.

If we take $K_p = 1 + R_p/R_e$ eqn. (23.1) can be rewritten as:

$$R = K_p R_e \tag{23.2}$$

In this formula the coefficient K_p characterizes the extent to which the process takes place without symptoms of polarization and deleterious reactions, not affecting the process of anodic dissolution but absorbing part of the energy. Hence, when $K = 1$, the process takes place properly, without losses. The more K_p is in excess of 1, the greater the perturbations in the process which are mainly due to the process of polarization. The coefficient K_p will be called a *loss factor*.

It is known that R_e can be represented by

$$R_e = \varrho \frac{s}{F} = \frac{s}{wF} \tag{23.3}$$

where: s—the electrode separation, i.e. the tool-work gap, F—the active anode area, i.e. the working area.

Moreover, the resistivity ϱ of the electrolyte, which is temperature-dependent, is defined by

$$\varrho = \frac{\varrho_0}{1 + \alpha T + \beta T^2} \tag{23.4}$$

and the conductivity w, which is equal in value to the reciprocal of the resistivity ($w = 1/\varrho$), is defined by

$$w = w_0(1 + \alpha T + \beta T^2) \tag{23.5}$$

Using eqns. (23.2) and (23.3) and Ohm's law, we can write the formula for the electric current in the form

$$I = \frac{V}{K_p S} wF \tag{23.6}$$

The current density j, being the ratio of the electric current to the active anode area is defined by

$$j = \frac{V}{K_p s} w \tag{23.7}$$

Hence, the current density j is directly proportional to the voltage of the current source and the conductivity at the given temperature [according to eqn. (23.5)]. It is inversely proportional to the size of the interelectrode gap and the coefficient of polarization and deleterious reactions.

By Faraday's laws, for a given pure metal the volume of machined material that is removed is given by

$$V = \frac{60C}{96\,540 n\gamma} It \quad \text{cm}^3 \tag{23.8}$$

Denoting

$$q_c = \frac{60C}{96\,540 n\gamma} \quad \frac{\text{cm}^2}{\text{A min}}, \tag{23.9}$$

we obtain

$$V = q_c I t \tag{23.10}$$

where: q_c—the electrochemical volumetric equivalent for the given metal (or specific loss of the given metal); the values of q_c for some metals are listed in Table 23.2, C—the atomic weight of the given metal, γ—the specific gravity of the given metal, t—elapsed time in minutes, n—the valency of the given metal.

Table 23.2 *Values of specific metal removal rates q_c of some metals*

Metal	Valency	q_c mm³/(A min)	Metal	Valency	q_c mm³/(A min)
Aluminium	3	2·06	Nickel	2	2·11
Chromium	2	2·25		3	1·36
	3	1·50	Tantalum	5	1·35
	6	0·76	Titanium	3	2·19
Cobalt	2	2·05		4	1·66
	3	1·37	Vanadium	3	1·74
Steel	2	2·21		5	1·05
	3	1·48	Tungsten	1	1·98
Molybdenum	3	1·95	Zinc	2	2·85
	4	1·48			
	6	0·99			

Equation 23.10 refers to the case when a pure metal is being machined. In the case of metal alloys, with two or more components, the total volume is calculated as the sum

$$V_m = I t \sum_{i=1}^{i=z} q_{ci} \tag{23.11}$$

where z—the number of components of the alloy.

Equation (23.6) can be used to rewrite (23.10) in a new form,

$$V_m = q_c \frac{V}{K_p s} w F t \tag{23.12}$$

The volume erosion rate of the electrochemical machining process is thus expressed by

$$Q_V = q_c F \frac{w}{K_p s} V \tag{23.13}$$

whereas the specific volume erosion rate is given by

$$q_V = \frac{Q_V}{I V} = q_c \frac{w}{K_s s j} \tag{23.14}$$

If the specific loss q_c expressed in units of cm³/A min, is multiplied by the current and divided by the active surface area of the anode, the result is the average material removal rate v_t

$$v_t = q_c j \quad \text{cm/min} \tag{23.15}$$

Of the equations above, the one for the linear rate of anode loss (23.15) is of particular importance for electrochemical hole sinking. This equation states that the rate of linear anode loss is proportional to the specific loss q and the current density. The first quantity is independent, whereas the second can be controlled. It is thus obvious that the higher the current density (and hence the hole-sinking speed), the greater the machining rate, since

$$V_m = v_t F t \quad \text{cm}^3 \tag{23.16}$$

and

$$Q_V = v_t F \quad \text{cm}^3/\text{min} \tag{23.17}$$

Directly dependent on the machining rate is the machining time

$$t_m = \frac{\Delta l}{v_t} \quad \text{min} \tag{23.18}$$

where Δl—the machining allowance in the direction of the hole sinking, in mm.

As follows from eqn. (23.7), the current density can be increased by:

— raising the supply voltage V,
— increasing the specific conductivity w,
— reducing the coefficient of polarization and deleterious reactions, K_p,
— decreasing the tool-work gap s between the electrodes.

Implementation of these directives is the principal objective pursued in the development of electrochemical machining.

23.3.2 The importance of the electrolyte and related factors

The electrolyte is of great importance in the electrochemical machining process.

Experience shows that the actual value of the erosion rate and the hole-sinking speed is most frequently smaller than the theoretical value i.e. when the total electrical energy is utilized to remove metal.

Results of the interdependence of the tool-work gap size and theoretical and measured erosion rates, v_t and v_{tm} respectively, are shown in Fig. 23.6. It is seen from the graph that the measured, i.e. the real, erosion rate is lower than the theoretical value throughout the entire range studied. Equations (23.15) and (23.7) can be used to write

$$v_{tm} = q_c \frac{V}{K_p s} w$$

The ratio v_t/v_{tm} thus is equal to the value of the loss factor

$$K_p = \frac{v_t}{v_{tm}} \tag{23.19}$$

The value of this coefficient for this example is also given in Fig. 23.6. It is seen from

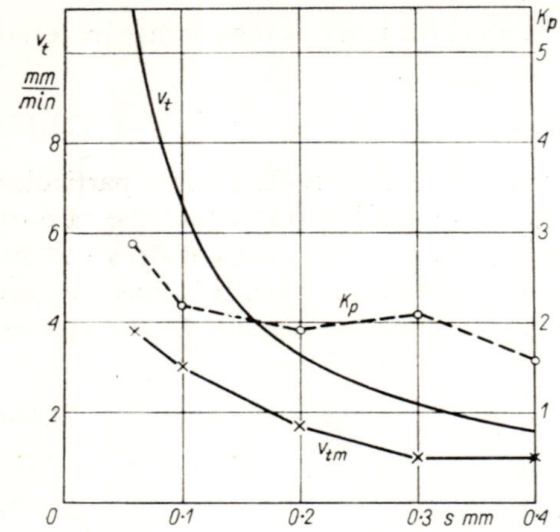

Fig. 23.6 Theoretical and real hole-sinking speeds, v_t and v_{tm}, and the loss factor K_p as functions of the tool-work gap. The values of v_{tm} were measured for diesinking at a supply voltage $V = 15$ V, and with a chloride electrolyte of specific conductivity $w = 0{\cdot}2\ (\Omega\ \mathrm{cm})^{-1}$ [23/2]

this graph that the value of K_p varies from 1·5 to 3. This is indicative of extremely large losses in erosion rate which may occur as a result of a failure to choose the best conditions.

Inhibition of the hole-sinking process is a major source of increased losses during the process. This may happen as a result of changes in the composition of the electrolyte itself because of electrolyte products inhibiting the process of dissolution. According to the Guldberg–Waage law, this may lead to the reaction stopping altogether. This case is, however, rare and is easily remedied by a renewal of the electrolyte, which is done as a matter of principle.

Another phenomenon which may inhibit the process is that of anode polarization. In this case, owing to an incorrect choice of electrolyte there forms on the anode surface either a thin layer of evolved gases or a thin layer of chemical compounds, most often oxides, which are difficult to remove. Consequently, the polarization resistance R_p increases.

This was the subject of special investigations [23/5]. By way of example Fig. 23.7 gives part of the results of investigations on the choice of electrolytes for machining iron (in steel alloys) and sintered carbides (SC) for various choices of cathode materials and various electrolyte compositions and concentrations. Three examples of electrolytes are given in the figure. The first (Fig. 23.7a), a 20% aqueous solution of NaCl, displays a slower or faster increase in polarization resistance when sintered carbides are machined, regardless of the kind of cathode (Fe, Cu, brass). On the other hand, in the case when steel is being machined, this electrolyte exhibits a relatively low resistance, regardless of the machining time. It can thus be seen that this electrolyte is suitable for machining iron and steel, but is misapplied when used for sintered carbides.

The plot in Fig. 23.7c shows an electrolyte that is good for machining sintered carbides, but is unsuitable for machining iron and steel. This is a mixture of sodium chromate (16%) and sodium chloride (4%).

The plot in Fig. 23.7b represents the case of a universal electrolyte which gives good results in machining both iron and steel, and sintered carbides. This is a 20% concentration of sodium chromate.

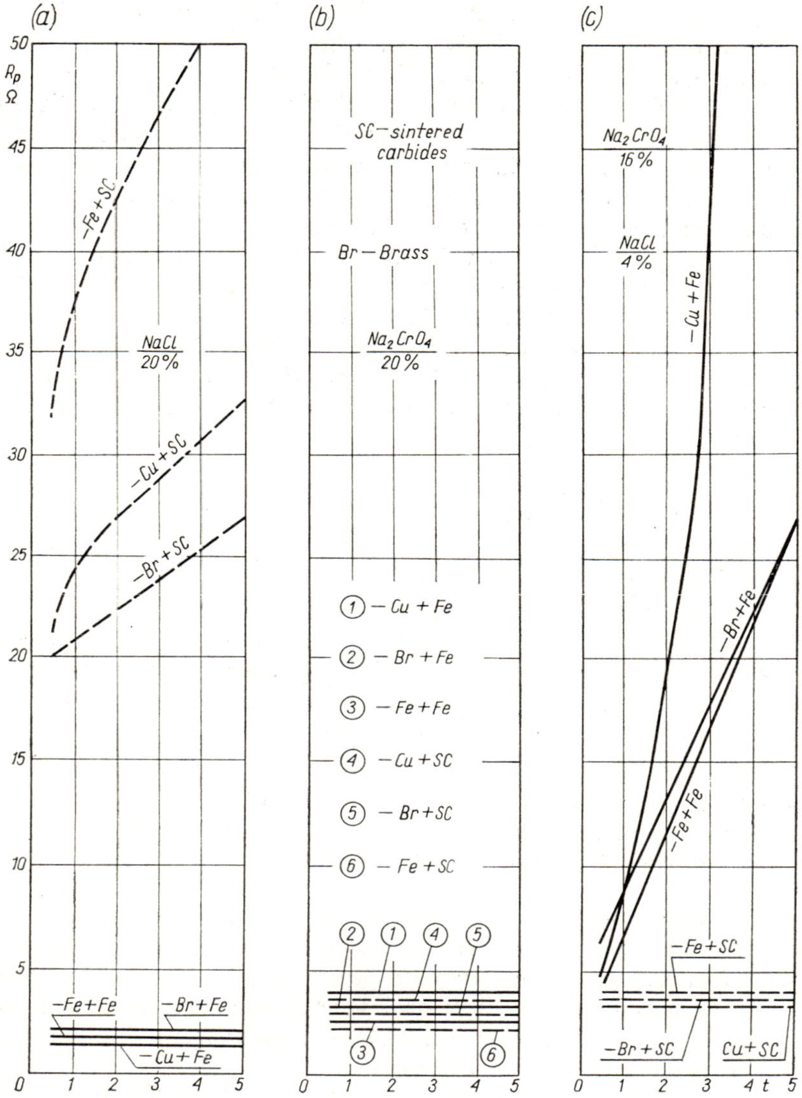

Fig. 23.7 Variations of the electrical polarization resistance R_p as a function of the duration of the process for various combinations of cathode (−) and anode (+) [23/5]

It is seen from the foregoing that in order to ensure an erosion rate corresponding to the nearest possible theoretical value, it is necessary to choose the right electrolyte for the anode material—the workpiece.

Many good, tried and tested electrolytes are now recommended by the manufacturers of electrochemical machine tools. Each of these electrolytes, however, has a limited range of optimal applicability. A change in the chemical composition of the workpiece material may entail a reduction in the efficiency of the machining process.

For this reason, it is recommended that the efficiency of the given electrolyte be checked whenever a new anode material is used. When the loss factor K_p is much greater than 1, a more suitable electrolyte should be found.

The choice of electrolyte which is optimal for the given range of applications is restricted by the deleterious properties of some electrolytes. These include the harmful effects they have on man, e.g. by having a corrosive or irritant effect on the human skin, or else by releasing toxic gases, and tending to corrode the machine. The injurious effects on man and the corrosive effects (which cannot be neutralized by a passivator) on the machine are factors which disqualify a given electrolyte.

In addition to the chemical composition, the concentration of a given electrolyte is extremely important. As illustrated in Fig. 23.8, with an increase in the concentration of electrolyte, there is an increase in the specific conductivity [23/3]. This increase, however, is only high in the low-concentration range. For this reason it has been found that it is usually pointless and uneconomical to increase the concentration above 10 moles.

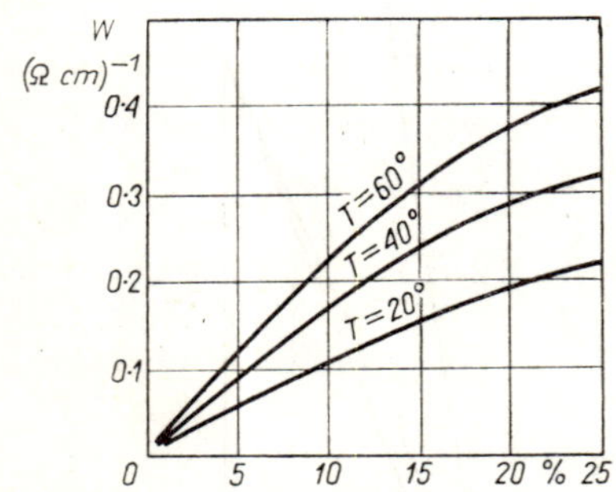

Fig. 23.8 Specific conductivity of aqueous solution of sodium chloride as a function of concentration at various temperatures [23/3]

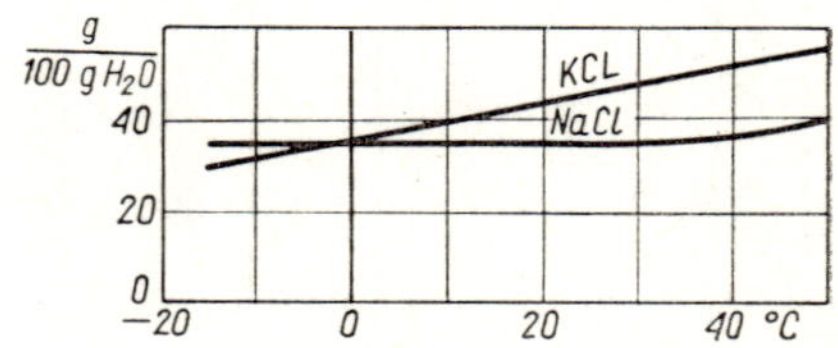

Fig. 23.9 Variation of solubility of sodium chloride and potassium chloride with the solution temperature [23/1]

In some cases, the process rate can be enhanced by heating the electrolyte since the specific conductivity of the electrolyte increases with the temperature. This can also be seen from Fig. 23.8 where the curves for the higher temperatures run higher.

It is also advantageous to raise the electrolyte temperature because the solubility of the salt increases. By way of example Fig. 23.9 gives the plot of the solubility of potassium chloride as a function of the solution temperature [23/1].

The elevation of electrolyte temperature is restricted by the thermal strain of the electrodes and machine tool, inasmuch as it may cause errors in the machining precision. In practice, the electrolyte temperature does not exceed 60–70°C. Special cooling equipment is used with any further increase in temperature.

Stabilization of the electrolyte temperature is of even greater importance, since the electrolyte temperature rises as a result of the heat released during the reaction. Temperature regulators are recommended for precision machining.

Circulation and replenishment of electrolyte constitute another aspect of the rational use of electrolyte. Suspensions which appear in the electrolyte as a result of the machining process alter the electrolyte properties.

The undesirable influence of suspensions and sediments is eliminated either by draining off the used electrolyte and replacing it by fresh electrolyte (system of open circulation), or by regenerating and cleansing it (closed system).

The electrolyte may be cleansed of suspensions by:

— filtration, which requires a relatively large surface if it is to be sufficiently effective and rapid; the cost of the installation is then relatively low, but the operating costs are higher,

— centrifugal separation, which gives a very good cleansing effect; the equipment is expensive, especially when it comes to automatic equipment, and so is operation,

— cyclone separation, of poor effectiveness, when the impurities are fine and differ little from the electrolyte as to specific gravity,

— sedimentation, which is second only to centrifugal separation as far as effectiveness is concerned; the equipment is relatively inexpensive and is very inexpensive to operate, but requires a large space,

— flotation, which gives good separation of impurities; the equipment is relatively expensive, but operating costs are low; moreover, it requires less space than sedimentation.

Any one of these methods may be economical, depending on the conditions, the machining rate and type and size of the impurities. As far as effectiveness and rate are concerned, centrifugal separation is considered the best method.

A diagram depicting the flow of electrolyte in a closed system is given in Fig. 23.10. From the machining area the electrolyte flows into a settling tank 2, from which a pump 5 sends it on to the filtering unit (or filtering and regenerating unit) 7. The electrolyte then flows into the main tank 8, from which the pump 6 sends it to the machining area.

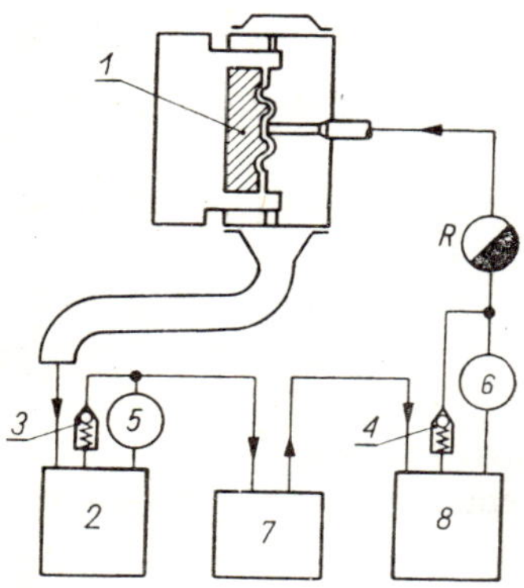

Fig. 23.10 Electrolyte flow diagram: 1—workpiece, 2—first settling tank, 3, 4—overload valves, 5, 6—pumps, 7—filter or device to regenerate and replenish electrolyte, 8—tank of regenerated or purified electrolyte

In the course of chemical dissolution itself no forces appear, if one ignores the forces stemming from the removal of layers in which permanent stresses might have existed. Forces due to the pressure of the pumped electrolyte do, however, act. The higher the pressure of the flow and the larger the surface area of the workpiece, the greater these forces. In practice, pressures of up to 50 atm are used. Accordingly, whenever electrochemical machine tools with a high electrolyte pressure in the tool-work gap are being designed, the forces resulting from this pressure must be taken into account.

The most important properties of the electrolyte include:

— breakdown resistance, which is evaluated by the limiting voltage V_{lim} at which electrical discharges occur,

— lack of corrosive properties and inoffensiveness to the skin, which is evaluated by, among other things, the acid number pH.

The electrolytes used in practice make it possible to employ working voltages of up to 20 V. At higher voltages discharges occur and these have a harmful effect on the surface smoothness and reduce the machining rate.

From the point of view of the acid number, one should avoid both highly acidic and highly alkaline electrolytes. The convention is that electrolytes are acidic when pH < 4, neutral $4 \leqq$ pH $\leqq 10$, and alkaline when pH > 10.

23.3.3 Optimization of the machining process by automation of the eroder feed

The principal factor affecting the process of electrochemical machining is the tool-work gap size s. The current density [eqn. (23.7)] depends on it, as does the machining speed, i.e. the rate of material removal [eqn. (23.15)]. The smaller the gap s, the better the results of machining should be from the point of view of metal removal rate. The extent to which s can be reduced is limited by the drop in the limiting voltage and the increase in the resistance to the flow of electrolyte.

Accordingly, for given machining conditions there is always an optimal tool-work gap size s_{opt}, which is usually determined by trial and error. It is extremely important to maintain an unvarying optimum s_{opt} during the machining process. This boils down to satisfying, in every interval of time, the condition

$$\Delta l_E = \Delta l_M \tag{23.20}$$

where Δl_E—the linear relative feed of the eroder in the direction of the linear material loss Δl_M.

This condition is fulfilled when the relative eroder velocity v_E is equal to the rate of linear material loss

$$v_E = \frac{\mathrm{d}l_E}{\mathrm{d}t} = q_c j = v_t \tag{23.21}$$

The condition $v_E = v_t$ can be attained in practice for cavity-sinking:

— with constant eroder feed chosen so that its speed is equal to the material removal rate,

— with eroder feed automatically regulated on the basis of a constant current. taking into account every instantaneous variation in the state of the electrolyte in the tool-work gap.

The first method in general offers insufficient precision and leads to the hole-sinking speeding up or slowing down. In the second case, discharges occur. The second method, i.e. automatic regulation of eroder feed, is better. This regulation must display a high degree of sensitivity and low inertia, i.e. good follow-up capacity.

Regulators now manufactured commercially, usually of the electro-hydraulic type, satisfy the follow-up requirements.

The general eroder feed control system can be explained with the aid of the design shown in Fig. 23.11. Voltage drops across the tool-work gap constitute the controlling quantity. These pulses are transmitted to control coils which are powered by a vacuum-valve amplifier from the comparison-measurement system *VM*. These coils control the anode voltage of the vacuum valves in the comparison-measurement system. The resultant voltage will thus be the difference between the voltage drop

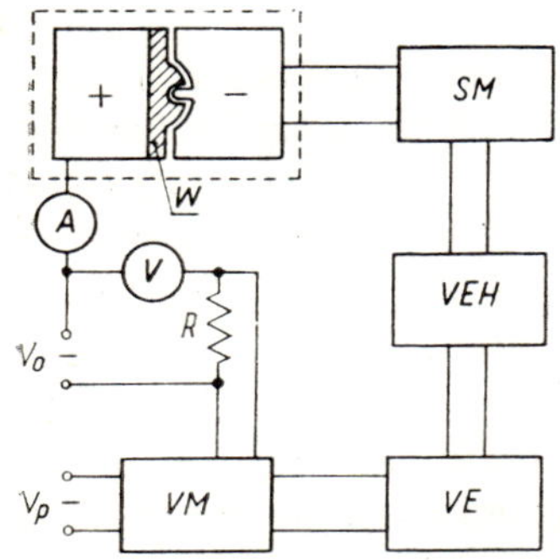

Fig. 23.11 Diagram showing circuit of working current and automatic system controlling eroder motion; *VM*—measuring and comparison system, *VE*—electronic amplifier, *VEH*—electrohydraulic amplifier, *SM*—servomotor

across the tool-work gap and the comparison voltage V_p, set by means of a potentiometer. The voltage differences are transmitted to the electronic amplifier *VE* and from there the amplified pulses go on to the second, electrohydraulic amplifier *VEH*. The stimuli are conveyed by this device to a mechanical executive component *SM*, directly connected with the electrode motion.

How effectively automatic regulators operate is best measured and evaluated by comparison of the machining rate for sinking holes to a depth of 1–1·5 times the eroder diameter. For, under these conditions frequent variations in the stage of the tool-work gap occur when there is a deterioration in the outflow of exhausted electrolyte and impurities (electrolysis products).*

Fig. 23.12 shows the results of investigations [23/2] into the effectiveness of a familiar and widely used thyratron regulator and an electrohydraulic regulator

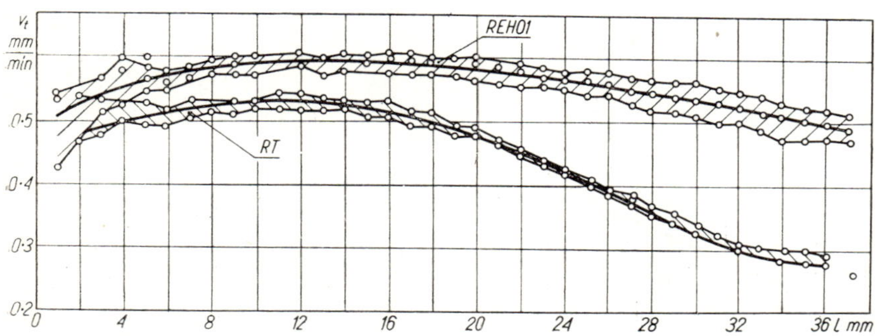

Fig. 23.12 Hole-sinking speed v_t as a function of the holesinking depth *l*, with eroder control by an electrohydraulic regulator and by a thyratron regulator [23/10]

* For this reason it is important to have the electrolyte flow in at a sufficiently high pressure so as to ensure a fast flow and among other things, to permit hydrodynamic depolarization to occur.

(developed at the Institute of Machining by Cutting in Cracow and used in Polish-made electric-spark and electrochemical machine tools). It is evident from the graph that by virtue of a better follow-up capability and lower losses for lost motion of electrode withdrawal and approach, that generator ensures better eroder working conditions and higher machining speeds. For example, the speed of an eroder controlled by that generator is about 40% higher than that of a thyratron-controlled eroder. Hence, the erosion rate increases and the machining time decreases correspondingly.

Development of sensitive control of eroder motion makes it possible to increase the machining rate also, because in practice some shaped workpieces can be machined from two sides at once. In the case of one-sided machining, the process rate is correspondingly lower since only one surface is machined. Two-sided machining entails many technical difficulties but it does permit an increase in production output.

23.3.4 Technical and economic indices

The principal factors affecting the electrolytic machining process have been discussed in the first three sections of this chapter. These factors have a close bearing and influence on the technical and economic indices of ECE.

The linear machining rate, i.e. the speed of material loss v_t, is expressed by

$$v_t = q_c \frac{w}{K_p} \frac{V}{s} = q_c j \tag{23.22}$$

Thus, if the machining conditions were to be totally stabilized, i.e. if $K_p = 1$, then the material removal rate, conventionally referred to as the machining speed, would be proportional to the voltage and specific conductivity of the electrolyte in the tool-work gap and inversely proportional to the gap size.

In fact, these relations depart from the theoretical calculations to a greater or lesser extent. The range of machining speeds obtained is illustrated in Fig. 23.13[23/2]. This graph concerns steel for which the theoretical specific loss is $q_c = 2{\cdot}0–2{\cdot}2\ \text{mm}^3/\text{A min}$.

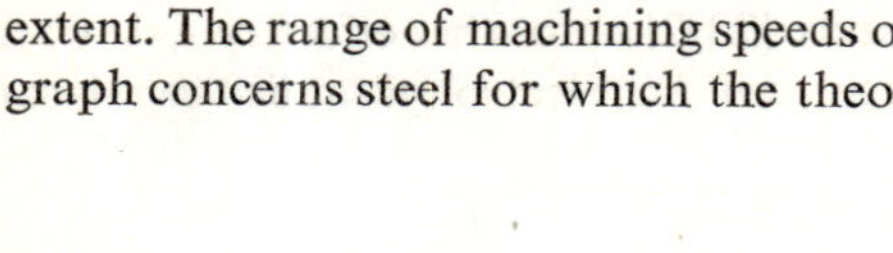

Fig. 23.13 Range of hole-sinking speeds (v_t) as a function of attainable current density j (for carbon and alloy steels). The shaded area denotes the scatter of v_t values obtained in practice

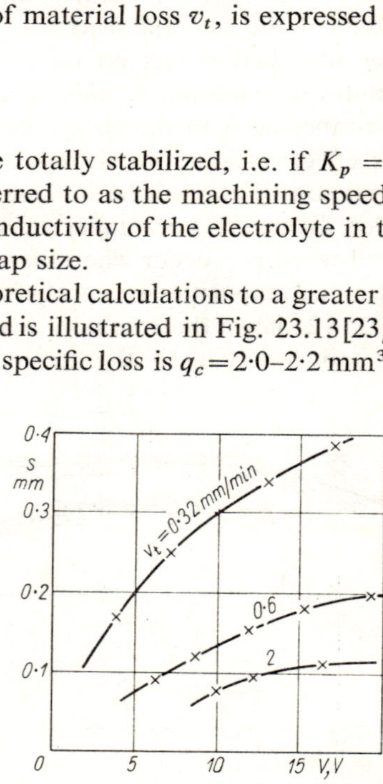

Fig. 23.14 Tool-work gap as a function of the working voltage (after [23/8])

The upper curve refers to the maximum theoretical value of the specific loss, and the shaded area indicates the range of values obtained in practice.

The dependence of the linear machining rate is not, in fact, directly proportional to the voltage even at the limits, when $V \leqq V_{\text{lim}}$. The reason is that when the voltage increases, the tool-work gap must also be increased for protection against harmful discharges. The results of investigations [23/3] on the size of the tool-work gap required as a function of the voltage for different machining speeds are plotted in Fig. 23.14.

An approximation of the relation depicted in Fig. 23.14 is expressed by

$$s = H \frac{V^m}{v_t^z} \tag{23.23}$$

where: H, m, z—numbers obtained from the approximation of experimental relations.

On substitution of (23.23) into (23.22), we obtain

$$v_t = \left(\frac{q_c}{H} \frac{w}{K_p} \right)^n V^c \tag{23.24}$$

where $n = \dfrac{1}{1-z}$ and $c = \dfrac{1-m}{1-z}$.

Equation (23.24) has been verified experimentally, i.e. the machining speed increases less than proportionally ($c < 1$) as the voltage increases.

Under industrial conditions and with the present state of technology, material loss speeds of $v_t = 0{\cdot}5$–5 mm/min are attainable for steel, and $v_t = 0{\cdot}1$–$1{\cdot}5$ mm/min for sintered carbides [23/3].

The precision of geometrically simple cavities or holes made by this method may be very high.

The taper (conicity) of holes is insignificant if the eroder has its side walls protected by insulation and works only with its end surface. In such cases a taper of less than 1 in 200 may be attained. However, when the side walls of the electrode are not insulated, the taper of the hole becomes much more pronounced and the shape of the hole becomes similar to that of a paraboloid.

The hole shape change can be reduced almost to zero if the design of the eroder (side-insulated) takes account of the tool-work gap size proper for the given machining conditions. Some influence on the out-of-roundness and shape change of the hole is exerted by the pressure and flow conditions of the electrolyte in the tool-work gap as well as by the stability (dynamic stiffness) of the machine tool. An accuracy in classes 7–4 should be regarded as feasible for round holes under industrial conditions.

In the case of complex, three-dimensional shapes, a suitably high accuracy can be attained but this necessitates overcoming some difficulties. The greatest difficulty is that of shaping the eroder; the more sharp curvatures and sharp corners and edges the final shape of the cavity has, the less the shape of the eroder must deviate from it. During electrolytic machining strong curvatures are dissolved more rapidly and the material is dissolved more quickly at spots where the local speed of electrolyte flow is higher [23/4].

In view of the latter, the ultimate shape of the eroder for geometrically complex cavities is obtained by adjustments during operation*. This increases the cost of the eroder and the machining operations. Consequently, the electrolytic machining of geometrically complex cavities is economical in general only for lots comprising a considerable number of workpieces.

The fact that the machined surface has a very low degree of roughness (roughness numbers 7 to 12) is an extremely important and valuable property of electrolytic machining.

Even more important, the roughness decreases as the current density increases, i.e. as the machining rate increases.

The roughness parameter R_a is plotted as a function of the current density in Fig. 23.15. The curves in this figure concern 0·45 carbon steel, machined in an electrolyte of NaCl + H_2O. As is seen from the diagram, in the case of electrolytic machining there is no fundamental division into roughing and finishing operations. An increase in smoothness with an increase in machining rate is a specific valuable feature of electrochemical machining.

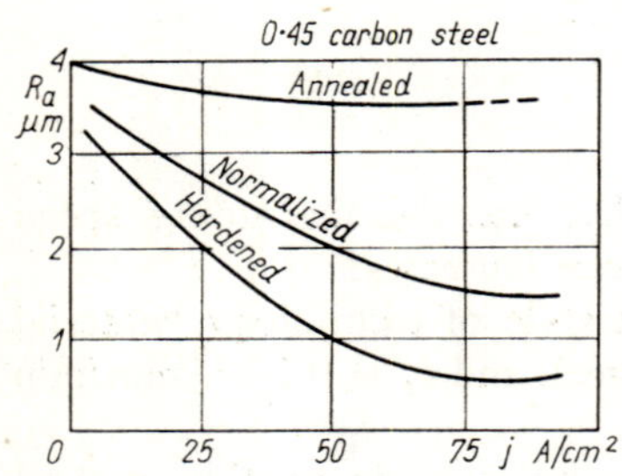

Fig. 23.15 Surface roughness parameter R_a as a function of the current density j and the state of the material (after [23/8])

It is seen from Fig. 23.15 that the state and structure of the material affects the roughness markedly. The more fine-grained and homogeneous the structure of the material, the better the smoothness. Hence, hardened steels, with a fine martensitic structure, always display a better smoothness than do normalized or annealed steels.

The geometric structure of the machined surface is specific. It exhibits traces of uneven etching of the particular chemical and structural components of the material. This effect is more pronounced at lower current densities (Fig. 23.16) [23/7].

Electrolytic machining leaves the surface layer unaltered (apart from geometric changes in the machined surface) which could be observed by etching and observation under a microscope at a magnification of 2000×.

The non-wearability of the eroder, if the machining conditions are chosen properly, is a very important technical and economic index of electrolytic machining. Discharges are among the factors which may chiefly be responsible for eroder wear. Moreover, an excessive rise in electrolyte temperature is undesirable.

* Theoretical determination of the eroder shape is feasible. The difficulty involved in the theoretical method is that so far there is no way of precisely determining the potential distribution of the electric field in the tool-work gap (which is characterized by the field gradient), and the accuracy of the proper shaping of the eroder depends on this factor. The electric field potential varies with time along with a change in the anode shape, and moreover also depends on the follow-up capacity of the eroder feed control system.

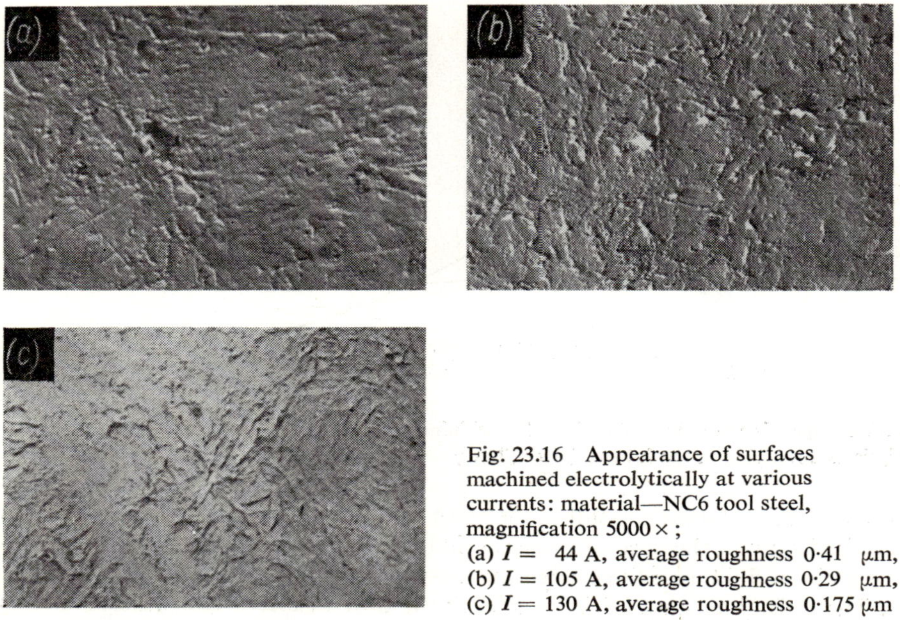

Fig. 23.16 Appearance of surfaces machined electrolytically at various currents: material—NC6 tool steel, magnification 5000×;
(a) $I =$ 44 A, average roughness 0·41 μm,
(b) $I =$ 105 A, average roughness 0·29 μm,
(c) $I =$ 130 A, average roughness 0·175 μm

The specific features of electrolytic machining determine its applications. Some typical applications of electrolytic machining are:
— machining of hemming dies after machining by cutting (so-called *electrolytic cleansing*),
— hole and cavity sinking in hardened high-alloy steels,
— die-sinking large numbers of simple dies,
— cutting off hardened high-strength steels and sintered carbides,
— shaping turbine vanes,
and similar operations.

Roughly speaking, the range of application of ECE as compared to EDM and machining by cutting and abrasion [23/3] with an eye to unit production costs k_j is illustrated in Fig. 23.17. The principal factor affecting the profitability of applications of electrolytic machining on large lots is that the eroder does not wear and that the machining conditions remain stable.

23.4 Electrochemical grinding

23.4.1 A general characteristic of chemical-abrasive machining

Chemical-abrasive machining is a combination of two concurrent processes: electrolytic machining and abrasive machining.

Depending on whether a bonded or loose abrasive is used in this combined process, we distinguish between:
— *electrochemical grinding*, when a grinding wheel is the tool, i.e. the eroder,
— *electrochemical lapping*, when the eroder is the cathode and the abrasive is in the form of a suspension in the electrolyte.

The electrochemical grinding method has been developed and has found applications.

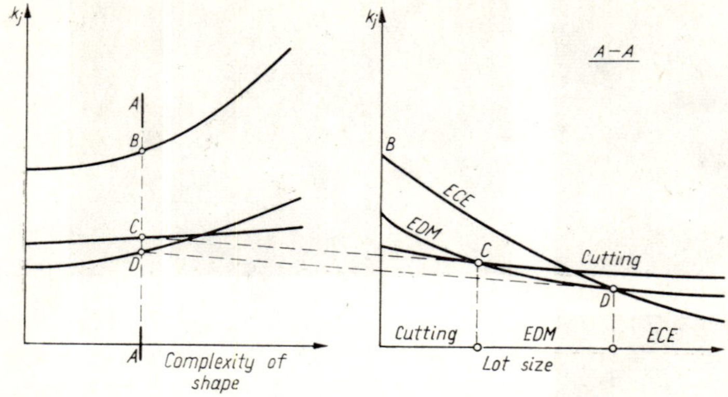

Fig. 23.17 Range of economic application of machining by cutting, electro-erosion, and electrolysis as a function of the lot size (after [23/8])

Electrochemical lapping is still in the stage of laboratory trials. In practice it has been realized only in the case of impact grinding where an electrolyte was used instead of an inert fluid, and the eroder and workpiece were incorporated into a D.C. circuit. The difficulties entailed in mastering the flow of electrolyte which contains an abrasive suspension are the main reason why the method of chemo-abrasive lapping has not hitherto found industrial application.

The introduction of chemo-abrasive machining is intended to enhance the yield of the process by adding the mechanical-abrasive work to the chemical work. Moreover, the abrasive grains can themselves automatically determine the tool-work gap size. The gap size should depend only on the grain size and the unit pressure, i.e. on the depth to which the grain enters the workpiece material.

The work of the abrasive grains also performs the function of mechanical depolarization, i.e. there is mechanical abrasion of any possible hard insoluble deposit (most often oxides) on the workpiece (anode) surface. This takes place especially in the case of alloys of many metals and sintered carbides of metals (tungsten, cobalt, titanium, molybdenum, tantalum).

For these reasons electrochemical grinding has gained application primarily to sharpening cutting tools with sintered carbide bits. To this end use is most often made of metal grinding wheels with loose abrasive, e.g. diamond dust. A diagram of an arrangement for electro-chemical tool sharpening is given in Fig. 23.18

The trend in electrochemical grinding has been towards mastering the production of grinding wheels which would conduct current but which could be sharpened and

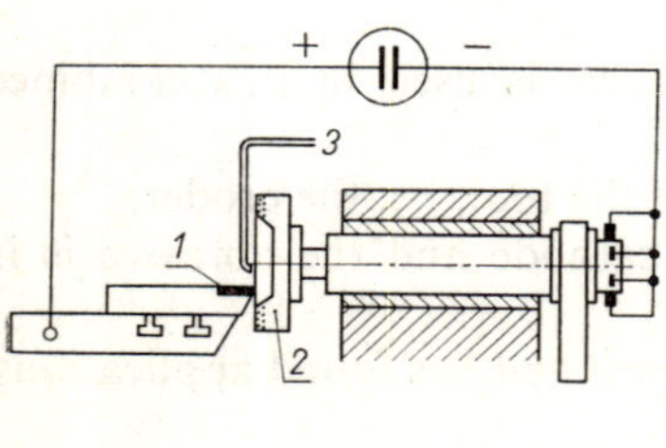

Fig. 23.18 Diagram of electrochemical grinding; 1—workpiece, 2—grinding wheel, 3—electrolyte

profiled as required. Attempts have been made in this connection to produce corundum abrasives with a graphite bond, ordinary grinding wheels with a ceramic bond with a conducting additive, and sintered grinding wheels made of powdered metals (bond) and abrasive mixed in a suitable proportion.

Mastery of the production of good and relatively inexpensive grinding wheels of this kind will be of fundamental importance for extending the applications of ECA to production grinding operations.

23.4.2 The contribution of mechanical and chemical work

Knowledge of the contribution of mechanical and chemical work to the total work done in electrochemical grinding serves as a basis for determining the rational range of application of ECA. This contribution is evaluated according to the proportion of the grinding rate accounted for by electrochemical machining, and by electrochemical grinding. This, in turn, is determined by experiment.

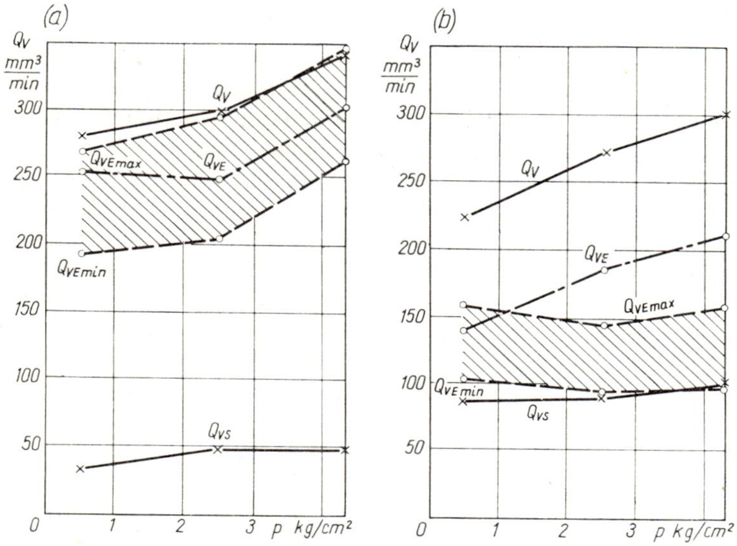

Fig. 23.19 Rates: of mechanical grinding Q_{VG}, electrochemical grinding (Q_V), electrochemical etching—maximum ($Q_{VE\max}$), minimum ($Q_{VE\min}$), and calculated as $Q_V - Q_{VG} = Q_{VI}$—as functions of the unit pressure. The values of Q_{VG} and Q_V were measured, those of $Q_{VE\max}$, $Q_{VE\min}$, and Q_{VE} were calculated by formulae. Plots drawn according to tests. Test conditions: electrolyte E2, $V = 12$ V, $v = 16$ m/s, $F = 1$ cm², grinding wheel with diamond grit; (a) grinding high-speed SW 18 steel (SS grinding wheel), (b) grinding sintered carbides S20 (SW grinding wheel)

Figure 23.19 shows some of the results of such investigations on grinding SW18 high-speed steel (Fig. 23.19a) and sintered carbides (Fig. 23.19b) at various unit pressures.

When high-speed steel is machined by electrochemical grinding (Fig. 23.19a) at various unit pressures, the curve Q_V indicates the rate of metal removal which increases with a rise in the unit pressure p. In the given case, the mechanical grinding rate Q_{VG} is about one-sixth as large, and an increase in the unit pressures causes it to rise only slightly.

The reason why the pressure has such little influence on variations in the grinding rate and on its generally very low value was, as found in the given case, that the grinding wheel became clogged with plastic steel and thus lost its cutting capabilities. Thus, the considerable increase in the electrochemical rate of metal removal with an increase in unit pressure should be attributed mainly to the reduction in the tool-work gap size, which accordingly increases the current density.

The same diagram gives two curves which illustrate the maximum and minimum theoretical rate of metal removal by electrochemical etching, $Q_{VE\,max}$ and $Q_{VE\,min}$, respectively. If the metal removal rate for mechanical grinding is deducted from the electrochemical grinding rate, the resultant curve Q_{VE} runs through the middle of the shaded area.

From the foregoing discussion it follows that in the case of electrochemical grinding of steel, the work of mechanical grinding plays a secondary role, since it accounts for only a small percentage of the overall rate of metal removal. This conclusion is confirmed by a number of other investigations, such as those of Pahlitsch [23/11].

On the other hand, a different situation arises in the case of the electrochemical grinding of sintered carbides. It is true that here, too, the influence of unit pressure is more pronounced on the electrochemical grinding rate than it is on the mechanical grinding rate, a fact which is also indicative of the dominant influence of changes in the tool-work gap and its ohmic resistance. Nevertheless, it is seen from Fig. 23.19b that the percentage contribution of the mechanical grinding work is much greater than in the case of steel. The shape of the curves $Q_{VE\,max}$ and $Q_{VE\,min}$ furthermore indicates that in the case of electrochemical etching the polarization resistance R_p probably increases since the curve $Q_{V\,max}$ runs below Q_{VE}.

The conclusion to be drawn from this is that in the case of the electrochemical grinding of sintered carbides the grinding wheel performs the extremely important function of mechanical depolarization and increases the overall metal removal rate by about 30%.

23.4.3 Technical and economic indices

The process and results of electrochemical grinding are affected by all those factors which influence grinding and electrolytic machining. Thus, these are parameters of the process of electrochemical grinding.

The domains of some of the principal parameters are as follows: operating voltage 5–15 V, current density j = 50–200 A/cm^2, unit pressures p up to 50 kg/cm^2, tool-work gap (using diamond dust) s = 0·01–0·1 mm, and most often s = 0·02–0·04 mm.

Of the factors which affect the machining rate, there has already been mention (in the chapter on electrolytic machining) of the material machined, the electrolyte and current density. Of the other factors affecting the yield, we shall discuss the dependence on the unit pressure, the abrasive grain size, and the peripheral speed of the grinding wheel, and the feed.

Increasing the unit pressure causes the abrasive grains to penetrate more deeply. As a result, a thicker layer is abraded and thus the grinding rate should increase.

Concurrently, the tool-work gap decreases. When the voltage remains constant, the current and current density increase and this leads to a rise in the electrolytic machining rate.

When the two influences operate simultaneously, the overall electrochemical machining rate goes up, as shown in Fig. 23.19, by 10 to 30%. However, if a reduction of the gap does not lead to a rise in current density, i.e. when the voltage decreases correspondingly as the working gap is reduced, the overall metal removal rate grows much less, by no more than 5%.

The conclusion from this is that an increase in the unit pressure affects the ECA rate chiefly by changing the current density.

The importance of the current density is also confirmed in studies of how abrasive grain size affects the yield. On the basis of results from studies of grinding it may be concluded that larger grain size should mean a higher rate of metal removal. In actual fact, the opposite is true, and this should be attributed to the fact that as the grain size increases so does the tool-work gap and, along with this, the current density falls off. The ensuing loss of metal removal cannot be compensated by the partial increase in the grinding rate.

The influence of the peripheral speed of the grinding wheel on the metal removal rate can be predicted from the theory underlying the grinding process.

As is known,

$$Q_V = F_{vt} = F\,z\,a_z \tag{23.25}$$

where F—the area of contact between the grinding wheel and the workpiece material, z—the number of points along the contact length per unit time, a_z—the average thickness of the layer abraded by a single grain.

If it is taken into account that

$$z = z_0 n = \frac{1000}{d} z_0 v = A v \tag{23.26}$$

where: z—the number of grains on the grinding wheel periphery, n—the rotational speed of the grinding wheel, d—the grinding wheel diameter, $A = 1000 z_0/d$.

Moreover, as is known from the chapter on grinding, the average thickness of the layer abraded by a single grain is

$$a_z = \frac{v_p}{60v \pm v_p} g = \frac{1}{\dfrac{60}{g\,v_p} v \pm g} \tag{23.27}$$

After substitution of eqns. (23.26) and (23.27) into (23.25),

$$Q_v = F\,A \frac{1}{\dfrac{60}{g\,v_p} v \pm g} \tag{23.28}$$

or, noting that $F = \text{const}$,

$$v_t = \frac{Q_v}{F} = A \frac{v}{Bv \pm g} \tag{23.29}$$

where $B = \dfrac{60}{g\,v_p}$.

The graphical interpretation of eqn. (23.29) is given in Fig. 23.20. It is seen from the plot that there is an optimal cutting speed v for which the material removal rate v_t is greatest.

Studies by ENIMS* [23/9] on how the peripheral speed of the grinding wheel affects the electrochemical grinding rate confirm theoretical predictions. It should be noted, however, that the optimal value v_{opt} depends on the values taken for other parameters, viz. on feed, the grinding wheel size, and cutting depth, which depends on the unit pressure.

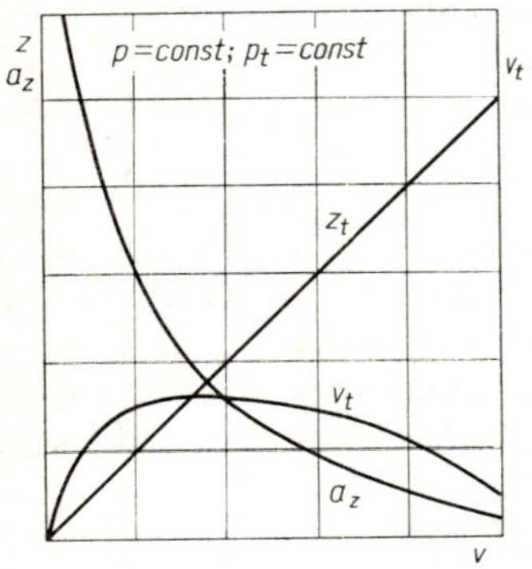

Fig. 23.20 Plots of thickness of layer a_z abraded by a single grain, total number z of abrading grains during contact with material, and speed v_t of material penetration

The recommended values of the mechanical parameters of electrochemical grinding must encompass interlinked groups of quantities: v, v_p, d, and p or g.

The influence of feed on the machining rate may be considered in analogous manner. Analysing eqns. (23.26) to (23.29), but taking v_p as an independent variable, one can establish that there is a certain value of v_p for which the machining rate is highest. This has also been confirmed by investigations [23/9].

The depletion of the abrasive in the case of electrochemical grinding, as referred to the amount of machined material removed, i.e. the relative depletion, averages $m_V = 0{\cdot}1$–$0{\cdot}4\%$ (when diamond is the abrasive used). The average overall unit diamond consumption is from 10 to 20% less than in the case of mechanical grinding. The lifetime of diamond abrasive in the ECA process is 20 to 50 times that in ordinary grinding.

The roughness of an electrochemically ground surface is in general of the same order as after electrochemical machining and somewhat less than after grinding. No lay in the geometric structure of the surface or grinding scratches can be seen after electrochemical grinding. Nevertheless, accurate measurement of the roughness longitudinally and transversely with respect to the direction of grinding does reveal some slight differences.

The properties of the surface layer after electrochemical grinding are, from the point of view of abrasion resistance, almost equal to those of a mechanically ground layer. The slight differences in wearability which do occur (of the order of 2–8% to the disadvantage of EAM) are detectable only by statistical means. By way of

* ENIMS—Eksperymental'nyi Nauchno-Issledovatel'skii Institut Metallorezhushchikh Stankov (Experimental Scientific Research Institute of Metal-cutting Machine Tools), Moscow.

illustration Fig. 23.21 presents the results of wearability tests [23/9] on layers machined by mechanical (ASM), electrochemical (ECA), and electrolytic (ECE) grinding.

Other tests have shown that the surface layer has similar properties after all three modes of machining, or even that an ECA-machined layer has slightly better properties [23/11].

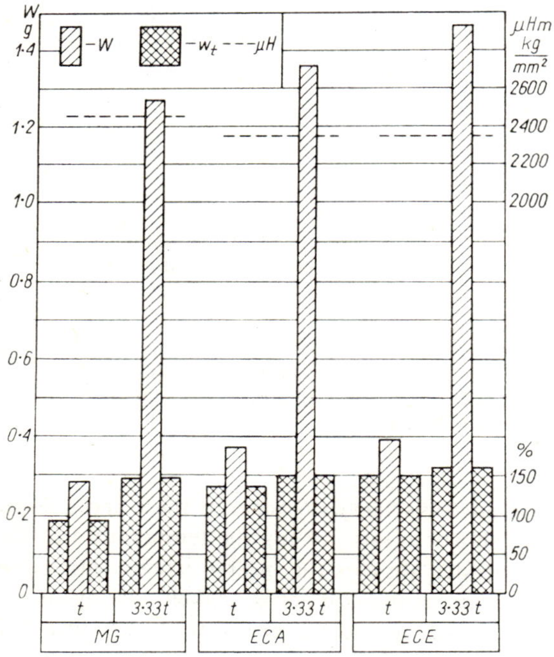

Fig. 23.21 Weight loss W, unit weight loss w_t, and microhardness μHm of surfaces machined by mechanical grinding MG, electrochemical grinding ECA, and electrochemical etching ECE during process of abrasion by a carborundum-powder suspension [23/10]

23.5 Chemical and spark machining

23.5.1 The principle of chemical and spark machining

The metal machining process called *chemical and spark machining* takes place as a result of simultaneous electrochemical (electrolytic) dissolution, local melting due to spark erosion, and mechanical abrasion of surface irregularities.

The decisive role of electrolytic and thermal processes has been emphasized in the terminology which employs the term thermoelectrolytic machining in parallel with chemical and spark machining.

The principle of chemical and spark machining can be explained by means of Fig. 23.22.

The d.c. generator 1 energizes a circuit in which the cathode 2 is the tool (by analogy with other machining methods, it might be called an eroder) and the anode 3 is the workpiece. Electrolyte 4 is fed between the eroder and the workpiece. The rheostat 5 in this circuit serves to regulate the voltage and current.

A process of electrolytic machining occurs as a result of the flow of current across the tool-work gap. The electrolyte is chosen so that on the surface of the workpiece 2 (Fig. 23.23) there forms a membrane, most often an oxide membrane 3, which increases the resistance, thus reducing the efficiency of the electrolytic process.

The eroder 1, which travels at some speed relative to the workpiece, is subjected to some pressure. Consequently, at the tips of the irregularities on the surface the

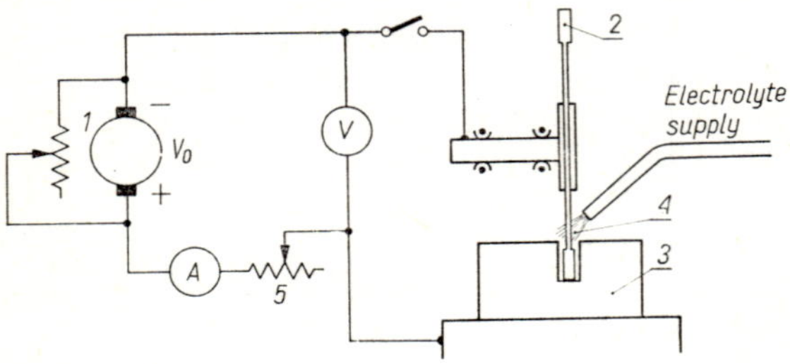

Fig. 23.22 Diagram of electrical circuit for chemical-spark machining

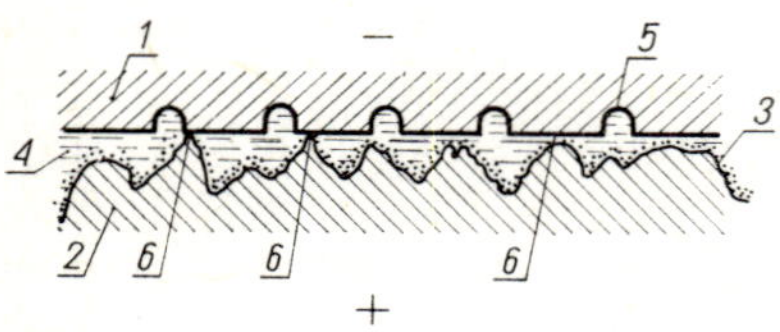

Fig. 23.23 Diagrammatic representation of loss of anode material during process of chemical-spark machining

polarization membrane is ruptured and microdischarges 6 occur at these spots. Heat evolves locally as a result of these brief short-circuits, raising the temperature of the contacts above the melting point. The eroder abrades the molten irregularities and thus breaks off the contact. Sparking occurs as the contact is broken. Owing to the good deionization properties of the electrolyte, no electric arc is produced. In order to enhance the abrasive action of the eroder and to provide storage space for the abrasion products and the gases formed, the eroder is provided with indentations 5.

The process of chemical and spark machining can proceed in a stable manner only if all the machining conditions have been appropriately chosen.

23.5.2 Conditions and applications of chemical and spark machining

The current source usually has a voltage range of 10–28 V. This voltage is regulated so that roughing takes place at the higher voltage, whereas finishing is done at 10–12 V.

With such voltages, the polarization almost disappears and the short-lived contacts do not produce enough heat to cause melting. Accordingly, under these conditions it is almost exclusively electrolytic machining which takes place.

The electrolyte for chemical and spark machining is a solution of water glass. Water glass is characterized by a modulus M which is defined by the formula

$$M = \frac{A}{B} 1{\cdot}032 \tag{23.30}$$

where: A—the percentage content of SiO, B—the percentage content of Na_2O, and 1·032—the value of the molecular weight ratio of sodium oxide to silica.

The modulus of sodium water glass is $M = 2{\cdot}6\text{–}3{\cdot}0$.

An aqueous solution of water glass is used for the machining. It has been found by experiment that the optimum concentration of the solution should be such that the specific gravity of the electrolyte be 1·16–1·19 g/cm³, and the modulus be 2·5–2·85.

The fact that it tends to dry is a very incovenient feature of the electrolyte. Layers which adhere very strongly to the surface form as a consequence.

To reduce this proneness to drying and to facilitate cleansing of the surfaces of the machine tool, the eroder, and the workpiece, 5 to 10% spindle or transformer oil is added to the electrolyte.

Another disadvantageous feature of the electrolyte is that the composition changes with time. Owing to evaporation, splashing, and drying of the water glass, the composition of the electrolyte must be checked and adjusted.

Apart from the voltage and the electrolyte, the eroder pressure on the surface being machined is the most important factor affecting the process. In practice, unit pressures from 0·25 to 3·0 kg/cm² are employed.

Of the other factors which affect the process of chemical and spark machining, mention should be made of the eroder speed relative to the workpiece. This speed in practice is 4 to 60 m/s, the range of speeds employed most often being $v = 15\text{–}25$ m/s.

The applications of chemical and spark machining changed during the 'fifties. At first, this mode of machining was used to grind tools, especially with sintered carbide bits. It was particularly recommended for profiling form tools. In addition, chemical and spark machining was used to shape complex profiles (instead of grinding) and also to cut off materials.

Once other erosion methods, e.g. electrochemical grinding, had been perfected, the use of chemical and spark machining for tool sharpening ceased almost completely. The principal reason for this was that it was difficult to control the process of surface fusion and sparking which could cause all sorts of damage to the surface layer (microfissures invisible to the naked eye).

At present, chemical and spark machining is used mainly to cut off materials, especially those which are difficult to machine by cutting.

23.5.3 Technical and economic indices

The metal removal rate attained by chemical and spark machining is relatively high, inasmuch as it may reach a value of the order of 500–600 mm³/min [23/15].

Figure 23.24 shows a typical dependence of the volume rate of metal removal on the unit pressure. The shape of the curve can be explained as follows.

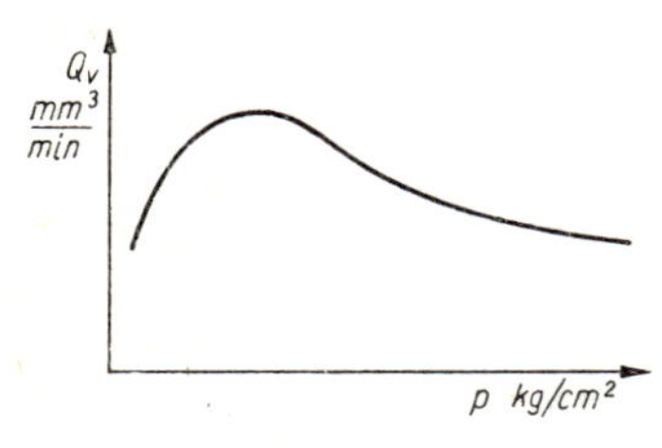

Fig. 23.24 Character of plot of the machining rate of chemical and spark machining as a function of the unit pressure

As the pressure increases from 0 to some value *p*, the tool-work gap is seen to reduce in size. Consequently, the current increases and along with it so does the metal removal rate of the process. A further rise in the pressure causes the number of short-lived short circuits to rise steadily. Under these conditions a considerable fraction of the current now flows across the metallic contacts. Hence, the work of the current causing melting and the chemical work decrease. As a consequence of this, the rise in the yield is observed to be inhibited. If the unit pressures continue to increase, this phenomenon is seen to intensify further. As a result of this, the yield continues to fall off gradually.

The character of the pressure-dependence of the yield was the reason why different opinions were held on the role of pressure. Depending on the pressure range, investigators found that with an increase in pressure, the yield rose, fell, or was unaffected [23/14].

A rise in the current density *j* causes the yield to increase almost in proportion [23/14]. A similar effect is also exerted by a voltage of up to 25 V, i.e. to the voltage at which sparking occurs with increased intensity.

The peripheral speed of the eroder also causes the yield to increase.

The roughness number of the machined surface ranges from 5 to 8. The current density and voltage are of decisive importance with regard to the smoothness of the surface machined.

In the range wherein the current density is raised to about 40 A/cm^2, the roughness is found to increase monotonically, almost linearly. Accordingly, any improvement in smoothness must be brought about by a decrease in the current density, hence at the cost of reducing the metal removal rate of the process.

The surface smoothness can be changed effectively by changing the peripheral speed of the eroder. A plot of such a relation is given in Fig. 23.25 as an example. It is seen from the plot that the dependence of the height of the irregularity on the peripheral speed of the eroder is hyperbolic in character [23/15].

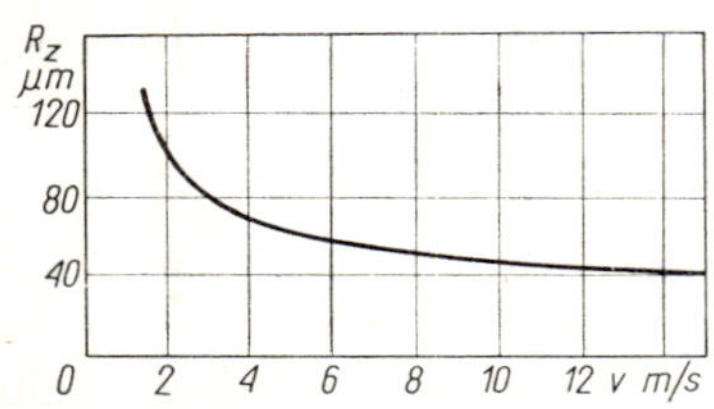

Fig. 23.25 Plot of height of surface roughness against the peripheral eroder velocity (after [23/15])

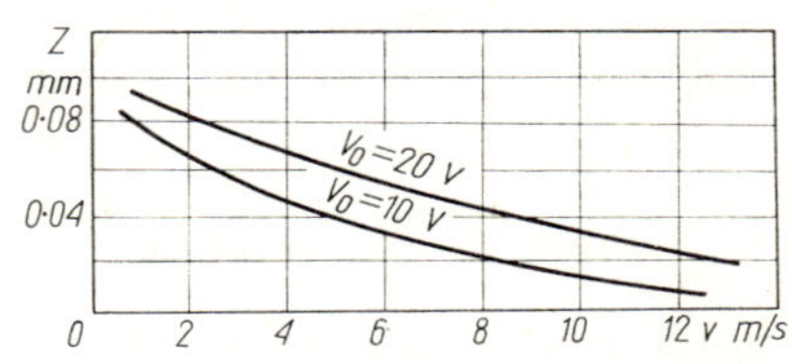

Fig. 23.26 Thickness *z* of altered layer versus the peripheral eroder velocity (after [23/13])

The thickness of the surface layer that is damaged or altered depends chiefly on the current conditions and on the speed of the eroder relative to the workpiece.

Figure 23.26 gives an example of the results of investigations [23/14] into how the speed *v* affects the thickness of the altered surface layer.

The foregoing examples provide justification for the use of the highest possible peripheral eroder speeds. For design and operating considerations (electrolyte splashing), the speed is restricted in practice to a value of 35–40 m/s. Moreover, eroder wear also provides an argument for restricting the speed.

Eroder wear during chemical and spark machining occurs as a result of friction and reduction of the strength owing to the eroder heating up. The latter effect is particularly pronounced at high current densities and high eroder speeds.

REFERENCES

23/1 ALLISON, C. R.: 'How electrolytes influence the ECM process', Technical Paper, Amer. Soc. of Tool and Mfg. Engineers, 1964

23/2 KACZMAREK, J.: 'Niektóre problemy rozwoju teorii i techniki obróbki elektrochemicznej' (Some aspects of the development of the theory and techniques of electrochemical machining), Prace IOS 1965, seria Referaty, No. 46

23/3 KUBETH, H.: 'Elektrochemische Bearbeitung—Grundlagen und Verfahrenskennwerte', Ind. Anz., 80, 1964

23/4 KURAFUJI, H., and SUDA, K.: 'Accuracy of electrolytic machining', Paper at 15th CIRP General Assembly, 1965

23/5 MAJSAK, Z.: 'Badania nad doborem elektrolitu dla elektrochemicznej obróbki metali' (Research on the choice of electrolyte for electrochemical machining of metals), IOS Research Report, No. B1C-11, Cracow 1963

23/6 PELLISIER, J., and HESSE, A.: 'L'usinage électrolytique', La Machine-Outil Française, 5, 6, 7, 8, 1963

23/7 THINAT, A., HESSE, A.: 'La Rectification Électrolytique', Assemblée Générale du CIRP, 1963

23/8 'Die abtragende Bearbeitung und ihre Abgrenzung gegenüber spanenden Verfahren', Ind. Anz., 69, 1965

23/9 'Elektrokhemicheskoe shlifovanie bystrorezhushchikh stali' (Electrochemical grinding of high-speed steels), R-18, ENIMS, Moscow 1964

23/10 KACZMAREK, J., and ZACHWIEJA, T.: 'Badania wydajności elektrochemicznego szlifowania materiałów narzędziowych w zależności od własności ściernicy' (Research on the metal removal rate of electrochemical grinding of tool materials as a function of the properties of the abrasive), Prace IOS, seria Referaty No. 61

23/11 PAHLITSCH, G.: 'Elektrolytisches Schleifen von Hartmetall', Klepzig Fachberichte, 8, 1964

23/12 'Die abtragende Bearbeitung und ihre Abgrenzung gegenüber spanenden Verfahren, Elektrochemisches Schleifen', Ind. Anz. 87, 1965

23/13 KARASIK, G. A., and KOSOLANOV, I. I.: 'Konstruirovanie anodnomekhanicheskikh otreznykh i zatochnykh stankov' (Design of Chemical-spark Cutting-off and Tool-grinding Machines), Mashgiz, Moscow 1951

23/14 ALBIŃSKI, K.: 'Elektroerozyjna obróbka metali' (Electro-erosion Machining of Metals), PWT, Warsaw 1954

23/15 KOSMACHEV, I. G.: 'Obrabotka metallov anodno-mekhanicheskim sposobom' (Machining Metals by the Chemical and Spark Process), Mashgiz, Moscow 1961

CHAPTER 24

Electro-beam machining

24.1 Electron-beam machining

24.1.1 THE CHARACTERISTICS OF ELECTRON-BEAM MACHINING

24.1.1.1 *A general characterization of electron-beam machining*

Electron-beam machining, or *electro-beam machining by electrons* (EBE), is machining by evaporation by means of a beam of electrons emitted by a suitably heated electrode (cathode) under vacuum. An electric field can be employed to focus the electrons and direct them. Accordingly, when the electron beam strikes an obstacle, i.e. the material being machined, the high kinetic energy of the beam is converted into heat at the point of contact. The process of irradiation with a beam of electrons takes place on a very small section of the surface. As a result of this, within a brief time there is a local rise of temperature much above the boiling point (and in some cases, above the point of sublimation), even of the most refractory materials. As a consequence of these processes, at the spot irradiated by the electron beam the material is evaporated violently; in other words, in accordance with the conventional definition of erosion, there is *erosion by evaporation.*

The electron beam may thus be treated as a thermal tool.

24.1.1.2 *The operating principles of the electron-beam machine tool*

The first technical device, which was the prototype of electron-beam machine tools, was designed by Steigerwald [24/1] in 1949. Electron-beam machine tools built to this day are still based on the same principle, which is explained in Fig. 24.1.

The cathode 1 is heated to 2500–3000°C, at which temperature it begins to emit electrons. Thanks to the Wehnelt electrode 2 and the voltage V_0 applied across the cathode and the anode 3, the beam of electrons irradiated is accelerated and gets past the anode. The alignment system 4 enables the principal axis of the beam to be positioned so that only the central part of the beam passes through the diaphragm 5. The focusing system, the magnetic lens 6, refocuses the partially spread beam 7, which is then directed through the control system 8 into the desired spot of the workpiece 9. The entire equipment is in a vacuum of the order of 10^{-5} torr. A microscope 10 is built into the stereoscopic device for checking that the beam has been aimed at the right spot.

A small deflection of the beam, of the order of 4 mm on the surface being machined, is effected by means of the control system. If the beam must be moved more with respect to the workpiece, a table with a feed motion in three mutually perpendicular directions is used for this purpose.

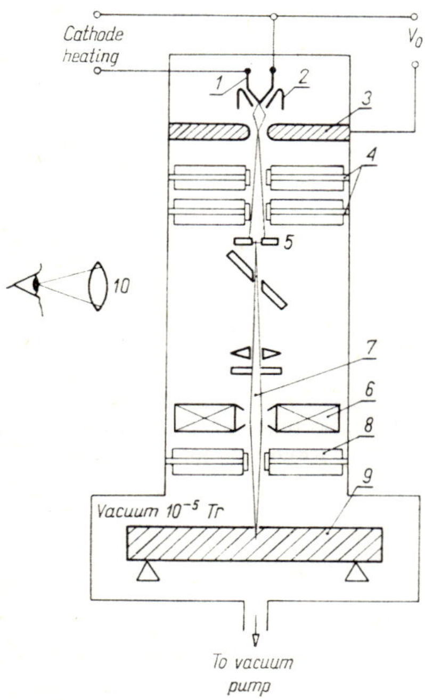

Fig. 24.1 Operating principle of electron-beam machine tool

24.1.2 The fundamental concepts and relationships of electron-beam machining

The accelerating potential V_0 is the voltage across the cathode and anode. The voltages used are up to 150 000 V. The high voltage causes the electrons to be accelerated to such an extent that X-rays may be produced if these electrons are allowed to bombard the workpiece. Accordingly, wherever voltages in excess of 30 000 V are employed, it is necessary to employ protective lead shielding. At voltages under 30 000 V the radiation is soft enough and of sufficiently low intensity for protection shielding not to be required.

The emission current I_e is the intensity of the stream of electrons emitted by the cathode. The emission current depends on the emissivity of the cathode, hence it depends on the type of cathode material, the accelerating potential, and the heating temperature. Within the range of voltages and heating temperature used, the emission current has a value of the order of milliamperes. The relation between the emission current and the accelerating potential, with a tungsten cathode heated to different temperatures, is given in Fig. 24.2.

The emission current density j_e indicates the extent of beam concentration and is expressed by the ratio

$$j_e = \frac{I_e}{F_e} \tag{24.1}$$

where F_e—the cross-sectional area of the beam.

The emission current density is usually given in A/mm^2 or A/cm^2.

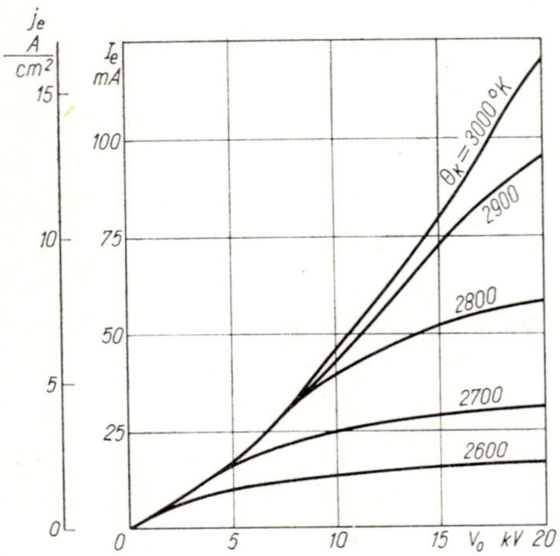

Fig. 24.2 Emission current I_e and emission current density j_e as functions of accelerating potential V_0 for various cathode temperatures [24/2]

The pulse charge L_p is defined as the product of the emission current and the pulse duration, i.e.

$$L_i = I_e t_p \tag{24.1a}$$

On account of the strength of the equipment and in order to reduce the temperature of the workpiece in the vicinity of the spot being machined (owing to conduction of heat), the irradiation is carried out in brief, repetitive pulses.

The energy E_p of a single pulse can be calculated by

$$E_p = V_0 I_e t_p \tag{24.2}$$

where t_p—the duration of a single pulse.

The pulse frequency f_p is defined by

$$f = \frac{1}{t_p + t_i} = \frac{1}{t_0} \tag{24.3}$$

where t_i—the pulse interval, t_0—the period.

The power density w, that is, the ratio of the beam power to the beam area, as defined by

$$w = \frac{I_e V_0}{F_e} \tag{24.4}$$

has an effect on the technical and economic indices of the process of electron erosion, since the rate and temperature of the local heating of the material depend on it.

The electron penetration depth a characterizes a specific property of electrons, that of penetrating into the materials encountered. The depth of electron penetration is computed by the formula [24/2]

$$a = H\frac{V_0^2}{\gamma}, \tag{24.5}$$

where γ—the specific gravity, and $H = 2{\cdot}1\times10^{-12}$.

In the zone of electron penetration the energy of the electrons is converted into heat. In the zone of heat concentration the material heats up, melts, and evaporates, and heat is conducted from this zone deeper into the material, thus setting up a temperature field with a high gradient.

While the pulse lasts, the electron beam erodes a mass of material to a depth g_e. The longitudinal cross section of the cavity after a single pulse in chromium-molybdenum steel is shown in Fig. 24.3 [24/3].

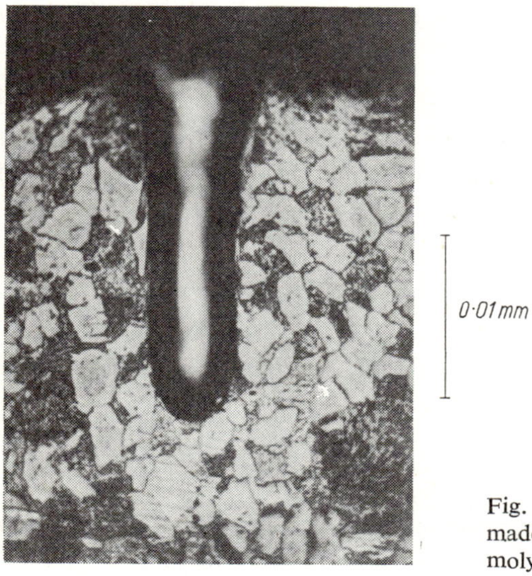

Fig. 24.3 Longitudinal cross-section of a cavity made by a single electron-beam pulse in chromium-molybdenum steel [24/3]

If the hole is round, the diameter d_B of the hole made is

$$d_B = d_e+2b \tag{24.6}$$

where d_e—the diameter of the electron beam at the point of contact with the machined surface, b—the hole shape change, caused when gas protuberances carry off a layer of the molten walls.

The hole made is surrounded by a layer of average thickness z formed mainly by solidification of that part of the molten walls of the hole which were not eroded. The approximate diameter d_R of this layer is

$$d_R = d_B+2z = d_e+2b+2z \tag{24.7}$$

If it is assumed that the energy of a pulse of electrons is converted completely into heat, the total amount of heat Q is

$$Q = K I_e V_0 t_p \tag{24.8}$$

where K—a coefficient of proportionality.

The heat Q is used as follows:

— for eroding a mass in the form of vapour—Q_v

$$Q_v = A\, m_e[C_s(\Theta_m-\Theta_0)+C_m+C_l(\Theta_b-\Theta_m)+C_b+ \\ +C_v(\Theta_v-\Theta_b)] = A\, m_e Q_A \tag{24.9}$$

where Q_A—the total amount of heat required to convert a unit mass to vapour;

— for eroding mass in liquid form (i.e. the part of the molten walls of the hole carried off by the expanding gas—Q_{lv}

$$Q_{lv} = B\, m_e[C_s(\Theta_m-\Theta_0)+C_m+C_l(\Theta_l-\Theta_m)] = B\, m_e Q_B \tag{24.10}$$

where Q_B—the total amount of heat required to convert a unit mass to liquid;

— for molten mass not eroded by the scattered electrons on the hole walls and on the walls of the zone of electron penetration—Q_{lk}

$$Q_{lk} = m_{lk}[C_s(\Theta_m-\Theta_0)+C_m+C_c(\Theta_c-\Theta_m)] = m_{lk} Q_B \tag{24.11}$$

— for heating the material in the solid state by conduction heat from the zone of electron penetration and from part of the surface of the hole walls (irradiated by scattered electrons) during the pulse—Q_s

$$Q_s = \lambda \frac{\mathrm{d}\Theta}{\mathrm{d}l} t_i F_l \tag{24.12}$$

The symbols used in eqns. (24.9)–(24.12) have the following connotations: A and B—positive coefficients, found empirically, such that $A+B = 1$; m_e—the average mass eroder per pulse

$$m_e = \pi \frac{d_B}{4} g_e \gamma \tag{24.13}$$

where g_e—the average erosion depth per pulse, and the other symbols are the same as before.

C_s—the specific heat of the material in the solid state; C_m—the heat of melting (fusion); C_l—the specific heat of the material in the liquid state; C_b—the boiling heat; C_v—the specific heat in the vapour state; Θ_0—the initial temperature of the material; Θ_m—the melting temperature; Θ_l—the temperature of the liquid; Θ_b—the boiling temperature; Θ_v—the vapour temperature; m_{lk}—the mass of material molten but not eroded

$$m_{lk} = \left(n_e g_e \pi d_B z + \pi \frac{d_B^2}{4} z\right)\gamma \tag{24.14}$$

where n_e—the number of pulses; and the other symbols as before.

λ—the heat conductivity factor; $\mathrm{d}\Theta/\mathrm{d}l$—the temperature gradient in the direction normal to the loss surface F_l (l—the distance in the direction normal to the surface loss); F_l—the loss surface, equal to the sum of the face surface F_B of the zone of electron penetration, the lateral surface F_W of the zone of electron penetration, and part of the lateral surface F_C of the entire hole

$$F_l = F_B + F_W + F_C \tag{24.15}$$

When the hole is round, then:

$$F_l = \frac{\pi d_B^2}{4} + 4\pi d_B a + \pi d_B M\, n_e g_e \tag{24.16}$$

where a—the depth of electron beam penetration [according to eqn. (24.5)]; M—a positive empirical coefficient, less than unity, for determining what part of the surface of the entire hole is irradiated by scattered electrons; the other symbols as before.

By assumption the total heat of a single pulse is

$$Q = K I_e V_0 t_p = Q_v + Q_{lv} + Q_{lk} + Q_s \tag{24.17}$$

If the summands, expressed by eqns. (24.9), (24.10), (24.11) and (24.12), and the auxiliary eqns. (24.13), (24.14), (24.16), and (24.5), are inserted into eqn. (24.17) we can derive an equation for the average depth g_e eroded by one pulse

$$g_e = \frac{\dfrac{K}{\pi r} I_e - r\gamma \dfrac{z}{B} P_B - \lambda \dfrac{d\Theta}{dl}\left(r + 2\dfrac{H}{\gamma} V_0^2\right)}{\gamma\left[r(P_A + P_B) + 2\dfrac{z}{B} P_B n_e + 2\lambda \dfrac{d\Theta}{dl} M\, n_e\right]} \tag{24.18}$$

In eqn. (24.18), in addition to symbols used hitherto, we have introduced $r = d_B/2$, i.e. the radius of the hole being machined, and $P_B = B\,Q_B/t_p$ and $P_A = A\,Q_A/t_p$. It may be noted that the ratios Q_A/t_p and Q_B/t_p denote the average rate of heat intake by a unit mass of material during a pulse.

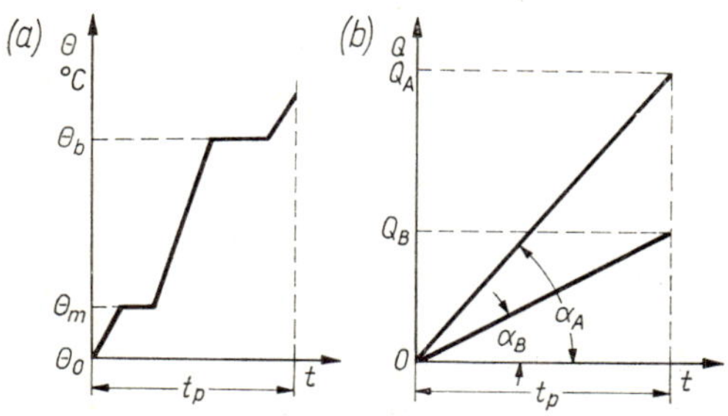

Fig. 24.4 Temperature of unit mass, and heat absorbed by unit mass for heating, melting, and boiling as functions of the pulse duration

The graphical interpretation of this expression is illustrated in Fig. 24.4. During a pulse a unit of mass must be melted and evaporated. Accordingly, there are two temperature arrests (Fig. 24.4a). On the other hand, the heat intake is continuous. If the actual course of the heat intake is replaced by a straight line (Fig. 24.4b), then the slope of this line is

$$\tan\alpha_A = \frac{Q_A}{t_p} \quad \text{and} \quad \tan\alpha_B = \frac{Q_B}{t_p} \tag{24.19}$$

From analysis of eqn. (24.18) for the value of g_e it follows that, among other things, there exists an optimal value of V_0 for which g_e attains a maximum. Moreover, as the number of pulses increases, the value of g_e falls off hyperbolically.

Calculating g_e by eqn. (24.18) and knowing F_e and γ, one can also compute the mass eroded by one pulse, since

$$m_e = g_e F_B \gamma \tag{24.20}$$

Equation (24.18) is also useful for finding an equation for the number of pulses n_e required to erode a hole to a depth of g. Knowing that

$$n_e = \frac{g}{g_e} \tag{24.21}$$

one replaces g_e by an expression from eqn. (24.18) and after appropriate manipulations, one obtains

$$n_e = \frac{gr\gamma(P_A+P_B)}{\frac{K}{\pi r} I_e V_0 - r\gamma \frac{z}{B} P_B - \lambda \frac{\mathrm{d}\Theta}{\mathrm{d}l}\left(r+2\frac{H}{\gamma}V_0^2\right) - 2g\left(\gamma \frac{z}{B} P_B + \lambda \frac{\mathrm{d}\Theta}{\mathrm{d}l} M\right)} \tag{24.22}$$

Equation (24.22) indicates that the number of pulses depends on several factors and makes it possible to analyse the influence of those factors.

By way of example it may be said that the number of pulses required to machine a material to a depth of g depends non-monotonically on the accelerating potential.

The graphical interpretation of eqn. (24.22), as shown in Fig. 24.5, treats n_e as a function of the accelerating potential V_0 with the other factors remaining fixed. On the basis of a graph, e.g. Fig. 24.5 one might formulate the conclusion that there is an optimum accelerating potential at which the number of pulses required becomes

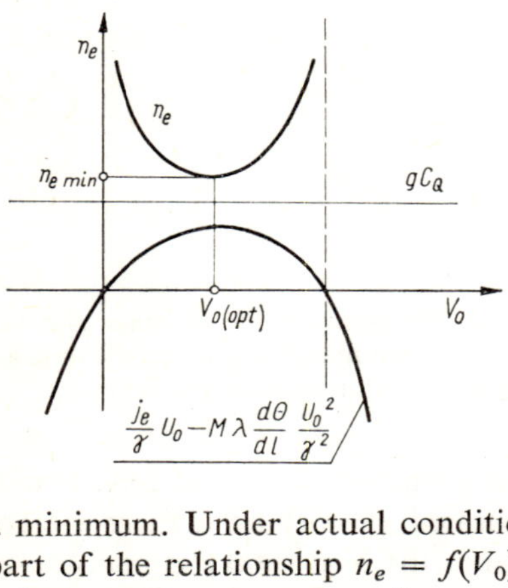

Fig. 24.5 Graphical interpretation of the dependence of the number n_e of pulses required on the accelerating potential V_0

a minimum. Under actual conditions, with the available techniques, the decreasing part of the relationship $n_e = f(V_0)$ comes under consideration, i.e. the number of pulses may be expected to decrease with an increase in V_0.

It also follows from eqn. (24.22) that with all the factors constant, and only the required depth of the hole or cavity changing, the number of pulses required n_e increases hyperbolically as the depth g increases. In practice this means that starting from a certain depth it is pointless to increase the depth any further since this would require a very large increase in the number of pulses.

Knowing the required number of pulses n_e, one can calculate the time needed for electron-beam machining.

The machining time for a hole or cavity, i.e. the hole or cavity sinking time, is expressed by

$$t_m = n_e t_0 = \frac{n_e}{f} \tag{24.23}$$

The machining time when cutting a slot to a depth g is found by

$$t_m = n_e t_0 \frac{L}{d_e} \tag{24.24}$$

where L—the length of the slot; d_e—the diameter or width of the electron beam at the point of contact with the workpiece.

24.1.3 THE TECHNICAL AND ECONOMIC INDICES OF ELECTRON-BEAM MACHINING

24.1.3.1 *The parameters and indices of electron-beam machining*

Just as in the case of other machining methods, so in electron-beam machining one distinguishes process parameters associated with:

— the material being machined, and especially with all of its thermal properties which enter the pulse energy balance, such as density, specific heat, transition temperatures, and thermal conductivity; moreover, the electrical properties of the material are also of some importance;

— the emitting cathode, i.e. the kind of material used to make the cathode, its emissivity as a function of the temperature (emission current), the shape and dimensions of the cathode, and its resistance to wear;

— the electrical circuitry, characterized by the accelerating potential, effectiveness in diaphragming and focusing the electron beam (which is expressed in the emission current density; emissive power, and pulse charge), depth of electron penetration, and also the pulse duration;

— the machining conditions, i.e. the position of the workpiece and the degree of vacuum.

The technical and economic indices of electron-beam machining are analogous to those of electro-erosion machining (Chapter 22).

Electron-beam machining is one of the newest machining processes, hence its technical and economic indices have not yet been examined as extensively as in the case of other machining methods. Of all the indices of electron-beam machining, fairly detailed studies have been made of erosion rate, machining precision, and, to a certain extent, the properties of the surface layer.

24.1.3.2 *Machining rate*

It has become the custom to take as a measure of the machining rate not the absolute magnitude of the material eroded per unit time but the number of pulses required to evaporate a particular volume or mass of material [24/3]. Such a convention is convenient in practical application of electron-beam machining since observation of an electron counter showing the number of pulses facilitates adjustment of the machining time. Hence, a distinction is made between two types of numbers of pulses required:

— volume or weight number

$$n_{ev} = \frac{V\gamma}{m_e} = \frac{m}{m_E} \tag{24.25}$$

— linear number

$$n_{eL} = \frac{g}{g_E} \tag{24.26}$$

where V—the volume of material eroded; g_E—the depth of cavity sunk by a single pulse; other symbols as before.

The volume coefficient of the required number of pulses is used when cutting off or cutting into the material, i.e. slotting, etc. The linear coefficient of the required number of pulses is used above all for hole sinking.

Studies on the machinability of different metals have shown that it depends to a large extent on the type of material. The results of measurements on the erosion of the same volume of Cd, Zn, Fe, Ti, Ta, Ni, Cu, and W are plotted in Fig. 24.6.

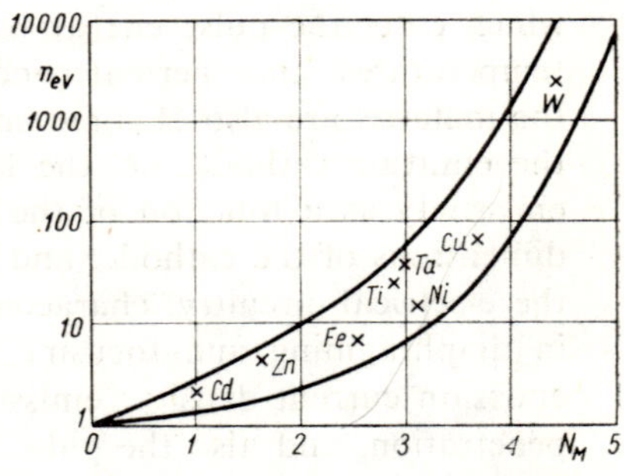

Fig. 24.6 Number of pulses required n_{eV} versus relative index of machinability N_M (after [24/3])

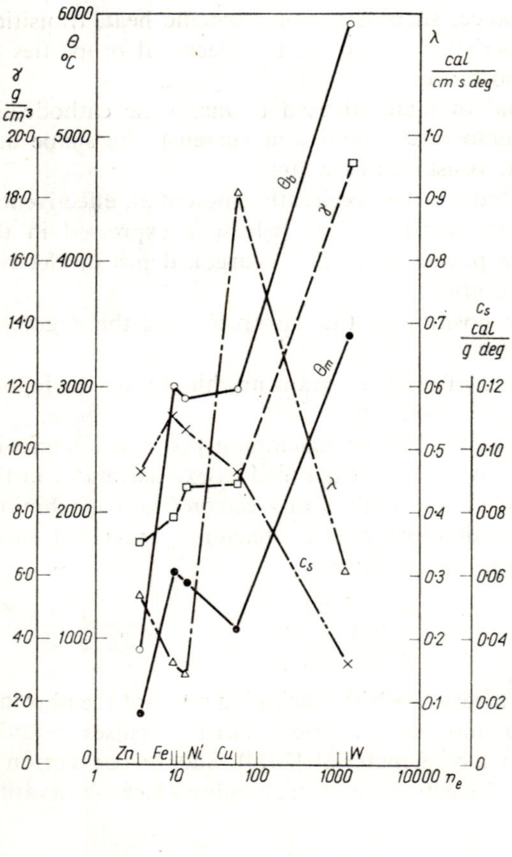

Fig. 24.7 The thermal properties of some metals (melting point Θ_m, boiling point Θ_b, thermal conductivity λ, and specific heat c_s) as functions of the number of pulses required, n_e

The number of pulses for a fixed volume of cadmium was taken as 1, and then the machinability number N_M was calculated as

$$N_M = \frac{n_{eM}}{n_{eCd}} \tag{24.27}$$

where n_{eM}—the required number of pulses for the given material, n_{eCd}—the required number of pulses for cadmium.

The order in which the machinability numbers are arranged may be explained by an analysis of material properties which affect the process of electron-beam machining. Two quantities, viz. boiling point and thermal conductivity, play a decisive role among these properties.

The melting and boiling points, the specific heat, and the thermal conductivity factor are plotted in Fig. 24.7 against the number of pulses required. The order in which the n_{eV}'s of the metals studied are arrayed corresponds to the order of the machinability numbers. On the other hand, the curves of the individual specific properties plotted as functions of n_{eV} are not monotonic. The same is also true of the dependence of the material densities on n_{eV}.

It is thus seen that the machinability number is affected jointly by all the thermal properties, as well as other properties of the material, such as density and electrical conductivity. It should be borne in mind that these properties vary with the temperature. Only the overall influence of all the aforementioned properties could make possible a quantitative treatment of the relationship between the kind of material and the number of pulses required.

Studies of how the accelerating potential affects the machining rate [24/3] have confirmed the theoretical predictions (interpreted graphically in Fig. 24.5). The experimental curves for the dependence of n_{eV} on V_0, shown in Fig. 24.8, have the same shape as the left-hand part of the curve in Fig. 24.5. It is seen from Fig. 24.8

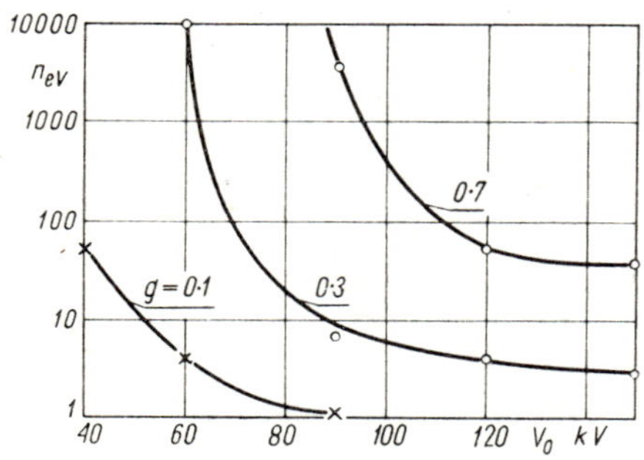

Fig. 24.8 Number of pulses required n_{eV} as a function of the accelerating voltage V_0 (after [24/3])

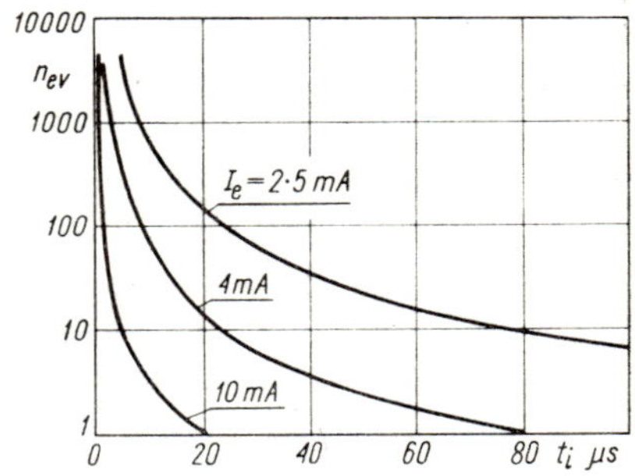

Fig. 24.9 Variation of number of pulses required n_{eV} with pulse duration t_p (after [24/3])

that increasing the accelerating potential above 120 000 V is quite ineffective (for given design parameters of the machine tool and under given machining conditions). This is explained by the fact that the electron-beam machine tools built thus operate at voltages not greater than 150 000 V.

It can also be seen from Fig. 24.8 that increasing the hole-sinking depth g entails a more than proportional rise in the number of pulses. This is due primarily to the increase in heat losses resulting from conduction and melting of the adjoining layer of material.

Increasing the pulse duration, which leads to an increase in the pulse energy E_p, leads to a reduction in the number of pulses required. This reduction occurs in accordance with a relation of the hyperbolic type. The results of studies on this relation are given in Fig. 24.9 [24/3].

The position of the focus of the electron beam relative to the depth of the material being machined is extremely important for the machining rate of the electron erosion process. Tests have shown that there is only one distance which enables the least number of pulses required n_e to be obtained. This is called the *optimum working distance* (by analogy to abrasive-blast treatment).

Figure 24.10 presents, after Pahlitsch and Visser [24/3], a simplified plot of the dependence of the number of pulses required on the position of the upper surface of the workpiece relative to the focus. The thickness of the material machined was 2 mm in the given case. It is seen from the plot that a focal position slightly below the upper surface was most advantageous in the given case.

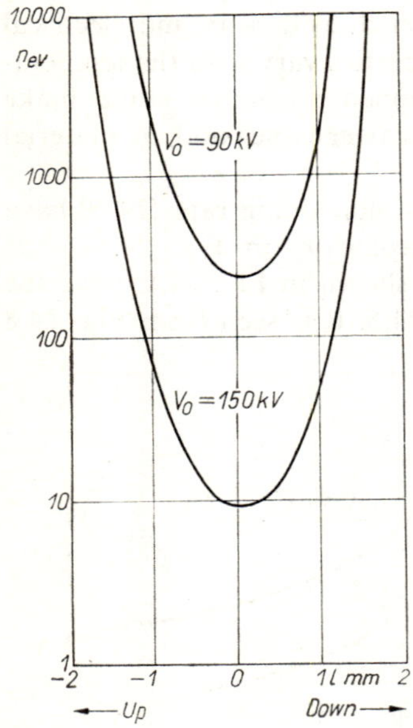

Fig. 24.10 Dependence of number of pulses required n_{eV} on displacement l of focal length of beam relative to upper surface of workpiece (after [24/3])

24.1.3.3 *Machining accuracy*

The accuracy with which holes, cavities, and cuts are executed is of particularly great importance for the technical and economic evaluation of electron-beam machining. The accuracy of electron-beam machining is appraised on the basis of:

— the sharpness or jaggedness of the edge of the workpiece or its surface roughness; the sharpness or jaggedness of the edge is indicated by the letter S,

— changes of exterior angles (K_{ext}) and interior angles (K_{int}), which is also determined by photomicrographic comparison with a standard, measuring the shift of the edge in micrometres,

— taper of cavities and holes (conicity) and size changes in slots and holes.

Studies of edge sharpness or jaggedness have shown that they depend to a large extent on the type of material machined. The reason for this is that the jaggedness is due mainly to local pitting of the material, and, other conditions remaining constant, this depends on the thermal properties of the material. Moreover, in accordance with theoretical predictions, an influence on the edge sharpness and jaggedness is exerted by the pulse charge (proportional to the pulse energy).

The jaggedness of the edges of workpieces made of several materials (C, W, Ag, Ti, Ni) is plotted in Fig. 24.11 against the pulse charge [24/4]. With an increase in charge, the jaggedness also increases, as a result of the growth of the pulse energy.

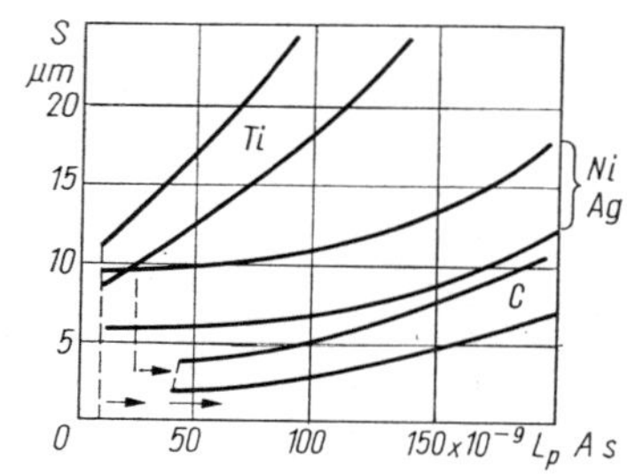

Fig. 24.11 Edge sharpness S versus pulse charge L_p (after [24/4])

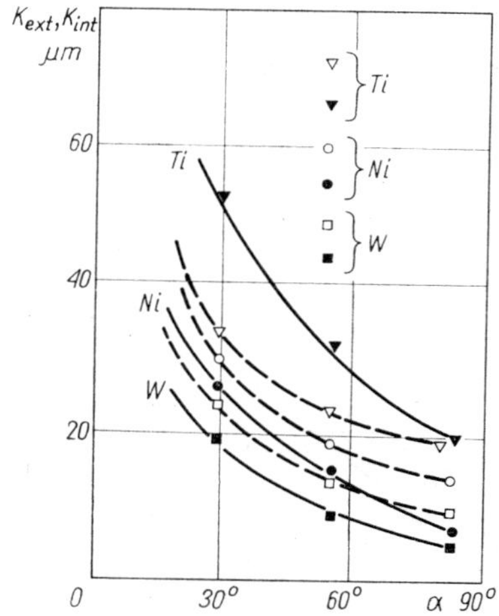

Fig. 24.12 Interior and exterior angle errors, K_{ext} and K_{int}, versus magnitude of angle α (after [24/3])

The angle changes also depend on the kind of material, the extent of the changes depending on the thermal properties of the material just as the machinability numbers do. Moreover, the smaller the angle made, the greater the change in the angles. This is due to the fact that a greater concentration of heat occurs when sharp edges are being machined.

The results of studies [24/4] on how these two parameters depend on the magnitude of the change of exterior and interior angles are presented in Fig. 24.12. In the case of tungsten and nickel, when internal corners (dashed line) are made the errors are greater than when external corners (solid line) are executed. On the other hand, titanium displayed the reverse relationship, this being attributed to, among other things, the thermal properties and the conditions which arise during the machining process [24/4].

The change in the width of slots and notches cut depends first and foremost on what material is being machined and on the pulse energy, i.e. on the pulse charge, when the accelerating potential is constant. An example of this kind of dependence obtained on the basis of investigations [24/4] is given in Fig. 24.13.

The diameter of holes or cavities obtained as a result of pulses depends on the material being machined and on the pulse energy, i.e. the pulse charge L_p(Fig. 24.14). Experience shows that under the conditions employed in practice cavity diameters ranging from 35 to 40 μm are most often obtained. For this range of cavity or hole diameters the out-of-roundness varies from 3 to 6 μm in more than 60% of the cases studied [24/4].

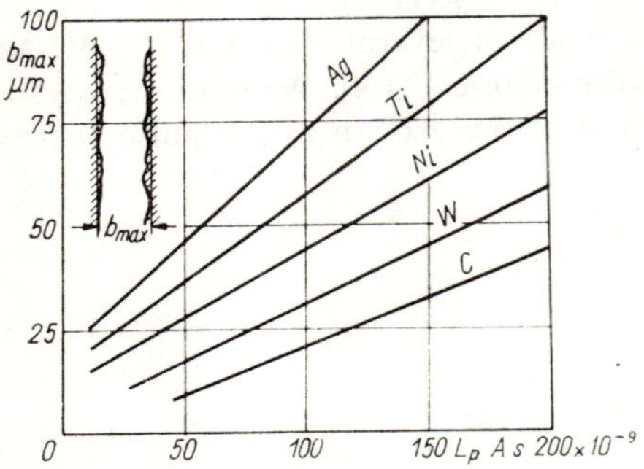

Fig. 24.13 Maximum groove width b_{max} as a function of the pulse charge l_p (after [24/4])

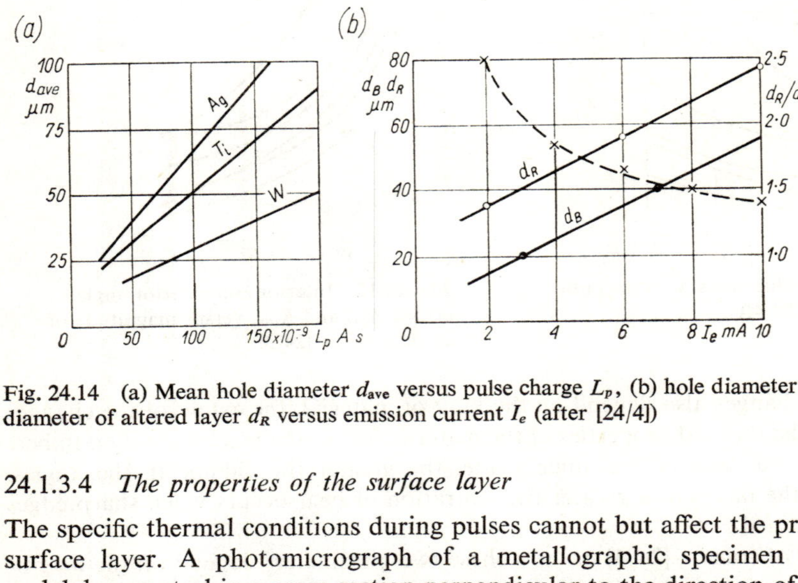

Fig. 24.14 (a) Mean hole diameter d_{ave} versus pulse charge L_p, (b) hole diameter d_B and diameter of altered layer d_R versus emission current I_e (after [24/4])

24.1.3.4 *The properties of the surface layer*

The specific thermal conditions during pulses cannot but affect the properties of the surface layer. A photomicrograph of a metallographic specimen of chromium-molybdenum steel in a cross section perpendicular to the direction of electron-beam machining is given in Fig. 24.15. Visible in the photomicrograph is a white ring of metal (d_R), with altered properties, which surrounds a hole of diameter d_B. Studies of the white ring of the surface layer have shown that decarbonization of the steel had not taken place and that the carbon had gone over into an unetchable solid solution. It is characteristic that with the pulse duration fixed, the width of the white ring does not change. This is indicated by the results of studies [24/3] illustrated by way of example in Fig. 24.14b and 24.16b. In both these cases the pulse duration was constant and in the case of the investigations illustrated in Fig. 24.16b, the emission current was also constant.

The distinct influence of the pulse duration is confirmed by the graph in Fig. 24.16a where the results of studies on the behaviour of d_B and d_R have been plotted. As can be seen from this graph, both the hole diameter and the diameter of the altered layer increase as the pulse duration is extended. This indicated that some time is required both for the material to be evaporated and for microstructural changes to occur in the surface layer. This is consistent with the results presented in Fig. 24.9.

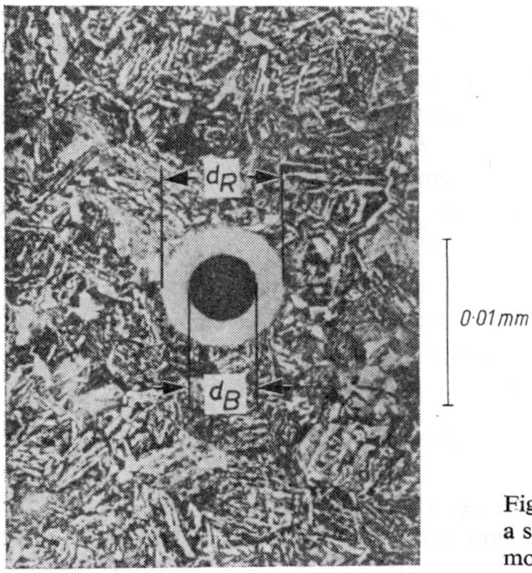

Fig. 24.15 Cross-section of cavity made by a single electron-beam pulse in chromium-molybdenum steel [24/3]

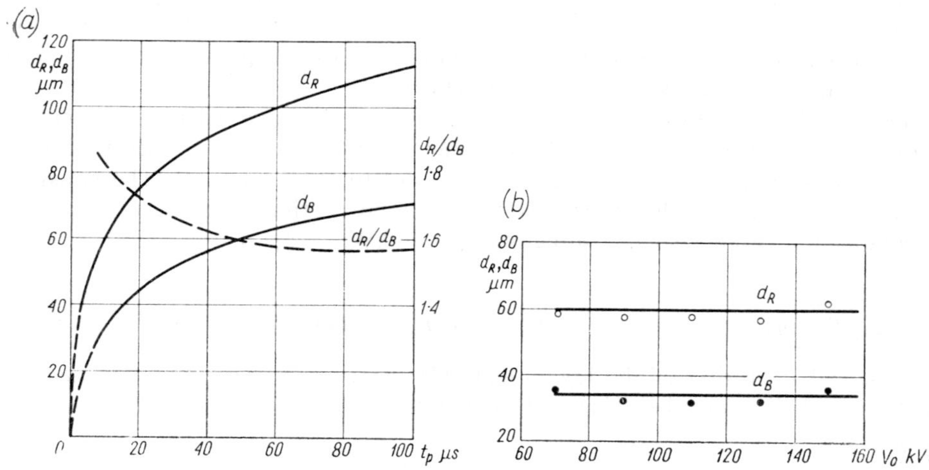

Fig. 24.16 Hole diameter and diameter of altered layer d_R as functions of the pulse duration t_p and voltage V_0 (after [24/3])

24.1.3.5 *Range of applications of electron-beam machining*

The properties of electron-beam machining indicate that it should be earmarked for working materials that are very difficult to machine. Hitherto, electron-beam machining has found practical application to electron-beam hole-sinking and milling of recesses and profiles.

In the case of drilling, holes made thus far have most often been in the 50–200 μm range. The pulse frequency is chosen so that, on the one hand, the machining rate be sufficient and, on the other hand, changes be prevented from occurring in the surface layer. In practice, the pulse frequency does not exceed 10^4 Hz.

Electron-beam machining has found particularly valuable application in precision engineering and in electrical engineering, especially in electronics. One example of the applications of electron-beam machining is that of cutting circuitry in conducting semiconductor films evaporated into microcomponents made of insulating materials.

The expectations are that as technical means develop and as the properties of electron-beam machining are learned, the range of economical applications will be extended.

24.2 Ion-beam machining

24.2.1 FORMATION OF PLASMA AND AN ION BEAM

24.2.1.1 *Plasma generation systems* [24/2]

Plasma is taken to mean a state of gas at high temperature which is characterized by total or partial ionization.

Electric fields can be applied to ionized gas to cause segregation and an outflow* of a stream of ions; moreover, this ion beam can be accelerated, just as an electron beam can. According to the accelerating potential, a distinction is made between two systems for generating plasma and ion beams:

— the first system (high-voltage) is characterized by high voltages of up to 25 000 V, which allows spontaneous ionization of the gas to occur. In this system emission currents of up to 2 A are attainable, which makes possible powers of the order of 40 kW (at the voltages employed in practice);

— the second system (low-voltage) is characterized by a low accelerating potential, of the order of 30–50 V. In this system, an additional source of energy, usually radio-frequency waves, must be used in order to ionize the gas. The emission current obtained in this system is several hundred times higher, so that it is possible to operate also at a rating of about 40 kW.

24.2.1.2 *The operating principle of the plasma generator* (*plasmatron*)

The mode of operation of the low-voltage plasma generator used for engineering purposes, including metal cutting and hole sinking, can be elucidated by means of Fig. 24.17.

* Since no hot cathode is required in the case of machining by a beam of ions extracted from plasma, the process whereby the plasma is obtained is referred to as cold emission.

Gas 2 flows into a hollow cathode 1, and in the lower part of the cathode the gas is subjected to the action of the energy of radio-frequency waves from source 3 operating at a voltage of up to 12 V and a current of up to 100 A. Under the applied electric field, segregation and the outflow of ions or electrons also occur in the ionized gas 4. The ion beam 5 is accelerated by the potential impressed across the cathode and anode.

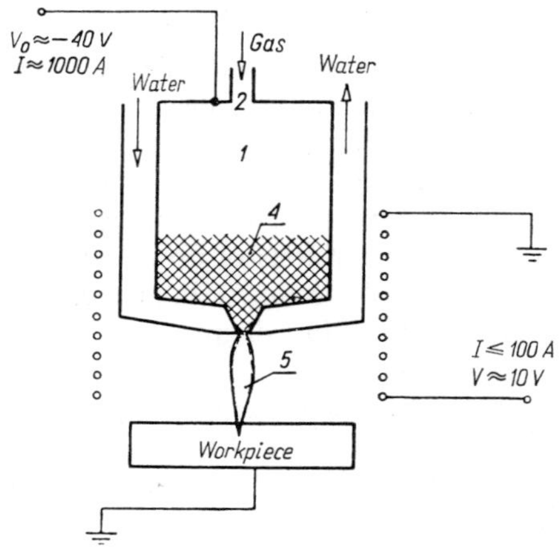

Fig. 24.17 Operating principle of plasma generator

The ion beam could be controlled just as electrons from a hot cathode are. Most frequently, however, the fact that the beam has relatively little divergence is exploited and the beam is used directly as a thermal tool.

The dimensions of cathodes hitherto produced lie within the following ranges:

— for low and intermediate powers—diameters up to 30 mm and length up to 40 mm,

— for high powers—diameters up to 75 mm and length up to 100 mm.

24.2.2 The range of applications of ion-beam machining

Compared with the electron-beam process, ion-beam machining has a number of disadvantages as well as some advantages.

The advantages include a capability for operation without a vacuum, or only a slight vacuum (0·01–0·1 torr). Moreover, the simplicity of the equipment and the simplicity of the beam generation should be considered an advantage. In view of the presence of negative and positive charges, the equipment can be operated at voltages of either polarity.

The disadvantages include the substantially poorer machining accuracy (because the beam cannot be controlled accurately), the necessity of continuously monitoring the gas pressure and the voltage, and the relatively large dimensions.

With the present state of the technical art in this method, ion-beam machining has found practical application to low-precision but high-efficiency cutting off (up to 70 mm thickness) and to removal of damaged surface layers (e.g. scales). After ion-beam machining, however, the pitting and changes in the surface layer are much greater than in the case of electron-beam machining.

As a result of these properties, ion-beam machining is being put to work more in welding and metal melting than in profiling.

24.3 Photon-beam machining*

24.3.1 The characteristics of photon-beam machining

Electro-beam machining using photons is a method of erosion shaping of holes, cavities, and slits, which is based on the principle of exploitation of the energy of a beam of light, i.e. a stream of photons.

Thanks to the use of almost completely coherent monochromatic light, which admits a very high degree of focusing, it is possible to obtain power densities never before achieved by any other method. Maximum power densities of up to 10^8–10^{13} W/cm^2, that is, 10^2–10^7 MW/cm^2, have been successfully attained thus far. This corresponds to a rate of temperature rise of 10 to 20 million degrees per second [24/3].

The temperatures of the order of 8000–10 000°C attained, because of the high concentration of the photon beam, make it possible to effect the lightning-fast, often explosive, evaporation of all solids known to man. Owing to this, a suitably focused beam of photons constitutes a new kind of thermal tool which can also be utilized, among other engineering applications, for erosion machining of precision and very minute cavities, holes, and slits.

24.3.2 A simplified explanation of the operation of the laser [24/7]

To produce a photon beam with a high concentration of power there is need for ideally coherent light, i.e. having the same wavelength, the same direction, and almost identical phase (differences of no more than 1/20th of a wavelength).

Light of this kind can be produced by amplification of controlled emission of light radiation. The device employing this method of producing coherent light is called a *laser*, the acronym for light amplification by stimulated emission of radiation.

The essence of directed stimulation and emission, which is the basis of the laser may be presented in simplified terms as follows.

Atoms can be at different energy levels, this being possible only in a quantum manner, which is to say that there are distinct electron energy levels (orbits). By a suitable influx of energy, in the given case by irradiation with photons, electrons attain higher energy levels. This phenomenon is called *optical stimulation* or *pumping*. If the inflow of photon energy were halted, atoms have a natural inclination to return to the original energy state, thus spontaneously radiating photons of different energies in various directions and at different intervals of time.

If further pumping is employed in the excitation state, the effect continues in a different way. Each additional photon which strikes the excited atom causes two photons to be emitted. These in turn strike the next excited atoms, thus causing the emission of four photons in all. In this way an avalanche emission of photons of

* Also called *laser machining*.

almost identical frequency and direction of radiation comes about. This is an amplified flash of light which can be focused by means of an appropriate optical system.

The basic condition for obtaining an amplified, coherent flash of light is that of preventing a state of energy equilibrium, i.e. the additional optical pumping of atoms must be capable of exciting the overwhelming majority of atoms. For only in this state can more photons be emitted than are received. In order to sustain the emission of an amplified light pulse for some time a large number of atoms must be kept at a higher energy level, i.e. the pumping must proceed at a faster rate than the emission.

The foregoing model of the effect can be implemented only in the following manner. A flash tube producing pulses with a duration of the order of milliseconds can be used for stimulation. Chromium atoms, say, which are dispersed (about 0·5% content) in a ruby medium, can be stimulated. The ruby most often in the form of a rod, has reflecting ends, one of them being completely reflecting and the other semi-transparent. Stimulation of the chromium atoms is indicated by a green glow, and the initial phase of the emission, by red radiation. The red light bounces off the mirrored ends of the ruby rod, further stimulating atoms. Then all the atoms surrender the absorbed photons in an avalanche, which are transmitted through the semi-transparent end of the ruby rod. Rays which are not parallel to the ruby rod axis, i.e. are not perpendicular to the ends, leave by way of the cylindrical surface of the rod. In consequence, only parellel rays of little divergence get out via the end.

The low divergence of the beam, coupled with an almost identical frequency, enable a simple optical system to produce an extremely high photon concentration yielding an energy density hitherto unattainable by any other method.

24.3.3 Scheme of construction of laser machine tool

The optics of a laser machine tool are indicated in Fig. 24.18. The light source 1 in the form of a flash tube can be moved in order to regulate the amount of energy supplied. In designs known so far the power input is about 12·5 kW, whereas the average power output (for the total operating time of the laser) is about 10 W. The laser head 2 consists of a mirror shield and a chromium-doped ruby. The end planes of the ruby rod should not be out of parallel by more than 1/20th of a wavelength. The dimensions of the ruby rod affect the amount of energy radiated during light pulses. In 1960, for example, the ruby rods used to generate 3-joule pulses had average dimensions of 8·5 mm dia. and 76 mm length. Further advances in 1961–1969 in the design of laser heads made it possible to obtain pulse energies of the order of 30–100 J.*

On leaving the laser head (Fig. 24.18) the pulsed beam encounters a radiation divider 3 in which most of the radiation is deflected downwards, and only about 5% is bent upwards, to the observation system. The latter system consists of a heat shield 4, objective 5, another radiation divider 6, prism 7, and ocular 8. Part of the radiation is directed from the second radiation divider 6 through the convergent lens 9 to the mercury vapour lamp 10 for visual observation of the pulses. The bulk of the radiation is directed onto the workpiece 14 by one of the optical systems 12 or 13, rotated about the axis 15. The distance between the spherical lens 13 and the

* The Anglo-American firm Corads Corp. has built a giant laser with a maximum pulse power of the order of 50 MW. Owing to the large amount of heat liberated, a frequency of only 20 pulses per hour could be achieved [24/2].

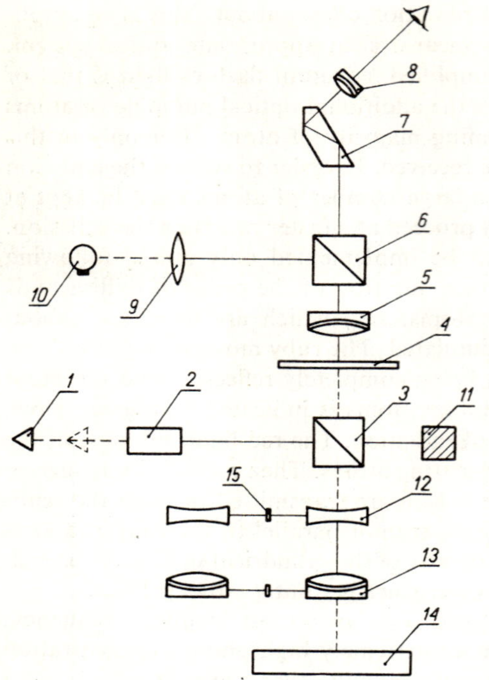

Fig. 24.18 Optics of laser, or photon-beam, machine tool

workpiece surface is called the working distance. This distance ranges from 5 to 70 mm with a divergence angle of 1 to 300 milliradians.

24.3.4 The properties of photon-beam machining

A photon beam does not have the ability to penetrate into a material. Consequently, heating practically on the surface ensues with an attendant increase in temperature. The temperature can be computed from

$$Q = k\frac{w}{f^2\theta^2} \tag{24.28}$$

where w—the energy density, f—the focal length, θ—divergence angle, k—an empirical constant.

It follows from eqn. (24.18) that the temperature rise remains within a distinct range since an increase in the power density causes the divergence angle to grow.

Figure 24.19 shows the successive phases in the process of irradiation of a workpiece (titanium) and the development of the process of liquation, evaporation, and protuberance of photon erosion products. The consecutive photographs in Fig. 24.19 were taken at intervals of 250 μs.

With the aid of observation of the photographs* the process of photon erosion may be depicted as follows.

* These photographs were taken with a high-speed camera taking pictures at a rate of up to 40 000 a second [24/8].

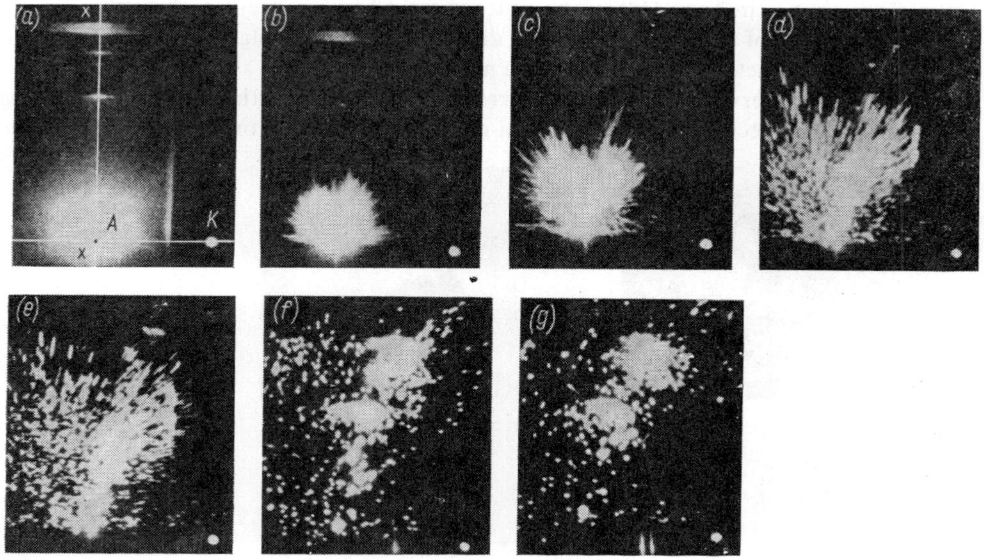

Fig. 24.19 Consecutive phase in protuberance of photon erosion products [24/8]

As soon as the laser pulse is produced a very thin layer of the material around point *A* (Fig. 24.19a) is subjected to thermal action. The position of this point is determined by the intersection of the *x*-axis of the photon beam with a straight line perpendicular to it and passing through the luminous control point *K*, lying on the surface of the workpiece.

The position of the objective is best seen in the photograph in Fig. 24.19a, where the set of the most intensely luminous points are still almost on the very surface.

The intensity of the light gradually falls off as the centre of the glow (and of the highest temperature) moves deeper into the material. This effect lasts for about 200–250 μs. After this time, protuberance of luminous erosion products (photograph in Fig. 24.19b) make their first appearance. Clouds of erosion products are already indicative of the beginnings of orientation. After a further 250 μs (photograph in Fig. 24.19c) a developed cloud of erosion products is visible and individual luminous particles of matter begin to appear. This process continues for some time (photographs in Figs. 30.19d, e, f). The luminous cloud of photon erosion products begins to disappear after about 1 ms has elapsed and has completely disappeared after about 2 ms.

The cloud takes about 750 μs to traverse the distance from the workpiece to the objective, which was 50 mm in the studies described here.

Studies of erosion products (by precipitating them on a glass plate) have shown that in the overwhelming majority of cases these are particles of molten machined material ejected by the pressure of gases produced in the cavity being sunk. Moreover, the dark spots on the glass are indicative of machined material being deposited by evaporation. There is reason to believe that the cloud of erosion products may also contain minute solid particles carried off by the expanding vapour of the material.

Examples of particles of various materials deposited on glass from the vapour phase are shown in Fig. 24.20. The bottom right-hand corner of each photograph gives the symbol of the material and the distance of the glass plate (in mm), on which the erosion products solidified, from the machined surface.

It may be observed that there is a correlationship between the melting and boiling points and the specific heat and thermal conductivity on the one hand, and the size

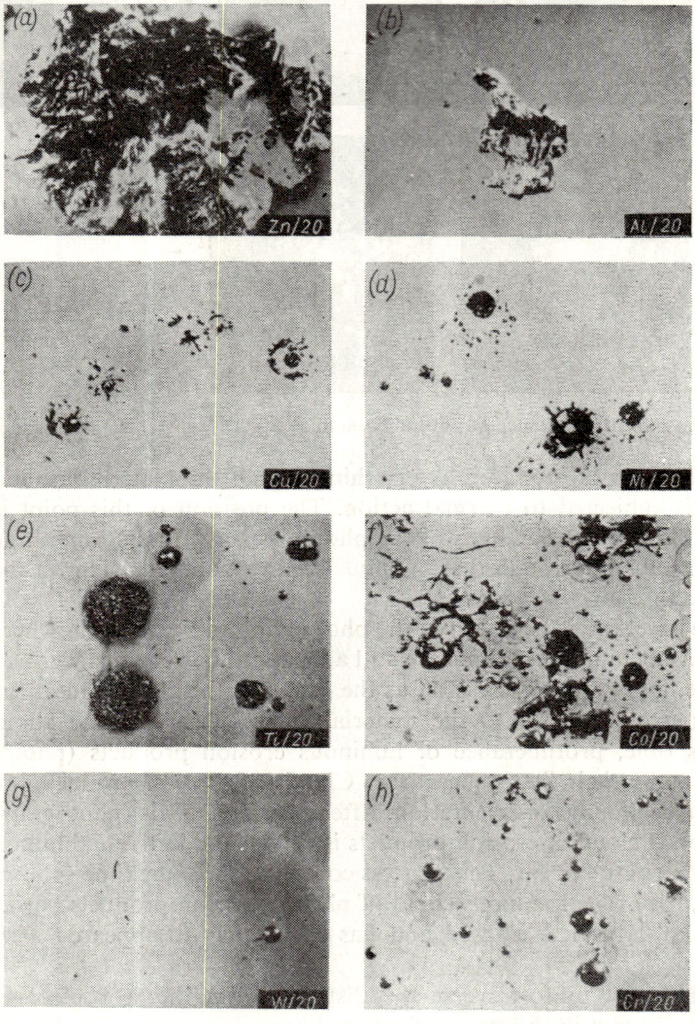

Fig. 24.20 Appearance of particles eroded from different materials machined (all photographs to the same scale) [24/8]

of the erosion particles on the other. This would point to a great similarity between electron-beam and photon-beam erosion.

This similarity is also evident in the appearance and shape of the cavities. Figure 24.21 shows the longitudinal cross sections of cavities produced by pulses of different durations. A hole right through the material is shown in Fig. 24.21a. It was made by means of a pulse of 1·2 ms duration. The photograph in Fig. 24.21b shows a partially plugged hole executed by a pulse of 1·8 ms duration. The analogous hole in Fig. 24.21c has been completely plugged by a white (in photo) layer of fused material. This hole also displays more taper. It was made by a pulse of 2·8 ms duration.

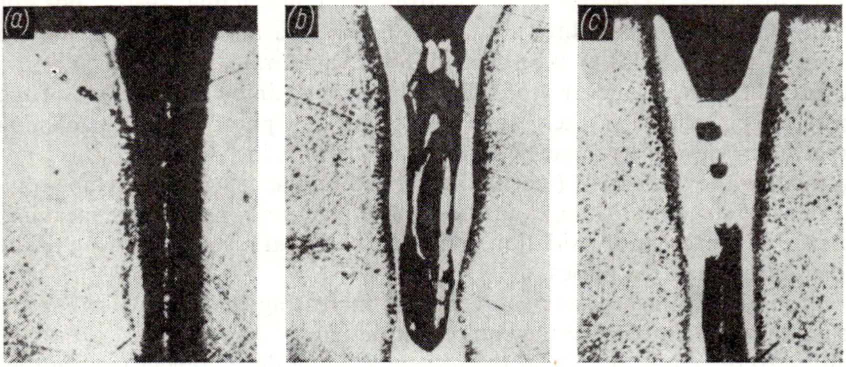

Fig. 24.21 Longitudinal cross section of cavities eroded by a single photon-beam pulse of varying duration: (a) 1·2 ms, (b) 1·8 ms, (c) 2·8 ms [24/8]

It follows from the foregoing discussion that there exists a certain optimal pulse duration for which the best results are obtained from the point of view of machining rate and accuracy.

Studies of the structure of the surface layer after photon-beam machining revealed the existence of a zone of thermal influences consisting of a white layer (Fig. 24.22) and a transition layer.

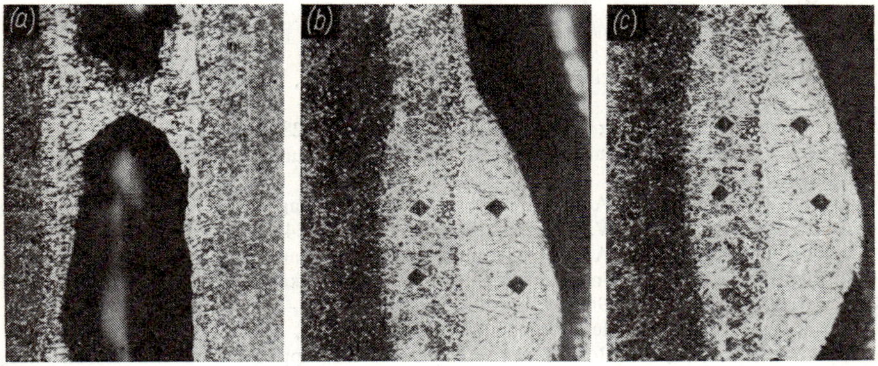

Fig. 24.22 Microsection of longitudinal cross section of top layer of cavities made by a photon-beam pulse of varying duration: (a) 1·2 ms, (b) 1·8 ms, (c) 2·8 ms [24/8]

The white layer is residual austenite with visible martensite spines with a hardness of 800–900 μHV/50. The transition layer consists of helical spines of martensite and residual austenite with carbides of chromium and iron. The hardness of the transition layer ranges from 950 to 1500 μHV/50.

The conclusion may thus be drawn that the structure of the surface layer is shaped by thermal influences, and the shorter the pulse duration, but with a power density sufficient for the process of evaporation and liquation, the smaller the depth to which these changes extend.

The technical and economic indices of photon-beam machining are, however, only now being determined by experiment. It may nevertheless be assumed that the heat balance, discussed in reference to electron-beam machining, also holds for photon-beam machining. Accordingly, the machining rate and its dependence on the properties of the materials may be defined in the same way.

At present, the energy efficiency of photon-beam machining is very low, of the order of several per cent. A further rapid development of photon-beam machining is to be expected in view of its important advantages. These include:

— ease in moving the workpiece (mechanically) and in operating the machine tool,
— small extent of structural changes in the surface layer and, practically speaking, the absence of a process of oxidation and other chemical reactions, owing to the very brief duration of the pulses,
— no need for high-voltage power sources, thus increasing the operational safety and reliability and simplifying the equipment design.

In addition to these advantages, at the present stage of development the photon-beam machine tool present a number of difficulties. These include:

— the aforementioned very low efficiency,
— the difficulties entailed in cooling the laser unit and the pumping lamp,
— restriction of application to only shallow machining.

The properties listed above have determined the hitherto limited application of photon-beam machining to very accurate and shallow cavity sinking and other special machining applications, especially in fabricating semiconductors. However, this method of electro-beam machining does have a great potential for a much broader range of economical application.

REFEREUCES

24/1 STEIGERWALD, K. H.: 'Materialbearbeitung mit Elektronen', Physikalische Verhandlungen, 4, 1953
STEIGERWALD, K. H.: Drilling by electrons, U.S. Patent 2793281, May 21, 1957.
24/2 CREMOSNIK, G.: 'Materialbearbeitung mit Ladungsträger und Laserstrahlen', Technische Rundschau, 3, 4, 5, 1965
24/3 PAHLITSCH, G., and VISSER, A.: 'Studien zum Werkstoffabtrag durch energiereiche Elektronenstrahlen', Vortrag auf XIV. Kongress, Stresa—Italien 1964
24/4 PAHLITSCH, G., and VISSER, A.: 'Studien zum Werkstoffabtrag durch energiereiche Elektronenstrahlen. Systematische Untersuchung der Arbeitsgenaugkeit', Mitteilung auf XV. CIRP Kongress, Liège (Belgium) 1965
24/5 MICROSA, R.: Laser in der Metallbearbeitung. Microtecnic 2, 1965
24/6 SCHMIDT, A., HAM, L., and HOSI, T.: 'Laser Features and Applications to Micro-Machining and Micro-Welding', Int. Journal of M.T.D.R., 1, 1965
24/7 KLEJMAN, H.: 'Masery i lasery—Nowe zdobycze elektroniki' (Masers and Lasers—New Achievements of Electronics), WNT, Warsaw 1965
24/8 KACZMAREK J., and KREHLIK R.: 'Badania procesu fotonowej erozji ewaporacyjnej i likwacyjnej' (Research on the process of photon-beam erosion by evaporation and liquation), Prace IOS—Seria Referaty, No. 104, 1967

Subject index